SYSTEM DYNAMICS

SYSTEM DYNAMICS

KATSUHIKO OGATA
University of Minnesota

PRENTICE-HALL, INC., Englewood Cliffs, New Jersey 07632

Library of Congress Cataloging in Publication Data

OGATA, KATSUHIKO.
  System dynamics.

  Includes bibliographies and index.
    1. Engineering—Mathematical models. 2. Dynamics.
3. System analysis. I. Title.
TA342.035   620'. 72   77–20180
ISBN 0–13–880385–4

© 1978 by Prentice-Hall, Inc., Englewood Cliffs, New Jersey 07632

Printed in the United States of America

10 9 8 7 6 5

PRENTICE-HALL INTERNATIONAL, INC., *London*
PRENTICE-HALL OF AUSTRALIA PTY. LIMITED, *Sydney*
PRENTICE-HALL OF CANADA, LTD., *Toronto*
PRENTICE-HALL OF INDIA PRIVATE LIMITED, *New Delhi*
PRENTICE-HALL OF JAPAN, INC., *Tokyo*
PRENTICE-HALL OF SOUTHEAST ASIA PTE. LTD., *Singapore*
WHITEHALL BOOKS LIMITED, *Wellington, New Zealand*

CONTENTS

APPENDICES

PREFACE

In recent years, courses in system dynamics that deal with the mathematical modeling and response analysis of dynamic systems are required in most engineering curricula. This book is intended for use in such courses. It presents a comprehensive treatment of the dynamics of physical systems in several media and is written for the junior engineering student.

The outline of the book is as follows: Chapter 1 presents a general introduction to systems. Mathematical modeling and analysis of mechanical systems are described in Chapter 2. Electrical systems are covered in Chapter 3; and Chapters 4 and 5 deal with hydraulic and pneumatic systems, respectively, and include discussions of the linearization process in mathematical modeling of nonlinear systems. The Laplace transform method for analyzing linear systems is given in Chapter 6, and linear systems analysis is described in Chapter 7. Finally, Chapter 8 presents control systems analysis.

Throughout the book, carefully chosen examples are presented at strategic points so that the reader will have a better understanding of the subject matter discussed. And, with the exception of Chapter 1, numerous solved problems are provided at the end of each chapter. These problems, many of which depict realistic situations encountered in engineering practice, represent an integral part of the text. Therefore, it is suggested that the reader study all such problems carefully. In addition, many unsolved problems of various degrees of difficulty are included to test the reader's ability to apply the theory involved.

Physical quantities are introduced in SI units (modernized metric system of units). In some examples and problems, however, physical quantities are introduced in SI units *and* other systems of units so that the reader will be able to convert freely from one system to another.

Most of the material presented here has been tested for several years in junior-level courses in system dynamics at the Department of Mechanical Engineering, University of Minnesota.

For the convenience of the instructor using this book in the classroom, the more challenging sections (and the corresponding solved and unsolved problems) are marked with an asterisk in order that he or she may include or omit these sections, depending on course objectives.

If this book is used for quarter-length courses, most of the material in Chapters 1 through 7 (with the possible exception of starred sections) could be covered. After studying these seven chapters, the reader should be able to formulate mathematical models for physical systems with reasonable simplicity and should be able to determine the transient and frequency response of such systems. For semester-length courses, the entire book could be covered.

In conclusion, I would like to express my appreciation to my former students who solved many problems included in this book, anonymous reviewers who made many constructive comments, and to all those who, in one way or another, were helpful in its completion.

K. OGATA

SYSTEM DYNAMICS

1

INTRODUCTION

1-1 SYSTEMS

Systems. A *system* is a combination of components acting together to perform a specific objective. A *component* is a single functioning unit of a system. By no means limited to physical ones, the concept of a system can be extended to abstract dynamic phenomena, such as those encountered in economics, transportation, population growth, and biology.

A system is called *dynamic* if its present output depends on past input; if its current output depends only on current input, the system is known as *static*. Output of a static system remains constant if input does not change and it changes only when input changes. In a dynamic system output changes with time if it is not in a state of equilibrium. We are concerned in this book only with dynamic systems.

Mathematical models. Any attempt to design a system must begin with a prediction of its performance before the system itself can be designed in detail or actually built. Such prediction is based on a mathematical description of the system's dynamic characteristics. This mathematical description is called a *mathematical model*. For physical systems, most useful mathematical models are described in terms of differential equations.

System dynamics deals with mathematical modeling and response analy-

1

sis of dynamic systems. A careful study of this subject is necessary in present-day engineering design.

Linear and nonlinear differential equations. Linear differential equations may be classified as linear, time-invariant, differential equations and linear, time-varying, differential equations.

A *linear, time-invariant, differential equation* is one in which a dependent variable and its derivatives appear as linear combinations. An example of such an equation would be

$$\frac{d^2x}{dt^2} + 5\frac{dx}{dt} + 10x = 0$$

Since the coefficients of all terms are constant, a linear, time-invariant, differential equation is also called a *linear, constant coefficient, differential equation.*

In the case of a *linear, time-varying, differential equation*, the dependent variable and its derivatives appear as linear combinations, but a coefficient or coefficients of terms may involve the independent variable. An example of this type of differential equation is

$$\frac{d^2x}{dt^2} + (1 - \cos 2t)x = 0$$

It is important to remember that, in order to be linear, the equation must contain no powers or other functions or products of the dependent variables and its derivatives.

A differential equation is called *nonlinear* if it is not linear. Examples of nonlinear differential equations include

$$\frac{d^2x}{dt^2} + (x^2 - 1)\frac{dx}{dt} + x = 0$$

$$\frac{d^2x}{dt^2} + \frac{dx}{dt} + x + x^3 = \sin \omega t$$

Linear systems and nonlinear systems. For linear systems, the equations that constitute the model are linear. We shall deal in this book with linear systems that can be represented by linear, time-invariant, ordinary differential equations.

The most important property of linear systems is that the *principle of superposition* is applicable. This principle states that the response produced by simultaneous applications of two different forcing functions or inputs is the sum of two individual responses. Consequently, for linear systems, the response to several inputs can be calculated by dealing with one input at a time and then adding the results. As a result of the principle of superposition, complicated solutions to linear differential equations can be derived as a sum of simple solutions.

In an experimental investigation of a dynamic system, if cause and effect are proportional, thereby implying that the principle of superposition holds, the system can be considered linear.

Nonlinear systems, of course, are those that are represented by nonlinear equations.

Although physical relationships are often represented by linear equations, in many instances the actual relationships may not be quite linear. In fact, a careful study of physical systems reveals that so-called linear systems are actually linear only within limited operating ranges. For instance, many hydraulic systems and pneumatic systems involve nonlinear relationships among their variables.

For nonlinear systems, the most important characteristic is that the principle of superposition is not applicable. In general, procedures for finding the solutions of problems involving such systems are extremely complicated. Because of the mathematical difficulty involved, it is frequently necessary to linearize a nonlinear system near the operating condition. Once a nonlinear system is approximated by a linear mathematical model, however, a number of linear techniques may be used for analysis and design purposes.

1-2   MODELING

Mathematical modeling. By applying physical laws to a specific system, it may be possible to develop a mathematical model that describes the system. Such a model may include unknown parameters, which must then be evaluated through actual tests. Sometimes, however, the physical laws governing the behavior of a system are not completely defined, and formulating a mathematical model may be impossible. If so, an experimental modeling process can be used. In this process the system is subjected to a set of known inputs and its outputs are measured. Then a mathematical model is derived from these input-output relationships.

Simplicity versus accuracy. When attempting to build a model, a compromise must be made between the simplicity of the model and the accuracy of the results of the analysis. It is important to note that the results obtained from the analysis are valid only to the extent that the model approximates a given physical system.

The rapidity with which a digital computer can perform arithmetic operations allows us to include hundreds of equations to describe a system and to build an accurate but extremely complicated model. If extreme accuracy is not needed, however, it is preferable to develop only a reasonably simplified model.

In determining a reasonably simplified model, we must decide which physical variables and relationships are negligible and which are crucial to the accuracy of the model. In order to obtain a model in the form of linear differential equations, any distributed parameters and nonlinearities that may be present in the physical system should be ignored. If the effects that these ignored properties have on the response are small, then the results of the analysis of a mathematical model and the results of the experimental study of the physical system will be in good agreement. Whether any particular features are important may be obvious in some cases and may, in other instances, require physical insight and intuition. Experience is an important factor in this connection.

Usually, when solving a new problem, it is desirable first to build a simplified model to obtain a general idea about the solution. Afterward a more detailed mathematical model can be built and used for a more complete analysis.

Remarks on mathematical modeling. No mathematical model can represent any physical component or system precisely. Approximations and assumptions are always involved. Such approximations and assumptions restrict the range of validity of the mathematical model. (The degree of approximation can be determined only by experiments.) So in making a prediction about the system's performance, any approximations and assumptions involved in the model must be kept in mind.

Mathematical modeling procedure. The procedure for obtaining a mathematical model for a system can be summarized as follows.

1. Draw a schematic diagram of the system, and define variables.

2. Using physical laws, write equations for each component, combine them according to the system diagram, and obtain a mathematical model.

3. To verify the validity of the model, the performance prediction, obtained by solving the equations of the model, is compared with experimental results. (The question of the validity of any mathematical model can be answered only by experiment.) If the experimental results deviate from the prediction to a great extent, the model must be modified. A new model is then derived and a new prediction compared with experimental results. The process is repeated until a satisfactory agreement is obtained between prediction and experimental results.

1-3 ANALYSIS AND DESIGN OF DYNAMIC SYSTEMS

This section briefly explains what is involved in the analysis and design of dynamic systems.

Analysis. *System analysis* means the investigation, under specified conditions, of the performance of a system whose mathematical model is known.

The first step in analyzing a dynamic system is to derive its mathematical model. Since any system is made up of components, analysis must start by developing a mathematical model of each component and combining these models in order to build a model of the complete system. Once the latter model is obtained, the analysis may be formulated in such a way that system parameters in the model are varied to produce a number of solutions. The engineer then compares these solutions and interprets and applies the results of his analysis to his basic task.

It should always be remembered that deriving a reasonable model for the complete system is the basic part of the entire analysis. Once such a model is available, various analytical and computer techniques can be used for analysis. The manner in which analysis is carried out is independent of the type of physical system involved—mechanical, electrical, hydraulic, and so on.

Design. *System design* refers to the process of finding a system that accomplishes a given task. In general, the design procedure is not straightforward and will require trial and error.

Synthesis. By *synthesis* we mean using an explicit procedure to find a system that will perform in a specified way. Here the desired system characteristics are postulated at the outset and then various mathematical techniques are used to synthesize a system having these characteristics. Generally, such a procedure is completely mathematical from the start to the end of the design process.

Basic approach to system design. The basic approach to the design of any dynamic system necessarily involves trial-and-error procedures. Theoretically, synthesis of linear systems is possible, and the engineer can systematically determine the components necessary to perform the given objective. In practice, however, the system may be subjected to many constraints or may be nonlinear; in such cases, no synthesis methods are available at present. Moreover, the component features may not be precisely known. Thus trial-and-error techniques are almost always needed.

Design procedures. Frequently, the design of a system proceeds as follows. The engineer begins the design procedure knowing the specifications to be met and the dynamics of the components, the latter of which involve

5

design parameters. The specifications may be given in terms of both precise numerical values and vague qualitative descriptions. (Engineering specifications normally include statements on such factors as cost, reliability, space, weight, and ease of maintenance.) It is important to note that the specifications may be changed as the design progresses, for detailed analysis may reveal that certain requirements are impossible to meet. Next, the engineer will apply synthesis techniques if available, as well as other methods, to build a mathematical model of the system.

Once the design problem is formulated in terms of this model, he carries out a mathematical design that yields a solution to the mathematical version of the design problem. With the mathematical design completed, the engineer simulates the model on a computer in order to test the effect of various inputs and disturbances on the behavior of the resulting system. If the initial system configuration is not satisfactory, the system must be redesigned and the corresponding analysis completed. This process of design and analysis is repeated until a satisfactory system is found. Then a prototype physical system can be constructed.

Note that this process of constructing a prototype is the reverse of mathematical modeling. The prototype is a physical system that represents the mathematical model with reasonable accuracy. Once the prototype has been built, the engineer tests it to see whether it is satisfactory. If it is, the design is complete. If not, the prototype must be modified and tested. This process continues until a satisfactory prototype is obtained.

1-4   SUMMARY

From the point of view of analysis, a successful engineer must be able to obtain a mathematical model of a given system and predict its performance. (The validity of a prediction depends to a great extent on the validity of the mathematical model used in making the prediction.) And from the design standpoint, he must be able to carry out a thorough performance analysis before a prototype is constructed.

The objective of this book is to enable the reader (a) to obtain mathematical models that closely represent behaviors of physical systems and (b) to develop system responses to various deterministic inputs so that he or she can effectively analyze and design dynamic systems. (Deterministic inputs, as opposed to probabilistic inputs, are those inputs that are completely specified as functions of time. Probabilistic inputs, on the other hand, depend on unpredictable random variables.)

Outline of the text. Chapter 1 presented an introduction to dynamic systems. Chapters 2 through 5 are primarily concerned with the problem of mathematical modeling. Basically, a system model is formulated by applying physical laws to the components and considering the way in which they are connected. Specifically, Chapter 2 deals with mechanical systems and Chapter 3 with electrical systems. Chapters 4 and 5 present hydraulic and pneumatic components and systems, respectively. These two chapters include linearization techniques for nonlinear systems. In linearizing a nonlinear system, we limit the deviations of variables to small values and obtain a linear mathematical model. Although accurate only for small ranges of variables, such a model has a great deal of practical value because many nonlinear systems are designed to be maintained as close to certain operating conditions or equilibrium states as possible.

Chapter 6 describes the Laplace transform method, which is useful for analyzing linear, time-invariant systems. Then the transient and frequency response analyses of linear systems are discussed in Chapter 7, which covers the Laplace transform approach to the transient response analysis, transfer functions, frequency response analysis with applications to solving problems on vibration isolation, and analog computers. Finally, Chapter 8 presents introductory materials on control systems analysis. Included are discussions on automatic controllers, standard techniques to obtain various control actions through the use of pneumatic, hydraulic, and electronic components, transient response analysis of control systems, and a design problem.

2

MECHANICAL SYSTEMS

2-1 INTRODUCTION

Mathematical modeling and response analysis of mechanical systems are the subjects of this chapter. Introductory material appears in Sec. 2-1, where we begin with a brief description of systems of units followed by a review of such topics as mass, force, and Newton's laws, all of which are a normal part of college physics courses. In Sec. 2-2 mathematical modeling of mechanical systems and free vibrations of simple mechanical systems are described. Section 2-3 deals with free vibrations of systems having two or more degrees of freedom, and Sec. 2-4 describes mechanical systems with dry friction. The topic of static and sliding friction is discussed first, followed by an analysis of sliding and rolling motions of systems. Finally, d'Alembert's principle is introduced. Section 2-5 presents concepts of work, energy, and power. The section concludes with an energy method for obtaining equations of motion of systems. The final section, Sec. 2-6, deals with motion, energy, and power transformers, and here examples of motion, energy, and power transformers are presented and analyzed.

Systems of units. A clear understanding of the various systems of units is obviously necessary to the quantitative study of system dynamics. In the past most engineering calculations in the United States were based on the

British Engineering System (BES) of measurement. From now on, however, calculations will be made in International System (abbreviated SI) units.

The *International System of units* is a modified metric system, and as such it differs from conventional metric absolute or metric gravitational systems of units. Table 2-1 lists International System, conventional metric

Table 2-1. SYSTEMS OF UNITS

Systems of units / Quantity	Absolute systems				Gravitational systems	
	Metric			British	Metric engineering	British engineering
	SI	mks	cgs			
Length	m	m	cm	ft	m	ft
Mass	kg	kg	g	lb	$\dfrac{\text{kg}_f\text{-s}^2}{\text{m}}$	slug $= \dfrac{\text{lb}_f\text{-s}^2}{\text{ft}}$
Time	s	s	s	s	s	s
Force	N $= \dfrac{\text{kg-m}}{\text{s}^2}$	N $= \dfrac{\text{kg-m}}{\text{s}^2}$	dyn $= \dfrac{\text{g-cm}}{\text{s}^2}$	poundal $= \dfrac{\text{lb-ft}}{\text{s}^2}$	kg_f	lb_f
Energy	J $= \text{N-m}$	J $= \text{N-m}$	erg $= \text{dyn-cm}$	ft-poundal	$\text{kg}_f\text{-m}$	ft-lb$_f$ or Btu
Power	W $\dfrac{\text{N-m}}{\text{s}}$	W $\dfrac{\text{N-m}}{\text{s}}$	$\dfrac{\text{dyn-cm}}{\text{s}}$	$\dfrac{\text{ft-poundal}}{\text{s}}$	$\dfrac{\text{kg}_f\text{-m}}{\text{s}}$	$\dfrac{\text{ft-lb}_f}{\text{s}}$ or hp

systems, and British systems of units. (This table presents only those units necessary to describe behaviors of mechanical systems. For additional details on systems of units, refer to Appendix A.)

The chief difference between "absolute" systems of units and "gravitational" systems of units lies in the choice of mass or force as a primary dimension. In the absolute systems (SI and the metric and British absolute systems) mass is chosen as a primary dimension and force is a derived quantity. Conversely, in gravitational systems (metric engineering and British engineering systems) of units force is a primary dimension and mass is a derived quantity. Moreover, in the latter systems the mass of a body is defined as the ratio of the magnitude of the force to that of the acceleration. (Thus the dimension of mass is force/acceleration.)

A special effort has been made in this book to familiarize the reader with the various systems of measurement. In examples and problems, for instance, calculations are often made in SI units, conventional metric units, and BES units in order to illustrate how to convert from one system to another. Table 2-2 shows some convenient conversion factors among different systems of units. (Other detailed conversion tables are given in Appendix B.)

Abbreviations. Abbreviations of SI units are written in lowercase letters, such as m for meter and kg for kilogram. Abbreviations of a unit named after a person are usually capitablized—W for watt, N for newton, for instance. Commonly used units and their abbreviations are listed below.

ampere	A	newton	N
candela	cd	ohm	Ω
coulomb	C	pascal	Pa
farad	F	radian	rad
henry	H	second	s
hertz	Hz	steradian	sr
joule	J	tesla	T
kelvin	K	volt	V
kilogram	kg	watt	W
meter	m	weber	Wb

Multiples and submultiples in powers of 10 are indicated by abbreviated prefixes as follows.

	Prefix	Abbreviated prefix
10^{18}	exa	E
10^{15}	peta	P
10^{12}	tera	T
10^{9}	giga	G
10^{6}	mega	M
10^{3}	kilo	k
10^{2}	hecto	h
10	deka	da
10^{-1}	deci	d
10^{-2}	centi	c
10^{-3}	milli	m
10^{-6}	micro	μ
10^{-9}	nano	n
10^{-12}	pico	p
10^{-15}	femto	f
10^{-18}	atto	a

Table 2-2. CONVERSION TABLE

Length	1	1 m = 100 cm	
	2	1 ft = 12 in.	1 in. = 2.54 cm
	3	1 m = 3.281 ft	1 ft = 0.3048 m
Mass	4	1 kg = 2.2046 lb	1 lb = 0.4536 kg
	5	1 kg = 0.10197 kg$_f$-s^2/m	1 kg$_f$-s^2/m = 9.807 kg
	6	1 slug = 14.594 kg	1 kg = 0.06852 slug
	7	1 slug = 32.174 lb	1 lb = 0.03108 slug
	8	1 slug = 1.488 kg$_f$-s^2/m	1 kg$_f$-s^2/m = 0.6720 slug
Moment of inertia	9	1 slug-ft^2 = 1.356 kg-m^2	1 kg-m^2 = 0.7376 slug-ft^2
	10	1 slug-ft^2 = 0.1383 kg$_f$-s^2-m	1 kg$_f$-s^2-m = 7.233 slug-ft^2
	11	1 slug-ft^2 = 32.174 lb-ft^2	1 lb-ft^2 = 0.03108 slug-ft^2
Force	12	1 N = 10^5 dyn	
	13	1 N = 0.10197 kg$_f$	1 kg$_f$ = 9.807 N
	14	1 N = 7.233 poundals	1 poundal = 0.1383 N
	15	1 N = 0.2248 lb$_f$	1 lb$_f$ = 4.4482 N
	16	1 kg$_f$ = 2.2046 lb$_f$	1 lb$_f$ = 0.4536 kg$_f$
	17	1 lb$_f$ = 32.174 poundals	1 poundal = 0.03108 lb$_f$
Energy	18	1 N-m = 1 J = 1 W-s	1 J = 0.10197 kg$_f$-m
	19	1 dyn-cm = 1 erg = 10^{-7} J	1 kg$_f$-m = 9.807 N-m
	20	1 N-m = 0.7376 ft-lb$_f$	1 ft-lb$_f$ = 1.3557 N-m
	21	1 J = 2.389 × 10^{-4} kcal	1 kcal = 4186 J
	22	1 Btu = 778 ft-lb$_f$	1 ft-lb$_f$ = 1.285 × 10^{-3} Btu
Power	23	1 W = 1 J/s	
	24	1 hp = 550 ft-lb$_f$/s	1 ft-lb$_f$/s = 1.818 × 10^{-3} hp
	25	1 hp = 745.7 W	1 W = 1.341 × 10^{-3} hp

Mass. The *mass* of a body is the quantity of matter in it, which is assumed to be constant. Physically, mass is the property of a body that gives it inertia—that is, resistance to starting and stopping. A body is attracted by the earth, and the magnitude of the force that the earth exerts on it is called its *weight*.

In practical situations, we know the weight w of a body but not the mass m. We calculate mass m from

$$m = \frac{w}{g}$$

where g is the gravitational acceleration constant. The value of g varies slightly from point to point on the earth's surface. As a result, the weight of a body varies slightly at different points on the earth's surface, but its mass remains constant. For engineering purposes, g is taken as

$$g = 9.81 \text{ m/s}^2 = 981 \text{ cm/s}^2 = 32.2 \text{ ft/s}^2 = 386 \text{ in./s}^2$$

Far out in space, a body becomes weightless. Yet its mass remains constant and so the body possesses inertia.

The units of mass are kg, g, lb, kg_f-s^2/m, and slug, as shown in Table 2-1. If mass is expressed in units of kilogram (or pound), we call it kilogram mass (or pound mass) to distinguish it from the unit of force, which is termed kilogram force (or pound force). In this book kg is used to denote a kilogram mass and kg_f a kilogram force. Similarly, lb denotes a pound mass and lb_f a pound force.

A slug is a unit of mass such that, when acted on by 1-pound force, a 1-slug mass accelerates at 1 ft/s² (slug = lb_f-s^2/ft). In other words, if a mass of 1 slug is acted on by 32.2 pounds force, it accelerates at 32.2 ft/s² ($= g$). Hence the mass of a body weighing 32.2 lb_f at the earth's surface is 1 slug or

$$m = \frac{w}{g} = \frac{32.2 \text{ lb}_f}{32.2 \text{ ft/s}^2} = 1 \text{ slug}$$

Force. *Force* can be defined as the cause that tends to produce a change in motion of a body on which it acts. In order to move a body, force must be applied to it. Two types of forces are capable of acting on a body: *contact* forces and *field* forces. Contact forces are those that come into direct contact with a body, whereas field forces, such as gravitational force and magnetic force, act on a body but do not come into contact with it.

The units of force are newton (N), dyne (dyn), poundal, kg_f, and lb_f. In SI units and the mks system (a metric absolute system) of units the force unit is the newton. The newton is the force that will give a 1-kilogram mass an acceleration of 1 m/s² or

$$1 \text{ N} = 1 \text{ kg-m/s}^2$$

This means that 9.81 newtons will give a kilogram mass an acceleration of 9.81 m/s². Since the gravitational acceleration is $g = 9.81$ m/s² (as stated

earlier, for engineering calculations, the value of g may be taken as 9.81 m/s²
or 32.2 ft/s²), a mass of 1 kilogram will produce a force on its support of 9.81
newtons.

The force unit in the cgs system (a metric absolute system) is the dyne,
which will give a gram mass an acceleration of 1 cm/s² or

$$1 \text{ dyn} = 1 \text{ g-cm/s}^2$$

In the British absolute system of units the pound is the mass unit and
the poundal is the force unit. A poundal is the force that will give a pound
mass an acceleration of 1 ft/s² or

$$1 \text{ poundal} = 1 \text{ lb-ft/s}^2$$

Thus a force of 32.2 poundals will give a pound mass an acceleration of
32.2 ft/s². And so a mass of 1 pound will produce a force on its support of
32.2 poundals.

The force unit in the metric engineering (gravitational) system is kg_f,
which is a primary dimension in the system. Similarly, in the British engineering system the force unit is lb_f. It is also a primary dimension in this
system of units.

Comments. SI units for force, mass, and length are newton (N),
kilogram mass (kg), and meter (m). The mks units for force, mass, and length
are the same as the SI units. Similarly, the cgs units for force, mass, and
length are dyne (dyn), gram (g), and centimeter (cm) and those for BES units
are pound force (lb_f), slug, and foot (ft). Each of the unit systems is consistent
in that the unit of force accelerates the unit of mass 1 unit of length per second
per second.

Torque or moment of force. *Torque,* or moment of force, is defined as
any cause that tends to produce a change in rotational motion of a body on
which it acts. Torque is the product of a force and the perpendicular distance
from a point of rotation to the action line of the force. The units of torque are
units of force times length, such as N-m, dyn-cm, kg_f-m, and lb_f-ft.

Rigid body. When any real body is accelerated, internal elastic deflections are always present. If these internal deflections are negligibly small
relative to the gross motion of the entire body, the body is called a *rigid body.*
Thus a rigid body does not deform. In other words, for a pure translational
motion, every point in a rigid body has identical motion.

Moments of inertia. The *moment of inertia J* of a rigid body about an
axis is defined by

$$J = \int r^2 \, dm$$

where *dm* is an element of mass, *r* is distance from the axis to *dm*, and integration is performed over the body. In considering moments of inertia, we assume that the rotating body is perfectly rigid. Physically, the moment of inertia of a body is a measure of its resistance to angular acceleration.

Consider a rectangular (xyz) coordinate system in a body. The moment of inertia of a body about the x axis is

$$J_x = \int (y^2 + z^2)\, dm$$

Similarly, those about the y and z axes are

$$J_y = \int (z^2 + x^2)\, dm$$

$$J_z = \int (x^2 + y^2)\, dm$$

If the width (in the z direction) of a flat body, such as a disk, is negligible compared with the x or y dimension, then

$$J_x = \int y^2\, dm$$

$$J_y = \int x^2\, dm$$

$$J_z = \int (x^2 + y^2)\, dm = J_x + J_y$$

Table 2-3 gives moments of inertia for common shapes.

Example 2-1. Figure 2-1 shows a homogeneous cylinder of radius R and length L. The moment of inertia of this cylinder about axis AA' can be obtained as follows.

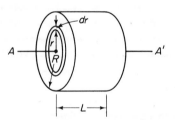

Fig. 2-1. Homogeneous cylinder.

Consider a ring-shaped mass element of infinitesimal width dr at radius r. The mass of this ring-shaped element is $2\pi r L \rho\, dr$, where ρ is the density of this cylinder. Thus

$$dm = 2\pi r L \rho\, dr$$

Consequently,

$$J = \int_0^R r^2 2\pi r L \rho\, dr = 2\pi L \rho \int_0^R r^3\, dr = \frac{\pi L \rho R^4}{2}$$

Since the entire mass m of the cylinder body is $m = \pi R^2 L \rho$, we obtain

$$J = \tfrac{1}{2} m R^2$$

Table 2-3. MOMENTS OF INERTIA

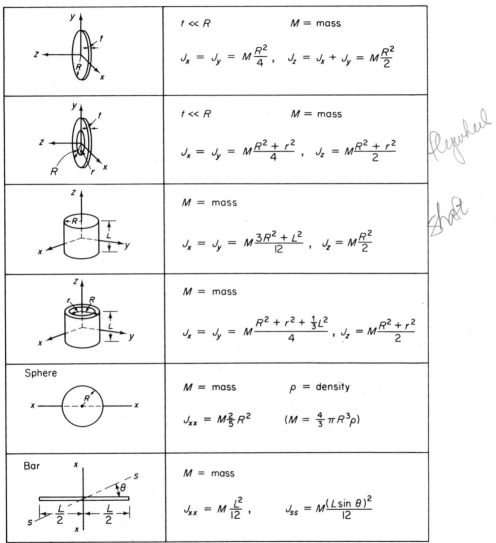

	$t \ll R$ $\qquad$ M = mass $$J_x = J_y = M\frac{R^2}{4}, \quad J_z = J_x + J_y = M\frac{R^2}{2}$$
	$t \ll R$ $\qquad$ M = mass $$J_x = J_y = M\frac{R^2 + r^2}{4}, \quad J_z = M\frac{R^2 + r^2}{2}$$
	M = mass $$J_x = J_y = M\frac{3R^2 + L^2}{12}, \quad J_z = M\frac{R^2}{2}$$
	M = mass $$J_x = J_y = M\frac{R^2 + r^2 + \frac{1}{3}L^2}{4}, \quad J_z = M\frac{R^2 + r^2}{2}$$
Sphere	M = mass $\qquad$ ρ = density $$J_{xx} = M\tfrac{2}{5}R^2 \qquad (M = \tfrac{4}{3}\pi R^3 \rho)$$
Bar	M = mass $$J_{xx} = M\frac{L^2}{12}, \qquad J_{ss} = M\frac{(L\sin\theta)^2}{12}$$

flywheel

shaft

Radius of gyration. The *radius of gyration* of a rigid body with reference to an axis is a length k, which, when squared and multiplied by the mass m of a rigid body, is equal to the moment of inertia J of the body with reference to the same inertia axis, or

$$mk^2 = J$$

Therefore

$$k = \sqrt{\frac{J}{m}}$$

Physically, if it were possible to concentrate all the mass of the rigid body at a point of some distance k from the axis in question, so that mk^2 is equal to the moment of inertia J of the body with respect to that axis, then the distance k is the radius of gyration.

Example 2-2. Suppose that a rigid body weighing 161 lb$_f$ has a moment of inertia about a particular axis of 125 slug-ft^2. What is the radius of gyration about this axis?

The mass of the rigid body is

$$m = \frac{w}{g} = \frac{161}{32.2} = 5 \text{ slugs}$$

The radius of gyration k is

$$k = \sqrt{\frac{J}{m}} = \sqrt{\frac{125}{5}} = 5 \text{ ft}$$

In this example we used units of the British engineering system. As an exercise, let us interpret this problem in terms of other systems of units.

1.   *SI units or mks units (refer to Tables 2-1 and 2-2).*

A rigid body weighs

$$w = 161 \text{ lb}_f = 161 \times 4.448 \text{ N} = 716 \text{ N} = 716 \text{ kg-m/s}^2$$

To get mass m, we divide this value of w by $g = 9.81$ m/s^2.

$$m = \frac{w}{g} = \frac{716 \text{ kg-m/s}^2}{9.81 \text{ m/s}^2} = 73.0 \text{ kg}$$

The moment of inertia J becomes

$$J = 125 \text{ slug-ft}^2 = 125 \times 1.356 \text{ kg-m}^2 = 169.5 \text{ kg-m}^2$$

Hence the radius of gyration k is

$$k = \sqrt{\frac{J}{m}} = \sqrt{\frac{169.5}{73.0}} = 1.52 \text{ m}$$

2.   *Metric engineering (gravitational) units (refer to Tables 2-1 and 2-2).*

A rigid body weighs

$$w = 161 \text{ lb}_f = 161 \times 0.4536 \text{ kg}_f = 73.0 \text{ kg}_f$$

Mass m is then obtained as

$$m = \frac{w}{g} = \frac{73.0 \text{ kg}_f}{9.81 \text{ m/s}^2} = 7.44 \text{ kg}_f\text{-s}^2/\text{m}$$

Also, the moment of inertia J is

$$J = 125 \text{ slug-ft}^2 = 125 \times 0.1382 \text{ kg}_f\text{-s}^2\text{-m} = 17.28 \text{ kg}_f\text{-s}^2\text{-m}$$

Consequently,

$$k = \sqrt{\frac{J}{m}} = \sqrt{\frac{17.28}{7.44}} = 1.52 \text{ m}$$

Moment of inertia about axis other than the geometrical axis. Sometimes it is necessary to calculate the moment of inertia of a homogeneous rigid body about an axis other than its geometrical axis. If the axes are parallel, this process can be done easily. The moment of inertia about an axis that is a distance x from the geometrical axis passing through the center of gravity is the sum of the moment of inertia about the geometrical axis and the moment of inertia of the body considered concentrated at the center of gravity about the new axis.

Example 2-3. Consider the system shown in Fig. 2-2, where a homogeneous cylinder of mass m and radius R rolls on a flat surface. Find the moment of inertia J_x of the cylinder about its line of contact (axis xx') with the surface.

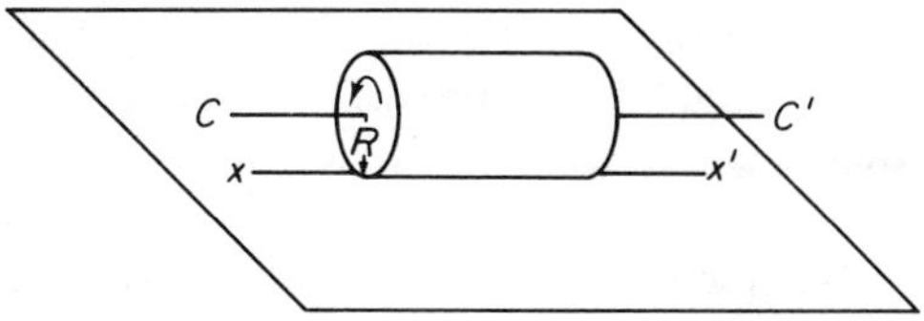

Fig. 2-2. Homogeneous cylinder rolling on a flat surface.

The moment of inertia of the cylinder about axis CC' is

$$J_C = \tfrac{1}{2}mR^2$$

The moment of inertia of the cylinder about axis xx' when mass m is considered concentrated at the center of gravity is mR^2. Thus the moment of inertia J_x of the cylinder about axis xx' is

$$J_x = J_C + mR^2 = \tfrac{1}{2}mR^2 + mR^2 = \tfrac{3}{2}mR^2$$

Displacement, velocity, and acceleration. *Displacement* x is a change in position from one point in a frame of reference to another. *Velocity* v is the time derivative of displacement or

$$v = \dot{x}$$

Acceleration a is the time derivative of velocity or

$$a = \dot{v} = \ddot{x}$$

It is important to distinguish speed from velocity. *Speed* is the rate at which a body moves along its path. Thus speed refers merely to rapidity of motion and is a scalar quantity. *Velocity* is the time rate of change of position. Therefore velocity represents rapidity, direction, and sense, and it is a vector quantity.

Angular displacement, angular velocity, and angular acceleration. For engineering purposes, *angular displacement* θ is measured in radians and is defined as positive if measured in the counterclockwise direction. *Angular* *velocity* ω is the time derivative of angular displacement or

$$\omega = \dot{\theta}$$

and *angular acceleration* α is the time derivative of angular velocity or

$$\alpha = \dot{\omega} = \ddot{\theta}$$

Note that angular speed refers to the rapidity of angular motion and that it is a scalar quantity. Angular velocity is the time rate of change of angular position, and so it represents rapidity of angular motion, direction, and sense. It is a vector quantity.

Furthermore, note that if velocity or angular velocity is measured with respect to a fixed reference, then it is called *absolute velocity* or absolute angular velocity. Otherwise it is called *relative velocity* or relative angular velocity. Similarly, if acceleration or angular acceleration is measured with respect to a nonaccelerating reference, it is called *absolute acceleration* or absolute angular acceleration; otherwise the term *relative acceleration* or relative angular acceleration is used.

Newton's laws. There are three well-known laws called *Newton's laws.* Newton's first law, which concerns conservation of momentum, states that the total momentum of a mechanical system is constant in the absence of external forces. Momentum is the product of mass m and velocity v, or mv, for translational or linear motion. For rotational motion, momentum is the product of moment of inertia J and angular velocity ω, or $J\omega$, and is called angular momentum.

Newton's second law gives the force-acceleration relationship of a rigid body or the torque-angular acceleration relationship of a rigid rotating body. The third law concerns action and reaction and, in effect, states that every action is always opposed by an equal reaction.

Newton's second law (for translational motion). For a translational motion, Newton's second law says that the acceleration of any rigid body is directly proportional to the force acting on it and is inversely proportional to the mass of the body. That is,

$$\text{Force} = (\text{mass})(\text{acceleration})$$

Suppose that forces are acting on a mass m. If $\sum F$ is the sum of all forces acting in a given direction, then

$$\sum F = ma \tag{2-1}$$

where a is the resulting absolute acceleration in that direction. The line of action of the force acting on a mass must pass through the center of mass. Otherwise rotational motion will also be involved. Rotational motion is not defined by Eq. (2-1) but is given by Eq. (2-2) below.

Newton's second law (for rotational motion). For a rigid body in pure rotation about a fixed axis, Newton's second law states that

$$\sum \text{Torques} = (\text{moment of inertia})(\text{angular acceleration})$$

or

$$\sum T = J\alpha \tag{2-2}$$

where $\sum T$ is the sum of all torques acting about a given axis, J is the moment of inertia of a body about that axis, and α is the angular acceleration.

Example 2-4. A ball of mass m is thrown vertically upward with an initial velocity of 10 m/s. The vertical displacement x is measured upward from the starting point. What is the maximum height that the ball will reach? Assume that air friction is negligible.

The only force acting on the ball is the gravitational force $-mg$. Applying Newton's second law, we have

$$m\ddot{x} = -mg$$

Noting that $\dot{x}(0) = 10$ m/s and $x(0) = 0$, we obtain

$$\dot{x}(t) = -gt + \dot{x}(0) = -gt + 10$$

$$x(t) = -\tfrac{1}{2}gt^2 + 10t$$

At the instant that the ball reaches the highest point the velocity is zero. Let us define this instant as t_p. Then

$$\dot{x}(t_p) = -gt_p + 10 = 0$$

or

$$t_p = \frac{10}{g} = \frac{10 \text{ m/s}}{9.81 \text{ m/s}^2} = 1.02 \text{ s}$$

The maximum height is thus obtained as $x(t_p)$ or

$$x(t_p) = -\tfrac{1}{2}gt_p^2 + 10t_p$$

$$= -\tfrac{1}{2} \times 9.81 \times 1.02^2 + 10 \times 1.02 = 5.10$$

And so the maximum height that the ball will reach is 5.10 m.

2-2　MATHEMATICAL MODELING

A mathematical model for any mechanical system can be developed by applying Newton's laws to the system. In this section we deal first with the problem of deriving mathematical models of mechanical systems, and then we present the response analysis of simple mechanical systems.

Three types of basic elements may be necessary in modeling mechanical systems—inertia elements, spring elements, and damper elements. We begin this section with a brief explanation of each of the three types.

Inertia elements. By *inertia elements* we mean masses and moments of inertia. Since masses and moments of inertia were presented in detail in Sec. 2-1, a short discussion will suffice here.

Inertia may be defined as the change in force (torque) required to make a unit change in acceleration (angular acceleration). That is,

$$\text{Inertia (mass)} = \frac{\text{change in force}}{\text{change in acceleration}} \quad \frac{\text{N}}{\text{m/s}^2} \text{ or kg}$$

$$\text{Inertia (moment of inertia)} = \frac{\text{change in torque}}{\text{change in angular acceleration}}$$

$$\frac{\text{N-m}}{\text{rad/s}^2} \text{ or kg-m}^2$$

Spring elements. A linear *spring* is a mechanical element that can be deformed by an external force such that the deformation is directly proportional to the force or torque applied to it.

Figure 2-3 is a schematic diagram of a spring. Here we consider translational motion only. The spring has been deflected from its initial position by a force applied to each end. Positions x_1 and x_2 of the ends of the spring are measured relative to the same frame of reference. The forces at both ends of the spring are on the same line and are equal in magnitude. Then force F and net displacement x of the ends of the spring are related by

$$F = kx = k(x_1 - x_2) \tag{2-3}$$

where k is a proportionality constant called a *spring constant*. The dimension of spring constant k is force/displacement.

Fig. 2-3. Spring.

For rotational motion, the torque applied to the ends of a torsional or rotational spring and the net change in the angular displacements of the ends are related by

$$T = k\theta = k(\theta_1 - \theta_2) \tag{2-4}$$

where

T = torque applied to the ends of torsional spring
θ_1 = angular displacement of one end
θ_2 = angular displacement of other end
$\theta = \theta_1 - \theta_2$ = net angular displacement of ends
k = torsional spring constant

The dimension of torsional spring constant k is torque/angular displacement, where angular displacement is measured in radians.

When a linear spring is stretched, a point is reached in which the force per unit displacement begins to change and the spring becomes a nonlinear spring. If stretched farther, a point is reached in which the material will either break or yield. For practical springs, therefore, the assumption of linearity may be good only for relatively small net displacements. Figure 2-4 shows the force-displacement characteristic curves for linear and nonlinear springs.

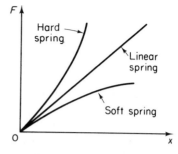

Fig. 2-4. Force-displacement characteristic curves for linear and nonlinear springs.

For practical springs, spring constant k (or an equivalent spring constant for a nonlinear spring) may be defined as

Spring constant k (for translational spring)

$$= \frac{\text{change in force}}{\text{change in displacement of spring}} \quad \frac{\text{N}}{\text{m}}$$

Spring constant k (for torsional spring)

$$= \frac{\text{change in torque}}{\text{change in angular displacement of spring}} \quad \frac{\text{N-m}}{\text{rad}}$$

Spring constants indicate stiffness; a large value of k corresponds to a hard spring and a small value of k to a soft spring. The reciprocal of spring constant k is called *compliance* or *mechanical capacitance* C. Thus $C = 1/k$. Compliance or mechanical capacitance indicates softness of a spring.

Compliance or mechanical capacitance C (for translational spring)

$$= \frac{\text{change in displacement of spring}}{\text{change in force}} \quad \frac{\text{m}}{\text{N}}$$

Compliance or mechanical capacitance C (for torsional spring)

$$= \frac{\text{change in angular displacement of spring}}{\text{change in torque}} \quad \frac{\text{rad}}{\text{N-m}}$$

Note that in terms of compliance or mechanical capacitance C, Eqs. (2-3) and (2-4) become

$$x = CF \quad \text{and} \quad \theta = CT$$

Practical spring versus ideal spring. All practical springs have inertia and damping. In our analysis in this book, however, we assume that the effect of the mass in a spring is small—that is, the inertia force due to acceleration of the spring is negligibly small compared to the spring force. Also, we assume that the damping effect of the spring is negligibly small.

An ideal spring, in comparison to a practical spring, will have neither mass nor damping (internal friction) and will obey linear force-displacement law as given by Eq. (2-3) or linear torque-angular displacement law as given by Eq. (2-4).

Damper elements. A *damper* is a mechanical element that dissipates energy in the form of heat instead of storing it. Figure 2-5(a) shows a schematic diagram of a translational damper or dashpot. It consists of a piston and an oil-filled cylinder. Any relative motion between piston rod and cylinder is resisted by oil because oil must flow around the piston (or through orifices provided in the piston) from one side to the other. Essentially, the damper absorbs energy and the absorbed energy is dissipated as heat that flows away to the surroundings.

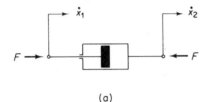

(a)

(b)

Fig. 2-5. (a) Translational damper; (b) torsional (or rotational) damper.

In Fig. 2-5(a) velocities $\dot{x}_1$ and $\dot{x}_2$ are taken relative to the same frame of reference. The forces at the ends of the translational damper are on the same line and are equal in magnitude. In the damper, the force F acting on it is proportional to the velocity difference $\dot{x}$ of both ends or

$$F = b\dot{x} = b(\dot{x}_1 - \dot{x}_2) \tag{2-5}$$

where proportionality constant b relating external force F and velocity difference $\dot{x}$ is called a *viscous friction coefficient* or *viscous friction constant*. The dimension of the viscous friction coefficient is force/velocity. Note that the initial positions of both ends of the damper do not appear in the equation.

For the torsional or rotational damper shown in Fig. 2-5(b), the torque applied to the ends of the damper is proportional to the angular velocity difference of both ends or

$$T = b\dot{\theta} = b(\dot{\theta}_1 - \dot{\theta}_2) \tag{2-6}$$

where

$T =$ torque applied to the ends of torsional damper
$\dot{\theta}_1 =$ angular velocity of one end
$\dot{\theta}_2 =$ angular velocity of other end
$\dot{\theta} = \dot{\theta}_1 - \dot{\theta}_2 =$ relative angular velocity of ends
$b =$ torsional viscous friction coefficient (torsional viscous friction constant)

The dimension of the torsional viscous friction coefficient b is torque/ angular velocity.

Note that a damper is an element that provides resistance in mechanical motion, and as such its effect on the dynamic behavior of a mechanical system is similiar to that of an electrical resistor on the dynamic behavior of an electrical system. Consequently, a damper is often referred to as a *mechanical resistance element* and the viscous friction coefficient as the *mechanical resistance*.

Mechanical resistance b (for translational damper)

$$= \frac{\text{change in force}}{\text{change in velocity}} \quad \frac{\text{N}}{\text{m/s}}$$

Mechanical resistance b (for torsional damper)

$$= \frac{\text{change in torque}}{\text{change in angular velocity}} \quad \frac{\text{N-m}}{\text{rad/s}}$$

Practical damper versus ideal damper. All practical dampers produce inertia and spring effects. In this book, however, we assume that these effects are negligible.

An ideal damper is massless and springless, dissipates all energy, and obeys linear force-velocity law or linear torque-angular velocity law as given by Eq. (2-5) or Eq. (2-6), respectively.

Nonlinear friction. Friction that obeys linear law is called *linear friction*, whereas friction that does not is described as *nonlinear*. Examples of nonlinear friction include static friction, sliding friction, and square-law friction. The topics of static friction and sliding friction are discussed in Section 2-4.

Square-law friction occurs when a solid body moves in a fluid medium. Here the friction force is essentially proportional to velocity at low speeds and becomes proportional to the square of velocity at high speeds. Figure 2-6 shows a characteristic curve for square-law friction.

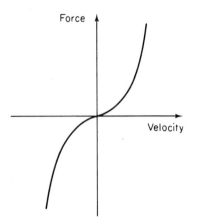

Fig. 2-6. Characteristic curve for square-law friction.

Forced response and natural response. The behavior determined by a forcing function is called a *forced response* and that due to initial conditions (initial energy storages) is called a *natural response*. The period between the initiation of a natural response and the ending is referred to as the *transient period*. After the natural response has become negligibly small, conditions are said to have reached a steady state.

Rotational systems. A schematic diagram of a rotor mounted in bearings is shown in Fig. 2-7. The moment of inertia of the rotor about the

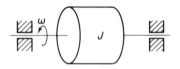

Fig. 2-7. Rotor mounted in bearings.

axis of rotation is J. Let us assume that at $t = 0$ the rotor is rotating at the angular velocity of $\omega(0) = \omega_0$. We also assume that the friction in the bearings is viscous friction and that no external torque is applied to the rotor. Then the only torque acting on the rotor is the friction torque $b\omega$ in the bearings.

Applying Newton's second law, Eq. (2-2), we obtain the equation of motion

$$J\dot{\omega} = -b\omega, \qquad \omega(0) = \omega_0$$

or

$$J\dot{\omega} + b\omega = 0 \tag{2-7}$$

Equation (2-7) is the equation of motion, as well as a mathematical model, of the system. A system governed by a first-order differential equation, such as this one, is called a *first-order system*.

In solving Eq. (2-7), we can assume

$$\omega(t) = \omega_0 e^{\lambda t} \tag{2-8}$$

Differentiating both sides of Eq. (2-8) with respect to t gives

$$\dot{\omega} = \omega_0 \lambda e^{\lambda t}$$

Next, we substitute this $\dot{\omega}$ and Eq. (2-8) into Eq. (2-7) and obtain

$$J\omega_0 \lambda e^{\lambda t} + b\omega_0 e^{\lambda t} = 0$$

Since $\omega_0 e^{\lambda t} \neq 0$, this last equation yields

$$J\lambda + b = 0$$

This result is called the *characteristic equation* for the system. The characteristic equation determines the value of λ.

$$\lambda = -\frac{b}{J}$$

Hence from Eq. (2-8) we have

$$\omega(t) = \omega_0 e^{-(b/J)t}$$

The angular velocity decreases exponentially, as shown in Fig. 2-8.

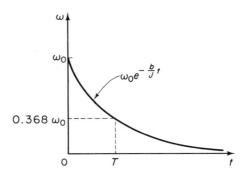

Fig. 2-8. Angular velocity curve for the rotor system shown in Fig. 2-7.

Since the exponential factor $e^{-(b/J)t}$ approaches zero as t increases without limit, mathematically the response lasts forever. When dealing with such an exponentially decaying response, it is convenient to depict the response in terms of a time constant. A *time constant* is that value of time that makes the exponent equal to -1. For this system, time constant T is equal to J/b, or $T = J/b$. When $t = T$, the value of the exponential factor is

$$e^{-T/T} = e^{-1} = 0.368$$

In other words, when time t in seconds is equal to the time constant, the exponential factor is reduced to approximately 37% of its initial value, as shown in Fig. 2-8.

Spring-mass system. Figure 2-9 depicts a system consisting of a mass and a spring. Here the mass is suspended by the spring. For the vertical motion, two forces are acting on the mass—spring force ky and gravitational

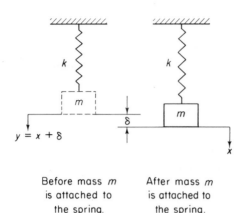

$$\delta = \text{Static deflection}$$

$$y = x + \delta$$

Before mass m
is attached to
the spring.

After mass m
is attached to
the spring.

Fig. 2-9. Spring-mass system.

force mg. In the diagram the positive direction of displacement y is defined downward. Note that the gravitational force pulls the mass downward. If the mass is pulled downward by an external force and then released, the spring force acts upward and tends to pull the mass upward. And so by applying Newton's second law to this system, we obtain the equation of motion

$$m\ddot{y} = \sum \text{forces} = -ky + mg$$

or

$$m\ddot{y} + ky = mg \tag{2-9}$$

The gravitational force is opposed statically by the equilibrium spring deflection δ. If we measure the displacement from this equilibrium position, then the term mg can be dropped from the equation of motion. Since $k\delta = mg$, by substituting $y = x + \delta$ into Eq. (2-9) and noting that $\delta = \text{constant}$, we have

$$m\ddot{x} + kx = 0 \tag{2-10}$$

which is a mathematical model for the system. Such a system is called a *second-order system*—that is, it is governed by a second-order differential equation.

Unless otherwise stated, when writing equations of motion for systems involving the gravitational force in this book, we measure the displacement of the mass from the equilibrium position in order to eliminate the term mg and simplify the mathematical model.

Free vibration. For the spring-mass system of Fig. 2-9, suppose that the mass is pulled downward and then released with arbitrary initial conditions $x(0)$ and $\dot{x}(0)$. In this case, the mass will oscillate and the motion will be a periodic one. The periodic motion, observed as the system is displaced from its static equilibrium position, is called *free vibration*. It is a natural response due to the initial conditions.

In order to find the mathematical form of the periodic motion, let us solve Eq. (2-10). A useful method in finding the solution is to assume that $x(t)$ is in either an exponential or sinusoidal form. Both approaches, exponential and sinusoidal, will be demonstrated in this chapter.

For a solution of the present problem, we assume that $x(t)$ is in the exponential form.

$$x(t) = Ke^{\lambda t} \tag{2-11}$$

If this equation is substituted into Eq. (2-10), then

$$mK\lambda^2 e^{\lambda t} + kKe^{\lambda t} = 0$$

Dividing both sides by $Ke^{\lambda t}$ gives

$$m\lambda^2 + k = 0$$

which is the characteristic equation for the system. From this characteristic equation we obtain

$$\lambda^2 = -\frac{k}{m}$$

Hence two roots λ_1 and λ_2 are

$$\lambda_1 = j\sqrt{\frac{k}{m}}, \qquad \lambda_2 = -j\sqrt{\frac{k}{m}}$$

and these two values of λ satisfy the assumed solution, Eq. (2-11). Since the second-order differential equation must have two arbitrary constants in its solution, we can write the general solution $x(t)$ as

$$x(t) = K_1 e^{j(\sqrt{k/m})t} + K_2 e^{-j(\sqrt{k/m})t}$$

Next, we can use the following equations (Euler's formula) to express exponential functions in terms of sine and cosine functions:

$$\begin{aligned} e^{j\omega t} &= \cos \omega t + j \sin \omega t \\ e^{-j\omega t} &= \cos \omega t - j \sin \omega t \end{aligned} \tag{2-12}$$

Then the general solution $x(t)$ becomes

$$\begin{aligned} x(t) &= K_1\left(\cos \sqrt{\frac{k}{m}}\, t + j \sin \sqrt{\frac{k}{m}}\, t\right) + K_2\left(\cos \sqrt{\frac{k}{m}}\, t - j \sin \sqrt{\frac{k}{m}}\, t\right) \\ &= j(K_1 - K_2) \sin \sqrt{\frac{k}{m}}\, t + (K_1 + K_2) \cos \sqrt{\frac{k}{m}}\, t \\ &= A \sin \sqrt{\frac{k}{m}}\, t + B \cos \sqrt{\frac{k}{m}}\, t \end{aligned} \tag{2-13}$$

where

$$A = j(K_1 - K_2), \qquad B = K_1 + K_2$$

A and B are now arbitrary constants that depend on the initial conditions $x(0)$ and $\dot{x}(0)$. Equation (2-13) can also be written

$$x(t) = C \cos \left(\sqrt{\frac{k}{m}}\, t + \phi \right)$$

where

$$C = \sqrt{A^2 + B^2}, \qquad \phi = -\tan^{-1} \frac{A}{B}$$

To determine constants A and B in terms of the initial conditions $x(0)$ and $\dot{x}(0)$, we substitute $t = 0$ into Eq. (2-13). Then

$$x(0) = B$$

As a result of differentiating both sides of Eq. (2-13) with respect to t, we have

$$\dot{x}(t) = A\sqrt{\frac{k}{m}} \cos \sqrt{\frac{k}{m}}\, t - B\sqrt{\frac{k}{m}} \sin \sqrt{\frac{k}{m}}\, t$$

Thus

$$\dot{x}(0) = A\sqrt{\frac{k}{m}}$$

It follows that $A = \sqrt{m/k}\, \dot{x}(0)$ and $B = x(0)$. In terms of the initial conditions Eq. (2-13) becomes

$$x(t) = \dot{x}(0)\sqrt{\frac{m}{k}} \sin \sqrt{\frac{k}{m}}\, t + x(0) \cos \sqrt{\frac{k}{m}}\, t \qquad (2\text{-}14)$$

The periodic motion described by this last equation is called a *simple harmonic motion*.

If the initial conditions were given as $x(0) = x_0$ and $\dot{x}(0) = 0$, then by substituting these initial conditions into Eq. (2-14), the displacement $x(t)$ would be given by

$$x(t) = x_0 \cos \sqrt{\frac{k}{m}}\, t$$

The period and frequency of a simple harmonic motion can now be defined as follows. The *period* T is the time required for a periodic motion to repeat itself. In the present case,

$$\text{Period } T = \frac{2\pi}{\sqrt{k/m}} \quad \text{seconds}$$

The *frequency* f of a periodic motion is the number of cycles per second and the standard unit of frequency is the hertz (Hz)—that is, one hertz is one cycle per second. In the present harmonic motion,

$$\text{Frequency } f = \frac{1}{T} = \frac{\sqrt{k/m}}{2\pi} \quad \text{hertz}$$

   The natural frequency or undamped natural frequency is the frequency in the free vibration of a system having no damping. If the natural frequency is measured in hertz (Hz) or cycles per second (cps), it is denoted by f_n. If it is measured in radians per second (rad/s), it is denoted by ω_n. In the present system

$$\omega_n = 2\pi f_n = \sqrt{\frac{k}{m}} \quad \text{rad/s}$$

   It is important to remember that when Eq. (2-10) is written in a form such that the coefficient of the $\ddot{x}$ term is unity,

$$\ddot{x} + \frac{k}{m}x = 0$$

the square root of the coefficient of the x term is the natural frequency ω_n. This means that a mathematical model for the system shown in Fig. 2-9 can be put in the form

$$\ddot{x} + \omega_n^2 x = 0$$

where $\omega_n = \sqrt{k/m}$.

   Experimental determination of moment of inertia. It is possible to calculate moments of inertia for homogeneous bodies having geometrically simple shapes. However, for rigid bodies with complicated shapes or those consisting of materials of various densities, such calculation may be difficult or even impossible; moreover, calculated values may not be accurate. In these instances, experimental determination of moments of inertia is preferable. The process is as follows. We mount a rigid body in frictionless bearings so that it can rotate freely about the axis of rotation around which the moment of inertia is to be determined. Next, we attach a torsional spring with known spring constant k to the rigid body (see Fig. 2-10). The spring is twisted

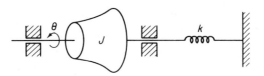

Fig. 2-10. A set-up for experimental determination of moment of inertia.

slightly, released, and the frequency of the resulting simple harmonic motion measured. Since the equation of motion for this system is

$$J\ddot{\theta} + k\theta = 0$$

or

$$\ddot{\theta} + \frac{k}{J}\theta = 0$$

the natural frequency ω_n is

$$\omega_n = \sqrt{\frac{k}{J}}$$

and the period of vibration is

$$T = \frac{2\pi}{\omega_n} = \frac{2\pi}{\sqrt{k/J}}$$

The moment of inertia J is then determined as

$$J = \frac{kT^2}{4\pi^2}$$

Similarly, in the spring-mass system of Fig. 2-9, if the spring constant k is known and the period T of the free vibration is measured, then mass m can be calculated from

$$m = \frac{kT^2}{4\pi^2}$$

Spring-mass-damper system. Most physical systems involve some type of damping—viscous damping, dry damping, magnetic damping, and so on. Such damping not only slows the motion, it also causes the motion to stop eventually. In the following discussion we shall consider a simple mechanical system involving viscous damping. Note that a typical viscous damping element is a damper or dashpot.

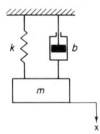

Fig. 2-11. Spring-mass-damper system.

Figure 2-11 is a schematic diagram of a spring-mass-damper system. Suppose that the mass is pulled downward and then released. If the damping is light, vibratory motion will occur. (The system is said to be underdamped.) If the damping is heavy, vibratory motion will not occur. (The system is said to be overdamped.) A critically damped system is one in which the degree of damping is such that the resultant motion is on the borderline between the underdamped and overdamped cases. Regardless of the system being underdamped, overdamped, or critically damped, because of the presence of damper the free vibration or free motion will diminish with time. This free vibration is called *transient*.

In the system shown in Fig. 2-11, for the vertical motion, three forces are acting on the mass: spring force, damping force, and gravitational force. As noted, if we measure the displacement of the mass from a static equilibrium position (so that the gravitational force is balanced by the equilibrium spring deflection), the gravitational force will not enter the equation of motion. So by measuring the displacement x from the static equilibrium position, we obtain the equation of motion

$$m\ddot{x} = \sum \text{forces} = -kx - b\dot{x}$$

or

$$m\ddot{x} + b\dot{x} + kx = 0 \qquad (2\text{-}15)$$

Equation (2-15), which describes the motion of the system, is also a mathematical model of the system.

Only the underdamped case is considered in our present analysis. (A more complete analysis of this system for the underdamped, overdamped, and critically damped cases is given in Chapter 7.)

Let us solve Eq. (2-15) for a particular case. Suppose that $m = 0.1$ slug, $b = 0.4$ lb_f-s/ft, and $k = 4$ lb_f/ft. Then Eq. (2-15) becomes

$$0.1\ddot{x} + 0.4\dot{x} + 4x = 0$$

or

$$\ddot{x} + 4\dot{x} + 40x = 0 \qquad (2\text{-}16)$$

Let us assume that

$$x(t) = Ke^{\lambda t} \qquad (2\text{-}17)$$

When Eq. (2-17) is substituted into Eq. (2-16), the result is

$$K\lambda^2 e^{\lambda t} + 4K\lambda e^{\lambda t} + 40Ke^{\lambda t} = 0$$

Dividing this last equation by $Ke^{\lambda t}$ gives

$$\lambda^2 + 4\lambda + 40 = 0$$

This quadratic equation is the characteristic equation for the system considered. As such, it has two roots

$$\lambda_1 = -2 + j6, \qquad \lambda_2 = -2 - j6$$

These two values of λ satisfy the assumed solution, Eq. (2-17). Consequently, we assume that the solution contains two terms of the form shown in Eq. (2-17) and write the general solution $x(t)$ as

$$x(t) = K_1 e^{(-2+j6)t} + K_2 e^{(-2-j6)t}$$
$$= e^{-2t}(K_1 e^{j6t} + K_2 e^{-j6t})$$
$$= e^{-2t}(A \sin 6t + B \cos 6t) \qquad (2\text{-}18)$$

where K_1 and K_2 are arbitrary constants and $A = j(K_1 - K_2)$, $B = K_1 + K_2$. In obtaining Eq. (2-18), we used Euler's formula, given in Eq. (2-12).

Let us obtain the motion $x(t)$ when the mass is pulled downward at $t = 0$, such that $x(0) = x_0$, and released with zero velocity, $\dot{x}(0) = 0$. Then the arbitrary constants A and B can be determined as follows. First, substitution of $t = 0$ into Eq. (2-18) yields

$$x(0) = B = x_0$$

Then differentiating Eq. (2-18) with respect to t,

$$\dot{x}(t) = -2e^{-2t}(A \sin 6t + B \cos 6t) + e^{-2t}(6A \cos 6t - 6B \sin 6t)$$
$$= -2e^{-2t}[(A + 3B) \sin 6t + (B - 3A) \cos 6t]$$

Hence

$$\dot{x}(0) = -2(B - 3A) = 0$$

and so

$$A = \tfrac{1}{3}x_0, \qquad B = x_0$$

The solution $x(t)$ becomes

$$x(t) = e^{-2t}(\tfrac{1}{3} \sin 6t + \cos 6t)x_0$$

Besides depicting a damped sinusoidal vibration (Fig. 2-12), this equation represents the free vibration of the spring-mass-damper system with the given numerical values.

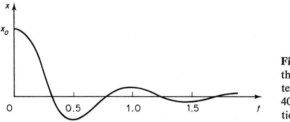

Fig. 2-12. Free vibration of the spring-mass-damper system described by $\ddot{x} + 4\dot{x} + 40 x = 0$ with initial conditions $x(0) = x_0$ and $\dot{x}(0) = 0$.

Comments. The numerical values in the preceding problem were stated in BES units. Let us convert these values into units of other systems.

1. *SI units or mks units (refer to Tables 2-1 and 2-2).*

    $m = 0.1$ slug $= 1.459$ kg

    $b = 0.4$ lb$_f$-s/ft $= 0.4 \times 4.448/0.3048$ N-s/m $= 5.837$ N-s/m

    $k = 4$ lb$_f$/ft $= 4 \times 4.448/0.3048$ N/m $= 58.37$ N/m

Hence Eq. (2-15) becomes

$$1.459\ddot{x} + 5.837\dot{x} + 58.37x = 0$$

or

$$\ddot{x} + 4\dot{x} + 40x = 0$$

which is the same as Eq. (2-16).

2. *Metric engineering (gravitational) units (refer to Tables 2-1 and 2-2).*

    $m = 0.1$ slug $= 0.1488$ kg$_f$-s^2/m

    $b = 0.4$ lb$_f$-s/ft $= 0.4 \times 0.4536/0.3048$ kg$_f$-s/m $= 0.5953$ kg$_f$-s/m

    $k = 4$ lb$_f$/ft $= 4 \times 0.4536/0.3048$ kg$_f$/m $= 5.953$ kg$_f$/m

Therefore Eq. (2-15) becomes

$$0.1488\ddot{x} + 0.5953\dot{x} + 5.953x = 0$$

or

$$\ddot{x} + 4\dot{x} + 40x = 0$$

which is the same as Eq. (2-16).

Note that as long as we use consistent units, the differential equation (mathematical model) of the system remains the same.

Summary on general procedure for obtaining responses. A general procedure for determining the natural behavior of mechanical systems may be summarized as below.

1. Obtain a mathematical model of the system. (Write the differential equation for the system, using Newton's second law.)
2. If the system involves damping, it is convenient to assume that the solution is in the form of an exponential function with undetermined constants. (If damping is not involved, we can assume that the solution is in a sinusoidal form with undetermined constants. See Section 2-3.)
3. Determine the exponent from the characteristic equation. (For the sinusoidal solution, determine the natural frequency from the characteristic equation.)
4. Evaluate the undetermined constants, using the initial conditions.

2-3 MECHANICAL SYSTEMS WITH TWO OR MORE DEGREES OF FREEDOM

In real-life situations, motion of a mechanical system may be simultaneously translational and rotational in three-dimensional space, and parts of the system may have path constraints on where they can move. The geometrical description of such motions can become complicated, but the fundamental physical laws, Eqs. (2-1) and (2-2), still apply.

For the spring-mass-damper system discussed in Section 2-2, only one coordinate x was needed to specify the motion of the system. However, more than one coordinate is necessary in describing the motion of complicated systems. The term used to describe the minimum number of independent coordinates required to specify this motion is *degrees of freedom*.

Degrees of freedom. The number of degrees of freedom that a mechanical system possesses is the minumim number of independent coordinates required to specify the positions of all its elements. For instance, if only one independent coordinate is needed to specify the geometric location of the mass of a system in space completely, it is a one-degree-of-freedom system. That is, a rigid body rotating on an axis has one degree of freedom, whereas a rigid body in space has six degrees of freedom—three translational and three rotational.

It is important to note that, in general, neither the number of masses nor any other obvious quantity will always lead to a correct assessment of the number of degrees of freedom.

In terms of the number of equations of motion and the number of constraints, the degrees of freedom may be written

The number of degrees of freedom

$$= \text{(number of equations of motion)}$$

$$- \text{(number of equations of constraint)}$$

Example 2-5. Referring to the systems shown in Fig. 2-13, let us find the degrees of freedom for each.

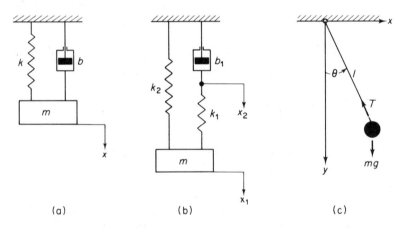

(a)   (b)   (c)

Fig. 2-13. Mechanical systems.

(a) We begin with the system shown in Fig. 2-13(a). If mass m is constrained to move vertically, only one coordinate x is required to define the location of the mass at any time. Thus the system shown in Fig. 2-13(a) has one degree of freedom.

We can verify this statement by counting the number of equations of motion and the number of equations of constraint. This system has one equation of motion

$$m\ddot{x} + b\dot{x} + kx = 0$$

and no equation of constraint. Consequently,

Degree of freedom $= 1 - 0 = 1$

(b) Next, consider the system shown in Fig. 2-13(b). The equations of motion here are

$$m\ddot{x}_1 + k_1(x_1 - x_2) + k_2 x_1 = 0$$
$$k_1(x_1 - x_2) = b_1 \dot{x}_2$$

So the number of equations of motion is two. There is no equation of constraint. Therefore

$$\text{Degrees of freedom} = 2 - 0 = 2$$

(c) Finally, consider the pendulum system shown in Fig. 2-13(c). Define the coordinates of the pendulum mass as (x, y). Then the equations of motion are

$$m\ddot{x} = -T\sin\theta$$
$$m\ddot{y} = mg - T\cos\theta$$

Thus the number of equations of motion is two. The constraint equation for this system is

$$x^2 + y^2 = l^2$$

The number of equation of constraint is one. And so

$$\text{Degree of freedom} = 2 - 1 = 1$$

Note that when physical constraints are present, the most convenient coordinate system may not be the rectangular one. In the pendulum system of Fig. 2-13(c) the pendulum is constrained to move in a circular path. The most convenient coordinate system here would be a polar coordinate system. Then the only coordinate that is needed is the angle θ through which the pendulum has swung. The rectangular coordinates x, y and polar coordinates θ, l (where l is a constant) are related by

$$x = l\sin\theta, \qquad y = l\cos\theta$$

In terms of the polar coordinate system, the equation of motion becomes

$$ml^2\ddot{\theta} = -mgl\sin\theta$$

or

$$\ddot{\theta} + \frac{g}{l}\sin\theta = 0$$

Note that since l is a constant, the configuration of the system can be specified by one coordinate, θ. Consequently, it is a *one-degree-of-freedom system*.

Two-degrees-of-freedom system. A *two-degrees-of-freedom system* requires two independent coordinates to specify the system's configuration. Consider the system shown in Fig. 2-14, which illustrates the two-degrees-of-freedom case, and let us derive a mathematical model of it. By applying Newton's second law to mass m_1 and mass m_2, we have

$$m_1\ddot{x}_1 = -k_1 x_1 - k_2(x_1 - x_2)$$
$$m_2\ddot{x}_2 = -k_3 x_2 - k_2(x_2 - x_1)$$

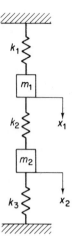

Fig. 2-14. Mechanical system.

Rearranging, the equations of motion become

$$m_1\ddot{x}_1 + k_1 x_1 + k_2(x_1 - x_2) = 0 \qquad (2\text{-}19)$$

$$m_2\ddot{x}_2 + k_3 x_2 + k_2(x_2 - x_1) = 0 \qquad (2\text{-}20)$$

These two equations represent a mathematical model of the system.

Free vibration of two-degrees-of-freedom system. Next, consider Fig. 2-15, which is a special case of the system given in Fig. 2-14. The equations of motion for the present system can be obtained by substituting $m_1 = m_2 = m$ and $k_1 = k_2 = k_3 = k$ into Eqs. (2-19) and (2-20) as follows.

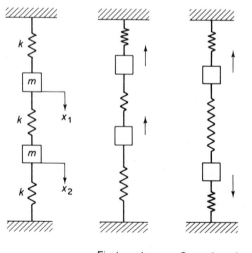

First mode   Second mode   **Fig. 2-15.** Mechanical system and its
of vibration   of vibration   two modes of vibration.

$$m\ddot{x}_1 + 2kx_1 - kx_2 = 0 \tag{2-21}$$

$$m\ddot{x}_2 + 2kx_2 - kx_1 = 0 \tag{2-22}$$

Let us examine the free vibration of this system. In order to find the natural frequencies of the free vibration, we assume that the motion is harmonic. That is, we assume

$$x_1 = A \sin \omega t, \qquad x_2 = B \sin \omega t$$

Then

$$\ddot{x}_1 = -A\omega^2 \sin \omega t, \qquad \ddot{x}_2 = -B\omega^2 \sin \omega t$$

If the preceding expressions are substituted into Eqs. (2-21) and (2-22), the resulting equations are

$$(-mA\omega^2 + 2kA - kB) \sin \omega t = 0$$

$$(-mB\omega^2 + 2kB - kA) \sin \omega t = 0$$

Since these equations must be satisfied at all times, and since $\sin \omega t$ cannot be zero at all times, the quantities in the brackets must be equal to zero. Thus

$$-mA\omega^2 + 2kA - kB = 0$$

$$-mB\omega^2 + 2kB - kA = 0$$

Rearranging, we have

$$(2k - m\omega^2)A - kB = 0 \tag{2-23}$$

$$-kA + (2k - m\omega^2)B = 0 \tag{2-24}$$

Equation (2-23) yields

$$\frac{A}{B} = \frac{k}{2k - m\omega^2}$$

and Eq. (2-24)

$$\frac{A}{B} = \frac{2k - m\omega^2}{k}$$

And so we obtain

$$\frac{k}{2k - m\omega^2} = \frac{2k - m\omega^2}{k}$$

or

$$\omega^4 - \frac{4k}{m}\omega^2 + 3\frac{k^2}{m^2} = 0$$

which can be rewritten as

$$\left(\omega^2 - \frac{k}{m}\right)\left(\omega^2 - \frac{3k}{m}\right) = 0$$

or

$$\omega^2 = \frac{k}{m}, \qquad \omega^2 = \frac{3k}{m}$$

Consequently, ω^2 has two values, the first representing the first natural frequency ω_1 (first mode) and the second representing the second natural frequency ω_2 (second mode).

$$\omega_1 = \sqrt{\frac{k}{m}}, \qquad \omega_2 = \sqrt{\frac{3k}{m}}$$

It should be remembered that in the one-degree-of-freedom system only one natural frequency exists, whereas the two-degrees-of-freedom system has two natural frequencies.

At either of the natural frequencies the two masses must vibrate at the same frequency. For instance, at the first (or lowest) natural frequency ω_1 the amplitude ratio A/B becomes unity, or $A = B$, which means that both masses move the same amount in the same direction—that is, the motions are in phase. However, at the second natural frequency ω_2 the amplitude ratio A/B becomes -1, or $A = -B$, and so the motions are opposite in phase. (In the present system the amplitude ratio A/B becomes equal to 1 or -1 when the masses vibrate at a natural frequency. It is important to point out that this situation occurs because we assumed that $m_1 = m_2$ and $k_1 = k_2 = k_3$. Without such assumptions the ratio A/B may not be equal to 1 or -1.)

Note also that the system may not always vibrate at one of the two frequencies but that two modes of vibration may occur simultaneously, depending on the initial conditions. That is, the vibration of m_1 may consist of the sum of two components—a harmonic motion of amplitude A_1 at the frequency ω_1 and a harmonic motion of amplitude A_2 at the frequency ω_2. In this case, the vibration of m_2 consists of the sum of two harmonic components of amplitude B_1 at the frequency ω_1 and of amplitude B_2 at the frequency ω_2.

Many-degrees-of-freedom system. Generally, an *n-degrees-of-freedom system* (such as that consisting of n masses and $n + 1$ springs) has n natural frequencies. If free vibration takes place at any one of its natural frequencies, all the n masses will vibrate at that frequency and the amplitude of any mass will bear a fixed value relative to the amplitude of any other mass. The system, however, may vibrate with more than one natural frequency. Then the resultant vibration may appear quite complicated and may seem to be a random vibration.

2-4 MECHANICAL SYSTEMS WITH DRY FRICTION

Sliding, rolling, and rubbing of various parts constitute some of the friction forces involved in mechanical systems. In most cases, actual friction forces are a combination of viscous friction, dry friction, and various other types.

In this section we are concerned with *dry friction*—the friction force observed when a body with an unlubricated surface slides over another unlubricated surface. We begin with static friction, sliding friction, and rolling friction. Afterward we derive mathematical models of mechanical systems with dry friction, followed by response analysis of such systems. Finally, d'Alembert's principle and its application to mathematical modeling are discussed.

Static friction and sliding friction. Whenever the surface of one body slides over that of another, each body exerts a frictional force on the other, parallel to the surfaces. The force on each body is opposite to the direction of its motion relative to the other.

Suppose that a body is placed on a rough surface and that a pulling force is exerted on it [see Fig. 2-16(a)]. If the body is pulled with an increasing force,

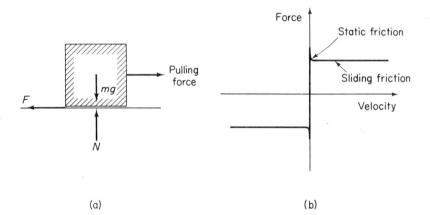

(a)   (b)

Fig. 2-16. (a) A body placed on a rough surface and exerted by a pulling force; (b) characteristic curve for static and sliding friction.

it will not move at first. But as the magnitude of the force increases and reaches a sufficient value to overcome the friction between two contacting surfaces, the body will begin to move. When two contacting surfaces are at rest relative to each other, the static friction force reaches a maximum when sliding between the two surfaces is imminent. Immediately after the motion begins, the magnitude of the friction force decreases slightly. The friction force acting on the body when it is moving with uniform motion is called *sliding* or *kinetic friction*. Sometimes it is also referred to as *coulomb friction*. A characteristic curve for static and sliding friction appears in Fig. 2-16(b).

For the system shown in Fig. 2-16(a), the forces acting on the body, other than the pulling and friction forces, are the gravitational force and the so-called normal force, which is created by the surface on which the body is resting or sliding and which is pushing up on the body. The latter force acts *normal* to the surface, hence its name.

The magnitude N of the normal force and the magnitude F_s of the maximum static friction force are proportional to each other. The ratio F_s/N is called the *coefficient of static friction* and is denoted by μ_s or

$$\mu_s = \frac{F_s}{N}$$

The actual force F of static friction can have any value between zero (when no applied force is parallel to the surface) and a maximum value $\mu_s N$ or

$$0 \leq F \leq \mu_s N$$

If the friction force is that observed for uniform motion of the body, the ratio F_k/N, where F_k is the magnitude of the friction force during uniform motion, is called the *coefficient of sliding friction* or *kinetic friction* and is denoted by μ_k or

$$\mu_k = \frac{F_k}{N}$$

Thus when the body is in motion, the force of sliding or kinetic friction is given by

$$F_k = \mu_k N$$

Note that the maximum static friction is larger than sliding friction—that is,

$$\mu_s > \mu_k$$

The coefficients of static and sliding friction depend primarily on the nature of the surfaces in contact.

Comments. The preceding facts on frictional forces give a macroscopic description of friction phenomena. They are based on experimental studies. That is, they are empirical relations but do not represent fundamental laws. In the following list we summarize the facts about static and sliding friction.

1. Friction force always acts opposite to the direction of actual or intended motion. Until motion of the body takes place, the magnitude of the static friction force is equal to the magnitude of the force acting in the direction of motion.
2. The magnitude of sliding friction is proportional to the magnitude of normal force and is nearly independent of the contact area.
3. The coefficient of sliding friction varies somewhat with the relative velocity but may be considered constant over a wide range of speeds.
4. For a given pair of surfaces, the maximum static friction is larger than sliding friction.

Example 2-6. For the system shown in Fig. 2-17, let us obtain the force F_i needed at the end of the lever in order to keep the brake drum from rotating. Assume that the coefficient of the static friction μ_s is 0.4.

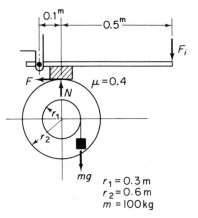

Fig. 2-17. Brake system.

$r_1 = 0.3\,\text{m}$
$r_2 = 0.6\,\text{m}$
$m = 100\,\text{kg}$

The torque due to the weight mg is clockwise and its magnitude T_1 is

$$T_1 = mgr_1 = 100 \times 9.81 \times 0.3 \,\frac{\text{kg-m}^2}{\text{s}^2} = 294.3 \text{ N-m}$$

The friction force acting on the brake drum is

$$F = \mu_s N = 0.4 \times 6F_i = 2.4F_i \text{ N}$$

The torque due to the brake force F is counterclockwise and its magnitude T_2 is

$$T_2 = Fr_2 = 2.4F_i \times 0.6 = 1.44F_i \text{ N-m}$$

If $T_2 > T_1$, then the brake drum will not rotate, and so

$$1.44F_i > 294.3$$

or

$$F_i > 204.4 \text{ N}$$

Thus the magnitude of force F_i needed to keep the brake drum from rotating must be greater than 204.4 newtons.

Rolling friction. The motion of one body rolling on another is opposed by a force called *rolling friction*, which results from the deformation of the two bodies where they make contact. Figure 2-18 shows a homogeneous cylinder being rolled over a smooth surface. Here the pulling force P is acting parallel to the surface. The gravitational force mg acting downward and the reaction force or normal force N acting upward, exerted on the cylinder by the plane's surface, constitute a couple of rolling friction. (A *couple* is a pair of forces equal in magnitude but in opposite direction, not having the same line

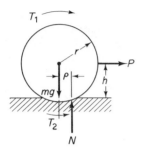

Fig. 2-18. Homogeneous cylinder being rolled over a smooth surface.

of action.) The action line of the gravitational force and that of the normal force are separated by the distance ρ due to deformation of the cylinder and surface.

The couple of rolling friction is a couple having for its axis the tangent to the surface of the plane about which the cylinder is rotating. Its maximum value—that is, the normal pressure multiplied by distance ρ—is generally very small, so that this couple is usually neglected. The direction in which the couple of rolling friction tends to turn the cylinder is opposite to that in which it is actually rotating. If the cylinder is at rest but is acted on by forces that tend to make it rotate, the couple of rolling friction tends to prevent rotation about a common tangent to the two surfaces.

In Fig. 2-18 the torque due to the pulling force is acting in the clockwise direction and its magnitude T_1 is

$$T_1 = Ph \doteq Pr$$

The torque T_2 that resists the rotation is caused by the couple due to mg and N. It is acting in the counterclockwise direction and its magnitude is

$$T_2 = N\rho$$

Suppose that a body is pulled with an increasing force. If the applied force reaches a value large enough to overcome the resisting torque T_2, the cylinder will begin to rotate. Therefore $T_1 = T_2$ or $Pr = N\rho$ gives the condition for imminent rotation. The distance ρ, where

$$\rho = \frac{Pr}{N}$$

is called the *coefficient of rolling friction*. Besides having the dimension of length, it is dependent on such factors as the nature of the contacting surfaces and contact pressure.

Since the rolling friction couple is generally very small as noted, it is usually neglected in the engineering analysis of mechanical systems. In the following analyses of mechanical systems with dry friction, we also neglect it.

Rolling and sliding motion. Consider the motion of a homogeneous cylinder of mass m and radius R rolling down an inclined plane. Questions of rolling and/or sliding immediately involve questions of friction between the surfaces in contact. Without friction the cylinder would slide. Without friction the translational motion of the center of mass and the rotational motion about the center of mass are independent of each other. On the other hand, if the cylinder rolls without sliding, a static friction force acts at all times; the magnitude and direction of the static friction force are such as to ensure $x = R\theta$, where x is the translational displacement of the center of mass and θ is the angle of rotation.

If the cylinder is sliding, the friction force F is equal to $\mu_k N$, where μ_k is the coefficient of sliding friction and N is the normal force. Note that if the cylinder is rolling without sliding, the friction force F becomes static friction with its magnitude unknown but F is smaller than $\mu_k N$. In other words, the condition for the cylinder to roll down without sliding is that F be smaller than $\mu_k N$.

The static friction force in the rolling cylinder is a nondissipative force that interchanges rotational and translational energy. This friction force is not applied over a displacement, for the point of contact between the rolling cylinder and the inclined plane changes continuously. Work done by the friction force in increasing the translational kinetic energy of the center of mass is matched by an equal but negative amount of rotational work done by the same friction force, thereby decreasing the rotational kinetic energy about the center of mass, and vice versa, and dissipating no energy.

Example 2-7. Consider a homogeneous cylinder of mass m and radius R initially at rest on a horizontal surface and suppose that a force P is applied to it (Fig. 2-19). Assuming that the cylinder rolls without sliding, find the magnitude and direction of the friction force F.

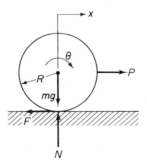

Fig. 2-19. Homogeneous cylinder subjected to a horizontal force P.

Let us assume that F is acting in the direction opposite to P. Applying Newton's second law to the system, we obtain for the translational motion of the center of mass

$$m\ddot{x} = P - F \tag{2-25}$$

and for the rotational motion about the center of mass

$$J\ddot{\theta} = FR \tag{2-26}$$

where J, the moment of inertia of the cylinder about the axis of rotation passing through the center of gravity, is equal to $\frac{1}{2}mR^2$ and x is equal to $R\theta$. (Note that the condition that the cylinder rolls without sliding is $x = R\theta$.) Equations (2-25) and (2-26) represent a mathematical model of the system.

Substituting $\theta = x/R$ into Eq. (2-26) gives

$$\tfrac{1}{2}mR\ddot{x} = FR$$

or

$$\tfrac{1}{2}m\ddot{x} = F \qquad\qquad (2\text{-}27)$$

By eliminating $\ddot{x}$ from Eqs. (2-25) and (2-27),

$$2F = P - F$$

or

$$F = \frac{P}{3}$$

Thus the magnitude of friction force F is $P/3$ and the direction of F is to the left, as shown in Fig. 2-19. (Had we chosen the direction of F to be to the right, F would have been $-P/3$.)

Note that if the external force P is zero, the static friction force F is also zero and we obtain $\ddot{x} = 0$. That is, if the cylinder is originally rolling, it will continue to roll with constant linear velocity $\dot{x} = v_0$ (where v_0 is the initial velocity) and constant angular velocity $\dot{\theta} = v_0/R$.

Example 2-8. A homogeneous cylinder of mass m and radius R moves down an inclined plane whose angle of inclination is α as shown in Fig. 2-20. Let us define

θ = angular displacement as defined in Fig. 2-20
x = linear displacement (along the inclined plane) of the center of gravity of the cylinder
F = friction force acting upward on the plane
N = normal force acting through the point of contact

Determine the angle α for which the cylinder will roll down without sliding. Assume that $x(0) = 0$, $\dot{x}(0) = 0$, $\theta(0) = 0$, and $\dot{\theta}(0) = 0$.

When the cylinder rolls down without sliding, the friction force F and the normal force N are related by

$$F < \mu_k N$$

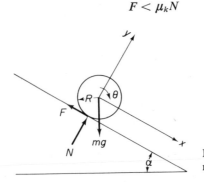

Fig. 2-20. Homogeneous cylinder moving down an inclined plane.

where μ_k is the coefficient of sliding friction. The equation of translational motion in the x direction is

$$m\ddot{x} = mg \sin \alpha - F \qquad (2\text{-}28)$$

and the equation of rotational motion is

$$J\ddot{\theta} = FR \qquad (2\text{-}29)$$

where $\theta = x/R$ (since the cylinder rolls without sliding) and J is the moment of inertia of the cylinder about the axis of rotation passing through the center of gravity. Equations (2-28) and (2-29) define a mathematical model for the system. (Note that, for the y direction, $m\ddot{y} = N - mg \cos \alpha = 0$.)

By eliminating F from Eqs. (2-28) and (2-29) and using the relationship $\theta = x/R$, we have

$$m\ddot{x} = mg \sin \alpha - \frac{J\ddot{x}}{R^2}$$

Noting that $J = \frac{1}{2}mR^2$, this last equation simplifies to

$$\ddot{x} = \frac{2}{3}g \sin \alpha \qquad (2\text{-}30)$$

And so from Eqs. (2-28) and (2-30) we obtain

$$F = mg \sin \alpha - m\ddot{x} = \frac{1}{3}mg \sin \alpha$$

The condition that the cylinder roll down without sliding is $F < \mu_k N$. Consequently,

$$\mu_s N > \mu_k N = \mu_k mg \cos \alpha > F = \frac{1}{3}mg \sin \alpha$$

where $\mu_s N$ is the maximum static friction. Thus

$$\mu_s > \mu_k > \frac{1}{3} \tan \alpha$$

If this condition is satisfied, the cylinder will roll down the inclined plane without sliding.

---

d'Alembert's principle. The following discussion will show that a slight rearrangement of Newton's second law sometimes leads to a simpler way of obtaining the equation of motion. When a force is acting on a mass, it accelerates. Instead of thinking of acceleration as resulting from the application of a force, we can convert this dynamic situation to an equilibrium situation in which the sum of all external forces is opposed by a fictitious inertia force.

Consider a motion of a particle. Newton's second law for a translational motion can be rewritten as

$$F - ma = 0$$

If a fictitious force $-ma$ is assumed to be acting on the particle, the latter can be treated as if it were in equilibrium. This fact is known as *d'Alembert's principle*.

The equation resulting from the application of d'Alembert's principle is one in which the sum of all forces, including fictitious inertia forces, is set equal to zero. The d'Alembert's approach applies both to translational and rotational systems and provides an important analytical simplification in complicated situations involving combined translation and rotation. It is important to note that in this method the inertia force is a fictitious force that is mentally added to the system for analysis purposes only rather than a real force capable of causing a body initially at rest to move.

The main advantage of d'Alembert's method over direct application of Newton's second law is that we need not consider the action of forces and torques about an axis through the center of gravity. Instead we can sum up such action around any axis that we consider convenient. We shall demonstrate this advantage through an example.

Example 2-9. Consider the same system discussed in Example 2-8. A cylinder rolls down an inclined plane whose angle of inclination is α (see Fig. 2-20). In this example we assume that there is no sliding. Let us obtain $x(t)$ as a function of time t by using d'Alembert's approach. Assume that $x(0) = 0$, $\dot{x}(0) = 0$, $\theta(0) = 0$, and $\dot{\theta}(0) = 0$.

In applying d'Alembert's principle to this problem, we treat the cylinder as if it were in equilibrium under the action of all forces and torques, including the fictitious inertia force $m\ddot{x}$ and the fictitious torque $J\ddot{\theta}$, as shown in Fig. 2-21. We can

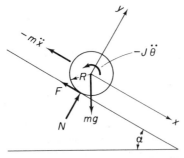

Fig. 2-21. Homogeneous cylinder on an inclined plane indicating all forces acting on the cylinder including fictitious inertia force and fictitious torque.

sum up torques about any axis in the cylinder. However, it is convenient to sum up torques about the axis at the point of contact between the cylinder and the inclined plane because doing so eliminates the forces F and N. Then we obtain

$$mgR \sin \alpha - J\ddot{\theta} - m\ddot{x}R = 0 \qquad (2\text{-}31)$$

Since there is no sliding, we have $x = R\theta$. The moment of inertia J is equal to $\frac{1}{2}mR^2$. Hence Eq. (2-31) can be simplified to

$$\ddot{x} = \tfrac{2}{3}g \sin \alpha \qquad (2\text{-}32)$$

Note that Eq. (2-32) is identical to Eq. (2-30). Integrating Eq. (2-32) twice with respect to t and noting that $x(0) = 0$ and $\dot{x}(0) = 0$, we have

$$x = \tfrac{1}{3}gt^2 \sin \alpha$$

This last equation gives the linear displacement of the cylinder's center of mass.

Let us compare d'Alembert's approach with Newton's approach to the solution. In Newton's method the equations of motion in terms of x and θ are given by two equations, Eqs. (2-28) and (2-29), whereas in d'Alembert's approach we obtained only one equation, Eq. (2-31). Thus d'Alembert's approach simplifies the process of obtaining the final result.

2-5   WORK, ENERGY, AND POWER

If force is considered a measure of effort, then *work* is a measure of accomplishment and *energy* is the ability to do work. The concept of work makes no allowance for a time factor. When a time factor is considered, the concept of power must be introduced. *Power* is work per unit time.

In the following pages the concepts of work, energy, and power (which are normally presented in college physics courses) are covered in some detail, after which a method, based on the law of conservation of energy, for obtaining equations of motion of systems is presented.

Work. The *work* done in a mechanical system is the product of force multiplied by the distance (or torque multiplied by the angular displacement) through which it is exerted, both force and distance being measured in the same direction. For instance, if a body is pushed with a horizontal force of F newtons along a horizontal floor for a distance of x meters, the work done W in pushing the body is

$$W = Fx \quad \text{N-m}$$

Example 2-10. For a translational spring with spring constant k, the work done by an infinitesimal displacement dx is given by

$$\text{Work done} = F \, dx = kx \, dx$$

The total work done over any displacement is the integral of the work done by an infinitesimal displacement dx. If the total displacement is x, then

$$\text{Total work done} = \int_0^x kx \, dx = \tfrac{1}{2}kx^2$$

Similarly, for a torsional spring,

$$\text{Total work done} = \int_0^\theta k\theta \, d\theta = \tfrac{1}{2}k\theta^2$$

where k is the spring constant of the torsional spring and θ is the angular displacement.

Units of work. Units of work for different systems of units are listed below.

SI units and mks (metric absolute) system of units. Force is measured in newtons and distance in meters. Thus the unit of work is the N-m. Note that

$$1 \text{ N-m} = 1 \text{ joule} = 1 \text{ J}$$

British engineering system of units. Force is measured in pounds and distance in feet. Thus the unit of work is the ft-lb$_f$.

$$1 \text{ ft-lb}_f = 1.356 \text{ J} = 1.285 \times 10^{-3} \text{ Btu}$$

$$1 \text{ Btu} = 778 \text{ ft-lb}_f$$

cgs (metric absolute) system of units. The unit of work is the dyn-cm or erg. Note that

$$10^7 \text{ erg} = 10^7 \text{ dyn-cm} = 1 \text{ J}$$

Metric engineering (gravitational) system of units. The unit of work is the kg$_f$-m. Note that

$$1 \text{ kg}_f\text{-m} = 9.81 \times 10^7 \text{ dyn-cm} = 9.81 \text{ J}$$

$$1 \text{ J} = 0.102 \text{ kg}_f\text{-m}$$

British absolute system of units. The unit of work is the foot-poundal (ft-pdl). Note that

$$1 \text{ ft-pdl} = 0.0421 \text{ J}$$

$$1 \text{ J} = 23.7 \text{ ft-pdl}$$

Energy. In a general way, *energy* can be defined as the capacity or ability to do work. It is found in many different forms and can be converted from one form into another. For instance, an electric motor converts electrical energy into mechanical energy, a battery converts chemical energy into electrical energy, and so forth.

A system is said to possess energy when it can work. When a system does mechanical work, the system's energy decreases by the amount equal to the energy required for the work done. Units of energy are the same as units for work—that is, newton-m, joule, kcal, Btu, and so on.

According to the law of conservation of energy, energy can be neither created nor destroyed. This means that the increase in the total energy within a system is equal to the net energy input to the system. So if there is no energy input, there is no change in total energy of the system.

The energy that a body possesses because of its position is called *potential* energy, whereas the energy that a body has as a result of its velocity is called *kinetic* energy.

Potential energy. In a mechanical system only mass and spring elements can store potential energy. The change in the potential energy stored in a system equals the work required to change the system's configuration.

Potential energy is always measured with reference to some chosen level and is relative to that level.

The potential energy is the work done by the external force. For a body of mass m in a gravity field, the potential energy U measured from some reference level is mg times the altitude h measured from the same reference level or

$$U = \int_0^h mg \, dx = mgh$$

Notice that the body, if dropped, has the capacity to do work, since the weight mg (force) will travel a distance h when released. Once the body is released, the potential energy decreases. The lost potential energy is converted into kinetic energy.

For a translational spring, the potential energy U is equal to the net work done on it by the forces acting on its ends as it is compressed or stretched. Since the spring force F is equal to kx, where x is the net displacement of the ends of the spring, the total energy stored is

$$U = \int_0^x F \, dx = \int_0^x kx \, dx = \tfrac{1}{2}kx^2$$

If the initial and final values of x are x_1 and x_2, respectively, then

Change in potential energy $\Delta U = \displaystyle\int_{x_1}^{x_2} F \, dx = \int_{x_1}^{x_2} kx \, dx = \tfrac{1}{2}kx_2^2 - \tfrac{1}{2}kx_1^2$

Note that the potential energy stored in a spring does not depend on whether it is compressed or stretched.

Similarly, for a torsional spring,

Change in potential energy $\Delta U = \displaystyle\int_{\theta_1}^{\theta_2} T \, d\theta = \int_{\theta_1}^{\theta_2} k\theta \, d\theta = \tfrac{1}{2}k\theta_2^2 - \tfrac{1}{2}k\theta_1^2$

Kinetic energy. Only inertia elements can store kinetic energy in mechanical systems. A mass m in pure translation at velocity v has kinetic energy $T = \tfrac{1}{2}mv^2$, whereas a moment of inertia J in pure rotation at angular velocity $\dot\theta$ has kinetic energy $T = \tfrac{1}{2}J\dot\theta^2$. A change in the kinetic energy of a mass m is equal to the work done on it by an applied force as it accelerates or decelerates. Thus a change in the kinetic energy T of a mass m moving in a straight line is

$$\text{Change in kinetic enegy} = \Delta T = \Delta W = \int_{x_1}^{x_2} F \, dx = \int_{t_1}^{t_2} F \frac{dx}{dt} \, dt$$

$$= \int_{t_1}^{t_2} Fv \, dt = \int_{t_1}^{t_2} m\dot v v \, dt = \int_{v_1}^{v_2} mv \, dv$$

$$= \tfrac{1}{2}mv_2^2 - \tfrac{1}{2}mv_1^2$$

where $x(t_1) = x_1$, $x(t_2) = x_2$, $v(t_1) = v_1$, and $v(t_2) = v_2$. Notice that the kinetic energy stored in the mass does not depend on the sign of velocity v.

A change in the kinetic energy of a moment of inertia in pure rotation at angular velocity $\dot{\theta}$ is

$$\text{Change in kinetic energy } \Delta T = \tfrac{1}{2}J\dot{\theta}_2^2 - \tfrac{1}{2}J\dot{\theta}_1^2$$

where J is the moment of inertia about the axis of rotation, $\dot{\theta}_1 = \dot{\theta}(t_1)$, and $\dot{\theta}_2 = \dot{\theta}(t_2)$.

Example 2-11. A car of mass 1500 kg is moving with a velocity of 50 km/h. What is the force required to stop the car at a distance of 100 m?

The velocity of 50 km/h is equal to 13.89 m/s. So

$$v = 50 \text{ km/h} = 13.89 \text{ m/s}$$

The kinetic energy T is

$$T = \tfrac{1}{2}mv^2 = \tfrac{1}{2} \times 1500 \times (13.89)^2$$

$$= 1.447 \times 10^5 \, \frac{\text{kg-m}^2}{\text{s}^2}$$

$$= 1.447 \times 10^5 \text{ N-m}$$

The force F required to stop the car can be obtained by equating Fx (where x is the distance) and T or by equating the work done Fx with the kinetic energy T.

$$Fx = T$$

Thus

$$F = \frac{T}{x} = \frac{1.447 \times 10^5}{100} \frac{\text{N-m}}{\text{m}}$$

$$= 1447 \text{ N}$$

If the numerical values of this example are converted into British engineering system of units, then

$$\text{mass of car} = 1500 \times 0.0685 = 102.8 \text{ slugs}$$

$$(\text{weight of car} = mg = 3307 \text{ lb}_f)$$

$$v = 50 \text{ km/h} = 31.07 \text{ mi/h} = 45.57 \text{ ft/s}$$

$$T = \tfrac{1}{2}mv^2 = \tfrac{1}{2} \times 102.8 \times (45.57)^2$$

$$= 1.067 \times 10^5 \, \frac{\text{slug-ft}^2}{\text{s}^2} = 1.067 \times 10^5 \text{ ft-lb}_f$$

$$x = 100 \text{ m} = 328.1 \text{ ft}$$

Therefore the force required to stop the car is

$$F = \frac{T}{x} = \frac{1.067 \times 10^5}{328.1} \frac{\text{ft-lb}_f}{\text{ft}} = 325.3 \text{ lb}_f$$

Dissipated energy. Consider the damper shown in Fig. 2-22 in which one end is fixed and the other end is moved from x_1 to x_2. The dissipated energy ΔW of the damper is equal to the net work done on it.

$$\Delta W = \int_{x_1}^{x_2} F \, dx = \int_{x_1}^{x_2} b\dot{x} \, dx = b \int_{t_1}^{t_2} \dot{x} \frac{dx}{dt} \, dt = b \int_{t_1}^{t_2} \dot{x}^2 \, dt$$

The energy of the damper element is always dissipated regardless of the sign of $\dot{x}$.

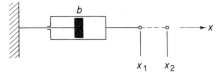

Fig. 2-22. Damper.

Power. *Power* is the time rate of doing work. That is,

$$\text{Power} = P = \frac{dW}{dt}$$

where dW denotes work done during time interval dt. The average power during time duration $t_2 - t_1$ seconds can be determined by measuring the work done in $t_2 - t_1$ seconds or

$$\text{Average power} = \frac{\text{work done in } (t_2 - t_1) \text{ seconds}}{(t_2 - t_1) \text{ seconds}}$$

In SI units or the mks (metric absolute) system of units the work done is measured in newton-meters and the time in seconds. The unit of power is the newton-meter per second or watt.

$$1 \text{ N-m/s} = 1 \text{ W}$$

For the British engineering system of units, the work done is measured in ft-lb$_f$ and the time in seconds. The unit of power is the ft-lb$_f$/s. The power of 550 ft-lb$_f$/s is called 1 horsepower (hp). Thus

$$1 \text{ hp} = 550 \text{ ft-lb}_f/\text{s} = 33\,000 \text{ ft-lb}_f/\text{min} = 745.7 \text{ W}$$

And in the metric engineering system of units the work done is measured in kg$_f$-m and the time in seconds. The unit of power is the kg$_f$-m/s.

$$1 \text{ kg}_f\text{-m/s} = 9.81 \text{ W}$$

$$1 \text{ W} = 1 \text{ J/s} = 0.102 \text{ kg}_f\text{-m/s}$$

Example 2-12. Find the power required to raise a body of mass 500 kg at a rate of 20 m/min.

Let us define displacement per second as x. Then

$$\text{Work done in one second} = mgx = 500 \times 9.81 \times \frac{20}{60} \frac{\text{kg-m}^2}{\text{s}^2} = 1635 \text{ N-m}$$

$$\text{Power} = \frac{\text{work done in one second}}{\text{one second}} = \frac{1635 \text{ N-m}}{1 \text{ s}} = 1635 \text{ W}$$

Thus the power required is 1635 W.

Power and energy. The power required in compressing or stretching a spring is

$$P = \frac{dW}{dt} = \frac{F\,dx}{dt} = F\dot{x} = kx\dot{x}$$

Since the potential energy of a spring compressed or stretched by an amount x is $U = \frac{1}{2}kx^2$, we obtain

$$P = kx\dot{x} = \dot{U}$$

Note that in the spring element the power P is the rate of change of potential energy U.

The power required to accelerate a mass in a straight line is

$$P = \frac{dW}{dt} = \frac{F\,dx}{dt} = F\dot{x} = m\ddot{x}\dot{x}$$

Since the kinetic energy of a mass m moving at velocity v is $T = \frac{1}{2}mv^2$,

$$P = m\ddot{x}\dot{x} = m\dot{v}v = \dot{T}$$

So for mass m moving in a straight line, power P is the rate of change of kinetic energy T.

The power dissipated in a damper or dashpot is

$$P = \frac{dW}{dt} = \frac{F\,dx}{dt} = F\dot{x}$$

Since $F = b\dot{x}$, where b is the viscous friction coefficient, we have

$$P = b\dot{x}^2$$

This power P is the rate at which energy is being dissipated in the damper. The total energy dissipated over a given time interval $t_2 - t_1$ is the time integral of $b\dot{x}^2$, or $\int_{t_1}^{t_2} b\dot{x}^2\,dt$.

Note that if the force applied by the external source and the velocity caused by it are in the same direction, power is supplied by the source to the system. If the force and velocity are opposed, the system is returning power to the source. For instance, a spring stores energy as a force is applied to compress it. If the force is gradually removed, the external force and the velocity will have opposite signs and the spring delivers power.

Passive elements and active elements. Some of the elements in a system—masses and springs, for instance—store energy. This energy can later be introduced into the system. The amount introduced, however, cannot exceed the amount that the element has stored, and unless such an element stored energy beforehand, it cannot deliver any to the system. As a result, such an element is called a *passive element*. That is, passive elements are nonenergy-producing elements. A system containing only passive elements is called a *passive system*. Examples of passive elements include masses, inertias, dampers, and springs in mechanical systems and inductors, resistors, and capacitors in electrical systems. It should be noted that, for passive systems, every term in the homogeneous system differential equation has the same sign.

A physical element that can deliver external energy into a system is called an *active element*. External forces and torques in mechanical systems and current and voltage sources in electrical systems are examples of active elements.

An energy method for deriving equations of motion. Earlier in this chapter we presented two basic methods for deriving equations of motion of mechanical systems. These methods are based on Newton's second law and d'Alembert's principle. Several other approaches for obtaining equations of motion are available, one of which is based on the law of conservation of energy. Here we derive such equations from the fact that the total energy of a system remains the same if no energy enters or leaves the system.

For mechanical systems, friction dissipates energy as heat. Systems that do not involve friction are called *conservative* systems. Consider a conservative system in which the energy is in the form of kinetic and/or potential energy. Since energy enters and leaves the conservative system in the form of mechanical work, we obtain

$$\Delta(T + U) = \Delta W$$

where $\Delta(T + U)$ is the change in the total energy and ΔW is the net work done on the system by the external force. If no external energy enters the system, then

$$\Delta(T + U) = 0$$

which results in

$$T + U = \text{constant}$$

Referring to the mechanical system shown in Fig. 2-23(a), if we assume no friction, this system can be considered a conservative one. The kinetic energy T and potential energy U are given by

$$T = \tfrac{1}{2}m\dot{x}^2, \qquad U = \tfrac{1}{2}kx^2$$

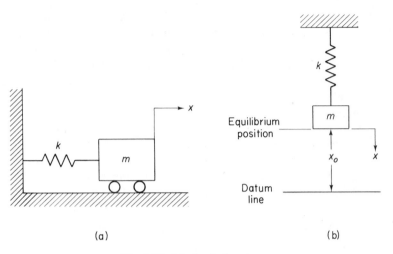

(a)   (b)

Fig. 2-23. Mechanical systems.

Consequently, in the absence of any external energy input,

$$T + U = \tfrac{1}{2}m\dot{x}^2 + \tfrac{1}{2}kx^2 = \text{constant}$$

The equation of motion for the system can be obtained by differentiating the total energy with respect to t and setting the result equal to zero.

$$\frac{d}{dt}(T + U) = m\dot{x}\ddot{x} + kx\dot{x} = (m\ddot{x} + kx)\dot{x} = 0$$

Since $\dot{x}$ is not always zero, we have

$$m\ddot{x} + kx = 0$$

which is the equation of motion for the system.

Let us look next at the mechanical system of Fig. 2-23(b). Here no damping is involved, therefore it is a conservative system. In this case, since the mass is suspended by a spring, the potential energy includes that due to the position of the mass element. At the equilibrium position, the potential energy U_0 of the system is

$$U_0 = mgx_0 + \tfrac{1}{2}k\,\delta^2$$

where x_0 is the equilibrium position in a gravitational field of the mass element above an arbitrary datum line and δ is the deformation of the spring when the system is in equilibrium position or $k\,\delta = mg$.

The instantaneous potential energy U is the instantaneous potential of the weight of the mass element plus the instantaneous elastic energy stored in the spring. Thus

$$U = mg(x_0 - x) + \tfrac{1}{2}k(\delta + x)^2$$
$$= mgx_0 - mgx + \tfrac{1}{2}k\,\delta^2 + k\,\delta x + \tfrac{1}{2}kx^2$$
$$= mgx_0 + \tfrac{1}{2}k\,\delta^2 - (mg - k\,\delta)x + \tfrac{1}{2}kx^2$$

Since $mg = k\,\delta$, if follows that

$$U = U_0 + \tfrac{1}{2}kx^2$$

Notice that the increase in the total potential energy of the system is due to the increase in the elastic energy of the spring that results from its deformation from the equilibrium position. In addition, note that since x_0 is the displacement measured from an arbitrary datum line, it is possible to choose the datum line such that $U_0 = 0$.

The kinetic energy of the system is $T = \tfrac{1}{2}m\dot{x}^2$. Since the total energy is constant, we obtain

$$T + U = \tfrac{1}{2}m\dot{x}^2 + U_0 + \tfrac{1}{2}kx^2 = \text{constant}$$

By differentiating the total energy with respect to t and noting that U_0 is a constant, we have

$$\frac{d}{dt}(T + U) = m\dot{x}\ddot{x} + kx\dot{x} = 0$$

or

$$(m\ddot{x} + kx)\dot{x} = 0$$

Since $\dot{x}$ is not always zero, it follows that

$$m\ddot{x} + kx = 0$$

This is the equation of motion for the system.

From this analysis we see that in the mechanical system where the motion of the mass is due only to a spring force, the increase in the total potential energy of the system is the elastic energy of the spring that results from its deformation from the configuration of the equilibrium position.

Example 2-13. Figure 2-24 shows a homogeneous cylinder of radius R and mass m that is free to rotate about the axis of rotation and that is connected to the wall through a spring. Assuming that the cylinder rolls on a rough surface without sliding, obtain the kinetic energy and potential energy of the system. Then derive the equations of motion from the fact that the total energy is constant.

The kinetic energy of the cylinder is the sum of the translational kinetic energy of the center of mass and the rotational kinetic energy about the axis of rotation.

$$\text{Kinetic energy} = T = \tfrac{1}{2}m\dot{x}^2 + \tfrac{1}{2}J\dot{\theta}^2$$

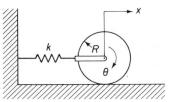

Fig. 2-24. Homogeneous cylinder connected to a wall through a spring.

The potential energy of the system is due to the spring deflection.

$$\text{Potential energy} = U = \tfrac{1}{2}kx^2$$

Since the total energy $T + U$ is constant in this conservative system (which means that the loss in potential energy equals the gain in kinetic energy), it follows that

$$T + U = \tfrac{1}{2}m\dot{x}^2 + \tfrac{1}{2}J\dot{\theta}^2 + \tfrac{1}{2}kx^2 = \text{constant}$$

The cylinder rolls without sliding, which means that $x = R\theta$. Rewriting this last equation and noting that the moment of inertia J is equal to $\tfrac{1}{2}mR^2$, we have

$$\tfrac{3}{4}m\dot{x}^2 + \tfrac{1}{2}kx^2 = \text{constant}$$

Differentiating both sides of this last equation with respect to t yields

$$\tfrac{3}{2}m\dot{x}\ddot{x} + kx\dot{x} = 0$$

or

$$(m\ddot{x} + \tfrac{2}{3}kx)\dot{x} = 0$$

Note that $\dot{x}$ is not always zero, and so $m\ddot{x} + \tfrac{2}{3}kx$ must be identically zero. Therefore

$$m\ddot{x} + \frac{2}{3}kx = 0$$

or

$$\ddot{x} + \frac{2k}{3m}x = 0$$

This equation describes the horizontal motion of the cylinder. For the rotational motion, we substitute $x = R\theta$ into this last equation to get

$$\ddot{\theta} + \frac{2k}{3m}\theta = 0$$

In either of the equations of motion the natural frequency of vibration is the same, $\omega_n = \sqrt{2k/(3m)}$ rad/s.

Comments. Using the law of conservation of energy to derive the equations of motion is easy for simple systems. This method, however, may not be convenient for systems unless they *are* simple. A more general approach based on the energy principle was developed by Lagrange. (For details, refer to Appendix C.) It can be used for most general systems. In fact, in some cases, Lagrange's method is more convenient to use than the conventional Newton's approach.

For a complicated mechanical system, it is advisable to derive the equations of motion by using two different methods to ensure that they are correct. Lagrange's method is useful in providing one such method—in addition to the methods based on Newton's second law, d'Alembert's principle, and so on.

2-6 MOTION, ENERGY, AND POWER TRANSFORMERS

Numerous motion, energy, and power transformers are found in mechanical systems. However, we shall discuss only mechanical motion transformers and mechanical-to-mechanical energy and power transformers here. We begin with a Scotch yoke system as an example of a motion transformer, followed by mechanical-to-mechanical energy transformers (such as a lever, a block and tackle, a chain hoist) that can be used to move a heavy load a short distance by applying a light force a long distance. Finally, gear train systems are described—that is, systems that act as motion transformers or power tranformers, depending on the applications.

Scotch yoke. Figure 2-25(a) shows a Scotch yoke mechanism, which produces a sinusoidal motion at the output from a crank rotating at a constant speed. A motion transformer, this device is used as a motion source in mechanical systems. This motion source, however, can be converted into a force source if it is connected to the load element through a soft spring, as shown in Fig. 2-25(b).

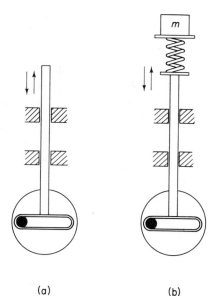

(a)   (b)

Fig. 2-25. Scotch yoke mechanisms.

Lever. A lever is a device that transmits energy from one part of a mechanical system to another. A simple lever system is shown in Fig. 2-26. The purpose of such a lever is to obtain a large mechanical advantage. The mechanical advantage of a machine is defined as the ratio between the force

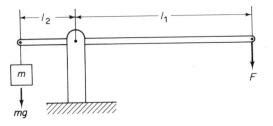

Fig. 2-26. Lever system.

exerted by the machine and the force put into the machine or

$$\text{Mechanical advantage} = \text{MA} = \frac{\text{force output}}{\text{force input}}$$

In the present lever system, if mg is the weight to be moved (or force to be balanced) and F is the input force (or balancing force), the mechanical advantage is mg/F. Since $Fl_1 = mgl_2$, the mechanical advantage is l_1/l_2. Note that the force input (force required to move the load or weight mg) is the force output (load or weight mg) divided by the mechanical advantage or

$$F = \frac{mg}{\text{MA}} = \frac{l_2}{l_1}mg$$

Block and tackle. Block and tackle refers to a device in which a heavy load is lifted by a comparatively small force. Essentially, the device is used to move a heavy load a short distance by applying a light load that moves a longer distance. Figure 2-27(a) shows a two-pulley hoist. Here a rope is attached to the upper pulley, passed around the lower pulley, and then back around the upper pulley to the lifting force F. A separate rope holding the weight mg is fastened to the lower pulley. In the present analysis, we include the weight of the pulleys and rope in the load weight mg. Also, we neglect any friction forces that may exist in the system. Notice that two ropes support the lower pulley and the weight in this system. Since F_1 and F_2 are forces (tension) in the same rope, they are equal, and so

$$F_1 = F_2 = \frac{mg}{2}$$

A four-pulley hoist is shown in Fig. 2-27(b). (In the actual hoist, two upper pulleys are of the same size and on the same shaft. The same is true for the lower two pulleys.) In this case, four ropes support the weight mg. Since tension in the rope is the same on its entire length, if follows that

$$4F = mg$$

where F is the tension in the rope and is the lifting force. Consequently,

$$F = \frac{mg}{4}$$

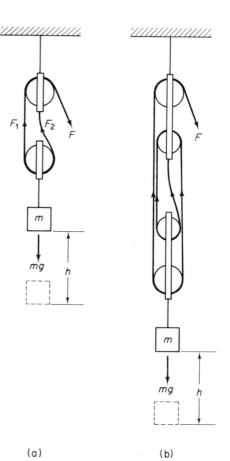

Fig. 2-27. (a) Two-pulley hoist; (b)
four-pulley hoist.

(a)   (b)

The mechanical advantage mg/F is 4. The work input by the pulling force is Fx, where x is the length of rope pulled by the lifting force, and it must equal the work output mgh. Therefore

$$Fx = mgh$$

Notice that height h to which the weight is raised is the distance x traveled by the lifting rope divided by the mechanical advantage, or $h = x/4$.

Because a large value of mechanical advantage is not obtained easily, the block and tackle is generally limited to small loads.

Chain hoist. Figure 2-28 shows a chain hoist, a device that differs from the block and tackle in that the chain is continuous. Use of a differential pulley provides a large mechanical advantage. In this figure the two upper pulleys are of different diameters but fastened to the same shaft, and thus they rotate through the same angle. If chain a is pulled, the upper pulleys turn

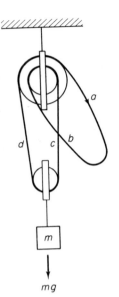

mg **Fig. 2-28.** Chain hoist.

clockwise. As chain *a* winds up the large pulley of radius *R*, chain *b* unwinds from the small pulley of radius *r*. The result is that the load *mg* is raised with a large mechanical advantage.

If the upper pulleys rotate an angle θ radians, chain *c* unwinds from the small pulley a distance $r\theta$ and, at the same time, chain *d* is winding up on the large pulley a distance $R\theta$. So the resulting shortening of the loop *cd* is $\theta(R-r)$. The load *mg* is raised a height equal to $\frac{1}{2}\theta(R-r)$.

The work input $FR\theta$, where *F* is the pulling force of the chain and $R\theta$ is the distance moved, must equal the work output $\frac{1}{2}mg\theta(R-r)$. Hence

$$FR\theta = \tfrac{1}{2}mg\theta(R-r)$$

The mechanical advantage is

$$\text{MA} = \frac{mg}{F} = \frac{2R}{R-r}$$

Since the mechanical advantage of the chain hoist can be made very large, this device is used for lifting heavy loads.

Note that if $R = r$, the mechanical advantage becomes infinite, but then the chain hoist will not lift the load at all. In practice, $R - r$ is made small enough to hold the weight in any position indefinitely with a very small pulling force on chain *a*. In the preceding analysis of the chain hoist, we did not consider the friction of the system. Actually, friction does exist, and it conveniently supplies the necessary small pulling force. As a result, no other pulling force is required to hold the weight in any position indefinitely.

Example 2-14. Consider the chain hoist shown in Fig. 2-28 and assume that the two upper pulleys have radii of 0.4 m and 0.38 m, respectively. What is the mechanical advantage? Also, find the pulling force required to lift a body of mass 500 kg.

The mechanical advantage is

$$\text{MA} = \frac{2R}{R-r} = \frac{2 \times 0.4}{0.4 - 0.38} = 40$$

The pulling force F is

$$F = \frac{mg}{\text{MA}} = \frac{500 \times 9.81}{40} \frac{\text{kg-m}}{\text{s}^2} = 122.6 \text{ N}$$

Gear train. Gear trains are frequently used in mechanical systems to reduce speed, to magnify torque, or to obtain the most efficient power transfer by matching the driving member to the given load. Figure 2-29

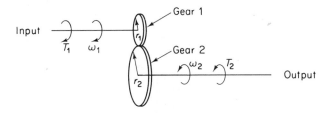

Fig. 2-29. Gear train system.

illustrates a simple gear train system in which the gear train transmits motion and torque from the input member to the output member. If the radii of gear 1 and gear 2 are r_1 and r_2, respectively, and the numbers of teeth on gear 1 and gear 2 are n_1 and n_2, respectively, then

$$\frac{r_1}{r_2} = \frac{n_1}{n_2}$$

Because the surface speeds at the point of contact of the two gears must be identical, we have

$$r_1 \omega_1 = r_2 \omega_2$$

where ω_1 and ω_2 are angular velocities of gear 1 and gear 2, respectively. Therefore

$$\frac{\omega_2}{\omega_1} = \frac{r_1}{r_2} = \frac{n_1}{n_2}$$

If we neglect friction loss, the gear train transmits the power unchanged. In other words, if the torque applied to the input shaft is T_1 and the torque transmitted to the output shaft is T_2, then

$$T_1\omega_1 = T_2\omega_2$$

Example 2-15. To illustrate the concept, consider the system shown in Fig. 2-30. Here a load is driven by a motor through the gear train. Assuming that the stiffness of the shafts of the gear train is infinite, that there is neither backlash nor elastic deformation, and that the number of teeth on each gear is proportional to the radius of the gear, find the equivalent inertia and equivalent friction referred to the motor shaft (shaft 1) and those referred to the load shaft (shaft 2). The numbers of teeth on gear 1 and gear 2 are n_1 and n_2, respectively. The angular velocities of shaft 1 and shaft 2 are ω_1 and ω_2, respectively, whereas the inertia and viscous friction coefficient of each gear train component are denoted by J_1, b_1, and J_2, b_2, respectively.

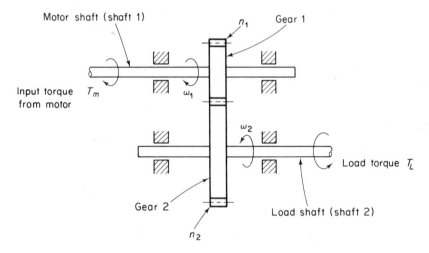

Fig. 2-30. Gear train system.

By applying Newton's second law to this system, the following two equations can be derived. For the motor shaft (shaft 1),

$$J_1\dot{\omega}_1 + b_1\omega_1 + T_1 = T_m \tag{2-33}$$

where T_m is the torque developed by the motor and T_1 is the load torque on gear 1 due to the rest of the gear train. For the load shaft (shaft 2),

$$J_2\dot{\omega}_2 + b_2\omega_2 + T_L = T_2 \tag{2-34}$$

where T_2 is the torque transmitted to gear 2 and T_L is the load torque. Since the work done by gear 1 is equal to that of gear 2,

$$T_1\omega_1 = T_2\omega_2$$

or

$$T_1 = T_2 \frac{\omega_2}{\omega_1} = T_2 \frac{n_1}{n_2}$$

If $n_1/n_2 < 1$, the gear ratio reduces the speed in addition to magnifying the torque. Elimination of T_1 and T_2 from Eqs. (2-33) and (2-34) yields

$$J_1 \dot{\omega}_1 + b_1 \omega_1 + \frac{n_1}{n_2}(J_2 \dot{\omega}_2 + b_2 \omega_2 + T_L) = T_m \qquad (2\text{-}35)$$

Since $\omega_2 = (n_1/n_2)\omega_1$, eliminating ω_2 from Eq. (2-35) gives

$$\left[J_1 + \left(\frac{n_1}{n_2}\right)^2 J_2\right]\dot{\omega}_1 + \left[b_1 + \left(\frac{n_1}{n_2}\right)^2 b_2\right]\omega_1 + \left(\frac{n_1}{n_2}\right)T_L = T_m \qquad (2\text{-}36)$$

Thus the equivalent inertia and equivalent viscous friction coefficient of the gear train referred to shaft 1 are given by

$$J_{1\ eq} = J_1 + \left(\frac{n_1}{n_2}\right)^2 J_2, \qquad b_{1\ eq} = b_1 + \left(\frac{n_1}{n_2}\right)^2 b_2$$

The effect of J_2 on the equivalent inertia $J_{1\ eq}$ is determined by the gear ratio n_1/n_2. For speed-reducing gear trains, the ratio n_1/n_2 is much smaller than unity. If $n_1/n_2 \ll 1$, then the effect of J_2 on the equivalent inertia $J_{1\ eq}$ is negligible. Similar comments apply to the equivalent friction of the gear train.

In terms of the equivalent inertia $J_{1\ eq}$ and equivalent viscous friction coefficient $b_{1\ eq}$, Eq. (2-36) can be simplified to give

$$J_{1\ eq}\dot{\omega}_1 + b_{1\ eq}\omega_1 + nT_L = T_m$$

where $n = n_1/n_2$.

The equivalent inertia and equivalent viscous friction coefficient of the gear train referred to shaft 2 are

$$J_{2\ eq} = J_2 + \left(\frac{n_2}{n_1}\right)^2 J_1, \qquad b_{2\ eq} = b_2 + \left(\frac{n_2}{n_1}\right)^2 b_1$$

So the relationship between $J_{1\ eq}$ and $J_{2\ eq}$ is

$$J_{1\ eq} = \left(\frac{n_1}{n_2}\right)^2 J_{2\ eq}$$

and that between $b_{1\ eq}$ and $b_{2\ eq}$ is

$$b_{1\ eq} = \left(\frac{n_1}{n_2}\right)^2 b_{2\ eq}$$

REFERENCES

2-1   Cannon, R. H., *Dynamics of Physical Systems*, New York: McGraw-Hill Book Company, Inc., 1967.

2-2   Den Hartog, J. P., *Mechanical Vibrations*, New York: McGraw-Hill Book Company, Inc., 1947.

2-3   Reswick, J. B., and C. K. Taft, *Introduction to Dynamic Systems*, Englewood Cliffs, N.J.: Prentice-Hall, Inc., 1967.

2-4   SEELY, S., *Dynamic Systems Analysis*, New York: Reinhold Publishing Corp., 1964.

2-5   SHEARER, J. L., A. T. MURPHY, AND H. H. RICHARDSON, *Introduction to System Dynamics*, Reading, Mass.: Addison-Wesley Publishing Company, Inc., 1967.

EXAMPLE PROBLEMS AND SOLUTIONS

PROBLEM A-2-1. A steel ball of mass 10 kg is supported by wall *ABC* (see Fig. 2-31). Find the reaction forces R_1 and R_2.

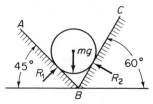

Fig. 2-31. Steel ball supported by wall.

Fig. 2-32. Force balance diagram.

Solution. The three forces mg, R_1, and R_2 must balance as shown in Fig. 2-32. From this figure we obtain

$$\frac{R_1}{\sin 60°} = \frac{R_2}{\sin 45°} = \frac{mg}{\sin 75°}$$

or

$$\frac{R_1}{0.866} = \frac{R_2}{0.707} = \frac{10 \times 9.81}{0.966}$$

Hence

$$R_1 = 87.9 \text{ N}, \qquad R_2 = 71.8 \text{ N}$$

PROBLEM A-2-2. In the system shown in Fig. 2-33 a steel bar is pivoted at point *A*. A mass of 10 kg is supported by wire *CBP*. Find the reaction force F_1 acting on bar *AB* and the tension F_2 in the wire.

Solution. The three forces mg, F_1, and F_2 must balance as in Fig. 2-34. Referring to the figure,

$$\frac{F_1}{\sin 120°} = \frac{F_2}{\sin 30°} = \frac{mg}{\sin 30°}$$

or

$$\frac{F_1}{0.866} = \frac{F_2}{0.5} = \frac{10 \times 9.81}{0.5}$$

So

$$F_1 = 169.9 \text{ N}, \qquad F_2 = 98.1 \text{ N}$$

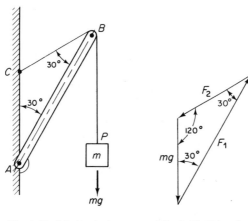

Fig. 2-33. Mechanical system.

Fig. 2-34. Force balance diagram.

PROBLEM A-2-3. For the homogeneous disk of mass m and radius r shown in Fig. 2-35(a), calculate the moments of inertia about the x axis and z axis. Assume that the origin of the xyz coordinate system is at the center of gravity and that the disk is symmetrical about each axis.

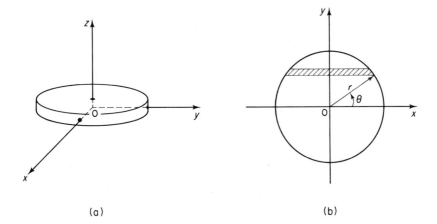

(a)   (b)

Fig. 2-35. (a) Homogeneous disk; (b) disk shown with rectangular and polar coordinate systems.

Solution. Define

$$\rho = \frac{m}{\pi r^2}$$

Then the moment of inertia J_x about the x axis is given by

$$J_x = \int_{-r}^{r} y^2 \rho (2\sqrt{r^2 - y^2})\, dy$$

Changing the rectangular coordinate system into a polar coordinate system, as in Fig. 2-35(b), we obtain $y = r \sin \theta$. Thus

$$J_x = \int_{-\frac{\pi}{2}}^{\frac{\pi}{2}} (r \sin \theta)^2 \, \rho(2r \cos \theta) r \cos \theta \, d\theta$$

$$= r^2 \rho 2r^2 \int_{-\frac{\pi}{2}}^{\frac{\pi}{2}} \sin^2 \theta \cos^2 \theta \, d\theta$$

$$= 4\rho r^4 \int_{0}^{\frac{\pi}{2}} \sin^2 \theta \cos^2 \theta \, d\theta$$

$$= 4\rho r^4 \left[\tfrac{1}{8}(\theta - \tfrac{1}{4} \sin 4\theta) \right]_{0}^{\frac{\pi}{2}}$$

$$= \tfrac{1}{4}\rho r^4 \pi$$

$$= \tfrac{1}{4}mr^2$$

Similarly, $J_y = \tfrac{1}{4}mr^2$. Therefore the moment of inertia J_z about the z axis is

$$J_z = J_x + J_y = \tfrac{1}{2}mr^2$$

PROBLEM A-2-4. A ball of mass m is thrown into the air in a direction 45° from the horizontal line. After 3 seconds the ball is seen in a direction 30° from the horizontal line (Fig. 2-36). Neglecting the friction of the air, find the initial velocity v of the ball.

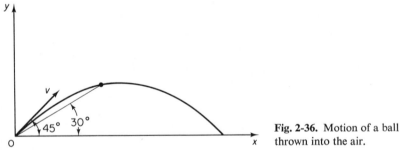

Fig. 2-36. Motion of a ball thrown into the air.

Solution. The equations of motion of the ball in the x and y directions are

$$m\ddot{x} = 0$$

$$m\ddot{y} = -mg$$

Integrating these equations, we find

$$\dot{x}(t) = v \cos 45°$$

$$x(t) = (v \cos 45°)t$$

and

$$\dot{y}(t) = -gt + v \sin 45°$$

$$y(t) = -\tfrac{1}{2}gt^2 + (v \sin 45°)t$$

where we used the initial conditions $x(0) = 0$, $\dot{x}(0) = v \cos 45°$, $y(0) = 0$, and $\dot{y}(0) = v \sin 45°$.

At $t = 3$ seconds

$$\frac{y(3)}{x(3)} = \tan 30° = \frac{1}{\sqrt{3}}$$

Since $x(3) = (0.707v) \times 3$ and $y(3) = -\frac{1}{2} \times 9.81 \times 3^2 + (0.707v) \times 3$, we obtain

$$\frac{-\frac{1}{2} \times 9.81 \times 9 + (0.707v) \times 3}{(0.707v) \times 3} = \frac{1}{1.732}$$

Solving this equation for v yields

$$v = 49.2 \text{ m/s}$$

PROBLEM A-2-5. A brake is applied to a car traveling at a constant speed of 90 km/h. If the deceleration α caused by the braking action is 5 m/s², find the time and distance before the car stops.

Solution. Note that

$$90 \text{ km/h} = 25 \text{ m/s}$$

The equation of motion for the car is

$$m\ddot{x} = -m\alpha$$

where m is the mass of the car and x is the displacement of the car measured from the point where the brake is first applied. Integrating this last equation, we have

$$\dot{x}(t) = -\alpha t + v(0)$$

and

$$x(t) = -\tfrac{1}{2}\alpha t^2 + v(0)t + x(0)$$

where $x(0) = 0$ and $v(0) = 25$ m/s.

Assume that the car stops at $t = t_1$. Then $\dot{x}(t_1) = 0$. The value of t_1 is determined from

$$\dot{x}(t_1) = -\alpha t_1 + v(0) = 0$$

or

$$t_1 = \frac{v(0)}{\alpha} = \frac{25}{5} = 5 \text{ s}$$

The distance traveled before the car stops is $x(t_1)$, where

$$x(t_1) = -\tfrac{1}{2}\alpha t_1^2 + v(0)t_1 = -\tfrac{1}{2} \times 5 \times 5^2 + 25 \times 5$$
$$= 62.5 \text{ m}$$

PROBLEM A-2-6. A homogeneous cylinder of radius 1 m has mass of 100 kg. What will be its angular acceleration if it is acted on by an external torque of 10 N-m about the axis of cylinder?

Solution. The moment of inertia J is

$$J = \tfrac{1}{2}mR^2 = \tfrac{1}{2} \times 100 \times 1^2 = 50 \text{ kg-m}^2$$

The equation of motion for this system is

$$J\ddot{\theta} = T$$

where $\ddot{\theta}$ is the angular acceleration. Therefore

$$\ddot{\theta} = \frac{T}{J} = \frac{10 \text{ N-m}}{50 \text{ kg-m}^2} = 0.2 \text{ rad/s}^2$$

(Note that in examining the units of this last equation, we see that the unit of θ is not s^{-2} but rad/s². This usage occurs because writing rad/s² indicates that the angle θ is measured in radians. The radian is a measure of angles and is the ratio of length of arc to radius. That is, in radian measure, the angle is a pure number. In the algebraic handling of units the unit "radian" is added as necessary.)

PROBLEM A-2-7. Obtain the equivalent spring constant for the system shown in Fig. 2-37.

Solution. For the springs in parallel, the equivalent spring constant k_{eq} is obtained from

$$k_1 x + k_2 x = F = k_{eq} x$$

or

$$k_{eq} = k_1 + k_2$$

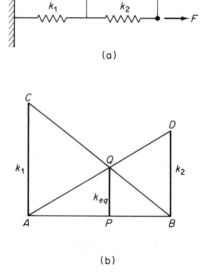

(a)

(b)

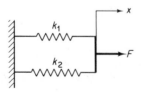

x

Fig. 2-37. System consisting of two springs in parallel.

Fig. 2-38. (a) System consisting of two springs in series; (b) diagram showing the equivalent spring constant.

PROBLEM A-2-8. Find the equivalent spring constant for the system shown in Fig. 2-38(a) and show that it can also be obtained graphically as in Fig. 2-38(b).

Solution. For the springs in series, the force in each spring is the same. Thus

$$k_1 y = F, \qquad k_2(x - y) = F$$

Elimination of y from these two equations results in

$$k_2\left(x - \frac{F}{k_1}\right) = F$$

or

$$k_2 x = F + \frac{k_2}{k_1}F = \frac{k_1 + k_2}{k_1}F$$

The equivalent spring constant k_{eq} for this case is then found as

$$k_{eq} = \frac{F}{x} = \frac{k_1 k_2}{k_1 + k_2} = \frac{1}{\dfrac{1}{k_1} + \dfrac{1}{k_2}}$$

For the graphical solution, notice that

$$\frac{\overline{AC}}{\overline{PQ}} = \frac{\overline{AB}}{\overline{PB}}, \qquad \frac{\overline{BD}}{\overline{PQ}} = \frac{\overline{AB}}{\overline{AP}}$$

from which

$$\overline{PB} = \frac{\overline{AB} \cdot \overline{PQ}}{\overline{AC}}, \qquad \overline{AP} = \frac{\overline{AB} \cdot \overline{PQ}}{\overline{BD}}$$

Since $\overline{AP} + \overline{PB} = \overline{AB}$, we have

$$\frac{\overline{AB} \cdot \overline{PQ}}{\overline{BD}} + \frac{\overline{AB} \cdot \overline{PQ}}{\overline{AC}} = \overline{AB}$$

or

$$\frac{\overline{PQ}}{\overline{BD}} + \frac{\overline{PQ}}{\overline{AC}} = 1$$

Solving for $\overline{PQ}$, we obtain

$$\overline{PQ} = \frac{1}{\dfrac{1}{\overline{AC}} + \dfrac{1}{\overline{BD}}}$$

So if lengths $\overline{AC}$ and $\overline{BD}$ represent the spring constants k_1 and k_2, respectively, then length $\overline{PQ}$ represents the equivalent spring constant k_{eq}. That is,

$$\overline{PQ} = \frac{1}{\dfrac{1}{k_1} + \dfrac{1}{k_2}} = k_{eq}$$

PROBLEM A-2-9. In Fig. 2-39 the simple pendulum shown consists of a sphere of mass m suspended by a string of negligible mass. Neglecting the elongation of the string, find the equation of motion of the pendulum. In addition, find the natural frequency of the system when θ is small. Assume no friction.

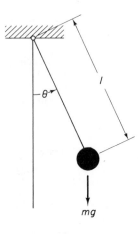

Fig. 2-39. Simple pendulum.

Solution. The gravitational force mg has the tangential component of $mg \sin \theta$ and the normal component of $mg \cos \theta$. The torque due to the tangential component of mg is $-mgl \sin \theta$. So the equation of motion is

$$J\ddot{\theta} = -mgl \sin \theta$$

where $J = ml^2$. Therefore

$$ml^2\ddot{\theta} + mgl \sin \theta = 0$$

or

$$\ddot{\theta} + \frac{g}{l} \sin \theta = 0$$

For small θ, $\sin \theta \doteq \theta$ and the equation of motion simplifies to

$$\ddot{\theta} + \frac{g}{l}\theta = 0$$

The natural frequency ω_n is then obtained as

$$\omega_n = \sqrt{\frac{g}{l}}$$

PROBLEM A-2-10. A mass m is suspended by two springs in Fig. 2-40. One spring is

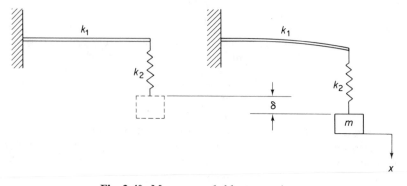

Fig. 2-40. Mass suspended by two springs.

a cantilever beam of spring constant k_1 if loaded transversely at its end. The other spring is a tension-compression spring of spring constant k_2. Determine the static deflection δ of the mass when measured from the position at which the springs carry no load. Determine also the natural frequency of the system.

Solution. Two springs are, in a sense, connected in series. Thus the equivalent spring constant k_{eq} is

$$k_{eq} = \frac{k_1 k_2}{k_1 + k_2}$$

Then the static deflection δ is

$$\delta = \frac{mg}{k_{eq}}$$

The natural frequency of the system is

$$\omega_n = \sqrt{\frac{k_{eq}}{m}}$$

PROBLEM A-2-11. Consider Fig. 2-41 where a homogeneous disk of radius R and mass m that can rotate about the center of mass is hung from the ceiling and is spring preloaded. (Two springs are connected by a wire such that it passes over a pulley as shown.) Each spring is stretched by an amount x. Assuming that the disk is initially rotated by a small angle θ and then released, obtain both the equation of motion of the disk and the natural frequency.

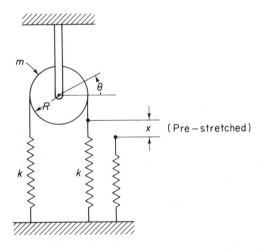

Fig. 2-41. Spring-pulley system.

Solution. If the disk is rotated by angle θ as in Fig. 2-41, the right-hand side spring is stretched by $x + R\theta$ and the left-hand side spring is stretched by $x - R\theta$. So application of Newton's second law to the rotational motion of the disk gives

$$J\ddot{\theta} = -k(x + R\theta)R + k(x - R\theta)R$$

where the moment of inertia J is $\frac{1}{2}mR^2$. Simplifying the equation of motion, we have

$$\ddot{\theta} + \frac{4k}{m}\theta = 0$$

Therefore the natural frequency is

$$\omega_n = \sqrt{\frac{4k}{m}}$$

PROBLEM A-2-12. For the spring-mass-pulley system of Fig. 2-42, the moment of inertia of the pulley about the axis of rotation is J and the radius is R. Assume that the system is initially at equilibrium. The gravitational force of mass m causes a static deflection of the spring such that $k\,\delta = mg$. Assuming that the displacement x of mass m is measured from the equilibrium position, what is the equation of motion of the system? In addition, find the natural frequency.

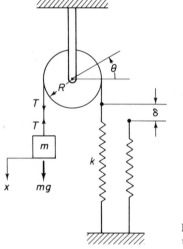

Fig. 2-42. Spring-mass-pulley system.

Solution. Applying Newton's second law, we obtain for mass m

$$m\ddot{x} = -T \tag{2-37}$$

where T is the tension in the wire. (Note that since x is measured from the static equilibrium position, the term mg does not enter the equation.) For the rotational motion of the pulley,

$$J\ddot{\theta} = TR - kxR \tag{2-38}$$

If we eliminate tension T from Eqs. (2-37) and (2-38), the result is

$$J\ddot{\theta} = -m\ddot{x}R - kxR$$

Noting that $x = R\theta$, this last equation is simplified to

$$(J + mR^2)\ddot{\theta} + kR^2\theta = 0$$

or

$$\ddot{\theta} + \frac{kR^2}{J + mR^2}\theta = 0$$

The natural frequency ω_n is

$$\omega_n = \sqrt{\frac{kR^2}{J + mR^2}}$$

PROBLEM A-2-13. For the mechanical system of Fig. 2-43, one end of the lever is connected to a spring and a damper and a force F is applied to the other end of the lever. What is the equation of motion for the system, assuming small displacement x? Assume also that the lever is rigid and massless.

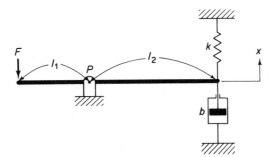

Fig. 2-43. Lever system.

Solution. Using Newton's second law, for small displacement x, the rotational motion about pivot P is given by

$$F l_1 - (b\dot{x} + kx)l_2 = 0$$

or

$$b\dot{x} + kx = \frac{l_1}{l_2}F$$

which is the equation of motion for the system.

PROBLEM A-2-14. A ladder is placed against a wall (see Fig. 2-44). Assuming that the coefficient of sliding friction of the wall is 0.2 and that of the floor is 0.5, find the critical angle θ that the ladder will start sliding down to the floor.

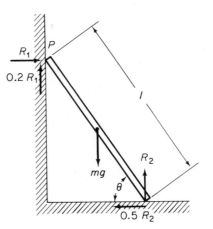

Fig. 2-44. Ladder placed against a wall.

Solution. The magnitudes and directions of the forces and reaction forces acting in the system appear in the figure. The critical angle θ can be found by considering the equations for force balance and moment balance. The equations for force balance are

$$R_1 = 0.5R_2$$
$$mg = 0.2R_1 + R_2$$

from which

$$mg = 1.1R_2$$

The equation for moment balance at point P is

$$R_2 l \cos \theta = mg\frac{l}{2} \cos \theta + 0.5R_2 l \sin \theta$$

which may be simplified to

$$R_2 = \tfrac{1}{2}mg + 0.5R_2 \tan \theta$$

Substituting $mg = 1.1R_2$ into this last equation, we have

$$R_2 = 0.55R_2 + 0.5R_2 \tan \theta$$

which can be simplified to

$$\tan \theta = 0.9$$

or

$$\theta = 42°$$

Hence the critical angle θ is 42°.

PROBLEM A-2-15. Consider a homogeneous disk of radius R and mass m standing on a horizontal plane. Assume that at $t = 0$ the center of mass of the disk is given an initial velocity v_0 in the x direction (but no initial angular velocity) as shown in Fig. 2-45—that is, $x(0) = 0$, $\dot{x}(0) = v_0 > 0$, $\theta(0) = 0$, and $\dot{\theta}(0) = 0$, where θ is the angular displacement of the disk about the axis of rotation. Determine the motion of the disk for $t > 0$. Assume that there is no viscous friction acting on the disk.

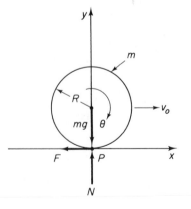

Fig. 2-45. Homogeneous disk standing on a horizontal plane subjected to an initial velocity v_0 in the x direction.

Solution. When initial velocity v_0 is given, the disk begins to slide with rolling. (Point P, the point of contact of the disk and the plane, has velocity $\dot{x} - R\dot{\theta} > 0$.) The friction force F with its magnitude $\mu_k N$, where N is the normal force acting through the point of contact, acts until such time as the cylinder starts rolling without sliding.

The equations of motion for this system during the time interval for which $\dot{x} - R\dot{\theta} > 0$ (which means that the disk slides with rolling) are

$$m\ddot{x} = -F$$
$$J\ddot{\theta} = FR$$

where $F = \mu_k N = \mu_k mg$ and J is the moment of inertia of the disk about the rotating axis, or $J = \frac{1}{2}mR^2$. These two equations can be simplified to

$$\ddot{x} = -\mu_k g$$
$$\ddot{\theta} = 2\mu_k \frac{g}{R}$$

Integrating them with respect to time, we have

$$\dot{x} = v_0 - \mu_k gt \qquad (2\text{-}39)$$

$$\dot{\theta} = 2\mu_k \frac{g}{R}t \qquad (2\text{-}40)$$

Note that Eqs. (2-39) and (2-40) hold true for the time interval for which $\dot{x} - R\dot{\theta} > 0$.

Referring to Eqs. (2-39) and (2-40), we see that velocity $\dot{x}$ decreases and that angular velocity $\dot{\theta}$ increases as time t increases. So at some time $t = t_1$, $\dot{x}$ becomes equal to $R\dot{\theta}$. This particular instant t_1 can be determined by equating $\dot{x}$ and $R\dot{\theta}$ and solving for t_1 as follows:

$$v_0 - \mu_k gt_1 = 2R\mu_k \frac{g}{R}t_1$$

or

$$t_1 = \frac{v_0}{3\mu_k g}$$

For $t < t_1$, the disk rolls and slides. At $t = t_1$ the disk starts to roll without sliding at velocity $\dot{x}(t_1) = R\dot{\theta}(t_1) = 2v_0/3 = $ constant.

Once the velocity of the point of contact (point P) becomes zero, the friction force no longer acts on the system and sliding ceases. Thus $F = \mu_k N$ no longer applies. In fact, friction force F becomes equal to zero for $t \geq t_1$, and the disk continues to roll without sliding at the constant linear velocity $2v_0/3$ and the constant angular velocity $2v_0/(3R)$.

PROBLEM A-2-16. Referring to Problem A-2-15, consider a homogeneous disk of radius R and mass m standing on a horizontal plane. At $t = 0$ the center of mass of the disk is given an initial velocity v_0 in the x direction and at the same time the disk is given an initial angular velocity ω_0 about the axis of rotation (Fig. 2-46). Determine the motion of the disk for $t > 0$.

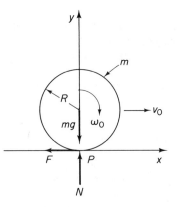

Fig. 2-46. Homogeneous disk standing on a horizontal plane subjected to an initial velocity v_0 in the x direction and an initial angular velocity ω_0 about the axis of rotation.

Solution. The motion of the disk depends on whether the velocity of the point of contact (point P) is positive, negative, or zero. The velocity of the point of contact is

$$u_0 = v_0 - R\omega_0$$

In the following material we shall consider the three cases ($v_0 > R\omega_0$, $v_0 < R\omega_0$, $v_0 = R\omega_0$) separately.

Case 1. $v_0 > R\omega_0$: There is sliding initially. The equations of motion are

$$m\ddot{x} = -F = -\mu_k N = -\mu_k mg$$
$$J\ddot{\theta} = FR$$

where $J = \frac{1}{2}mR^2$. Simplifying, these equations become

$$\ddot{x} = -\mu_k g$$
$$\ddot{\theta} = 2\mu_k \frac{g}{R}$$

Hence

$$\dot{x} = v_0 - \mu_k gt$$
$$\dot{\theta} = \omega_0 + 2\mu_k \frac{g}{R}t$$

At some time $t = t_1$ velocity $\dot{x}$ becomes equal to $R\dot{\theta}$. This particular time t_1 is found from

$$v_0 - \mu_k gt_1 = R\left(\omega_0 + 2\mu_k \frac{g}{R}t_1\right)$$

or

$$t_1 = \frac{1}{3\mu_k g}(v_0 - R\omega_0)$$

At $t = t_1$ the point of contact (point P) will have zero velocity and sliding ceases. For $t \geq t_1$, the disk rolls without sliding at the velocity

$$\dot{x} = R\dot{\theta} = \frac{1}{3}(2v_0 + R\omega_0) = \text{constant}$$

Case 2. $v_0 < R\omega_0$: In this case, the point of contact (point P) has the velocity $u_0 = v_0 - R\omega_0 < 0$. Thus there is sliding initially. The friction force acts in the direction opposite to Case 1. So by changing F to $-F$ in the equations of motion for Case 1, the equations of motion for the present case are found to be

$$m\ddot{x} = F$$
$$J\ddot{\theta} = -FR$$

Simplifying,

$$\ddot{x} = \mu_k g$$
$$\ddot{\theta} = -2\mu_k \frac{g}{R}$$

It follows that

$$\dot{x} = v_0 + \mu_k g t$$
$$\dot{\theta} = \omega_0 - 2\mu_k \frac{g}{R} t$$

We see that $\dot{x}$ increases and $\dot{\theta}$ decreases as t increases. Therefore we can expect that, at some time $t = t_2$, $\dot{x}(t_2) = R\dot{\theta}(t_2)$ or

$$v_0 + \mu_k g t_2 = R\left(\omega_0 - 2\mu_k \frac{g}{R} t_2\right)$$

Solving for t_2, we find

$$t_2 = \frac{1}{3\mu_k g}(R\omega_0 - v_0)$$

For $t < t_2$, the disk rolls and slides. At $t = t_2$ the point of contact (point P) will have zero velocity; for $t \geq t_2$, the disk rolls without sliding at a constant velocity that is

$$\dot{x} = R\dot{\theta} = \frac{1}{3}(2v_0 + R\omega_0)$$

Case 3. $v_0 = R\omega_0$: In this case, initially the point of contact (point P) has the velocity $u_0 = v_0 - R\omega_0 = 0$. Consequently, there is no sliding in the motion and the equations of motion are

$$m\ddot{x} = -F$$
$$J\ddot{\theta} = FR$$

where $x = R\theta$. By eliminating F from the equations of motion, we obtain

$$m\ddot{x} + \tfrac{1}{2}mR\ddot{\theta} = 0$$

Then substituting $\theta = x/R$ into this last equation gives

$$m\ddot{x} + \tfrac{1}{2}m\ddot{x} = 0$$

or

$$\ddot{x} = 0$$

So for $t \geq 0$ we have

$$\dot{x} = v_0 = \text{constant}, \qquad \dot{\theta} = \frac{\dot{x}}{R} = \omega_0 = \text{constant}$$

PROBLEM A-2-17. Consider the system shown in Fig. 2-47(a), where the cylinder of radius R and mass m is pulled through a massless spring with spring constant k. Assume that the cylinder rotates freely about its axis and that the input displacement x_i is known. What is the natural frequency of the system? Assume that there is no sliding.

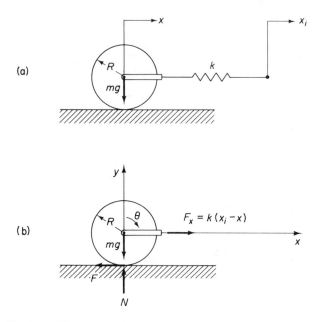

Fig. 2-47. (a) Cylinder being pulled through a spring; (b) force diagram for the cylinder system.

Solution. The forces acting on the system are the pulling force $F_x = k(x_i - x)$, gravitational force mg, friction force F, and normal force N as shown in Fig. 2-47(b). (The normal force N balances with mg—that is, $N = mg$.)

The equation of motion in the x direction is

$$m\ddot{x} = F_x - F = k(x_i - x) - F \qquad (2\text{-}41)$$

For the rotational motion about the center of mass,

$$J\ddot{\theta} = FR \qquad (2\text{-}42)$$

Since there is no sliding, $x = R\theta$. Hence Eq. (2-42) becomes

$$J\ddot{x} = FR^2$$

Noting that $J = \frac{1}{2}mR^2$, we obtain

$$F = \frac{J}{R^2}\ddot{x} = \frac{1}{2}m\ddot{x}$$

Elimination of F from Eq. (2-41) results in

$$m\ddot{x} = k(x_i - x) - \frac{1}{2}m\ddot{x}$$

or

$$\ddot{x} + \frac{2k}{3m}x = \frac{2k}{3m}x_i$$

The natural frequency of the system is $\sqrt{2k/(3m)}$.

PROBLEM A-2-18. A box is placed on a cart as shown in Fig. 2-48. Assuming that the cart is accelerated in the x direction, find the condition for the box to slide and the condition for the box to fall down. Also assume that μ_s, the coefficient of maximum static friction between the box and the floor of the cart, is 0.3. If the acceleration of the cart is increased from zero to $0.4g$, where g is the gravitational acceleration constant, will the box fall down to the floor?

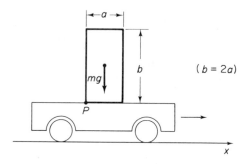

Fig. 2-48. Box placed on a cart.

Solution. The box will start to slide if

$$\text{(inertia force)} > \text{(static friction force)}$$

The box will fall down if

$$\left(\begin{matrix}\text{moment due to inertia force}\\ \text{about point } P\end{matrix}\right) > \left(\begin{matrix}\text{moment due to } mg\\ \text{about point } P\end{matrix}\right)$$

If the cart has an acceleration α, the magnitude of the inertia force is $m\alpha$. The magnitude of the maximum static friction force is $\mu_s mg$. Thus the condition for the box to start to slide is

$$m\alpha > \mu_s mg$$

or

$$\alpha > \mu_s g = 0.3g$$

The condition for the box to fall down is

$$m\alpha \frac{b}{2} > mg\frac{a}{2}$$

or

$$\alpha > \frac{a}{b}g = 0.5g$$

Since the condition for the box to start to slide is $\alpha = 0.3g$ and the condition for the box to start to fall down is $\alpha = 0.5g$, the box will start to slide as the acceleration of the cart is increased from zero to $0.4g$. The box will not fall down.

PROBLEM A-2-19. A box of mass m is put to slide on a smooth horizontal surface. If the initial velocity of the box is 5 m/s, how far will it slide? Assume that μ_k, the coefficient of sliding friction between the box and the surface, is 0.2.

Solution. The equation of motion for the box is

$$m\ddot{x} = -F \tag{2-43}$$

where F, the friction force, is

$$F = \mu_k mg = 0.2mg$$

Hence Eq. (2-43) simplifies to

$$\ddot{x} = -0.2g$$

By integrating this last equation and noting that the initial velocity is 5 m/s, we have

$$\dot{x}(t) = -0.2 \times 9.81t + 5 = -1.962t + 5 \tag{2-44}$$

The velocity $\dot{x}$ becomes zero at $t = t_1$, where

$$t_1 = \frac{5}{1.962}$$

Then integrating Eq. (2-44) with respect to t gives

$$x(t) = -0.981t^2 + 5t$$

where we assumed $x(0) = 0$. The distance that the box travels before it stops is obtained by substituting $t = t_1$ into this last equation.

$$x(t_1) = -0.981\left(\frac{5}{1.962}\right)^2 + 5\left(\frac{5}{1.962}\right) = 6.37$$

Thus the distance traveled is 6.37 meters.

PROBLEM A-2-20. A box of mass m is placed on an inclined plane with the angle of inclination equal to 30° (see Fig. 2-49). Find the work done when the box is moved

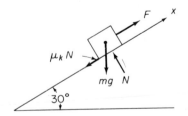

Fig. 2-49. Box placed on an inclined plane.

upward at a constant speed along the inclined plane for a distance of 5 m. Assume that mass m is 10 kg and that the coefficient of sliding friction μ_k is 0.3.

Solution. The work done is

$$W = Fl$$

where F is the pulling force along the inclined plane and l is the distance moved. The equation of motion for the system is

$$m\ddot{x} = -mg \sin 30° - \mu_k N + F$$

where m is the mass of the box and $N = mg \cos 30°$. When the box is moved along the inclined plane at a constant speed, $\ddot{x} = 0$. Hence

$$F = mg \sin 30° + 0.3mg \cos 30° = mg(0.5 + 0.3 \times 0.866)$$

$$= 10 \times 9.81 \times 0.760 = 74.5 \text{ N}$$

Therefore the work done is

$$W = 74.5 \times 5 = 373 \text{ N-m}$$

PROBLEM A-2-21. A circular water tank of radius 2 m and height 5 m is placed at 20 m above the ground as shown in Fig. 2-50. Obtain the potential energy of water filled in this tank. Assume that the density of water is 1000 kg/m³.

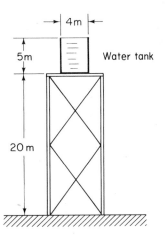

Fig. 2-50. Water tank.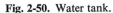

Solution. The mass m of water in the tank is

$$2 \times 2 \times 3.14 \times 5 \times 1000 = 62.8 \times 10^3 \text{ kg}$$

The height h of the center of gravity of water in the tank is $20 + \frac{1}{2} \times 5 = 22.5$ m. Hence the potential energy U with the ground level as the datum line is

$$U = mgh$$

$$= 62.8 \times 10^3 \times 9.81 \times 22.5$$

$$= 13.86 \times 10^6 \text{ N-m}$$

PROBLEM A-2-22. A body of mass m is thrown vertically upward in the air with an initial velocity of 20 m/s. How far up will it go? Find the vertical distance by using the law of conservation of energy.

Solution. Let us measure the vertical distance from the point where the body is thrown upward. At the instant that the body is thrown upward the potential energy U_1 is zero and the kinetic energy T_1 is

$$T_1 = \tfrac{1}{2}mv^2(0)$$

At the instant that the body reaches the maximum height h the potential energy U_2 is

$$U_2 = mgh$$

and the kinetic energy T_2 is zero.

By applying the law of conservation of energy, we obtain

$$T_1 + U_1 = T_2 + U_2$$

or

$$\tfrac{1}{2}mv^2(0) = mgh$$

from which

$$h = \frac{1}{2}\frac{v^2(0)}{g} = \frac{1}{2}\frac{20^2}{9.81} = 20.39 \text{ m}$$

PROBLEM A-2-23. The output of an engine is measured by a Prony brake as shown in Fig. 2-51. The angular speed measured is 4 Hz. Find the output power of the engine in watts.

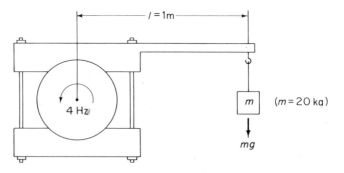

Fig. 2-51. Prony brake system.

Solution. The engine torque T is balanced with the braking torque mgl. Hence

$$T = mgl = 20 \times 9.81 \times 1 = 196.2 \text{ N-m}$$

Since the output power P is $T\omega$, we have

$$P = T\omega = 196.2 \times 4 \times 2\pi$$
$$= 4930 \text{ N-m/s}$$
$$= 4930 \text{ W}$$

PROBLEM A-2-24. Suppose that a car of mass 2000 kg is traveling at a constant speed of 90 km/h. When the car's brake is applied for a finite-time duration, the car's speed is reduced to 30 km/h. Find the energy absorbed by the brake.

Solution. Note that

$$90 \text{ km/h} = 25 \text{ m/s}, \qquad 30 \text{ km/h} = 8.33 \text{ m/s}$$

The energy absorbed by the brake is

$$W = \tfrac{1}{2}mv_1^2 - \tfrac{1}{2}mv_2^2$$
$$= \tfrac{1}{2} \times 2000 \times (25^2 - 8.33^2)$$
$$= 5.556 \times 10^5 \text{ N-m}$$
$$= 5.556 \times 10^5 \text{ J}$$

PROBLEM A-2-25. A homogeneous cylinder of radius R and mass m rolls down, without sliding, an inclined plane whose angle of inclination is ϕ (see Fig. 2-52). Assume that the cylinder is at rest initially. Applying the law of conservation of energy, find the linear speed of the center of mass of the cylinder when it has rolled down the plane a distance L.

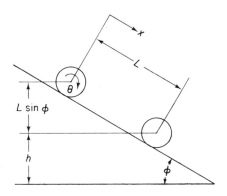

Fig. 2-52. Homogeneous cylinder rolling down an inclined plane.

Solution. At $t = 0$ the kinetic energy and potential energy of the cylinder are

Kinetic energy $T_1 = 0$

Potential energy $U_1 = mg(L \sin \phi + h)$

where the potential energy is measured from the reference horizontal line shown in Fig. 2-52.

At the instant that the cylinder has rolled down a linear distance L, the kinetic energy and potential energy are

Kinetic energy $T_2 = \tfrac{1}{2}m\dot{x}^2 + \tfrac{1}{2}J\dot{\theta}^2$

Potential energy $U_2 = mgh$

(The kinetic energy consists of the translational kinetic energy and the rotational kinetic energy.) By applying the law of conservation of energy, we have

$$T_1 + U_1 = T_2 + U_2$$

or

$$mg(L \sin \phi + h) = \tfrac{1}{2}m\dot{x}^2 + \tfrac{1}{2}J\dot{\theta}^2 + mgh \qquad (2\text{-}45)$$

Since the cylinder rolls without sliding, $x = R\theta$. Also, $J = \frac{1}{2}mR^2$. So Eq. (2-45) becomes

$$mgL \sin \phi = \frac{1}{2}m\dot{x}^2 + \frac{1}{4}m\dot{x}^2$$

or

$$\dot{x} = \sqrt{\frac{4}{3}gL \sin \phi}$$

This value of $\dot{x}$ gives the linear speed of the center of mass when it has rolled down the plane a distance L.

PROBLEM A-2-26. Consider the spring-mass-pulley system of Fig. 2-53(a). If mass m is pulled downward a short distance and released, it will vibrate. Obtain the natural frequency of the system by applying the law of conservation of energy.

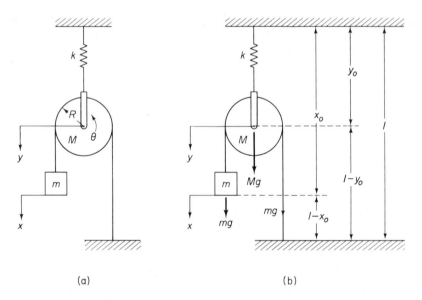

(a)   (b)

Fig. 2-53. (a) Spring-mass-pulley system; (b) diagram for figuring out potential energy of the system.

Solution. Define x, y, and θ as the displacement of mass m, the displacement of the pulley, and the angle of rotation of the pulley, measured respectively from their corresponding equilibrium positions. Note that $x = 2y$, $R\theta = x - y = y$, and $J = \frac{1}{2}MR^2$.

The kinetic energy T of the system is

$$T = \frac{1}{2}m\dot{x}^2 + \frac{1}{2}M\dot{y}^2 + \frac{1}{2}J\dot{\theta}^2$$

$$= \frac{1}{2}m\dot{x}^2 + \frac{1}{8}M\dot{x}^2 + \frac{1}{4}MR^2\left(\frac{\dot{y}}{R}\right)^2$$

$$= \frac{1}{2}m\dot{x}^2 + \frac{3}{16}M\dot{x}^2$$

Referring to Fig. 2-53(b), the potential energy U of the system can be obtained as follows. At the equilibrium state the potential energy U_0 is

$$U_0 = \tfrac{1}{2}ky_\delta^2 + Mg(l - y_0) + mg(l - x_0)$$

where y_δ is the static deflection of the spring due to the hanging masses M and m. When masses m and M are displaced by x and y, respectively, the instantaneous potential energy U can be obtained as

$$\begin{aligned} U &= \tfrac{1}{2}k(y_\delta + y)^2 + Mg(l - y_0 - y) + mg(l - x_0 - x) \\ &= \tfrac{1}{2}ky_\delta^2 + ky_\delta y + \tfrac{1}{2}ky^2 + Mg(l - y_0) - Mgy + mg(l - x_0) - mgx \\ &= U_0 + \tfrac{1}{2}ky^2 + ky_\delta y - Mgy - mgx \end{aligned}$$

Note that, referring to Fig. 2-53(b) again, the spring force ky_δ must balance with $Mg + 2mg$ or

$$ky_\delta = Mg + 2mg$$

Therefore

$$ky_\delta y = Mgy + 2mgy = Mgy + mgx$$

and

$$U = U_0 + \tfrac{1}{2}ky^2 = U_0 + \tfrac{1}{8}kx^2$$

where U_0 is constant.

By applying the law of conservation of energy to this conservative system,

$$T + U = \tfrac{1}{2}m\dot{x}^2 + \tfrac{3}{16}M\dot{x}^2 + U_0 + \tfrac{1}{8}kx^2 = \text{constant}$$

Then differentiating both sides of this last equation with respect to t yields

$$m\dot{x}\ddot{x} + \tfrac{3}{8}M\dot{x}\ddot{x} + \tfrac{1}{4}kx\dot{x} = 0$$

or

$$[(m + \tfrac{3}{8}M)\ddot{x} + \tfrac{1}{4}kx]\dot{x} = 0$$

Since $\dot{x}$ is not always zero, we must have

$$(m + \tfrac{3}{8}M)\ddot{x} + \tfrac{1}{4}kx = 0$$

or

$$\ddot{x} + \frac{2k}{8m + 3M}x = 0$$

The natural frequency of the system, therefore, is

$$\omega_n = \sqrt{\frac{2k}{8m + 3M}}$$

PROBLEM A-2-27. Consider the system shown in Fig. 2-54 in which a cylinder of radius r and mass m rolls without sliding on a cylindrical surface of radius R. Assuming the amplitude of oscillation to be small, find the frequency of oscillation of the cylinder by using the law of conservation of energy.

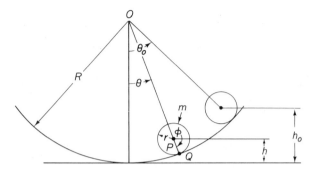

Fig. 2-54. Cylinder rolling on a cylindrical surface.

Solution. Let us define angle θ as the angle of rotation of line OP from the vertical position. Angle ϕ is the angle of rotation of line PQ. When the cylinder rolls without sliding

$$R|\theta| = r(|\theta| + |\phi|)$$

or

$$(R - r)|\theta| = r|\phi|$$

where $|\theta|$ and $|\phi|$ denote the absolute values of angles θ and ϕ. (The absolute values are used because the positive directions of angles θ and ϕ are opposite each other.)

Assume that the cylinder rolls up to height h_0 and then rolls down. Thus the potential energy at this extreme point is mgh_0. The kinetic energy is zero, since at this point the cylinder stops. Then at an arbitrary time t the potential energy that the cylinder possesses is mgh and the kinetic energy is $\frac{1}{2}mv^2 + \frac{1}{2}J\dot{\phi}^2$, where v is the velocity of the center of gravity of the cylinder and J is the moment of inertia of the cylinder about the axis of rotation at point P. So $J = \frac{1}{2}mr^2$.

By applying the law of conservation of energy to this system, we have

$$\frac{1}{2}mv^2 + \frac{1}{2}J\dot{\phi}^2 + mgh = mgh_0 \qquad (2\text{-}46)$$

Since

$$v^2 = (R - r)^2\dot{\theta}^2 = r^2\dot{\phi}^2$$

Eq. (2-46) can be simplified to

$$\tfrac{3}{4}mv^2 = \tfrac{3}{4}m(R - r)^2\dot{\theta}^2 = mg(h_0 - h) \qquad (2\text{-}47)$$

Noting that

$$h_0 = R - (R - r)\cos\theta_0$$
$$h = R - (R - r)\cos\theta$$

Eq. (2-47) can be written

$$\tfrac{3}{4}(R - r)\dot{\theta}^2 = g(\cos\theta - \cos\theta_0)$$

Differentiating this last equation with respect to t gives

$$\tfrac{3}{2}(R - r)\dot{\theta}\ddot{\theta} = g(-\sin\theta)\dot{\theta}$$

or

$$[\tfrac{3}{2}(R - r)\ddot{\theta} + g\sin\theta]\dot{\theta} = 0$$

Since $\dot{\theta}$ is not always zero, we must have

$$\tfrac{3}{2}(R - r)\ddot{\theta} + g \sin \theta = 0$$

For small values of θ, $\sin \theta \doteq \theta$, and this last equation can be simplified to

$$\ddot{\theta} + \frac{2}{3}\frac{g}{(R - r)}\theta = 0$$

which is the equation of motion for the system when θ is small. The frequency of this simple harmonic motion is

$$\omega_n = \sqrt{\frac{2}{3}\frac{g}{(R - r)}}$$

PROBLEM A-2-28. If, for the spring-mass system of Fig. 2-55, mass m_s of the spring is small but not negligibly small compared with suspended mass m, show that the inertia of the spring can be allowed for by adding one-third of its mass m_s to suspended mass m and then treating the spring as a massless spring.

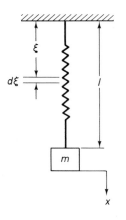

Fig. 2-55. Spring-mass system.

Solution. Consider the free vibration of the system. In the free vibration, displacement x can be written

$$x = A \cos \omega t$$

Since the mass of the spring is comparatively small, we can assume that the spring is stretched uniformly. Then the displacement of a point in the spring at distance ξ from the top can be given by $(\xi/l)A \cos \omega t$.

In the mean position where $x = 0$ and the velocity of mass m is maximum, the velocity of the suspended mass is $A\omega$ and that of the spring at distance ξ from the top is $(\xi/l)A\omega$. The maximum kinetic energy $T_{\max}$ is

$$T_{\max} = \frac{1}{2}m(A\omega)^2 + \int_0^l \frac{1}{2}\left(\frac{m_s}{l}\right)\left(\frac{\xi}{l}A\omega\right)^2 d\xi$$

$$= \frac{1}{2}mA^2\omega^2 + \frac{1}{2}\left(\frac{m_s}{l}\right)\left(\frac{A^2\omega^2}{l^2}\right)\frac{1}{3}l^3$$

$$= \frac{1}{2}\left(m + \frac{m_s}{3}\right)A^2\omega^2$$

Note that the mass of the spring does not affect the change in the potential energy of the system and that if the spring were massless, the maximum kinetic energy would have been $\frac{1}{2}mA^2\omega^2$. Therefore we conclude that the inertia of the spring can be allowed for simply by adding one-third of its mass to the suspensed mass and then treating the spring as a massless spring.

PROBLEM A-2-29. Find the acceleration of mass m_1 for the two-pulley hoist shown in Fig. 2-56.

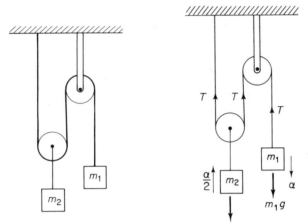

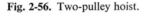
$m_2 = 1.2\ m_1$

Fig. 2-56. Two-pulley hoist.

Fig. 2-57. Forces acting in the two-pulley hoist system shown in Fig. 2-56.

Solution. Assume that the acceleration of mass m_1 is α. Referring to Fig. 2-57, the equation of motion for mass m_1 is

$$m_1\alpha = m_1g - T \tag{2-48}$$

For mass m_2,

$$m_2\frac{\alpha}{2} = 2T - m_2g$$

By substituting $m_2 = 1.2m_1$ into this last equation, we have

$$0.6m_1\alpha = 2T - 1.2m_1g \tag{2-49}$$

Then eliminating T from Eqs. (2-48) and (2-49) gives

$$2.6m_1\alpha = 0.8m_1g$$

from which

$$\alpha = \frac{0.8}{2.6}g = \frac{0.8 \times 9.81}{2.6} = 3.02 \text{ m/s}^2$$

PROBLEM A-2-30. A torque T is applied to shaft 1 in the gear train system of Fig. 2-58. Obtain the equation of motion of the system. Assume that the moments of

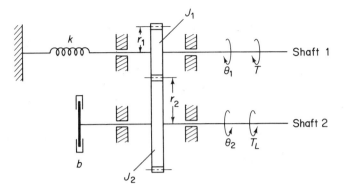

Fig. 2-58. Gear train system.

inertia of the gears are J_1 and J_2 as shown in the diagram and that the load torque is T_L.

Solution. For shaft 1, the equation of motion is

$$J_1\ddot{\theta}_1 = -k\theta_1 - T_1 + T \tag{2-50}$$

where T_1 is the torque transmitted to shaft 2. For shaft 2,

$$J_2\ddot{\theta}_2 = -b\dot{\theta}_2 - T_L + T_2 \tag{2-51}$$

where T_2 is the torque applied to shaft 2 through the gears.

From the geometrical constraints,

$$r_1\theta_1 = r_2\theta_2$$

Since the work done by gear 1 is equal to that of gear 2,

$$T_1\dot{\theta}_1 = T_2\dot{\theta}_2$$

or

$$\frac{T_1}{T_2} = \frac{\theta_2}{\theta_1} = \frac{r_1}{r_2} \tag{2-52}$$

Hence from Eqs. (2-51) and (2-52)

$$J_2\ddot{\theta}_2 + b\dot{\theta}_2 + T_L = T_2 = T_1\frac{r_2}{r_1} \tag{2-53}$$

If we eliminate T_1 from Eqs. (2-50) and (2-53), the result is

$$J_1\ddot{\theta}_1 + k\theta_1 + \frac{r_1}{r_2}(J_2\ddot{\theta}_2 + b\dot{\theta}_2 + T_L) = T$$

Then substituting $\theta_2 = (r_1/r_2)\theta_1$ into this last equation gives

$$\left[J_1 + \left(\frac{r_1}{r_2}\right)^2 J_2\right]\ddot{\theta}_1 + b\left(\frac{r_1}{r_2}\right)^2\dot{\theta}_1 + k\theta_1 = T - \frac{r_1}{r_2}T_L$$

This is the equation of motion of the system referred to shaft 1. In terms of angle

θ_2, this last equation becomes

$$\left[J_1 + \left(\frac{r_1}{r_2}\right)^2 J_2\right]\ddot{\theta}_2 + b\left(\frac{r_1}{r_2}\right)^2\dot{\theta}_2 + k\theta_2 = \frac{r_1}{r_2}T - \left(\frac{r_1}{r_2}\right)^2 T_L$$

which is the equation of motion of the system referred to shaft 2.

PROBLEMS

PROBLEM B-2-1. In the system of Fig. 2-59 a mass of 1 kg is suspended by two wires AB and BC. Find the tensions in the wires.

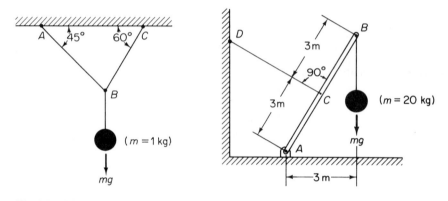

Fig. 2-59. Mass suspended by two wires.   **Fig. 2-60.** Mechanical system.

PROBLEM B-2-2. Figure 2-60 shows a steel bar pivoted at point A and a mass of 20 kg suspended from point B. The bar is supported by a wire CD. Find the tension in wire CD.

PROBLEM B-2-3. A homogeneous disk has a diameter of 1 m and mass of 100 kg. Obtain the moment of inertia of the disk about the axis perpendicular to the disk and passing through its center.

PROBLEM B-2-4. A ball is dropped from a point 100 m above the ground with an initial velocity of 20 m/s. How long will it take until the ball reaches the ground?

PROBLEM B-2-5. A flywheel of $J = 50$ kg-m² initially standing still is subjected to a constant torque. If the angular velocity reaches 20 Hz in 5 seconds, find the torque given to the flywheel.

PROBLEM B-2-6. A brake is applied to a flywheel rotating at an angular velocity of 100 rad/s. If the angular velocity reduces to 20 rad/s in 15 seconds, find the deceleration given by the brake and the total angle rotated in the 15-second period.

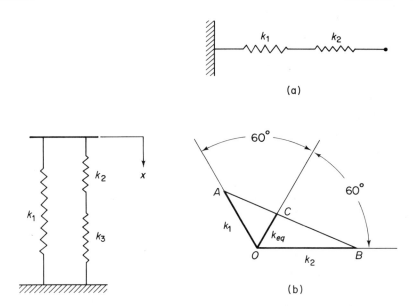

Fig. 2-61. System consisting of three springs.

Fig. 2-62. (a) System consisting of two springs in series; (b) diagram showing the equivalent spring constant.

PROBLEM B-2-7. Obtain the equivalent spring constant k_{eq} for the system shown in Fig. 2-61.

PROBLEM B-2-8. Consider the series-connected springs shown in Fig. 2-62(a). Referring to Fig. 2-62(b), show that the equivalent spring constant k_{eq} can be graphically obtained as the length $\overline{OC}$ if lengths $\overline{OA}$ and $\overline{OB}$ represent k_1 and k_2, respectively.

PROBLEM B-2-9. Find the natural frequency of the system shown in Fig. 2-63.

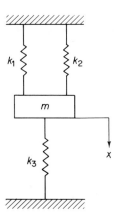

Fig. 2-63. Mechanical system.

PROBLEM B-2-10. The pendulum in Fig. 2-64 rotates freely under its own gravitational force. The moment of inerita of the body about the axis of rotation is J. Assuming that angle θ is small, obtain the frequency of oscillation.

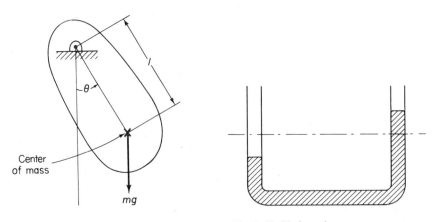

Fig. 2-64. Pendulum system.   Fig. 2-65. U-shaped manometer system.

PROBLEM B-2-11. Refer to the U-shaped manometer shown in Fig. 2-65, where liquid is partially filled in the U-shaped glass tube. Assuming that the total mass of the liquid in the tube is m, the total length of liquid in the tube is L, and the viscosity of the liquid is negligible, what is the equation of motion of the liquid? Find the frequency of oscillation.

PROBLEM B-2-12. In the mechanical system shown in Fig. 2-66 assume that the rod is massless, perfectly rigid, and pivoted at point P. The displacement x is measured from the equilibrium position. Assuming that displacement x is small, that weight mg at the end of the rod is 5 N, and that the spring constant k is 400 N/m, find the natural frequency of the system.

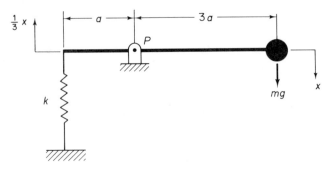

Fig. 2-66. Mechanical system.

PROBLEM B-2-13. Derive the equations of motion for the systems shown in Fig. 2-67(a), (b), and (c).

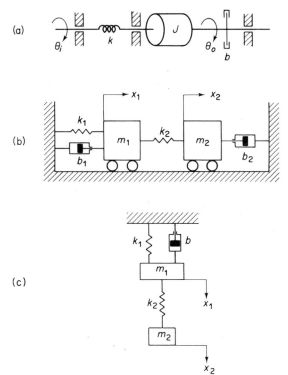

Fig. 2-67. Mechanical systems.

PROBLEM B-2-14. By applying Newton's second law to the spring-mass-pulley system of Fig. 2-53(a), obtain the motion of mass m when it is pulled down a short distance and then released. The displacement x of a hanging mass m is measured from the equilibrium position. (The mass, the radius, and the moment of inertia of the pulley are M, R, and $J = \frac{1}{2}MR^2$, respectively.)

PROBLEM B-2-15. A body of mass m (where $m = 10$ kg) is pulled by a force F in the direction shown in Fig. 2-68. If the coefficient of sliding friction between the body and the floor is 0.3, find the magnitude of force F necessary to keep moving the body at a constant speed.

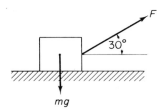

Fig. 2-68. Body placed on a floor and pulled by a force F.

PROBLEM B-2-16. A wood block of mass 2 kg placed on a rough horizontal surface is attached to a hanging mass of 1 kg through a massless wire (see Fig. 2-69). The wire passes horizontally over a frictionless pulley. Assume that the wire is inexten-

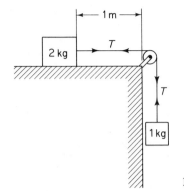

Fig. 2-69. Hanging-mass system.

sible. The wood block is initially held to stay at rest and the hanging mass is at rest. At $t = 0$ the wood block is released. Find the velocity of the block when it has moved 0.5 m. Assume that the coefficient of sliding friction between the wood block and the rough horizontal surface is 0.2.

PROBLEM B-2-17. A homogeneous cylinder of radius R and mass m rolls on a rough surface. It is connected to the wall through a spring as shown in Fig. 2-70. Assume that the cylinder rolls without sliding. Applying Newton's second law, obtain the frequency of oscillation of the system.

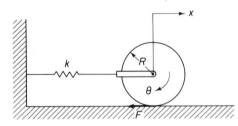

Fig. 2-70. Homogeneous cylinder connected to a wall through a spring.

PROBLEM B-2-18. Consider a homogeneous cylinder of radius R and mass m moving down an inclined plane whose angle of inclination is α as shown in Fig. 2-71. When the cylinder rolls down with sliding, $F = \mu_k N = \mu_k mg \cos \alpha$, and the equations of motion are

$$m\ddot{x} = mg \sin \alpha - \mu_k mg \cos \alpha$$

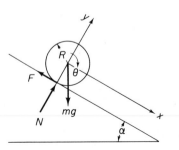

Fig. 2-71. Homogeneous cylinder rolling down an inclined plane.

$$m\ddot{y} = -mg \cos \alpha + N = 0$$
$$J\ddot{\theta} = FR$$

We assume that the initial conditions are $x(0) = 0$, $\dot{x}(0) = 0$, $\theta(0) = 0$, and $\dot{\theta}(0) = 0$. Show that when the cylinder rolls down with sliding, the solutions $x(t)$ and $\theta(t)$ of the equations of motion satisfy the relationship $x > R\theta$ or $\dot{x} > R\dot{\theta}$.

PROBLEM B-2-19. A homogeneous cylinder of radius R and mass m is initially at rest on a rough horizontal surface. An external force F is applied at the top rim of the cylinder (see Fig. 2-72). Assuming that the cylinder rolls without sliding, find the magnitude and direction of the static friction force.

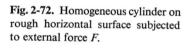

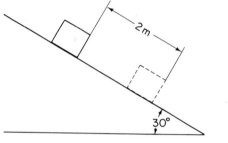

Fig. 2-72. Homogeneous cylinder on rough horizontal surface subjected to external force F.

Fig. 2-73. Body placed on an inclined plane.

PROBLEM B-2-20. A body of mass 1 kg is placed on an inclined plane whose angle of inclination is 30° as shown in Fig. 2-73. If the body slides down a linear distance 2 m along the inclined plane, how much work is done by the gravitational force and by the friction force? Assume that the coefficient of sliding friction is 0.2.

PROBLEM B-2-21. A body of mass 1 kg is at rest on an inclined plane whose angle of inclination is 30° (Fig. 2-74). If the body is pulled upward by a constant force F along the inclined surface, the speed reaches 5 m/s at the point where the body has moved 6 m. Find work done by the force F, by the gravitational force, and by the sliding friction force. Assume that the coefficient of sliding friction is 0.2.

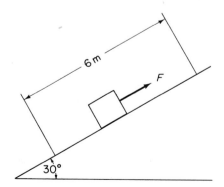

Fig. 2-74. Body placed on an inclined plane.

PROBLEM B-2-22. A spring requires a force of 50 N to be stretched by 5 cm. If the spring is stretched by 10 cm, how much potential energy does the spring possess due to its elongation?

PROBLEM B-2-23. A disk of radius 0.5 m and mass 10 kg is subjected to a tangential force of 50 N at its peripheral and is rotating at the angular velocity of 100 rad/s. Calculate the torque and power of the disk shaft.

PROBLEM B-2-24. Obtain the power needed to pull a body of mass 5 kg vertically upward by 10 m in 5 seconds.

PROBLEM B-2-25. Assuming that mass m of the rod of the pendulum shown in Fig. 2-75 is small but not negligible compared with mass M, find the natural frequency of the pendulum when angle θ is small.

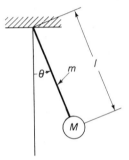

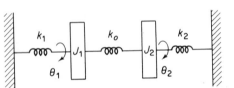

Fig. 2-75. Pendulum sys-   Fig. 2-76. Mechanical system.
tem.

PROBLEM B-2-26. Using the law of conservation of energy, derive the equation of motion for the system shown in Fig. 2-76.

PROBLEM B-2-27. A mass M placed on a smooth horizontal plane is attached to a hanging mass m through a massless wire as shown in Fig. 2-77. The wire passes

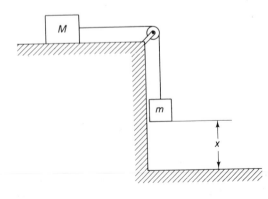

Fig. 2-77. Hanging-mass system.

horizontally over a pulley of radius r and mass m_p. The pulley is free to rotate about its axis of rotation. Assume that the wire is inextensible and does not slide when the pulley rotates. Assume also that mass M is initially held to stay at rest and that mass m is at rest at a vertical distance x from the floor. At $t = 0$ mass M is released to move. Using the law of conservation of energy, find the speed of mass m when it hits the floor. Neglect the friction between mass M and the horizontal plane.

PROBLEM B-2-28. A body of weight mg (where $m = 1000$ kg) is pulled vertically upward by a hoist as shown in Fig. 2-78. Find the force necessary to move the weight. How much power is needed to move the weight at a speed of 0.5 m/s?

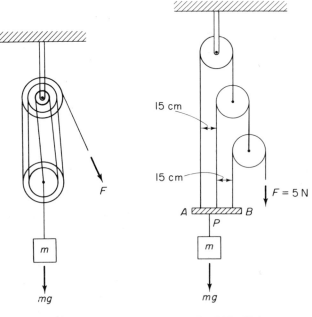

Fig. 2-78. Hoist.   Fig. 2-79. Hoist.

PROBLEM B-2-29. For the hoist shown in Fig. 2-79, find the maximum weight mg N that can be pulled upward by a force F of 5 N. Assume that bar AB weighs 2 N. To keep bar AB horizontal when pulling the weight mg, where should it be attached? (Determine the position of point P.)

PROBLEM B-2-30. A shaft is driven by an electric motor through a speed-reducing gear train. The output and speed of the motor are 1.5 kW and 60 Hz, respectively. Assuming the speed reduction ratio to be $1:30$, calculate the torque of the driven shaft; neglect losses in the transmission of power.

3

ELECTRICAL SYSTEMS

3-1 INTRODUCTION

This chapter is concerned with the mathematical modeling and response analysis of electrical systems. Since the basic materials of electrical circuits normally studied in college physics are covered in Sec. 3-1, this section should be regarded as review material. In Sec. 3-2 the basic laws for electrical circuits, such as Ohm's law and Kirchhoff's laws, are presented. Then Sec. 3-3 deals with mathematical modeling and analysis of circuit problems. Electrical power and energy are discussed in Sec. 3-4 and analogous systems in the final section, Sec. 3-5.

We begin with a brief review of voltage, charge, current, and current and voltage sources, followed by discussions of three types of basic elements of electrical systems—resistance, capacitance, and inductance elements.

Voltage. *Voltage* in electrical systems is analogous to pressure in hydraulic or pneumatic systems. It is the electromotive force needed to produce a flow of current in a wire, just as pressure is required to produce a flow of liquid or gas in a pipeline. The unit of voltage is the volt (V).

Charge. Electric *charge* is the integral of current with respect to time. The unit of charge is the coulomb (C). One coulomb is that amount of charge transferred in one second by a current of one ampere—that is,

$$\text{Coulomb} = \text{ampere-second}$$

In metric units, one coulomb is that amount of charge that experiences a force of one newton in an electric field of one volt per meter or

$$\text{Coulomb} = \text{newton-meter/volt}$$

Current. *Current* refers to the rate of flow of charge. Here the unit of current is the ampere. If charge dq coulombs crosses a given area in dt seconds, then the current i is

$$i = \frac{dq}{dt}$$

Thus, in a current of one ampere, charge is being transferred at the rate of one coulomb per second or

$$\text{Ampere} = \text{coulomb/second}$$

If positive charge flows from left to right (or negative charge from right to left), then the current flow is from left to right. Referring to Fig. 3-1, if i is positive, the current flow is from left to right. If i is negative, the current flow is from right to left.

$$i \longrightarrow$$

Fig. 3-1. Current flow.

Current sources and voltage sources. *Current source* means an energy source that produces a specified current, usually as function of time. It is capable of supplying a specified current independently of the voltage across the source. If a generator supplies the current relatively independently of the connected circuit, it is a *current generator*. Some commonly used current sources in electrical systems include transistors and other commercially available power supplies designed for constant current.

The *voltage source* is an energy source that supplies a specified voltage, usually as a function of time, completely independent of the current. In other words, it is a source of electrical power in which the voltage is independent of the current drawn. A generator that supplies a voltage output that is relatively independent of the circuit to which it is connected is called a *voltage generator*. Examples of voltage sources

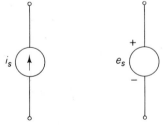

Current source   Voltage source

Fig. 3-2. Circuit symbols for current source and voltage source.

are rotating generators, batteries, and commercially available, constant voltage power supplies for electronic systems. The circuit symbols for a current source and voltage source are shown in Fig. 3-2.

It is possible for a battery to supply a nearly constant voltage at low currents. At high currents, however, the output voltage may drop considerably because part of the chemical energy is converted to heat. Thus a battery can be represented by a pure voltage source and an internal resistance, where the latter accounts for the heat loss.

Basic elements of electrical circuits. In Sec. 2-2 we discussed mathematical models of basic mechanical systems. An analogous situation exists for electrical systems. Three types of basic elements are found in electrical circuits—resistance elements, capacitance elements, and inductance elements—as discussed below.

Resistance elements. *Resistance* is defined as the change in voltage required to make a unit change in current or

$$\text{Resistance } R = \frac{\text{change in voltage}}{\text{change in current}} \quad \frac{\text{V}}{\text{A}}$$

The resistance R of a linear resistor can thus be given by

$$R = \frac{e_R}{i}$$

where e_R is the voltage across the resistor and i is the current through the resistor. The unit of resistance is the ohm (Ω), where

$$\text{Ohm} = \frac{\text{volt}}{\text{ampere}}$$

For the resistor shown in Fig. 3-3, a positive e_R causes a current i to flow from left to right. Consequently, we take the positive direction for i to the right.

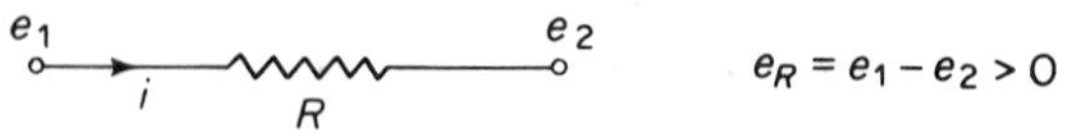

$$e_1 \;\xrightarrow{\;\;i\;\;}\; \text{—}\!\!\!\!\bigwedge\!\!\bigwedge\!\!\bigwedge\text{—} \; e_2 \qquad e_R = e_1 - e_2 > 0$$

Fig. 3-3. Resistor.

The reciprocal of resistance is called *conductance*. The unit of conductance is the siemens (S). ($1\,\text{S} = 1\,\text{A/V} = 1\,\Omega^{-1} = $ mho)

$$\text{Conductance} = \frac{1}{R} = G \quad \text{S}$$

Resistors do not store electric energy in any form but instead dissipate it as heat. Note that real resistors may not be linear and may also exhibit some capacitance and inductance effects.

Capacitance elements. Two conductors separated by a nonconducting medium (insulator or dielectric) form a capacitor. So two metallic plates separated by a very thin dielectric material form a capacitor. Sometimes the area is made variable, as in a tuning condenser in a radio.

Capacitance is defined as the change in the quantity of electric charge required to make a unit change in voltage or

$$\text{Capacitance } C = \frac{\text{change in quantity of electric charge}}{\text{change in voltage}} \quad \frac{C}{V}$$

Capacitance C is a measure of the quantity of charge that can be stored for a given voltage across the plates. (By bringing the plates close together, the capacitance increases and additional charge can be stored for a given voltage across the plates.) The capacitance C of a capacitor can thus be given by

$$C = \frac{q}{e_C}$$

where q is the quantity of charge stored and e_C is the voltage across the capacitor. The unit of capacitance is the farad (F), where

$$\text{Farad} = \frac{\text{ampere-second}}{\text{volt}} = \frac{\text{coulomb}}{\text{volt}}$$

Note that capacitance C is defined as a positive number. Therefore the algebraic sign of charge q is the same as that of voltage e_C across the capacitor. For the capacitor shown in Fig. 3-4, a positive e_C causes a current i to flow from left to right. As a result, we take the positive direction for i to the right as shown in the diagram.

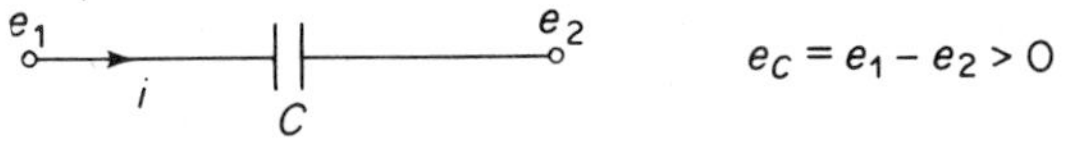

Fig. 3-4. Capacitor.

$$e_C = e_1 - e_2 > 0$$

Note that since $i = dq/dt$ and $e_C = q/C$, we have

$$i = C\frac{de_C}{dt}$$

or

$$de_C = \frac{1}{C}i\,dt$$

Therefore

$$e_C(t) = \frac{1}{C}\int_0^t i\,dt + e_C(0)$$

A capacitor having a capacitance of one farad is very large; those normally used in electronic devices are measured in microfarads (10^{-6} F). Some capacitors are even measured in picofarads (10^{-12} F).

Although a pure capacitor stores energy and can release all of it, real capacitors, on the other hand, exhibit various losses. These energy losses are indicated by a *power factor*, which is the ratio of energy lost per cycle of ac voltage to the energy stored per cycle. Thus a small-valued power factor is desirable.

Inductance elements. Surrounding a moving charge or current is a region of influence called a *magnetic field*. If a circuit lies in a time-varying magnetic field, an electromotive force is induced in the circuit. The ratio of the induced voltage to the rate of change of current (which means change in current per second) is defined as *inductance* or

$$\text{Inductance} = \frac{\text{change in induced voltage}}{\text{change in current per second}} \quad \frac{\text{V}}{\text{A/s}}$$

The inductive effects can be classified as self-inductance and mutual inductance.

Self-inductance is that property of a single coil that occurs when the magnetic field set up by the coil current links the coil itself. The magnitude of the induced voltage is proportional to the rate of change of flux linking the circuit. If the circuit does not contain ferromagnetic elements (such as iron core), the rate of change of flux is proportional to di/dt. Self-inductance, or simply inductance, L is the proportionality constant between the induced voltage e_L volts and the rate of change of current (or change in current per second) di/dt amperes per second—that is,

$$L = \frac{e_L}{di/dt}$$

The unit of inductance is the henry (H). (An electric circuit has an inductance of one henry when a rate of change of one ampere per second will induce an emf of one volt.)

$$\text{Henry} = \frac{\text{volt}}{\text{ampere/second}} = \frac{\text{weber}}{\text{ampere}}$$

For the inductor shown in Fig. 3-5, a positive e_L causes a current i to

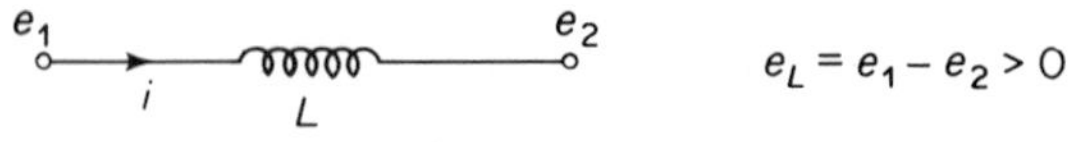

$$e_L = e_1 - e_2 > 0$$

Fig. 3-5. Inductor.

flow from left to right. So we take the positive direction for i to the right as in the diagram. Note that for the inductor shown

$$e_L = L\frac{di}{dt}$$

or

$$i_L(t) = \frac{1}{L}\int_0^t e_L\, dt + i_L(0)$$

Because most inductors are coils of wire, they have considerable resistance. The energy loss due to the presence of resistance is indicated by the *quality factor Q*, which shows the ratio of stored-to-dissipated energy. A high value of Q generally means that the inductor contains small resistance.

Mutual inductance refers to the influence between inductors that results from the interaction of their fields. If two inductors are involved in an electric circuit, each may come under the influence of the magnetic field of the other inductor. Then the voltage drop in the first inductor is related to the current flowing through the first inductor, as well as to the current flowing through the second inductor, whose magnetic field influences the first. The second inductor is also influenced by the first in exactly the same manner. When a change of current of one ampere per second in either of the two inductors induces an emf of one volt in the other inductor, their mutual inductance M is one henry. (Note that it is customary to use the symbol M to denote mutual inductance in order to distinguish it from self-inductance L.) We shall postpone further discussion of mutual inductance to Sec. 3-3.

3-2 BASIC LAWS FOR ELECTRICAL CIRCUITS

In this section we present Ohm's law and Kirchhoff's current and voltage laws. The former is fundamental in obtaining combined resistances in series and parallel circuits, currents and voltages in such circuits, and so on. Kirchhoff's laws are basic in formulating governing equations that characterize electrical circuits.

Ohm's law. *Ohm's law* states that the current in a circuit is proportional to the total electromotive force (emf) acting in the circuit and inversely proportional to the total resistance of the circuit. It can be expressed by

$$i = \frac{e}{R}$$

where i is the current (amperes), e the emf (volts), and R the resistance (ohms).

Series circuits. The combined resistance of series-connected resistors is the sum of the separate resistances. Figure 3-6 shows a simple series circuit. The voltage between points A and B is

$$e = e_1 + e_2 + e_3$$

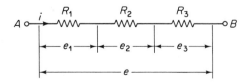

Fig. 3-6. Series circuit.

where

$$e_1 = iR_1, \qquad e_2 = iR_2, \qquad e_3 = iR_3$$

Thus

$$\frac{e}{i} = R_1 + R_2 + R_3$$

The combined resistance R is then given by

$$R = R_1 + R_2 + R_3$$

Parallel circuits. For the parallel circuit shown in Fig. 3-7,

$$i_1 = \frac{e}{R_1}, \qquad i_2 = \frac{e}{R_2}, \qquad i_3 = \frac{e}{R_3}$$

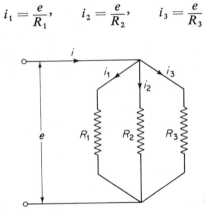

Fig. 3-7. Parallel circuit.

Since $i = i_1 + i_2 + i_3$, it follows that

$$i = \frac{e}{R_1} + \frac{e}{R_2} + \frac{e}{R_3} = \frac{e}{R}$$

where R is the combined resistance. Hence

$$\frac{1}{R} = \frac{1}{R_1} + \frac{1}{R_2} + \frac{1}{R_3}$$

or

$$R = \frac{1}{\dfrac{1}{R_1} + \dfrac{1}{R_2} + \dfrac{1}{R_3}} = \frac{R_1 R_2 R_3}{R_1 R_2 + R_2 R_3 + R_3 R_1}$$

Resistance of combined series and parallel resistors. Consider the circuit shown in Fig. 3-8(a). The combined resistance R_{BC} between points B and C is

$$R_{BC} = \frac{R_2 R_3}{R_2 + R_3}$$

Then the combined resistance R between points A and C is

$$R = R_1 + R_{BC} = R_1 + \frac{R_2 R_3}{R_2 + R_3}$$

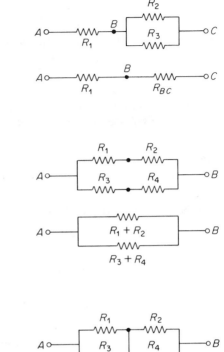

(a)

(b)

(c)

Fig. 3-8. Combined series and parallel resistors.

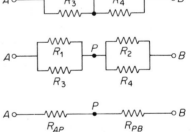

The circuit shown in Fig. 3-8(b) can be considered a parallel circuit consisting of resistances $(R_1 + R_2)$ and $(R_3 + R_4)$. So the combined resistance R between points A and B is

$$\frac{1}{R} = \frac{1}{R_1 + R_2} + \frac{1}{R_3 + R_4}$$

or

$$R = \frac{(R_1 + R_2)(R_3 + R_4)}{R_1 + R_2 + R_3 + R_4}$$

Next, consider the circuit shown in Fig. 3-8(c). Here R_1 and R_3 are parallel and R_2 and R_4 are parallel. These two parallel resistances are connected in series. Redrawing this circuit as shown in Fig. 3-8(c), therefore, we

obtain

$$R_{AP} = \frac{R_1 R_3}{R_1 + R_3}, \qquad R_{PB} = \frac{R_2 R_4}{R_2 + R_4}$$

As a result, the combined resistance R becomes

$$R = R_{AP} + R_{PB} = \frac{R_1 R_3}{R_1 + R_3} + \frac{R_2 R_4}{R_2 + R_4}$$

Currents and voltages in series and parallel circuits. In the circuit shown in Fig. 3-9, the voltage drop e_{BC} between points B and C is

$$e_{BC} = i R_{BC}$$

where

$$R_{BC} = \frac{R_1 R_2}{R_1 + R_2}$$

Hence the current i_1 flowing through resistance R_1 is

$$i_1 = \frac{e_{BC}}{R_1} = i \frac{R_{BC}}{R_1} = i \frac{R_2}{R_1 + R_2}$$

Similarly, the current i_2 flowing through resistance R_2 is

$$i_2 = \frac{e_{BC}}{R_2} = i \frac{R_{BC}}{R_2} = i \frac{R_1}{R_1 + R_2}$$

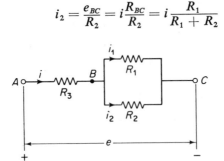

Fig. 3-9. Electrical circuit.

Looking next at the circuit shown in Fig. 3-10(a), the current i through the resistance $R_1 + R_2$ is

$$i = \frac{e}{R_1 + R_2}$$

where e is the voltage drop between points A and B. Thus voltage drops e_1 and e_2 are given by

$$e_1 = i R_1 = e \frac{R_1}{R_1 + R_2}, \qquad e_2 = i R_2 = e \frac{R_2}{R_1 + R_2}$$

from which

$$e_1 : e_2 = R_1 : R_2$$

Similarly, for the circuit shown in Fig. 3-10(b),

$$e_1 : e_2 : e_3 = R_1 : \frac{R_2 R_3}{R_2 + R_3} : R_4$$

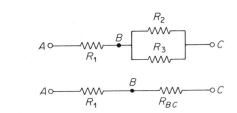

(a)

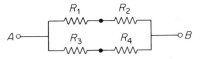

(b)

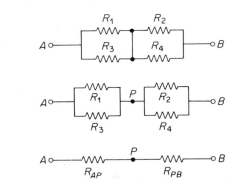

(c)

Fig. 3-8. Combined series and parallel resistors.

The circuit shown in Fig. 3-8(b) can be considered a parallel circuit consisting of resistances $(R_1 + R_2)$ and $(R_3 + R_4)$. So the combined resistance R between points A and B is

$$\frac{1}{R} = \frac{1}{R_1 + R_2} + \frac{1}{R_3 + R_4}$$

or

$$R = \frac{(R_1 + R_2)(R_3 + R_4)}{R_1 + R_2 + R_3 + R_4}$$

Next, consider the circuit shown in Fig. 3-8(c). Here R_1 and R_3 are parallel and R_2 and R_4 are parallel. These two parallel resistances are connected in series. Redrawing this circuit as shown in Fig. 3-8(c), therefore, we

obtain

$$R_{AP} = \frac{R_1 R_3}{R_1 + R_3}, \qquad R_{PB} = \frac{R_2 R_4}{R_2 + R_4}$$

As a result, the combined resistance R becomes

$$R = R_{AP} + R_{PB} = \frac{R_1 R_3}{R_1 + R_3} + \frac{R_2 R_4}{R_2 + R_4}$$

Currents and voltages in series and parallel circuits. In the circuit shown in Fig. 3-9, the voltage drop e_{BC} between points B and C is

$$e_{BC} = i R_{BC}$$

where

$$R_{BC} = \frac{R_1 R_2}{R_1 + R_2}$$

Hence the current i_1 flowing through resistance R_1 is

$$i_1 = \frac{e_{BC}}{R_1} = i \frac{R_{BC}}{R_1} = i \frac{R_2}{R_1 + R_2}$$

Similarly, the current i_2 flowing through resistance R_2 is

$$i_2 = \frac{e_{BC}}{R_2} = i \frac{R_{BC}}{R_2} = i \frac{R_1}{R_1 + R_2}$$

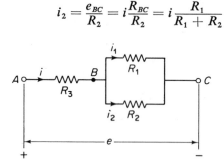

Fig. 3-9. Electrical circuit.

Looking next at the circuit shown in Fig. 3-10(a), the current i through the resistance $R_1 + R_2$ is

$$i = \frac{e}{R_1 + R_2}$$

where e is the voltage drop between points A and B. Thus voltage drops e_1 and e_2 are given by

$$e_1 = i R_1 = e \frac{R_1}{R_1 + R_2}, \qquad e_2 = i R_2 = e \frac{R_2}{R_1 + R_2}$$

from which

$$e_1 : e_2 = R_1 : R_2$$

Similarly, for the circuit shown in Fig. 3-10(b),

$$e_1 : e_2 : e_3 = R_1 : \frac{R_2 R_3}{R_2 + R_3} : R_4$$

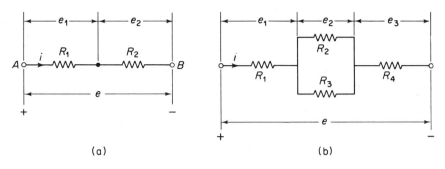

Fig. 3-10. Electrical circuits.

Kirchhoff's laws. In solving circuit problems that involve many electromotive forces, resistances, capacitances, inductances, and so on, it is sometimes necessary to use *Kirchhoff's laws*. There are two: the *current law* (node law) and the *voltage law* (loop law).

Kirchhoff's current law (node law). A *node* in an electrical circuit is a point where three or more wires are joined together. Kirchhoff's current law (node law) states that the algebraic sum of all currents entering and leaving a node is zero. (This law can also be stated as: the sum of currents entering a node is equal to the sum of currents leaving the same node.) In applying the law to circuit problems, the following rules should be observed. Currents going toward a node should be preceded by a plus sign. Currents going away from a node should be preceded by a minus sign. Referring to Fig. 3-11, Kirchhoff's current law states that

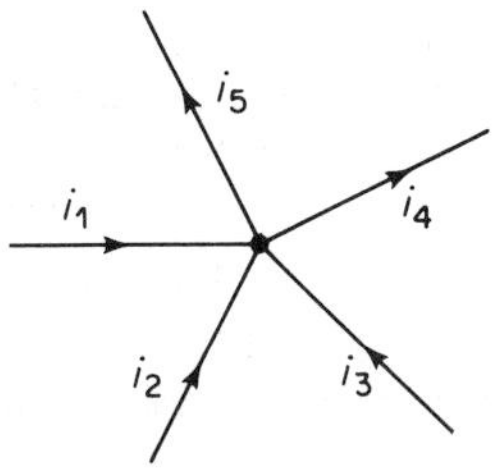

Fig. 3-11. Node.

$$i_1 + i_2 + i_3 - i_4 - i_5 = 0$$

Kirchhoff's voltage law (loop law). Kirchhoff's voltage law states that at any given instant of time the algebraic sum of the voltages around any loop in an electrical circuit is zero. This law can also be stated as: the sum of the voltage drops is equal to the sum of the voltage rises around a loop. When applying the law to circuit problems, the following rules should be observed. A rise in voltage [which occurs in going through a source of electromotive force from the negative to the positive terminal, as in Fig. 3-12(a), or in going through a resistance in opposition to the current flow, as in Fig. 3-12(b)] should be preceded by a plus sign. A drop in voltage [which occurs in going through a source of electromotive force from the positive to the negative

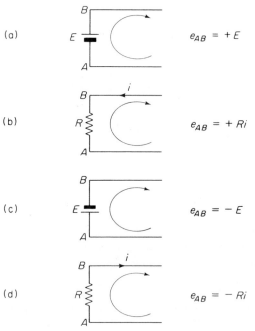

(a) $e_{AB} = +E$

(b) $e_{AB} = +Ri$

(c) $e_{AB} = -E$

(d) $e_{AB} = -Ri$

Fig. 3-12. Diagrams showing voltage rises and voltage drops in circuits.

terminal, as in Fig. 3-12(c), or in going through a resistance in the direction of the current flow, as in Fig. 3-12(d)] should be preceded by a minus sign.

Figure 3-13 shows a circuit that consists of a battery and an external resistance. Here E is the electromotive force, r the internal resistance of the battery, R the external resistance, and i the current. If we follow the loop in the clockwise direction $(A \rightarrow B \rightarrow C \rightarrow A)$ as shown, then

$$e_{\overrightarrow{AB}} + e_{\overrightarrow{BC}} + e_{\overrightarrow{CA}} = 0$$

or

$$E - iR - ir = 0$$

from which

$$i = \frac{E}{R + r}$$

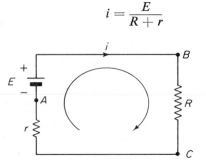

Figure 3-13. Electrical circuit.

A circuit consisting of two batteries and an external resistance appears in Fig. 3-14(a), where E_1 and r_1 (E_2 and r_2) are the electromotive force and internal resistance of battery No. 1 (battery No. 2), respectively, and R is the external resistance. By assuming the direction of the current i as shown and also following the loop clockwise as shown, the result is

$$E_1 - iR - E_2 - ir_2 - ir_1 = 0$$

or

$$i = \frac{E_1 - E_2}{r_1 + r_2 + R} \tag{3-1}$$

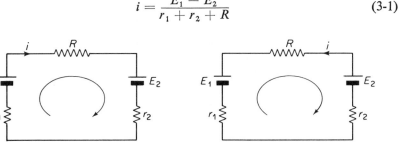

(a)   (b)

Fig. 3-14. Electrical circuits.

If we assume that the direction of the current i is reversed [Fig. 3-14(b)], then, by following the loop clockwise, we obtain

$$E_1 + iR - E_2 + ir_2 + ir_1 = 0$$

or

$$i = \frac{E_2 - E_1}{r_1 + r_2 + R} \tag{3-2}$$

Note that, in solving circuit problems, if we assume that the current flows to the right and if the value i is calculated and found to be positive, current i actually flows to the right. If the value of i is found to be negative, current i actually flows to the left. For the circuits shown in Fig. 3-14, suppose that $E_1 > E_2$. Then Eq. (3-1) gives $i > 0$, which means that the current i flows in the direction assumed. Equation (3-2), however, yields $i < 0$, which means that the current i flows opposite to the assumed direction.

It should be noted that the direction used to follow the loop is arbitrary, just as the direction of current flow can be assumed to be arbitrary. That is, the direction used in following the loop can be clockwise or counterclockwise. The final result is the same in either case.

Circuits with two or more loops. For circuits with two or more loops, both Kirchhoff's current law and voltage law may be applied. The first step in writing the circuit equations is to define the directions of the currents in each wire. The second is to determine the directions that we follow in each loop.

Consider the circuit shown in Fig. 3-15, which has two loops. Here we can assume the directions of currents as shown in the diagram. (Note that the directions of currents assumed are arbitrary and could differ from those shown in the diagram.) Suppose that we follow the loops clockwise as in Fig. 3-15. (Again, the directions could be either clockwise or counterclockwise.)

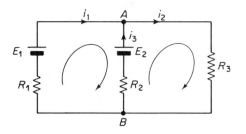

Fig. 3-15. Electrical circuit.

Then we obtain the equations

$$\text{At point } A: \quad i_1 + i_3 - i_2 = 0$$
$$\text{For the left loop:} \quad E_1 - E_2 + i_3 R_2 - i_1 R_1 = 0$$
$$\text{For the right loop:} \quad E_2 - i_2 R_3 - i_3 R_2 = 0$$

By first eliminating i_2 from the preceding three equations and then solving for i_1 and i_3, we find

$$i_1 = \frac{E_1(R_2 + R_3) - E_2 R_3}{R_1 R_2 + R_2 R_3 + R_3 R_1}$$

$$i_3 = \frac{E_2(R_1 + R_3) - E_1 R_3}{R_1 R_2 + R_2 R_3 + R_3 R_1}$$

Hence

$$i_2 = i_1 + i_3 = \frac{E_1 R_2 + E_2 R_1}{R_1 R_2 + R_2 R_3 + R_3 R_1}$$

Writing equations for loops by using cyclic currents. In this approach we assume that a cyclic current exists in each loop. For instance, in Fig. 3-16 we

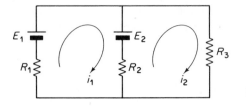

Fig. 3-16. Electrical circuit.

assume that clockwise cyclic currents i_1 and i_2 exist in the left and right loops, respectively, of the circuit.

Applying Kirchhoff's voltage law to the circuit results in the equations

$$\text{For left loop:} \quad E_1 - E_2 - R_2(i_1 - i_2) - R_1 i_1 = 0$$
$$\text{For right loop:} \quad E_2 - R_3 i_2 - R_2(i_2 - i_1) = 0$$

Note that the net current through resistance R_2 is the difference between i_1 and i_2. Solving for i_1 and i_2 gives

$$i_1 = \frac{E_1(R_2 + R_3) - E_2 R_3}{R_1 R_2 + R_2 R_3 + R_3 R_1}$$

$$i_2 = \frac{E_1 R_2 + E_2 R_1}{R_1 R_2 + R_2 R_3 + R_3 R_1}$$

(By comparing the circuits shown in Figs. 3-15 and 3-16, verify that i_3 in Fig. 3-15 is equal to $i_2 - i_1$ in Fig. 3-16.)

3-3 MATHEMATICAL MODELING AND CIRCUIT ANALYSIS

The first step in analyzing circuit problems is to obtain mathematical models for the circuits. (Although the terms *circuit* and *network* are sometimes used interchangeably, *network* implies a more complicated interconnection than *circuit*.) A mathematical model may consist of algebraic equations, differential equations, integrodifferential equations, and similiar ones. Such a model can be obtained by applying one or both of Kirchhoff's laws to a given circuit. The variables of interest in the circuit analysis are voltages and currents at various points along the circuit.

In this section we present mathematical modeling techniques and illustrate solutions of simple circuit problems. Although many significant problems can be solved by using the methods given, the discussions are intended as introductory and illustrative rather than comprehensive.

Node method for obtaining mathematical models. In the node method we write equations by applying Kirchhoff's current law (node law) to each node of the circuit.

Example 3-1. For the circuit shown in Fig. 3-17, assume that at $t = 0$ switch S is closed, so that $E = 12$ volts acts as the input to the circuit. Find voltages $e_A(t)$ and $e_B(t)$, where e_A and e_B are voltages at points A and B, respectively. Assume that the condenser is not charged initially.

In this problem we choose node D as a reference ($e_D = 0$) and measure the

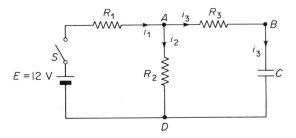

Fig. 3-17. Electrical circuit.

voltage of any node with respect to it. (In practice, many elements are connected to a metal chasis that is grounded to the earth. Such a ground, often shown as a common lead at the bottom of a circuit diagram, can serve as a convenient reference.)

At node A

$$i_1 - i_2 - i_3 = 0$$

where

$$i_1 = \frac{E - e_A}{R_1}, \qquad i_2 = \frac{e_A - e_D}{R_2} = \frac{e_A}{R_2}, \qquad i_3 = \frac{e_A - e_B}{R_3}$$

Then

$$\frac{E - e_A}{R_1} - \frac{e_A}{R_2} - \frac{e_A - e_B}{R_3} = 0 \tag{3-3}$$

At point B, i_3 is also equal to $C\,d(e_B - e_D)/dt = C\,de_B/dt$. Hence

$$\frac{e_A - e_B}{R_3} = C\frac{de_B}{dt} \tag{3-4}$$

Voltages $e_A(t)$ and $e_B(t)$ can be determined from Eqs. (3-3) and (3-4) as functions of time.

Let us solve Eqs. (3-3) and (3-4) for $e_A(t)$ and $e_B(t)$ for a special case where

$$2R_1 = R_2 = R_3, \qquad R_3C = 1$$

Then from Eq. (3-3) we obtain

$$2e_A - \tfrac{1}{2}e_B = E = 12 \tag{3-5}$$

From Eq. (3-4)

$$\dot{e}_B + e_B = e_A \tag{3-6}$$

By eliminating e_A from Eqs. (3-5) and (3-6), we have

$$\dot{e}_B + \tfrac{3}{4}e_B = 6 \tag{3-7}$$

To solve this last equation, let

$$\tfrac{3}{4}e_B - 6 = \tfrac{3}{4}x$$

or

$$x = e_B - 8$$

Then Eq. (3-7) can be written

$$\dot{x} + \tfrac{3}{4}x = 0$$

The solution of this last equation can be found by writing $x = Ke^{\lambda t}$ and substituting $\dot{x}$ and x into the equation or

$$K\lambda e^{\lambda t} + \tfrac{3}{4}Ke^{\lambda t} = 0$$

from which

$$\lambda = -\tfrac{3}{4} = -0.75$$

It follows that the solution of Eq. (3-7) can be written

$$e_B(t) = x(t) + 8 = Ke^{-0.75t} + 8$$

where K is to be determined by the initial condition. Since the condenser is not charged initially, $e_B(0) = 0$ or

$$e_B(0) = K + 8 = 0$$

This result gives us $K = -8$, and $e_B(t)$ is obtained as

$$e_B(t) = 8(1 - e^{-0.75t})$$

Then from Eq. (3-6)

$$e_A(t) = \dot{e}_B(t) + e_B(t) = 8 - 2e^{-0.75t}$$

Note that $e^{-4} = 0.0183$ and $e^{-6} = 0.00248$. Consequently, for $t > 8$, we have $e^{-0.75t} < 0.00248$ and approximately $e_A(t) = e_B(t) = 8$ volts. For $t > 8$, therefore, no power is dissipated by R_3. The power, however, is continuously dissipated by the resistances R_1 and R_2.

Loop method for obtaining mathematical models. In using this method, we first label the unknown currents and assume arbitrarily the directions of the currents around loops; then we write equations by applying Kirchhoff's voltage law (loop law).

Example 3-2. Assume that switch S is open for $t < 0$ and closed at $t = 0$ for the circuit of Fig. 3-18. Only one loop is involved here. By arbitrarily choosing the

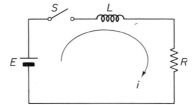

Fig. 3-18. Electrical circuit.

direction of the current around the loop as shown in the figure, we obtain the equation

$$E - L\frac{di}{dt} - Ri = 0$$

or

$$L\frac{di}{dt} + Ri = E \qquad (3-8)$$

This is a mathematical model for the circuit given. Note that at the instant that switch S is closed the current $i(0)$ is zero because the current in the inductor cannot change from zero to a finite value instantaneously. Thus $i(0) = 0$.

Let us solve Eq. (3-8) for the current $i(t)$. Note that by defining

$$x = i - \frac{E}{R}$$

Eq. (3-8) can be simplified to

$$L\frac{dx}{dt} + Rx = 0$$

Also, by assuming an exponential solution, as in Example 3-1, we obtain

$$x = Ke^{-(R/L)t}$$

or

$$i(t) = x(t) + \frac{E}{R} = Ke^{-(R/L)t} + \frac{E}{R}$$

where K is to be determined by the initial condition. Noting that $i(0) = 0$, we have

$$i(0) = K + \frac{E}{R} = 0$$

or

$$K = -\frac{E}{R}$$

Therefore the current $i(t)$ can be found as

$$i(t) = \frac{E}{R}[1 - e^{-(R/L)t}]$$

A typical plot of $i(t)$ versus t appears in Fig. 3-19.

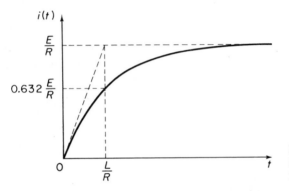

Fig. 3-19. Plot of $i(t)$ versus t for the circuit shown in Fig. 3-18 when switch S is closed.

Example 3-3. Again using the circuit shown in Fig. 3-18, assume that switch S is open for $t < 0$. It is closed at $t = 0$ and is opened again at $t = t_1 > 0$. Find the current $i(t)$ for $t \geq 0$.

The equation for the circuit for $t_1 > t \geq 0$ is

$$L\frac{di}{dt} + Ri = E, \qquad i(0) = 0$$

Referring to Example 3-2, the solution of this equation is

$$i(t) = \frac{E}{R}[1 - e^{-(R/L)t}] \qquad (t_1 > t \geq 0) \tag{3-9}$$

At $t = t_1$ the switch is opened. The equation for the circuit for $t \geq t_1$ is

$$L\frac{di}{dt} + Ri = 0 \tag{3-10}$$

where the initial condition at $t = t_1$ is given by

$$i(t_1) = \frac{E}{R}[1 - e^{-(R/L)t_1}] \tag{3-11}$$

(Note that the instantaneous value of the current at the switching instant $t = t_1$ serves as the initial condition for the transient response for $t \geq t_1$.)

The solution of Eq. (3-10) can be written

$$i(t) = Ke^{-(R/L)t} \tag{3-12}$$

where constant K is determined as follows. Substituting $t = t_1$ into Eq. (3-12) and equating the result to Eq. (3-11) yield

$$i(t_1) = Ke^{-(R/L)t_1} = \frac{E}{R}[1 - e^{-(R/L)t_1}]$$

from which

$$K = \frac{E}{R}[1 - e^{-(R/L)t_1}] e^{(R/L)t_1} \qquad (3\text{-}13)$$

Then using Eqs. (3-12) and (3-13), we have

$$i(t) = \frac{E}{R}[1 - e^{-(R/L)t_1}] e^{-(R/L)(t-t_1)} \qquad (t \geq t_1) \qquad (3\text{-}14)$$

And so, referring to Eqs. (3-9) and (3-14), the current $i(t)$ for $t \geq 0$ can be written

$$i(t) = \frac{E}{R}[1 - e^{-(R/L)t}] \qquad (t_1 > t \geq 0)$$

$$= \frac{E}{R}[1 - e^{-(R/L)t_1}] e^{-(R/L)(t-t_1)} \qquad (t \geq t_1)$$

A typical plot of $i(t)$ versus t for this case is given in Fig. 3-20.

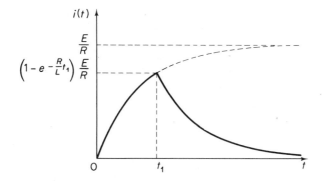

Fig. 3-20. Plot of $i(t)$ versus t for the circuit shown in Fig. 3-18 when switch S is closed at $t = 0$ and opened at $t = t_1$.

Example 3-4. Figure 3-21 shows a circuit consisting of a capacitor, a resistor, and a battery. The capacitor is charged to a voltage of 12 volts, and at $t = 0$ it is switched across the resistor. Thus $e_C(0) = 12$ volts. Obtain the current $i(t)$ as a function of time.

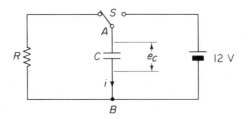

Fig. 3-21. Electrical circuit.

For $t > 0$, there is one loop in the circuit. Arbitrarily assuming the direction of the current as shown in the figure, we find

$$\frac{1}{C} \int i\, dt + Ri = 0$$

or

$$\frac{1}{C} \int_0^t i\, dt + \frac{1}{C} q(0) + Ri = 0$$

Note that $q(0)/C = e_A(0) - e_B(0) = e_C(0)$. So it follows that

$$\frac{1}{C} \int_0^t i\, dt + Ri = -e_C(0) \tag{3-15}$$

Differentiating both sides of Eq. (3-15) with respect to t results in

$$R \frac{di}{dt} + \frac{1}{C} i = 0$$

The solution of this last equation can be written

$$i(t) = Ke^{-t/RC} \tag{3-16}$$

where K is a constant to be determined by the initial condition, $e_C(0) = 12$ volts. Substituting $t = 0+$ into Eq. (3-15) gives

$$\frac{1}{C} \int_0^{0+} i\, dt + Ri(0+) = -e_C(0)$$

Since

$$\frac{1}{C} \int_0^{0+} i\, dt = 0$$

we obtain

$$i(0+) = -\frac{e_C(0)}{R}$$

From Eq. (3-16) we have

$$i(0+) = K$$

Hence

$$K = -\frac{e_C(0)}{R} = -\frac{12}{R}$$

The current $i(t)$, therefore, is obtained as

$$i(t) = -\frac{12}{R} e^{-t/RC}$$

Since the current $i(t)$ is found to be negative, the current actually flows opposite to the assumed direction.

Wheatstone bridge. The Wheatstone bridge shown in Fig. 3-22 consists of four resistors, a battery (or low voltage source of direct current), and a galvanometer. Resistance R_x is an unknown resistance. Resistances R_1 and

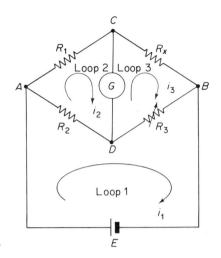

Fig. 3-22. Wheatstone bridge.

R_2 are ratio arms and can be made equal. The positions of the battery and galvanometer can be interchanged. By using such a bridge, resistors from a fraction of an ohm to 100 000 ohms or more can be measured precisely. Numerous modifications of the bridge are used to measure very low resistances and ac voltages.

In the diagram the Wheatstone bridge has three loops. Let us label them loop 1, loop 2, and loop 3 and label the cyclic current in each loop as i_1, i_2, and i_3. The condition for equilibrium in a Wheatstone bridge is that the current through the galvanometer is zero, or $i_2 - i_3 = 0$. This condition is equivalent to the condition that $e_C = e_D$. So if the bridge is in balance, we have

$$R_1 i_2 = R_2(i_1 - i_2)$$
$$R_x i_3 = R_3(i_1 - i_3)$$
$$i_2 = i_3$$

from which

$$\frac{R_1}{R_2} = \frac{R_x}{R_3} \quad \text{if the bridge is in balance}$$

If the bridge is not in balance, a current flows through the galvanometer. The equations for the bridge are as follows:

For loop 1: $E - R_2(i_1 - i_2) - R_3(i_1 - i_3) = 0$

For loop 2: $R_1 i_2 + R_G(i_2 - i_3) + R_2(i_2 - i_1) = 0$

For loop 3: $R_x i_3 + R_3(i_3 - i_1) + R_G(i_3 - i_2) = 0$

where R_G is the internal resistance of the galvanometer. Rewriting, we obtain

$$(R_2 + R_3)i_1 - R_2i_2 - R_3i_3 = E$$

$$-R_2i_1 + (R_1 + R_G + R_2)i_2 - R_Gi_3 = 0$$

$$-R_3i_1 - R_Gi_2 + (R_x + R_3 + R_G)i_3 = 0$$

Let us consider a special case where $R_1 = R_2 = 20\,\Omega$, $R_3 = 10\,\Omega$, $R_x = 5\,\Omega$, and $R_G = 50\,\Omega$ and find the current flowing through the galvanometer. Since $R_1/R_2 \neq R_x/R_3$, the bridge is not in balance. Substituting these numerical values into the last three equations gives

$$30i_1 - 20i_2 - 10i_3 = E$$

$$-20i_1 + 90i_2 - 50i_3 = 0$$

$$-10i_1 - 50i_2 + 65i_3 = 0$$

And solving for i_2, we have

$$i_2 = \frac{\begin{vmatrix} 30 & E & -10 \\ -20 & 0 & -50 \\ -10 & 0 & 65 \end{vmatrix}}{\begin{vmatrix} 30 & -20 & -10 \\ -20 & 90 & -50 \\ -10 & -50 & 65 \end{vmatrix}} = \frac{1800E}{45\,500} = 0.03956E$$

Similarly, solving for i_3,

$$i_3 = \frac{\begin{vmatrix} 30 & -20 & E \\ -20 & 90 & 0 \\ -10 & -50 & 0 \end{vmatrix}}{\begin{vmatrix} 30 & -20 & -10 \\ -20 & 90 & -50 \\ -10 & -50 & 65 \end{vmatrix}} = \frac{1900E}{45\,500} = 0.04176E$$

Hence the current flowing through the galvanometer is

$$i_2 - i_3 = 0.03956E - 0.04176E = -0.0022E$$

Since $i_2 - i_3$ is negative, the current $0.0022E$ flows in the direction from point D to point C.

Circuits with mutual inductance. Let us consider two coils (inductors) that are mutually coupled (Fig. 3-23). The induced voltage in coil 1 due to current change in coil 2 can either be added to or subtracted from the self-induced voltage in coil 1. Whether the induced voltage due to the mutual inductance M is to be added to the self-induced voltage depends on the direction of currents i_1 and i_2 and on the orientation of the coils. The direction of the currents can be chosen arbitrarily. The orientation of the coils is generally fixed, however, and it is customary to specify this orientation (based

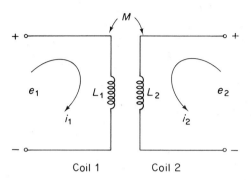

Fig. 3-23. Mutually coupled coils.   Coil 1   Coil 2

on experimental tests or physical arrangements) on the circuit diagram by placing a dot on one end of each coil of a mutual pair as shown in Fig. 3-24(a) and (b).

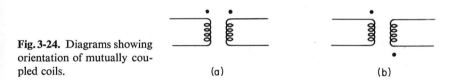

Fig. 3-24. Diagrams showing orientation of mutually coupled coils.   (a)   (b)

We will define L_1 and L_2 as the self-inductance of coils 1 and 2, respectively, and M as the mutual inductance. If both currents i_1 and i_2 enter (or leave) through a dot, then the voltage drop due to the mutual inductance will have the same sign as the voltage drop due to the self-inductance. For instance, for the circuit shown in Fig. 3-25(a),

$$e_1 = L_1 \frac{di_1}{dt} + M \frac{di_2}{dt}$$

$$e_2 = L_2 \frac{di_2}{dt} + M \frac{di_1}{dt}$$

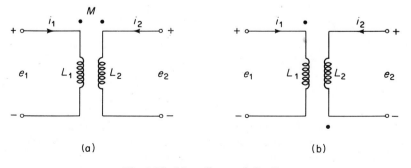

(a)   (b)

Fig. 3-25. Mutually coupled coils.

If one current enters by a dot and the other current leaves by a dot, the voltage drop due to the mutual inductance will have the opposite sign to the voltage drop due to the self-inductance. To illustrate, in the system of Fig. 3-25(b), i_1 enters by a dot and i_2 leaves by a dot. So for this case,

$$e_1 = L_1 \frac{di_1}{dt} - M \frac{di_2}{dt}$$

$$e_2 = L_2 \frac{di_2}{dt} - M \frac{di_1}{dt}$$

Example 3-5. The system of Fig. 3-26 is a network of two circuits coupled by the mutual inductance of a pair of coils with a common magnetic field. Assuming that switch S is closed at $t = 0$ and that there is no initial charge in the capacitor, find a mathematical model for the system.

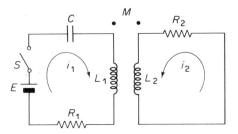

Fig. 3-26. Mutually coupled circuit.

Let us define cyclic currents i_1 and i_2 as in the diagram. Then for circuit 1 (left loop), we obtain

$$E - \frac{1}{C} \int i_1 \, dt - L_1 \frac{di_1}{dt} - M \frac{di_2}{dt} - R_1 i_1 = 0$$

For circuit 2 (right loop), we have

$$-L_2 \frac{di_2}{dt} - M \frac{di_1}{dt} - R_2 i_2 = 0$$

Rewriting the last two equations and noting that $q_1(0) = 0$ and thus that

$$\int i_1 dt = \int_0^t i_1 \, dt + q_1(0) = \int_0^t i_1 \, dt$$

we obtain

$$L_1 \frac{di_1}{dt} + M \frac{di_2}{dt} + R_1 i_1 + \frac{1}{C} \int_0^t i_1 \, dt = E$$

$$L_2 \frac{di_2}{dt} + M \frac{di_1}{dt} + R_2 i_2 = 0$$

These two equations constitute a mathematical model for the system.

If a mathematical model involves an integrodifferential equation, as in this case, it can be differentiated to obtain a differential equation. The solution of a set of differential equations can then be expressed in terms of exponential functions with undetermined constants. These constants can, in turn, be evaluated from the initial conditions.

3-4 POWER AND ENERGY

In Sec. 2-5 we discussed work, energy, and power in relation to mechanical systems. Here we deal with electrical energy and power. As stated earlier in Section 2-5.

Energy = ability to do work

Power = energy per unit time

The SI units of energy and power are the joule and the watt, respectively.

$$\frac{\text{joule}}{\text{second}} = \text{watt} = \text{volt-ampere} = \frac{\text{volt-coulomb}}{\text{second}} = \frac{\text{newton-meter}}{\text{second}}$$

Power and energy. Consider a two-terminal element as shown in Fig. 3-27. The input power, the rate at which energy is flowing into this element, is

$$P = \frac{dW}{dt}$$

where

$P =$ power, W
$dW =$ energy entering the element in dt second, J

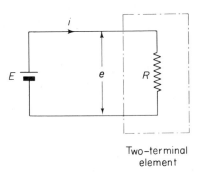

Fig. 3-27. Two-terminal element in a circuit.

Two-terminal element

Note that since voltage is the energy per unit charge (or $e = dW/dq$) and current is the rate of flow of charge (or $i = dq/dt$), we obtain

$$P = \frac{dW}{dq}\frac{dq}{dt} = ei \qquad (3\text{-}17)$$

Thus the resistance element R in Fig. 3-27 consumes power of ei watts.

In addition, note that Eq. (3-17) is analogous to the power equation for a mechanical system, rewritten thus

$$P = F\dot{x}$$

where F is the force and $\dot{x}$ is the velocity. In the SI system of units, power P is measured in watts, force F in newtons, and velocity $\dot{x}$ in meters per second. Thus

$$\text{Watt} = \frac{\text{newton-meter}}{\text{second}}$$

In the electrical system, if e and/or i are time varying, then power P becomes a function of time and it is called the *instantaneous power*. The amount of total energy that has entered the element during a time interval $t_0 \leq t \leq t_f$ is

$$W = \int_{t_0}^{t_f} P\,dt = \int_{t_0}^{t_f} ei\,dt$$

Energy dissipated by resistors. The energy dissipated or consumed by a resistor per unit time (second) is

$$ei = i^2 R = \frac{e^2}{R}$$

This dissipated energy becomes heat. If heat flow to the surroundings is prevented, the temperature of the resistor will rise whenever current flows through it until it burns out or melts. Resistance is a measure of the ability of a device to dissipate power irreversibly.

Example 3-6. For the circuit shown in Fig. 3-28, a load of 2 Ω is fed by two batteries. Assume that battery A has an open-circuit voltage of 12 V and an internal resistance of 3 Ω and that battery B has an open-circuit voltage of 6 V and an internal resistance of 2 Ω. Determine the power dissipated by the 2 Ω internal resistance of battery B.

The node equation for node P is $i_A + i_B - i = 0$, where $i_A = (12 - e_P)/3$,

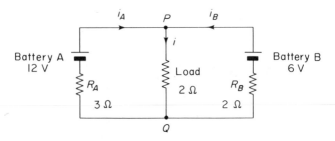

Fig. 3-28. Electrical circuit.

$i_B = (6 - e_P)/2$, and $i = (e_P - e_Q)/2 = e_P/2$. (We assume $e_Q = 0$.) Hence

$$\frac{12 - e_P}{3} + \frac{6 - e_P}{2} - \frac{e_P}{2} = 0$$

Solving for e_P yields

$$e_P = \frac{21}{4}$$

Then the current i_B is given as

$$i_B = \frac{6 - e_P}{2} = \frac{3}{8} \text{ A}$$

So the power dissipated in the 2 Ω internal resistance of battery B is

$$P = i_B^2 R_B = \left(\frac{3}{8}\right)^2 \times 2 = \frac{9}{32} \text{ W}$$

Energy stored in capacitors. Since an electrostatic field exists between the plates of a capacitor, energy is stored in it when a voltage is applied across the plates.

The work done to transfer a charge dq through a potential difference (voltage) e is $e\, dq$. The amount of energy stored in a capacitor during a time interval $t_0 \leq t \leq t_f$ is

$$\text{Energy stored} = \int_{t_0}^{t_f} e \frac{dq}{dt} dt = \int_{t_0}^{t_f} eC \frac{de}{dt} dt = \int_{e_0}^{e_f} Ce\, de$$

$$= \frac{1}{2} Ce_f^2 - \frac{1}{2} Ce_0^2$$

Capacitance is a measure of the ability of an element to store energy in the form of a separated charge or in the form of an electric field.

The energy supplied to the capacitor during the charging process is stored in the capacitor and can be recovered by connecting the charged capacitor to some energy-using device and letting the capacitor discharge into it. Because of various losses, not all energy supplied in actual capaciators can be released. (The wasted energy becomes heat.) In the discharging process, the voltage polarity remains the same as during charging, but the current is reversed. Thus the power into the capacitor becomes negative, which means that power is being taken from the capacitor.

Energy stored in inductors. An inductor stores electric energy as a result of the magnetic field that is produced when the current flows in it. The amount of power P stored in the inductor during a time interval $t_0 \leq t \leq t_f$ is

$$\text{Energy stored} = \int_{t_0}^{t_f} ei\, dt = \int_{t_0}^{t_f} L \frac{di}{dt} i\, dt = L \int_{i_0}^{i_f} i\, di$$

$$= \frac{1}{2} Li_f^2 - \frac{1}{2} Li_0^2$$

Inductance is a measure of the ability of an element to store energy in the form of moving charge or in the form of a magnetic field.

Power generated and power consumed. A charge of one coulomb receives or delivers an energy of one joule in moving through a voltage of one volt. Hence a current of one ampere (coulomb per second) consumes or generates a power of one watt (joule per second) in moving through a voltage of one volt.

Consider again the circuit of Fig. 3-27. If we neglect the internal resistance of the battery, then the power generated by the battery is $P_1 = Ei$ watts. The power consumed by the resistor is $P_2 = ei$ watts. Since $P_1 = P_2$ in this system, we have $E = e$. (Note that we neglected the internal resistance of the battery in this discussion.)

It should be noted that power P_1 differs from power P_2 in that, for the former, current i flows from a point of low voltage to a point of high voltage through the battery (the direction of voltage rise and the direction of the current flow are the same, which means that electric power is generated), whereas in the case of power P_2, current i flows from a point of high voltage to a point of low voltage through the resistor (the direction of voltage rise and the direction of the current flow are opposite, which means that electric power is consumed).

Example 3-7. In Fig. 3-29 the battery has an open-circuit voltage of E volts and an internal resistance of r ohms. Find the power dissipated by load resistance R. If resistance R is variable, at what value of R does the power dissipated by R become maximum?

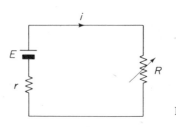

Fig. 3-29. Electrical circuit.

The current i is

$$i = \frac{E}{R + r}$$

The power dissipated by resistance R is

$$P = i^2 R = \left(\frac{E}{R + r}\right)^2 R$$

To find the value of R for which power P becomes maximum, let us rewrite the power expression as follows:

$$P = \left(\frac{E}{R+r}\right)^2 R = \frac{E^2}{[\sqrt{R} + (r/\sqrt{R})]^2} \tag{3-18}$$

The value of P becomes maximum when the denominator of the right-hand side of Eq. (3-18) becomes minimum.

Note that for two positive numbers a and b, if $ab =$ constant, then the sum $a + b$ becomes minimum when $a = b$, since

$$(a + b) = \sqrt{(a - b)^2 + 4ab}$$

In Eq. (3-18) note that

$$\sqrt{R} \cdot \frac{r}{\sqrt{R}} = r = \text{constant}$$

Therefore the denominator of the right-hand side of Eq. (3-18) becomes minimum when

$$\sqrt{R} = \frac{r}{\sqrt{R}}$$

or

$$R = r$$

Consequently, when the load resistance R equals the internal resistance r of the battery, the power dissipated by R becomes maximum. The maximum power dissipated is

$$P_{\text{max}} = \left(\frac{E}{2r}\right)^2 r = \frac{E^2}{4r}$$

Conversion between units of electric energy and thermal energy. Electric energy is measured in J (joule), W-s (watt-second), kWh (kilowatthour), and so on. Thermal energy is measured in J, kcal, Btu, and similiar units. These units are related to each other as follows:

1 J $= 0.2389$ cal $= 9.480 \times 10^{-4}$ Btu
1 kcal $= 4186$ J $= 3.968$ Btu
1 W-s $= 1$ J
1 kWh $= 1000$ Wh $= 1000 \times 3600$ W-s $= 1000 \times 3600$ J $= 860$ kcal
1 kcal $= 1.163$ Wh

3-5 ANALOGOUS SYSTEMS

Systems that can be represented by the same mathematical model but that are different physically are called *analogous* systems. Thus analogous systems are described by the same differential or integrodifferential equations or set of equations.

The concept of analogous systems is very useful in practice for the following reasons.

1. The solution of the equation describing one physical system can be directly applied to analogous systems in any other field.

2. Since one type of system may be easier to handle experimentally than another, instead of building and studying a mechanical system (or hydraulic system, pneumatic system, etc.), we can build and study its electrical analog, for electrical or electronic systems are, in general, much easier to deal with experimentally. (In particular, electronic analog computers are quite useful for simulating mechanical as well as other physical systems. For electronic analog computer simulation, see Sec. 7-7.)

This section presents analogies between mechanical and electrical systems. The concept of analogous systems, however, is applicable to any other system, and analogies among mechanical, electrical, hydraulic, pneumatic, and thermal systems are discussed in Chapters 4, 5, and 7.

Mechanical-electrical analogies. Mechanical systems can be studied through the use of their electrical analogs, which may be more easily constructed than models of the corresponding mechanical systems. There are two electrical analogies for mechanical systems—the force-voltage analogy and the force-current analogy.

Force-voltage analogy. Consider the mechanical system of Fig. 3-30(a) and the electrical system of Fig. 3-30(b). The system equation for the former is

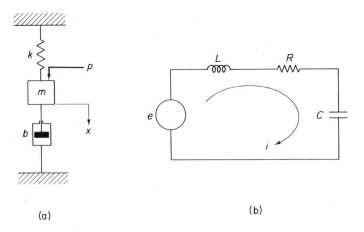

(a)   (b)

Fig. 3-30. Analogous mechanical and electrical systems.

$$m\frac{d^2x}{dt^2} + b\frac{dx}{dt} + kx = p \qquad (3\text{-}19)$$

whereas the system equation for the electrical system is

$$L\frac{di}{dt} + Ri + \frac{1}{C}\int i\,dt = e$$

In terms of electric charge q this last equation becomes

$$L\frac{d^2q}{dt^2} + R\frac{dq}{dt} + \frac{1}{C}q = e \tag{3-20}$$

Comparing Eqs. (3-19) and (3-20), we see that the differential equations for the two systems are of identical form. Thus these two systems are analogous systems. The terms that occupy corresponding positions in the differential equations are called *analogous quantities*, a list of which appears in Table 3-1. The analogy here is called the *force-voltage analogy* (or mass-inductance analogy).

Table 3-1. Force-voltage analogy

Mechanical systems	Electrical systems
Force p (torque T) Mass m (moment of inertia J) Viscous-friction coefficient b Spring constant k Displacement x (angular displacement θ) Velocity $\dot{x}$ (angular velocity $\dot{\theta}$)	Voltage e Inductance L Resistance R Reciprocal of capacitance, $1/C$ Charge q Current i

Force-current analogy. Another analogy between electrical and mechanical systems is based on the force-current analogy. Consider the mechanical system shown in Fig. 3-31(a). The system equation can be obtained as

$$m\frac{d^2x}{dt^2} + b\frac{dx}{dt} + kx = p \tag{3-21}$$

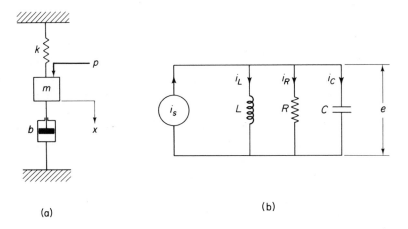

(a)   (b)

Fig. 3-31. Analogous mechanical and electrical systems.

Consider next the electrical system shown in Fig. 3-31(b). Application of Kirchhoff's current law gives

$$i_L + i_R + i_C = i_s \tag{3-22}$$

where

$$i_L = \frac{1}{L} \int e\, dt, \qquad i_R = \frac{e}{R}, \qquad i_C = C\frac{de}{dt}$$

Equation (3-22) can be written

$$\frac{1}{L} \int e\, dt + \frac{e}{R} + C\frac{de}{dt} = i_s \tag{3-23}$$

Since the magnetic flux linkage ψ is related to voltage e by the equation

$$\frac{d\psi}{dt} = e$$

in terms of ψ, Eq. (3-23) can be written

$$C\frac{d^2\psi}{dt^2} + \frac{1}{R}\frac{d\psi}{dt} + \frac{1}{L}\psi = i_s \tag{3-24}$$

Comparing Eqs. (3-21) and (3-24), we find that the two systems are analogous. The analogous quantities are listed in Table 3-2. The analogy here is called the *force-current analogy* (or mass-capacitance analogy).

Table 3-2. FORCE-CURRENT ANALOGY

Mechanical systems	Electrical systems
Force p (torque T) Mass m (moment of inertia J) Viscous-friction coefficient b Spring constant k Displacement x (angular displacement θ) Velocity $\dot{x}$ (angular velocity $\dot{\theta}$)	Current i Capacitance C Reciprocal of resistance, $1/R$ Reciprocal of inductance, $1/L$ Magnetic flux linkage ψ Voltage e

Comments. It should be noted that analogies between two systems break down if the regions of operation are extended too far. In other words, since the mathematical models (i.e., differential equations) on which the analogies are based are only approximations to the dynamic characteristics of physical systems, the analogy may break down if the operating region of one system is very wide. Nevertheless, if the operating region of a given mechanical system is wide, it can be divided into two or more subregions, and analogous electrical systems can be built for each subregion.

REFERENCES

3-1   D'AZZO, J. J., AND C. J. HOUPIS, *Principles of Electrical Engineering*, Columbus, Ohio: Charles E. Merrill Publishing Co., 1968.

3-2   GUILLEMIN, E. A., *Introductory Circuit Theory*, New York: John Wiley & Sons, Inc., 1953.

3-3 RESWICK, J. B., AND C. K. TAFT, *Introduction to Dynamic Systems*, Englewood Cliffs, N.J.: Prentice-Hall, Inc., 1967.

3-4 SEELY, S., *Dynamic Systems Analysis*, New York: Reinhold Publishing Corp., 1964.

3-5 SMITH, R. J., *Circuits, Devices, and Systems*, New York: John Wiley & Sons, Inc., 1966

EXAMPLE PROBLEMS AND SOLUTIONS

PROBLEM A-3-1. The circuit shown in Fig. 3-32 consists of a battery with emf E, resistor R, and switch S. The internal resistance of the battery is indicated as resistance r. Find the voltage e that will appear between the resistor terminals when switch S is closed.

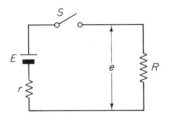

Fig. 3-32. Electrical circuit.

Solution. Since the combined resistance of the circuit is $R + r$, the current i when switch S is closed is

$$i = \frac{E}{R + r}$$

Hence

$$e = Ri = \frac{R}{R + r}E$$

PROBLEM A-3-2. Obtain resistance between points A and B of the circuit given in Fig. 3-33.

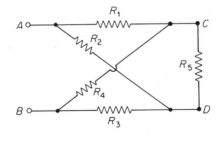

$$R_1 = R_4 = 10\ \Omega, \quad R_2 = R_3 = 20\ \Omega$$

Fig. 3-33. Electrical circuit.

$$R_5 = 100\ \Omega$$

Solution. This circuit is equivalent to the one shown in Fig. 3-34(a). Since $R_1 = R_4 = 10\ \Omega$ and $R_2 = R_3 = 20\ \Omega$, the voltages at points C and D are equal, and so there is no current flowing through R_5. Since resistance R_5 does not affect the value

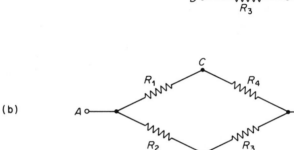

(a)

(b)

Fig. 3-34. Equivalent circuits for the one shown in Fig. 3-33.

of the total resistance between points A and B, it may be removed from the circuit as shown in Fig. 3-34(b). Then R_{AB} is obtained from

$$\frac{1}{R_{AB}} = \frac{1}{R_1 + R_4} + \frac{1}{R_2 + R_3} = \frac{1}{20} + \frac{1}{40} = \frac{3}{40}$$

as

$$R_{AB} = \frac{40}{3} = 13.3\ \Omega$$

PROBLEM A-3-3. Given the circuit of Fig. 3-35, calculate currents i_1, i_2, and i_3.

Fig. 3-35. Electrical circuit.

Solution. The circuit can be redrawn as in Fig. 3-36. The combined resistance R of the path in which current i_2 flows is

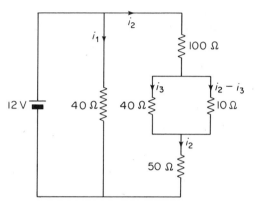

Fig. 3-36. Equivalent circuit for the one shown in Fig. 3-35.

$$R = 100 + \cfrac{1}{\cfrac{1}{10} + \cfrac{1}{40}} + 50 = 158\ \Omega$$

The combined resistance R_0 as seen from the battery is

$$\frac{1}{R_0} = \frac{1}{40} + \frac{1}{158}$$

or

$$R_0 = 31.92\ \Omega$$

Consequently,

$$i_1 + i_2 = \frac{12}{R_0} = \frac{12}{31.92} = 0.376\ \text{A}$$

Noting that $40i_1 = 158i_2$, we obtain

$$i_1 = 0.300\ \text{A}, \qquad i_2 = 0.076\ \text{A}$$

To determine i_3, note that

$$40i_3 = 10(i_2 - i_3)$$

Then

$$i_3 = \frac{10}{50}i_2 = 0.0152\ \text{A}$$

PROBLEM A-3-4. Obtain the combined resistance between points A and B of the circuit shown in Fig. 3-37, which consists of an infinite number of resistors connected in the form of a rudder.

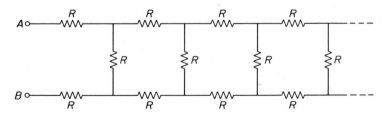

Fig. 3-37. Electrical circuit consisting of an infinite number of resistors connected in the form of a rudder.

Solution. Define the combined resistance between points A and B as R_0. Let us separate the first three resistors from the rest [see Fig. 3-38(a)]. Since the circuit consists of an infinite number of resistors, the removal of the first three resistors

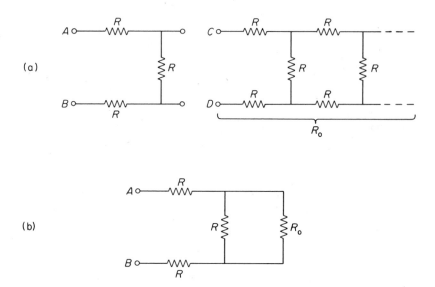

(a)

(b)

Fig. 3-38. Equivalent circuits for the one shown in Fig. 3-37.

does not affect the combined resistance value. Therefore, the combined resistance between points C and D is the same as R_0. Then the circuit shown in Fig. 3-37 may be redrawn as in Fig. 3-38(b) and R_0, the resistance between points A and B, can be obtained as

$$R_0 = 2R + \frac{1}{\dfrac{1}{R} + \dfrac{1}{R_0}} = 2R + \frac{RR_0}{R_0 + R}$$

Rewriting,

$$R_0^2 - 2RR_0 - 2R^2 = 0$$

Solving for R_0, we find

$$R_0 = R \pm \sqrt{3}\ R$$

Finally, by neglecting the negative value for resistance, we obtain

$$R_0 = R + \sqrt{3}\ R = 2.732R$$

PROBLEM A-3-5. Find currents i_1, i_2, and i_3 for the circuit shown in Fig. 3-39.

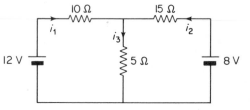

Fig. 3-39. Electrical circuit.

Solution. By applying Kirchhoff's voltage law and current law to the circuit, we have

$$12 - 10i_1 - 5i_3 = 0$$

$$8 - 15i_2 - 5i_3 = 0$$

$$i_1 + i_2 - i_3 = 0$$

Solving for i_1, i_2, and i_3 gives

$$i_1 = \frac{8}{11} \text{ A}, \qquad i_2 = \frac{12}{55} \text{ A}, \qquad i_3 = \frac{52}{55} \text{ A}$$

Since all i values are found to be positive, the currents flow in the directions shown in the digram.

PROBLEM A-3-6. Given the circuit shown in Fig. 3-40, obtain a mathematical model. Here currents i_1 and i_2 are cyclic currents.

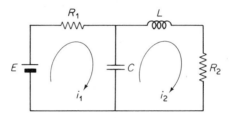

Fig. 3-40. Electrical circuit.

Solution. Application of Kirchhoff's voltage law gives

$$R_1 i_1 + \frac{1}{C} \int (i_1 - i_2)\, dt = E$$

$$L \frac{di_2}{dt} + R_2 i_2 + \frac{1}{C} \int (i_2 - i_1)\, dt = 0$$

These two equations constitute a mathematical model for the circuit.

PROBLEM A-3-7. In the circuit of Fig. 3-41, assume that, for $t < 0$, switch S is connected to voltage source E and the current in coil L is in steady state. At $t = 0$ switch S disconnects the voltage source and simultaneously short-circuits the coil. What is the current $i(t)$ for $t > 0$?

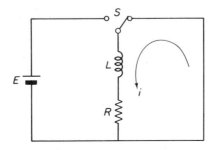

Fig. 3-41. Electrical circuit.

Solution. For $t > 0$, the equation for the circuit is

$$L\frac{di}{dt} + Ri = 0, \qquad i(0) = \frac{E}{R}$$

[Note that there is a nonzero initial current $i(0-) = E/R$. Since inductance L stores energy, the current in the coil cannot be changed instantaneously. Hence $i(0+) = i(0-) = i(0) = E/R$.]

Assuming an exponential solution

$$i(t) = Ke^{\lambda t}$$

we obtain the characteristic equation

$$L\lambda + R = 0$$

from which λ is determined as

$$\lambda = -\frac{R}{L}$$

Noting that $i(0) = E/R$,

$$i(0) = K = \frac{E}{R}$$

Therefore

$$i(t) = \frac{E}{R}e^{-(R/L)t}$$

PROBLEM A-3-8. Consider the circuit shown in Fig. 3-42 and assume that capacitor C is initially charged to q_0. At $t = 0$ switch S is disconnected from the battery and simultaneously connected to inductor L. The capacitance has a value of 50 μF.

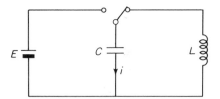

Fig. 3-42. Electrical circuit.

Calculate the value of the inductance L that will make the oscillation occur at the frequency of 200 Hz.

Solution. The equation for the circuit for $t > 0$ is

$$L\frac{di}{dt} + \frac{1}{C}\int i\,dt = 0$$

or by substituting $i = dq/dt$ into this last equation,

$$L\frac{d^2q}{dt^2} + \frac{1}{C}q = 0$$

where $q(0) = q_0$ and $\dot{q}(0) = 0$. The frequency ω_n of oscillation is

$$\omega_n = \sqrt{\frac{1}{LC}}$$

Since

$$200 \text{ Hz} = 200 \text{ cps} = 200 \times 6.28 \text{ rad/s} = 1256 \text{ rad/s}$$

we obtain

$$\omega_n = 1256 = \sqrt{\frac{1}{LC}} = \sqrt{\frac{1}{L \times 50 \times 10^{-6}}}$$

Thus

$$L = \frac{1}{1256^2 \times 50 \times 10^{-6}} = 0.0127 \text{ H}$$

PROBLEM A-3-9. In Fig. 3-43(a) suppose that switch S is open for $t < 0$ and that the system is in steady state. Switch S is closed at $t = 0$. Find the current $i(t)$ for $t \geq 0$.

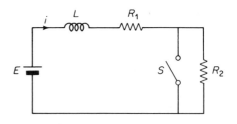

(a)

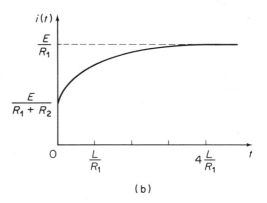

Fig. 3-43. (a) Electrical circuit; (b) plot of $i(t)$ versus t of the circuit when switch S is closed at $t = 0$.

(b)

Solution. Notice that, for $t < 0$, the circuit resistance is $R_1 + R_2$, and so there is a nonzero initial current $i(0-)$, where

$$i(0-) = \frac{E}{R_1 + R_2}$$

For $t \geq 0$, the circuit resistance becomes R_1. Note that because of the presence of inductance L, when switch S is closed, there is no instantaneous change in the current in the circuit. Hence

$$i(0+) = i(0-) = \frac{E}{R_1 + R_2} = i(0)$$

Therefore the equation for the circuit for $t \geq 0$ is

$$L\frac{di}{dt} + R_1 i = E, \qquad i(0) = \frac{E}{R_1 + R_2} \tag{3-25}$$

Let us define

$$x = i - \frac{E}{R_1}$$

Then Eq. (3-25) becomes

$$L\frac{dx}{dt} + R_1 x = 0$$

where

$$x(0) = i(0) - \frac{E}{R_1} = -\frac{R_2}{(R_1 + R_2)R_1}E \tag{3-26}$$

Assuming an exponential solution

$$x(t) = Ke^{\lambda t}$$

we obtain the characteristic equation as

$$L\lambda + R_1 = 0$$

which gives

$$\lambda = -\frac{R_1}{L}$$

It follows that

$$x(t) = Ke^{-(R_1/L)t}$$

where K is determined by use of Eq. (3-26) as follows:

$$x(0) = K = -\frac{R_2}{(R_1 + R_2)R_1}E$$

Therefore

$$x(t) = -\frac{R_2 E}{(R_1 + R_2)R_1}e^{-(R_1/L)t}$$

and

$$i(t) = x(t) + \frac{E}{R_1}$$

$$= \frac{E}{R_1}\left[1 - \frac{R_2}{R_1 + R_2}e^{-(R_1/L)t}\right]$$

A typical plot of $i(t)$ versus t is shown in Fig. 3-43(b).

PROBLEM A-3-10. For the circuit of Fig. 3-44(a), suppose that switch S is closed for $t < 0$ and that the system is in steady state. The switch is opened at $t = 0$. Obtain the current $i(t)$ for $t > 0$.

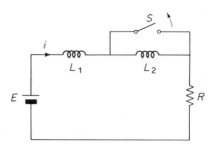

(a)

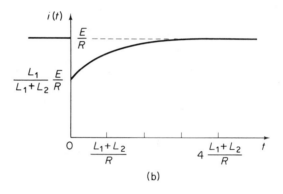

Fig. 3-44. (a) Electrical circuit;
(b) plot of $i(t)$ versus t of the
circuit when switch S is opened
at $t = 0$.

(b)

Solution. The equation for the circuit is

$$L \frac{di}{dt} + Ri = E \tag{3-27}$$

where $L = L_1$ for $t < 0$ and $L = L_1 + L_2$ for $t > 0$. Integrating both sides of Eq. (3-27) between $t = 0-$ and $t = 0+$, we have

$$\int_{0-}^{0+} L \frac{di}{dt}\, dt + \int_{0-}^{0+} Ri\, dt = \int_{0-}^{0+} E\, dt$$

Noting that

$$\int_{0-}^{0+} (E - Ri)\, dt = 0$$

we obtain

$$(L_1 + L_2)i(0+) - L_1 i(0-) = 0$$

Hence

$$i(0+) = \frac{L_1}{L_1 + L_2} i(0-)$$

We see that the steady-state current for $t < 0$ is $i = E/R$, and so we have $i(0-) = E/R$. Therefore

$$i(0+) = \frac{L_1}{L_1 + L_2} \frac{E}{R}$$

The equation for the circuit for $t \geq 0+$ can now be written

$$(L_1 + L_2)\frac{di}{dt} + Ri = E, \qquad i(0+) = \frac{L_1}{L_1 + L_2} \frac{E}{R} \qquad (3\text{-}28)$$

In order to solve Eq. (3-28), define a new variable x, where

$$x = i - \frac{E}{R} \qquad (3\text{-}29)$$

In terms of the new variable x, Eq. (3-28) becomes

$$(L_1 + L_2)\frac{dx}{dt} + Rx = 0, \qquad x(0+) = -\frac{L_2}{L_1 + L_2}\frac{E}{R}$$

By assuming the solution $x(t)$ to be

$$x(t) = Ke^{\lambda t}$$

the characteristic equation is given as

$$(L_1 + L_2)\lambda + R = 0$$

or

$$\lambda = -\frac{R}{L_1 + L_2}$$

Hence

$$x(t) = Ke^{-[R/(L_1+L_2)]t}$$

Since

$$x(0+) = K = -\frac{L_2}{L_1 + L_2}\frac{E}{R}$$

it follows that

$$x(t) = -\frac{L_2}{L_1 + L_2}\frac{E}{R}e^{-[R/(L_1+L_2)]t}$$

Then from Eq. (3-29) we obtain

$$i(t) = x(t) + \frac{E}{R}$$

$$= \frac{E}{R}\left\{1 - \frac{L_2}{L_1 + L_2}e^{-[R/(L_1+L_2)]t}\right\}$$

A typical curve $i(t)$ versus t appears in Fig. 3-44(b).

PROBLEM A-3-11. Consider the schematic diagram of a d'Arsonval moving-coil dc voltmeter shown in Fig. 3-45(a). Assume that when voltage E_0 is applied to the meter, it exhibits full-scale deflection with a current of i_0, where

$$i_0 = \frac{E_0}{r + R} = \frac{E_0}{R_v} \qquad (R_v = r + R)$$

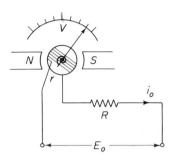

(a)

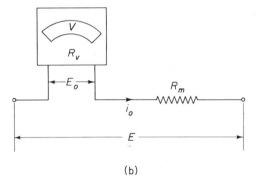

Fig. 3-45. (a) Schematic diagram of a
d'Arsonval moving-coil dc voltmeter;
(b) voltmeter connected in series with
a resistor.

(b)

The measurement range can be extended if resistance R_m is connected in series as
in Fig. 3-45(b). Determine the resistance R_m so that the meter can be used to measure
voltages up to E volts, where $E = mE_0$ and $m > 1$.

Solution. Note that the maximum current flow through the voltmeter must be
limited to i_0. Hence

$$\frac{E}{R_v + R_m} = \frac{mE_0}{R_v + R_m} = i_0 = \frac{E_0}{R_v}$$

from which

$$mR_v = R_v + R_m$$

or

$$R_m = (m - 1)R_v$$

PROBLEM A-3-12. Figure 3-46(a) shows a schematic diagram of a dc ammeter.
Assume that current I_0 is the maximum current that can be applied to this ammeter.
A large part of current I_0 is shunted through R_s, and only a small part of the current
flows through the moving coil. (For instance, for full-scale deflection with a meter

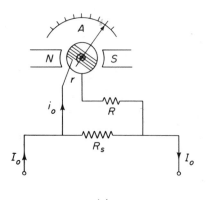

(a)

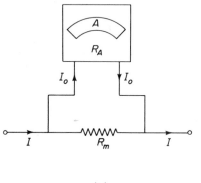

Fig. 3-46. (a) Schematic diagram of a dc ammeter; (b) ammeter connected in parallel with a resistor.

(b)

current of 3 A, the current in the shunt R_s may be 2.999 A and the current through the moving coil may be 1 mA.) Define the maximum allowable current through the moving coil (which corresponds to full-scale deflection) as i_0. Thus from the diagram

$$i_0 = \frac{R_s}{r + R + R_s} I_0$$

The measurement range can be extended if resistance R_m is connected in parallel as shown in Fig. 3-46(b).

Determine the resistance R_m so that the meter can be used to measure currents up to I amperes, where $I = mI_0$ and $m > 1$.

Solution. Let us define the resistance of the ammeter as R_A. Then

$$\frac{1}{r + R} + \frac{1}{R_s} = \frac{1}{R_A}$$

or

$$R_A = \frac{(r + R)R_s}{r + R + R_s}$$

The maximum current to the ammeter must be limited to I_0. So

$$I_0 R_A = (I - I_0)R_m = (m - 1)I_0 R_m$$

or

$$R_m = \frac{1}{m - 1} R_A$$

If this resistance R_m is connected in parallel to the ammeter as in Fig. 3-46(b), the meter can be used to measure up to mI_0 amperes, where $m > 1$.

PROBLEM A-3-13. Current I flows through a resistor whose resistance is R ohms (see Fig. 3-47). In order to measure the dc current I, the voltage drop across resis-

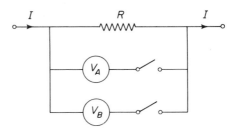

Fig. 3-47. Circuit for measuring dc current.

tance R is measured alternatively by two dc voltmeters. The voltage reading is 61 V when measured by voltmeter A, whose internal resistance is 15 000 Ω. The voltage reading is 60 V when measured by voltmeter B, whose internal resistance is 10 000 Ω. Determine current I and resistance R. Assume that current I does not change when a voltmeter is connected.

Solution. Let us redraw the system diagram as in Fig. 3-48(a) and (b). Define currents through resistance R and voltmeter A as i_1 and i_2, respectively, as in Fig.

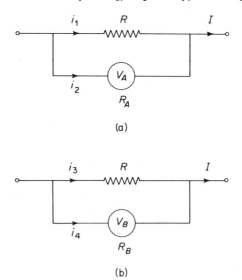

(a)

(b)

Fig. 3-48. Circuits for measuring dc currents.

3-48(a). Similarly, define currents through resistance R and voltmeter B as i_3 and i_4, respectively, as in Fig. 3-48(b). Then

$$i_1 R = i_2 R_A = 61, \qquad i_1 + i_2 = I$$
$$i_3 R = i_4 R_B = 60, \qquad i_3 + i_4 = I$$

Hence

$$i_1 = \frac{61}{R}, \qquad i_3 = \frac{60}{R}$$

Since $R_A = 15\,000\ \Omega$ and $R_B = 10\,000\ \Omega$, we have

$$i_2 = \frac{61}{15\,000}, \qquad i_4 = \frac{60}{10\,000}$$

Therefore

$$i_1 + i_2 = \frac{61}{R} + \frac{61}{15\,000} = I$$

$$i_3 + i_4 = \frac{60}{R} + \frac{60}{10\,000} = I$$

Solving these two equations for I and R, we obtain

$$I = 0.122\text{ A}, \qquad R = 517.2\ \Omega$$

PROBLEM A-3-14. A voltage of 6 V is applied between points A and B in the circuit of Fig. 3-49. Suppose that current i is 0.5 A, regardless of whether the switch is open or closed. Find the values of resistances R_1 and R_2.

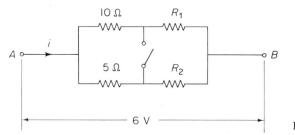

Fig. 3-49. Electrical circuit.

Solution. Notice that this circuit is a Wheatstone bridge. The fact that the current i is constant, regardless of whether the switch is open or closed, implies that the bridge is in balance. So

$$\frac{R_1}{R_2} = \frac{10}{5}$$

or

$$R_1 = 2R_2$$

The combined resistance R between points A and B is $6/0.5 = 12\ \Omega$. In terms of resistances R_1 and R_2 the combined resistance R can be written

$$R = \frac{1}{\dfrac{1}{10} + \dfrac{1}{5}} + \frac{1}{\dfrac{1}{R_1} + \dfrac{1}{R_2}} = \frac{50}{15} + \frac{R_1 R_2}{R_1 + R_2} = \frac{10}{3} + \frac{2}{3} R_2$$

Consequently,

$$R = 12 = \frac{10}{3} + \frac{2}{3}R_2$$

from which

$$R_2 = 13\ \Omega$$

Since $R_1 = 2R_2$, we have

$$R_1 = 26\ \Omega$$

PROBLEM A-3-15. In Fig. 3-50 two identical 6 V batteries in parallel are connected to a load resistance of 0.6 Ω. The internal resistance of each battery is 0.3 Ω. Assuming that the batteries supply a nearly constant voltage of 6 V for a limited time period, find the power consumed by the load resistance during this time period.

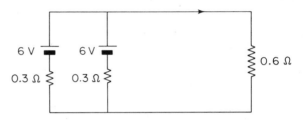

Fig. 3-50. Electrical circuit.

Solution. The current i is obtained as

$$i = \frac{6}{0.6 + (0.3/2)} = 8\ \text{A}$$

Hence the power consumed by the load resistance is

$$P = i^2 R = 8^2 \times 0.6 = 38.4\ \text{W}$$

PROBLEM A-3-16. The resistance R_1 is variable for the circuit of Fig. 3-51. At what value of R_1 does the power consumed by this resistance become maximum?

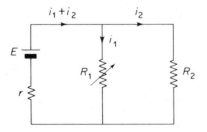

Fig. 3-51. Electrical circuit.

Solution. The current i_1 can be obtained as follows. First, noting that the equivalent resistance R for the parallel resistances R_1 and R_2 is

$$R = \frac{R_1 R_2}{R_1 + R_2}$$

we obtain

$$i_1 + i_2 = \frac{E}{r + R} = \frac{E}{r + [R_1 R_2/(R_1 + R_2)]}$$

Since $i_1 R_1 = i_2 R_2$, we have

$$i_1 = \frac{E R_2}{(r + R_2)R_1 + r R_2}$$

The power consumed by resistance R_1 is

$$P = i_1^2 R_1 = \left[\frac{E R_2}{(r + R_2)R_1 + r R_2} \right]^2 R_1$$

$$= \frac{E^2 R_2^2}{[(r + R_2)\sqrt{R_1} + (r R_2/\sqrt{R_1})]^2}$$

The power P becomes maximum when the denominator of the right-hand side of this last equation is minimum. Noting that the product of $(r + R_2)\sqrt{R_1}$ and $r R_2/\sqrt{R_1}$ is constant, the sum of these two terms becomes minimum when they are equal—that is,

$$(r + R_2)\sqrt{R_1} = \frac{r R_2}{\sqrt{R_1}}$$

Hence

$$R_1 = \frac{r R_2}{r + R_2} = \frac{1}{\dfrac{1}{r} + \dfrac{1}{R_2}}$$

Therefore we see that when R_1 is equal to the combined resistance of two parallel resistances r and R_2, the power consumed by resistance R_1 becomes maximum. The maximum power consumed by resistance R_1 is

$$P_{\max} = \frac{E^2 R_2^2}{[2(r + R_2)\sqrt{R_1}]^2} = \frac{E^2 R_2}{4r(r + R_2)}$$

PROBLEM A-3-17. Show that the mechanical and electrical systems given in Fig. 3-52 are analogous. Assume that the displacement x in the mechanical system is measured from the equilibrium position and that mass m is released from the initial displacement $x(0) = x_0$ with zero initial velocity, or $\dot{x}(0) = 0$. Assume also that in

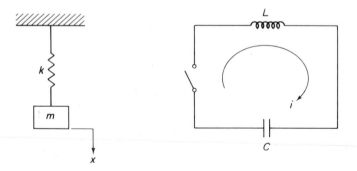

Fig. 3-52. Analogous mechanical and electrical systems.

the electrical system the capacitor has the initial charge $q(0) = q_0$ and that the switch is closed at $t = 0$. Note that $\dot{q}(0) = i(0) = 0$. Obtain $x(t)$ and $q(t)$.

Solution. The equation of motion for the mechanical system is

$$m\ddot{x} + kx = 0 \tag{3-30}$$

For the electrical system,

$$L\frac{di}{dt} + \frac{1}{C}\int i\,dt = 0$$

or by substituting $i = dq/dt = \dot{q}$ into this last equation,

$$L\ddot{q} + \frac{1}{C}q = 0 \tag{3-31}$$

Since Eqs. (3-30) and (3-31) are of the same form, these two systems are analogous. (Analogy here is based on the force-voltage analogy.)

The solution of Eq. (3-30) with the initial condition $x(0) = x_0$, $\dot{x}(0) = 0$ is a simple harmonic motion given by

$$x(t) = x_0 \cos\sqrt{\frac{k}{m}}t$$

Similarly, the solution of Eq. (3-31) with the initial condition $q(0) = q_0$, $\dot{q}(0) = 0$ is

$$q(t) = q_0 \cos\sqrt{\frac{1}{LC}}t$$

PROBLEM A-3-18. Obtain mathematical models for the systems shown in Fig. 3-53(a) and (b) and show that they are analogous systems.

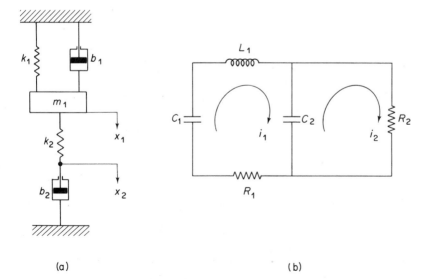

(a)   (b)

Fig. 3-53. Analogous mechanical and electrical systems.

Solution. For the system in Fig. 3-53(a), the equations of motion are

$$m_1\ddot{x}_1 + b_1\dot{x}_1 + k_1x_1 + k_2(x_1 - x_2) = 0$$

$$b_2\dot{x}_2 + k_2(x_2 - x_1) = 0$$

These two equations constitute a mathematical model for the mechanical system.
For the system in Fig. 3-53(b), the loop-voltage equations are

$$L_1\frac{di_1}{dt} + \frac{1}{C_2}\int (i_1 - i_2)\, dt + R_1i_1 + \frac{1}{C_1}\int i_1\, dt = 0$$

$$R_2i_2 + \frac{1}{C_2}\int (i_2 - i_1)\, dt = 0$$

Let us write $i_1 = \dot{q}_1$ and $i_2 = \dot{q}_2$. Then in terms of q_1 and q_2, the preceding two
equations can be written

$$L_1\ddot{q}_1 + R_1\dot{q}_1 + \frac{1}{C_1}q_1 + \frac{1}{C_2}(q_1 - q_2) = 0$$

$$R_2\dot{q}_2 + \frac{1}{C_2}(q_2 - q_1) = 0$$

The last two equations constitute a mathematical model for the electrical system.
By comparing these two mathematical models, we see that the two systems
are analogous. (The analogy is based on the force-voltage analogy.)

PROBLEM A-3-19. Using the force-voltage analogy, obtain an electrical analog of
the mechanical system shown in Fig. 3-54.

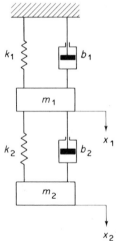

x_2 **Fig. 3-54.** Mechanical system.

Solution. The equations of motion for the mechanical system are

$$m_1\ddot{x}_1 + b_1\dot{x}_1 + k_1x_1 + b_2(\dot{x}_1 - \dot{x}_2) + k_2(x_1 - x_2) = 0$$

$$m_2\ddot{x}_2 + b_2(\dot{x}_2 - \dot{x}_1) + k_2(x_2 - x_1) = 0$$

By use of the force-voltage analogy, the equations for an analogous electrical system may be written

$$L_1\ddot{q}_1 + R_1\dot{q}_1 + \frac{1}{C_1}q_1 + R_2(\dot{q}_1 - \dot{q}_2) + \frac{1}{C_2}(q_1 - q_2) = 0$$

$$L_2\ddot{q}_2 + R_2(\dot{q}_2 - \dot{q}_1) + \frac{1}{C_2}(q_2 - q_1) = 0$$

Then substituting $\dot{q}_1 = i_1$ and $\dot{q}_2 = i_2$ into the last two equations gives

$$L_1\frac{di_1}{dt} + R_1 i_1 + \frac{1}{C_1}\int i_1\,dt + R_2(i_1 - i_2) + \frac{1}{C_2}\int (i_1 - i_2)\,dt = 0 \quad (3\text{-}32)$$

$$L_2\frac{di_2}{dt} + R_2(i_2 - i_1) + \frac{1}{C_2}\int (i_2 - i_1)\,dt = 0 \quad (3\text{-}33)$$

These two equations are loop-voltage equations. From Eq. (3-32) we obtain the diagram shown in Fig. 3-55(a). Similarly, from Eq. (3-33) we obtain the one given in Fig. 3-55(b). Combining these two diagrams produces the desired analogous electrical system (Fig. 3-56).

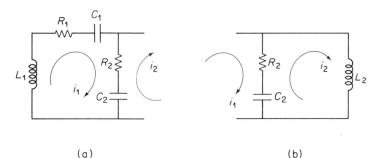

(a)   (b)

Fig. 3-55. (a) Electrical circuit corresponding to Eq. (3-32); (b) electrical circuit corresponding to Eq. (3-33).

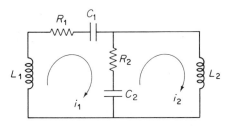

Fig. 3-56. Electrical system analogous to the mechanical system shown in Fig. 3-54. (Force-voltage analogy.)

PROBLEM A-3-20. Consider the mechanical system shown in Fig. 3-54 again. Using the force-current analogy, obtain an analogous electrical system.

Solution. The equations of motion for the mechanical system are

$$m_1\ddot{x}_1 + b_1\dot{x}_1 + k_1 x_1 + b_2(\dot{x}_1 - \dot{x}_2) + k_2(x_1 - x_2) = 0$$
$$m_2\ddot{x}_2 + b_2(\dot{x}_2 - \dot{x}_1) + k_2(x_2 - x_1) = 0$$

Using the force-current analogy, we can find the equations for an analogous electrical system.

$$C_1\ddot{\psi}_1 + \frac{1}{R_1}\dot{\psi}_1 + \frac{1}{L_1}\psi_1 + \frac{1}{R_2}(\dot{\psi}_1 - \dot{\psi}_2) + \frac{1}{L_2}(\psi_1 - \psi_2) = 0$$

$$C_2\ddot{\psi}_2 + \frac{1}{R_2}(\dot{\psi}_2 - \dot{\psi}_1) + \frac{1}{L_2}(\psi_2 - \psi_1) = 0$$

Noting that $\dot{\psi} = e$, the last two equations yield

$$C_1\dot{e}_1 + \frac{1}{R_1}e_1 + \frac{1}{L_1}\int e_1 \, dt + \frac{1}{R_2}(e_1 - e_2) + \frac{1}{L_2}\int (e_1 - e_2) \, dt = 0 \qquad (3\text{-}34)$$

$$C_2\dot{e}_2 + \frac{1}{R_2}(e_2 - e_1) + \frac{1}{L_2}\int (e_2 - e_1) \, dt = 0 \qquad (3\text{-}35)$$

Equations (3-34) and (3-35) are node equations. Corresponding to the first node equation, Eq. (3-34), we obtain the diagram shown in Fig. 3-57(a). Similarly, from the second node equation, Eq. (3-35), we derive the diagram shown in Fig. 3-57(b). If we combine the two diagrams, the result is the desired analogous electrical system (Fig. 3-58).

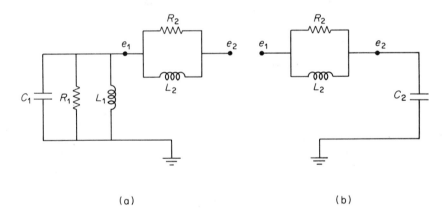

(a)   (b)

Fig. 3-57. (a) Electrical circuit corresponding to Eq. (3-34); (b) electrical circuit corresponding to Eq. (3-35).

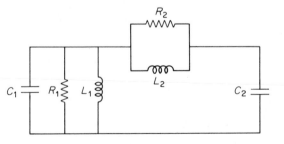

Fig. 3-58. Electrical system analogous to the mechanical system shown in Fig. 3-54. (Force-current analogy.)

PROBLEMS

PROBLEM B-3-1. A voltage source $E = 12$ V is applied between points A and C in Fig. 3-59. Find voltage E_0 between points B and C.

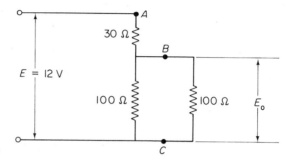

Fig. 3-59. Electrical circuit.

PROBLEM B-3-2. Three resistors R_1, R_2, and R_3 are connected in a triangular shape (Fig. 3-60). Obtain the combined resistance between points A and B.

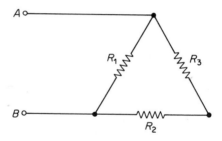

Fig. 3-60. Three resistors connected in a triangular shape.

PROBLEM B-3-3. Calculate the resistance between points A and B for the circuit in Fig. 3-61.

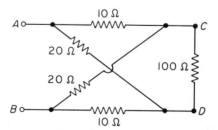

Fig. 3-61. Electrical circuit.

PROBLEM B-3-4. A voltage source $E = 12$ V is connected to a resistor as shown in Fig. 3-62(a). Voltage between terminals B and C is found to be 4 V. When a 40 Ω resistor is connected between terminals B and C as in Fig. 3-62(b), we observe that the voltage between terminals B and C is 2.4 V. If terminals B and C are short-circuited as shown in Fig. 3-62(c), what will be the value of current i?

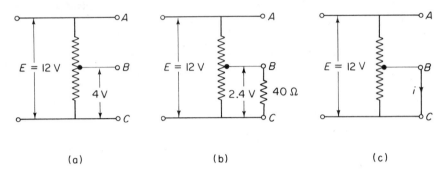

(a)   (b)   (c)

Fig. 3-62. Electrical circuits.

PROBLEM B-3-5. In the circuit of Fig. 3-63, assume that a voltage E is applied between points A and B and that current i is i_0 when switch S is open. When switch S is closed, current i becomes equal to $2i_0$. Find the value of resistance R.

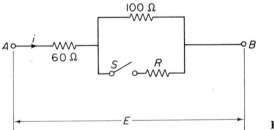

Fig. 3-63. Electrical circuit.

PROBLEM B-3-6. Referring to the circuit shown in Fig. 3-64, calculate the currents in resistors R_1, R_2, and R_3. Neglect the internal resistances of the batteries.

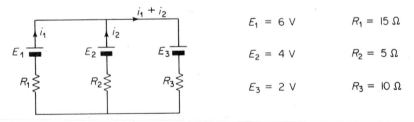

$E_1 = 6$ V $R_1 = 15$ Ω

$E_2 = 4$ V $R_2 = 5$ Ω

$E_3 = 2$ V $R_3 = 10$ Ω

Fig. 3-64. Electrical circuit.

PROBLEM B-3-7. Obtain a mathematical model for the circuit shown in Fig. 3-65.

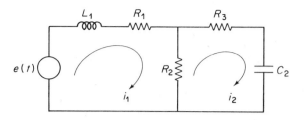

Fig. 3-65. Electrical circuit.

PROBLEM B-3-8. Consider the circuit shown in Fig. 3-66. Assume that switch S is open for $t < 0$ and that capacitor C is initially charged so that the initial voltage $q(0)/C = e_0$ appears on the capacitor. Calculate cyclic currents i_1 and i_2 when switch S is closed at $t = 0$.

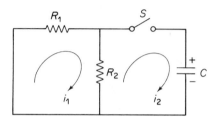

Figure 3-66. Electrical circuit.

PROBLEM B-3-9. In the circuit shown in Fig. 3-67, assume that the capacitor is initially not charged and switch S is closed at $t = 0$. Determine cyclic currents $i_1(t)$ and $i_2(t)$ for a special case where

$$2R_1 = R_2 = R_3, \qquad R_3C = 1$$

Also, determine $e_B(t)$, the voltage at point B. (Assuming $E = 12$ V, compare this result with that of Example 3-1.)

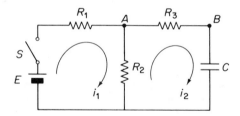

Fig. 3-67. Electrical circuit.

PROBLEM B-3-10. The circuit shown in Fig. 3-68 is in steady state with switch S closed. Switch S is then opened at $t = 0$. Obtain $i(t)$.

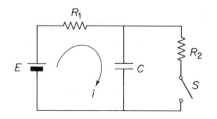

Fig. 3-68. Electrical circuit.

PROBLEM B-3-11. Assume that switch S is open for $t < 0$ and the system is in steady state (Fig. 3-69). At $t = 0$ the switch is closed. Obtain current $i(t)$ for $t > 0$. Plot a typical curve $i(t)$ versus t.

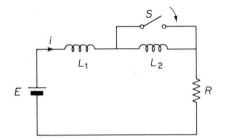

Fig. 3-69. Electrical circuit.

PROBLEM B-3-12. A dc voltmeter measures voltages between 0 and 150 V. Assume that the internal resistance of this voltmeter is 15 000 Ω. This dc voltmeter is to be used to measure voltages up to 400 V as shown in Fig. 3-70. Determine the necessary resistance R_m to be connected in series.

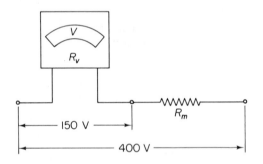

Fig. 3-70. Voltmeter connected in series with a resistor.

PROBLEM B-3-13. The circuit in Fig. 3-71 has a dc ammeter in parallel with resistance R and another dc ammeter connected in series. Assume that the readings of ammeter A_1 and ammeter A_2 are 20 A and 30 A, respectively. The resistance R is 0.1 Ω. Determine the internal resistance of ammeter A_1.

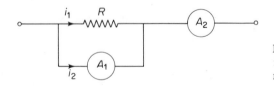

Fig. 3-71. Electrical circuit consisting of a resistor and two dc ammeters.

PROBLEM B-3-14. Two dc voltmeters V_1 and V_2 have internal resistances $R_1 = 15\,000\ \Omega$ and $R_2 = 13\,000\ \Omega$, respectively. Both are designed to measure voltages between 0 and 150 V. If these two dc voltmeters are connected in series as shown in Fig. 3-72, what is the maximum voltage that can be measured?

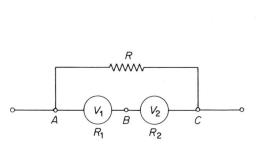

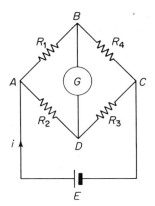

Fig. 3-72. Electrical circuit consisting of a resistor and two dc voltmeters.

Fig. 3-73. Wheatstone bridge.

PROBLEM B-3-15. Figure 3-73 shows a Wheatstone bridge. Find current i when the bridge is in balance.

PROBLEM B-3-16. Consider the bridge circuit shown in Fig. 3-74 and suppose that the bridge is in balance so that no current flows through the galvanometer, or $i_g = 0$. Obtain E_2 in terms of E_1, R_1, and R_2.

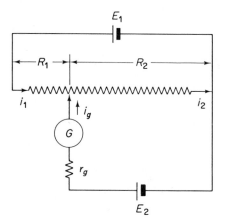

Fig. 3-74. Bridge circuit.

PROBLEM B-3-17. A resistor consumes a power of 500 W if 110 V is applied to the terminals. Calculate the power consumed by this resistor when the voltage applied to the terminals is 100 V.

PROBLEM B-3-18. Three 10 Ω resistors are connected in different ways as shown in Fig. 3-75(a), (b), (c), and (d). Calculate the combined resistance values for these four cases. If 12 V is applied between terminals A and B, what is the power dissipated? Calculate power dissipated for the four circuits shown.

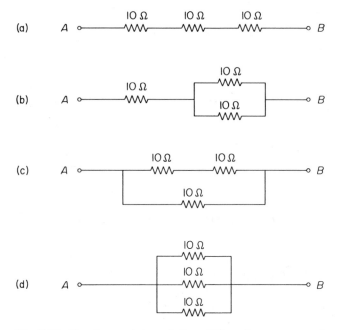

Fig. 3-75. Circuits consisting of three 10Ω resistors connected differently.

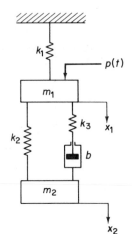

Fig. 3-76. Mechanical system.

PROBLEM B-3-19. Determine an analogous electrical system of the mechanical system shown in Fig. 3-76, where $p(t)$ is the force input to the system.

PROBLEM B-3-20. Obtain an analogous mechanical system of the electrical system
shown in Fig. 3-77.

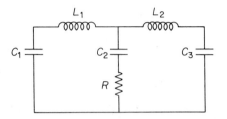

Fig. 3-77. Electrical system.

4

HYDRAULIC SYSTEMS

4-1 INTRODUCTION

As the most versatile medium for transmitting signals and power, fluids, either as liquids or gases, have wide usage in industry. Liquids and gases can be distinguished basically by their relative incompressibilities and the fact that a liquid may have a free surface, whereas a gas expands to fill its vessel. In the engineering field the term *hydraulic* describes fluid systems that use liquids and *pneumatic* applies to those using air or gases. Hydraulic systems are discussed in this chapter and pneumatic systems in Chapter 5.

Because of their prevalence in industry, hydraulic circuitry and hydraulic systems are a necessary part of an engineer's background. Most hydraulic systems are nonlinear. Sometimes, however, it is possible to linearize nonlinear systems so as to reduce their complexity and permit solutions that are sufficiently accurate for most purposes. Consequently, a useful linearization technique for dealing with nonlinear systems is presented here. A detailed analysis of linearized hydraulic control systems, however, is postponed to Chapter 8.

Before proceeding, let us define the units of pressure and gage and absolute pressures.

Units of pressure. *Pressure* is defined as force per unit area. The units of pressure include N/m^2, kg_f/cm^2, $lb_f/in.^2$, and so on. In the SI system the unit of pressure is N/m^2. The name *pascal* has been given to this unit (abbreviated Pa).

$$1 \text{ Pa} = 1 \text{ N/m}^2$$

Kilopascals (10^3 Pa = kPa) and megapascals (10^6 Pa = MPa) may be used in expressing hydraulic pressure. Note that

$$1 \text{ lb}_f/\text{in.}^2 = 6895 \text{ Pa}$$

$$1 \text{ kg}_f/\text{cm}^2 = 14.22 \text{ lb}_f/\text{in.}^2 = 0.9807 \times 10^5 \text{ N/m}^2 = 0.09807 \text{ MPa}$$

Gage pressure and absolute pressure. The standard barometer reading at sea level is 760 mm of mercury at 0°C (29.92 in. of mercury at 32°F). *Gage pressure* refers to that pressure that is measured with respect to atmospheric pressure. It is the pressure indicated by a gage above atmospheric. *Absolute pressure* is the sum of gage and barometer pressures. Note that in engineering measurement pressure is expressed in gage pressure. In theoretical calculations, however, absolute pressure must be used. Note also that

$$760 \text{ mm Hg} = 1.0332 \text{ kg}_f/\text{cm}^2 = 1.0133 \times 10^5 \text{ N/m}^2 = 14.7 \text{ lb}_f/\text{in.}^2$$

$$0 \text{ N/m}^2 \text{ gage} = 1.0133 \times 10^5 \text{ N/m}^2 \text{ abs}$$

$$0 \text{ kg}_f/\text{cm}^2 \text{ gage} = 1.0332 \text{ kg}_f/\text{cm}^2 \text{ abs}$$

$$0 \text{ lb}_f/\text{in.}^2 \text{ gage} = 0 \text{ psig} = 14.7 \text{ lb}_f/\text{in.}^2 \text{ abs} = 14.7 \text{ psia}$$

Hydraulic systems. The widespread use of hydraulic circuitry in machine tool applications, aircraft control systems, and similiar operations occurs because of such factors as positiveness, accuracy, flexibility, high horsepower-to-weight ratio, fast starting, stopping, and reversal with smoothness and precision, and simplicity of operations.

In many machine tool applications, for instance, the traverse and feed cycles required are best handled by hydraulic circuits. These cycles—where the piston advances rapidly on the work stroke until the work is contacted, advances slowly under pressure while the work is done, and then retracts rapidly at the end of the slow tool feed stroke—are easily handled by the use of two pumps (one large-capacity, low-pressure pump and one small-capacity, high-pressure pump) and flow control devices. The large-capacity, low-pressure pump is used only during the rapid advance and return of the cylinder. The small-capacity, high-pressure pump supplies hydraulic fluid for the compression stroke. An unloading valve maintains high pressure while the low-pressure pump is unloaded to the reservoir. (The unloading valve unloads the delivery of the large-capacity, low-pressure pump during the small-capacity, high-pressure phase of a cycle.) Such an unloading valve is designed for the rapid discharge of hydraulic fluid at near atmospheric pressure after permitting the buildup of pressure to a preset value.

Generally, the operating pressure in hydraulic systems is somewhere between 10^6 N/m² (1 MPa) and 35×10^6 N/m² (35 MPa) (approximately between 10 kg$_f$/cm² and 350 kg$_f$/cm²; or approximately between 145 lb$_f$/in.² and 5000 lb$_f$/in.²). In some special applications, the operating pressure may go up to 70×10^6 N/m² (70 MPa) (approximately 700 kg$_f$/cm² or 10,000 lb$_f$/in.²). For the same power requirement, the weight and size of the hydraulic unit can be made smaller by increasing the supply pressure.

Outline of the chapter. Our purpose is not to give a complete analysis of hydraulic systems but rather to present a brief outline of such systems, as well as the mathematical modeling techniques for them. After the introductory material of Sec. 4-1, a brief outline of the components of hydraulic systems without mathematical analysis follows in Sec. 4-2. Then the properties of hydraulic fluids are discussed in Sec. 4-3 and the basic laws and equations of fluid flow in Sec. 4-4. Mathematical modeling of hydraulic systems is covered in Sec. 4-5. Finally, Sec. 4-6 deals with a linearization technique for deriving linearized mathematical models of nonlinear components by using a hydraulic valve as an example of nonlinear components.

The material given in this chapter constitutes the absolute minimum required by an engineer. The reader who desires additional details of hydraulic systems should refer to specialized books—for instance, those appearing in References 4-4 and 4-7.

4-2 HYDRAULIC SYSTEMS

In this section, which introduces general concepts of hydraulic systems, we present brief descriptions of hydraulic circuits, power units, actuators, valves, and similiar devices.

Hydraulic circuits. Hydraulic circuits are capable of producing many different combinations of motion and force. All, however, are fundamentally the same regardless of the application. Such circuits involve four basic components: a reservoir to hold the hydraulic fluid, a pump or pumps to force the fluid through the circuit, valves to control fluid pressure and flow, and an actuator or actuators to convert hydraulic energy into mechanical energy to do the work. Figure 4-1 shows a simple circuit that involves a reservoir, a pump, valves, a hydraulic cylinder, and so on.

Hydraulic power unit. A *hydraulic power unit* includes such components as a reservoir, strainers, an electric motor to drive a hydraulic pump or pumps, and a maximum pressure control valve.

The reservoir, which functions as a source of hydraulic fluid, should be large enough to store more than the largest volume of fluid that the system will require. In addition, it should be completely enclosed in order to keep the

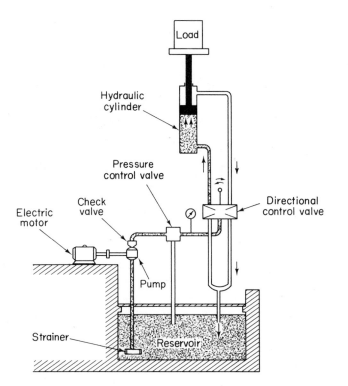

Fig. 4-1. Hydraulic circuit.

fluid clean. Frequently, the electric motor, pump, and valves are mounted on the reservoir.

Strainers, filters, and magnetic plugs are used to remove foreign particles from the hydraulic fluid, thereby ensuring the long life and trouble-free performance of the hydraulic system. (The magnetic plugs usually located in the reservoir will trap iron or steel particles from the hydraulic fluid.)

Used to convert mechanical energy into hydraulic energy, hydraulic pumps can be classified as *positive displacement* and *nonpositive displacement pumps*. Figure 4-2(a) and (b) shows schematic diagrams of each. One characteristic of the positive displacement pump is that the output is unaffected by variations in system pressure because of the presence of a positive internal seal against leakage. Almost all pumps used in power hydraulic systems are of the positive displacement type. (Because of the absence of a positive internal seal against leakage, the output of a nonpositive displacement pump varies considerably with the pressure.)

Positive displacement pumps can be classified as *fixed* or *variable displacement units*. In the former, in order to vary the volumetric output, the

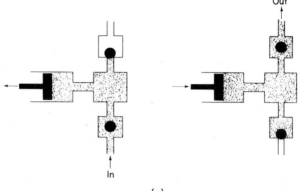

(a)

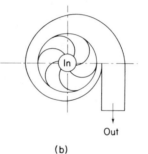

(b)

Fig. 4-2. (a) Positive displacement pump; (b) nonpositive displacement pump.

pump speed must be varied. Volumetric output can be varied in a variable-displacement pump by adjusting the physical relationship of the operating parts of the pump.

There are four basic types of positive displacement pumps commonly used in hydraulic systems.

1. Axial piston pumps
2. Radial piston pumps
3. Vane pumps
4. Gear pumps

Because of the similiarity in their mechanical construction, hydraulic pumps can be used as hydraulic motors. Table 4-1 shows the performance characteristics of hydraulic pumps and motors.

Table 4-1. PERFORMANCE CHARACTERISTICS
OF HYDRAULIC PUMPS AND MOTORS

	Pressure (MPa)	Output (m³/s)	Overall efficiency
Axial piston pumps	$7 \sim 70$	$3 \times 10^{-5} \sim 2 \times 10^{-2}$	$85 \sim 95$
Radial piston pumps	$5 \sim 50$	$3 \times 10^{-4} \sim 1.2 \times 10^{-2}$	$80 \sim 90$
Vane pumps	$2 \sim 18$	$3 \times 10^{-5} \sim 1.6 \times 10^{-2}$	$80 \sim 90$
Gear pumps	$2 \sim 18$	$1 \times 10^{-4} \sim 1 \times 10^{-2}$	$75 \sim 90$
	Pressure (MPa)	Angular frequency (Hz)	Overall efficiency
Axial piston motors	$1 \sim 70$	$0.2 \sim 50$	$85 \sim 95$
Radial piston motors	$1 \sim 50$	$0.2 \sim 30$	$80 \sim 90$
Vane motors	$1 \sim 18$	$2 \sim 50$	$80 \sim 90$
Gear motors	$1 \sim 18$	$2 \sim 50$	$70 \sim 90$

1 MPa $= 10^6$ Pa $= 10^6$ N/m^2 $= 10.197$ kg$_f$/cm^2 $= 145$ lb$_f$/in.^{2}
1 m^3/s $= 10^6$ cm^3/s $= 10^3$ ℓ/s $= 6 \times 10^4$ ℓ/min
1 Hz $= 1$ cps $= 60$ cpm $= 60$ rpm

Axial piston pumps. Figure 4-3 is a schematic diagram of an *axial piston pump*. The rotating cylinder block contains pistons that are free to move in and out of their bores. The drive shaft is located at an angle to the cylinder block. Rotation of the drive shaft causes rotation of the pistons and the cylinder block at the same speed. As each piston moves in and out within its

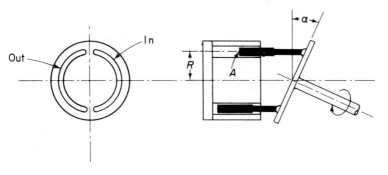

Fig. 4-3. Axial piston pump.

bore, the length of travel is $2R \tan \alpha$. (R is defined in Fig. 4-3.) This length depends on the angle α, the tilt angle of the cylinder block. As each piston moves outward, hydraulic fluid is drawn in through the valve. On the return stroke, fluid is forced out through the valve under pressure. For one cycle, the volumetric flow is $2ZAR \tan \alpha$, where Z is the number of pistons (a typical pump has nine pistons) and A is the area of piston.

Radial piston pumps. A *radial piston pump* is illustrated in Fig. 4-4(a). It consists of a stationary pintle that ports inlet and outlet flow, a cylinder block that revolves around the pintle and houses the pistons, and a rotor that controls piston stroke. The rotor centerline is offset from the cylinder block centerline.

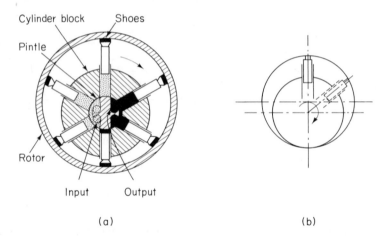

Fig. 4-4. (a) Radial piston pump; (b) schematic diagram of a radial piston pump.

Figure 4-4(b) is a schematic diagram of a radial piston pump in which only one piston is shown. Here, as the drive shaft rotates the cylinder block, the centrifugal force drives the plunger outward so that it presses against the rotor. Since the rotor centerline is offset from the cylinder block centerline, the piston moves in during one-half of a cylinder block revolution and out during the other half. The pintle includes inlet and outlet ports that connect with the open ends of the cylinder bores. During rotation the piston draws hydraulic fluid into the cylinder bore as it passes the inlet side of the pintle and forces this fluid out of the bore as it passes the outlet side of the pintle. The volumetric output depends on the amount of offset between rotor and cylinder centerlines.

Vane pumps. A schematic diagram of a simple *vane pump* appears in Fig. 4-5(a). A cylindrical rotor with movable vanes in radial slots rotates in a

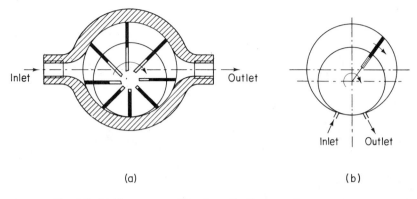

Inlet

Outlet

Inlet | Outlet

(a)   (b)

Fig. 4-5. (a) Vane pump; (b) schematic diagram of a vane pump.

circular housing. The diagram in Fig. 4-5(b) illustrates the principle of operation. To simplify the discussion, only one vane is shown. As the rotor turns, the centrifugal force drives the vane outward so that it is always in contact with the inner surface of the housing. The vane divides the area between the rotor and housing into two chambers. (The actual vane pump involves many vanes, and they divide the area between the rotor and housing into a series of chambers that vary in size, depending on their position around the housing.) The pump inlet is located at a point where the chamber is expanding in size. A partial vacuum caused by the expansion draws hydraulic fluid into the pump. Fluid is then carried to the outlet side of the pump, where the chamber contacts and forces it through the outlet port. This pump is called an *unbalanced* vane pump because high pressure is generated on only one side of the rotor and shaft.

A *balanced* vane pump has an elliptical housing that forms two separate pumping chambers on opposite sides of the rotor so that the side loads cancel out. Such a balanced pump is shown in Fig. 4-6. Advantages of this type are that it increases bearing life and permits higher operating pressure.

Gear pumps. Figure 4-7 presents a schematic diagram of a *gear pump*, which consists of a drive gear and a driven gear, enclosed in a closely fitted housing. The drive and driven gears rotate in opposite directions and mesh at a point in the housing between the inlet and outlet ports. Hydraulic fluid is drawn into inlet chamber A as the teeth of the drive and driven gears separate. It is trapped between the gear teeth and housing and carried through two separate paths around to outlet chamber B. As the teeth remesh, fluid is forced through the outlet port. Note that close fit of the gear teeth within the housing is necessary to minimize internal leakage.

Summary on positive displacement pumps. Because of lower cost, simpler maintenance, and greater tolerance for fluid contamination, the gear

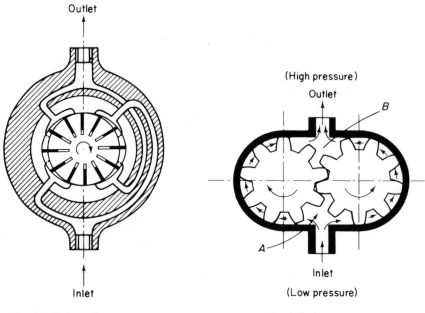

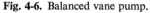

Fig. 4-6. Balanced vane pump.   **Fig. 4-7.** Gear pump.

pump is most frequently used in industries. The vane pump has wide industrial applications, such as machine tools and automation machinery. The axial and radial piston pumps are used mostly where higher pressures are needed.

Accumulators. The *accumulator* stores pressured fluid from a hydraulic pump. This component is often used in hydraulic circuits to make pressured fluid available on demand as well as to smooth out pulsating flows.

Actuators. *Hydraulic actuators* perform the opposite function of hydraulic pumps in that they convert hydraulic energy back to mechanical energy in order to permit useful work. Mechanically linked to the work load, this device is actuated by pressured fluid from the pump. Actuators can be classified as linear or rotary.

Linear actuators. *Linear actuators* come in the form of a ram or cylinder. Figure 4-8(a) and (b) shows double-acting cylinders. In a double-acting cylinder hydraulic pressure can be applied on either side of the piston. (The piston can be moved in either direction.) The type shown in Fig. 4-8(a) is called a *differential* cylinder because the piston area at the left is larger, thus providing a slower, more powerful work stroke when fluid pressure is applied to the left side. The return stroke is faster due to the smaller piston area.

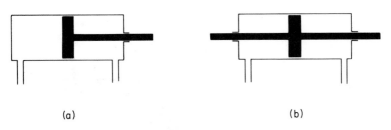

Fig. 4-8. Double-acting cylinders.

Figure 4-8(b) shows a *nondifferential* type of cylinder. Equal forces in both directions are required.

Rotary actuators. The *rotary actuators* include piston motors, vane motors, and gear motors. Many hydraulic pumps (such as piston pumps, vane pumps, and gear pumps) can be used as motors with little or no modification.

Figure 4-9 is a schematic diagram of an axial piston motor. The piston at the high-pressure side is pushed out with force Ap, where A is the piston

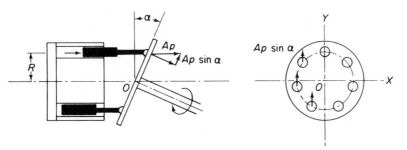

Fig. 4-9. Axial piston motor.

area and p the fluid pressure. This force may be resolved into forces normal and parallel to the driven plate. For each piston, the force parallel to the plate is $Ap \sin \alpha$. Therefore torque T acting on the shaft is

$$T = \sum_i Ap \sin \alpha \cdot R \sin \theta_i$$

where θ_i is the angle between line OY and the line connecting point O and the center of the ith plunger and R is defined in Fig. 4-9.

In a radial piston motor, pressured fluid enters half of the cylinder block bores, forcing the respective pistons radially from the cylinder block axis. These pistons can only move radially by revolving to a point where the rotor contour is farther from the pintle. Thus, driving the pistons radially causes the cylinder block and pistons to rotate. This principle of operation is illustrated in Fig. 4-10. The cylinder block is connected to the output shaft.

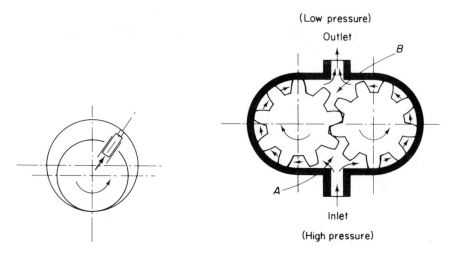

Fig. 4-10. Radial piston motor.   **Fig. 4-11.** Gear motor.

Figure 4-11 illustrates a gear motor. Since this device is a motor, both gears are driven gears, but only one is connected to the output shaft. Operation is essentially the reverse of a gear pump. Hydraulic fluid from the pump enters chamber *A* and flows in either direction around the inside surface of the housing to chamber *B*, forcing the gears to rotate as indicated. This rotary motion is then available for work at the output shaft.

Hydraulic control valves. The *hydraulic control valve* is a device that uses mechanical motion to control the direction of fluid flow to the actuator. Commonly used hydraulic control valves can be divided into four types: sliding spool, flapper, jet pipe, and poppet valves.

Sliding spool valves. Extensively used in hydraulic systems, *sliding spool valves* are usually classified by the number of ways that flow can enter and leave the valve.

A sliding-spool four-way valve (or four-way spool valve) connected to a power (or actuator) cylinder is shown in Fig. 4-12. The operating principle is as follows. The spool may be shifted to either direction. If shifted to the right, as shown, port *B* is open to the pressure port *P* and port *A* is open to the drain. The power (or actuator) piston moves to the left. Similarly, if the spool is shifted to the left, port *A* is open to the pressure port *P*, port *B* is open to the drain, and the power piston moves to the right.

This four-way valve has two lands on the spool. (In spool valves, the number of lands on a spool varies from one to three or four.) If the width of the land is smaller than the port in the valve sleeve, the valve is said to be *underlapped*. *Overlapped* valves have a land width greater than the port width

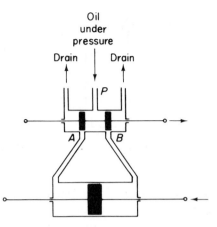

Fig. 4-12. Sliding-spool four-way valve connected to a power cylinder.

when the spool is at neutral. A zero-lapped valve has a land width that is identical to the port width.

Figure 4-13 shows a three-way valve connected to a power cylinder. It requires a bias pressure acting on one side of an unequal-area power piston for direction reversal.

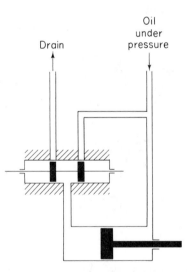

Fig. 4-13. Three-way valve connected to a power cylinder.

Flapper valves. *Flapper valves* are also called *nozzle-flapper valves*. A flapper is placed between two opposing nozzles (Fig. 4-14). If the flapper is moved slightly to the right, the pressure unbalance occurs in the nozzles and the power piston moves to the left, and vice versa.

These devices are frequently used in hydraulic servos as the first-stage valve in two-stage servovalves. This usage occurs because considerable force may be needed to stroke larger spool valves that result from the steady-state

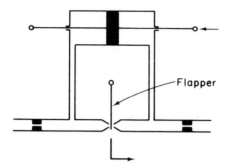

Fig. 4-14. Flapper valve connected to a power cylinder.

flow force. To reduce or compensate this force, two-stage valve configuration is often employed; a flapper valve is used as the first-stage valve to provide a necessary force to stroke the second-stage spool valve.

Jet pipe valves. Figure 4-15 shows a *jet pipe valve* connected to a power cylinder. Hydraulic fluid is ejected from the jet pipe. If the jet pipe is shifted to the right from the neutral position, the power piston moves to the left and vice versa. The jet pipe valve is not used as much as the flapper valve because of large null flow, slower response, and rather unpredictable characteristics. Its main advantage lies in its insensitivity to dirty fluids.

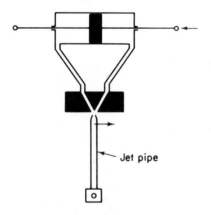

Fig. 4-15. Jet pipe valve connected to a power cylinder.

Poppet valves. Basically, *poppet valves* are two-way valves. Typical poppet valves are found in check valves and relief valves, where reversal in flow direction is not required.

The check valve is a one-way directional valve in that it permits flow in one direction and prevents it in the other.

The purpose of the relief valve is to provide protection against overloading of circuit components or to limit the force that can be exerted by an actuator. Such valves are required in almost all hydraulic circuits in order to control pressure. Figure 4-16 shows a simple relief valve in which one port is connected to the pressure line and the other to the reservoir. The spring force holds the valve on its seat. The adjusting screw controls the operating pressure.

The relief valve operates as follows. When pressure at the inlet exceeds the spring force, the valve is forced off its seat and fluid flows from the pressure line through the valve to the reservoir. When pressure drops below the spring force, the valve reseats and the flow is stopped. The pressure at which the valve is first forced off its seat and begins to pass fluid is called the *cracking pressure*. As flow through the valve increases, the valve is pushed farther off the seat and full flow pressure becomes higher than the cracking pressure. This phenomenon of pressure increase in the line as the flow through the relief valve increases is called the *pressure override*.

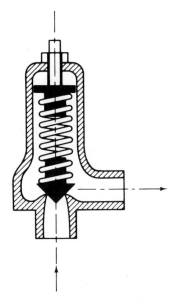

Fig. 4-16. Relief valve.

Advantages and disadvantages of hydraulic systems. There are certain advantages and disadvantages in using hydraulic systems rather than other systems. Some of the advantages are listed below.

1. Hydraulic fluid acts as a lubricant, in addition to carrying away heat generated in the system to a convenient heat exchanger.
2. Comparatively small-size hydraulic actuators can develop large forces or torques.
3. Hydraulic actuators have a higher speed of response with fast starts, stops, and speed reversals.
4. Hydraulic actuators can be operated under continuous, intermittent, reversing, and stalled conditions without damage.
5. Availability of both linear and rotary actuators gives flexibility in design.
6. Because of low leakages in hydraulic actuators, speed drop when loads are applied is small.

On the other hand, several disadvantages tend to limit their use.

1. Hydraulic power is not readily available compared to electric power.

2. Cost of a hydraulic system may be higher than a comparable electrical system performing a similar function.

3. Fire and explosion hazards exist unless fire-resistant fluids are used.

4. Because it is difficult to maintain a hydraulic system that is free from leaks, the system tends to be messy.

5. Contaminated oil may cause failure in the proper functioning of a hydraulic system.

6. As a result of the nonlinear and other complex characteristics involved, the design of sophisticated hydraulic systems is quite involved.

7. Hydraulic circuits have generally poor damping characteristics. If a hydraulic circuit is not designed properly, some unstable phenomena may occur or disappear, depending on the operating condition.

Comments. Particular attention is necessary to ensure that the hydraulic system is stable and satisfactory under all operating conditions. Since viscosity of hydraulic fluid can greatly affect damping and friction effects of the hydraulic circuits, stability tests must be carried out at the highest possible operating temperature.

It should also be noted that certain undesirable phenomena may occur in hydraulic systems, two of which are oil hammer and cavitation. Although nonexistent in well-designed systems, they do occur in some; therefore, designing hydraulic systems that avoid these phenomena is advisable.

Oil hammer. When oil or water flowing in a pipe is suddenly stopped by an instantaneous closure of the valve at the end of the pipeline, a severe pressure surge may result, thereby causing a series of shocks that sound like hammer blows. This phenomenon is known as *oil hammer* or *water hammer*, depending on the fluid medium involved.

The phenomenon of water hammer can occur in home plumbing systems. For instance, when faucets are turned off quickly or when the flow of water is automatically shut off by a water-using applicance like a dishwasher or clothes washer, hammering sound may occur. Such water hammer is caused by the fact that water flowing through a pipe develops momentum. When the flow is suddenly cut off by closing a faucet or by the action of an electric valve inside a washing machine, the water is still moving because of this momentum, and since water can hardly be compressed, it slams around inside the pipe. (It should be noted that pipes may make hammering noises for quite a different reason, such as being insecurely mounted so that water rushing through causes them to move about and bang against beams or other pipes next to them.)

In any hydraulic system, if the valve at the end of the pipeline is suddenly closed, the kinetic energy of the arrested colomn of hydraulic fluid is expended in compressing the fluid and stretching the pipe walls. By stopping the hydraulic fluid flow, the kinetic energy is transformed into potential energy. (The maximum pressure at the instant of valve closure may be obtained by equating kinetic energy and potential energy.) The increase of pressure is found to be proportional to the arrested velocity of hydraulic fluid flow. So in order to reduce the pressure surge, it is advisable to have low fluid velocity by making pipe areas sufficiently large. (A rule of thumb is to limit the velocity of hydraulic fluid to 5 m/s.)

Note that the pressure surge results only when the valve is closed in less than one round trip of the pressure wave. If the fluid flow is not stopped rapidly, so that the pressure wave has time to travel to the end of the hydraulic line and back again several times while the stopping is in progress, the excess pressure is greatly reduced. (The pressure wave continues to travel back and forth until the energy involved is lost by friction.) Consequently, to avoid severe pressure surge and hammering noise, the use of slow-closing valves on long pipelines and the installation of relief devices or antihammer devices, such as surge tanks, at proper sites to absorb the shock of pressure surge are advisable. An antihammer device, which basically consists of a chamber with trapped air, functions as a pneumatic cushion to absorb the shock created when the fast-flowing hydraulic fluid is slammed to a halt. Instead of the hydraulic fluid banging against the pipes and fittings, it forces its way into the air chamber of the antihammer device. Air compresses easily, and so the rushing hydraulic fluid starts to compress the air inside, thus absorbing the extra energy that would otherwise cause hammering. When the hydraulic fluid is at rest again, the compressed air inside expands and is ready for the next round.

In home plumbing systems, where washing machines almost always create this type of problem because of the frequent on-off cycles of the automatic valves, an antihammer device (such as a pipe or cylinder with one end sealed off and the other end connected to the water pipe so that air is trapped inside the pipe or cylinder) is usually installed next to the faucets that supply water to these appliances.

Cavitation. When the velocity of the liquid flow is increased locally and the liquid flows into a region where the pressure is reduced to vapor pressure, it boils and vapor pockets develop. In this situation, the vapor bubbles are carried along with the liquid until a region of higher pressure is reached and they suddenly collapse. When vapor pockets collapse, the forces exerted by the liquid rushing into the cavitation create very high localized pressure and cause pitting of the solid surface, a process that is accompanied by noise and vibration. This process of vaporization and subsequent collapse of vapor bubbles in a rapid flow of liquid is called *cavitation*. In a centrifugal pump, for instance, if cavitation arises because of the inlet pressure drop, noise and vibration occur and efficiency drops. Moreover, the pump may be damaged. Since the process causes such undesirable effects as lowered efficiency, damage to flow passages, and noise and vibration, hydraulic systems should be designed to avoid cavitation by eliminating local low-pressure regions and/or by using special cavitation-resistance materials or coatings.

4-3 PROPERTIES OF HYDRAULIC FLUIDS

The properties of hydraulic fluid have an important effect on the performance of hydraulic systems. Besides serving as a power-transmitting medium, hydraulic fluid must keep wear of moving parts to a minimum by providing satisfactory lubrication. In practice, petroleum-based oils with proper additives are the most commonly used hydraulic fluids because they give good lubrication for moving parts in the system and are almost imcompressible. The use of a clean, high-quality oil is required for satisfactory operation of the hydraulic system.

The following pages present those physical characteristics of hydraulic fluids that are necessary in discussing hydraulic systems.

Density and specific volume. The *mass density* ρ of a substance is its mass per unit volume. Commonly used units include kg/m³, lb/ft³, slug/ft³, and so on. For water at standard atmospheric pressure (1.0133×10^5 N/m² abs, which is equal to 1.0332 kg$_f$/cm² abs or 14.7 lb$_f$/in.² abs) and standard temperature (277.15 K which is equal to $4°C$ or $39.2°F$), the mass density is

$$\rho = 1000 \text{ kg/m}^3 = 62.43 \text{ lb/ft}^3 = 1.94 \text{ slug/ft}^3$$

For petroleum-based oils, the mass density is approximately

$$\rho = 820 \text{ kg/m}^3 = 51.2 \text{ lb/ft}^3 = 1.59 \text{ slug/ft}^3$$

The *specific volume* v is the reciprocal of the density ρ. It is the volume occupied by unit mass of fluid or

$$v = \frac{1}{\rho}$$

Specific weight and specific gravity. The *specific weight* γ of a substance is its weight per unit volume. The commonly used units are N/m², kg$_f$/m³, lb$_f$/ft³, and so forth. For water at standard atmospheric pressure and temperature,

$$\gamma = 9.807 \times 10^3 \text{ N/m}^3 = 1000 \text{ kg}_f/\text{m}^3 = 62.43 \text{ lb}_f/\text{ft}^3$$

For petroleum-based oils, the specific weight is approximately

$$\gamma = 8.04 \times 10^3 \text{ N/m}^3 = 820 \text{ kg}_f/\text{m}^3 = 51.2 \text{ lb}_f/\text{ft}^3$$

The specific weight γ and mass density ρ are related by

$$\gamma = \rho g$$

where g is the acceleration of gravity.

The *specific gravity* of a substance is the ratio of its weight to the weight of an equal volume of water at standard atmospheric pressure and temperature.

The density ρ of a liquid is a function of pressure and temperature. It may be written

$$\rho = \rho_0[1 + a(p - p_0) - b(\theta - \theta_0)]$$

where ρ, p, and θ are the mass density, pressure, and temperature, respectively. (The density of the liquid is assumed to be ρ_0 when the pressure is p_0 and the temperature is θ_0.) The values of a and b are positive. Thus the mass density of a liquid increases as pressure is increased and decreases with temperature increase. The coefficients a and b are called the *modulus of compressibility* and *cubical expansion coefficient*, respectively.

Compressibility and bulk modulus. The compressibility of a liquid is expressed by its bulk modulus. The bulk modulus of a liquid and the modulus of compressibility are related inversely. If the pressure of a liquid of volume V is increased by dp, it will cause a decrease in volume, dV. The bulk modulus K is defined by

$$K = \frac{dp}{-dV/V}$$

(Note that dV is negative, so that $-dV$ is positive.) The bulk modulus of water at ordinary temperature and pressure is approximately 2.1×10^9 N/m², which is equal to 2.1 GPa (gigapascal), 2.14×10^4 kg$_f$/cm², or 3×10^5 lb$_f$/in.²

It is important to note that all hydraulic fluids will combine with air to a certain extent. So in the experimental determination of the bulk modulus, the value of the bulk modulus of any given liquid depends on the amount of air in it.

Viscosity. *Viscosity*, the most important property of the hydraulic fluid, is a measure of the internal friction or the resistance of fluid to flow. Low viscosity means an increase of leakage losses, and high viscosity implies sluggish operation. In hydraulic systems allowable viscosities are limited by the operating characteristics of the pump, motor, and valves as well as by ambient and operating temperatures. The viscosity of a liquid decreases with temperature.

Viscosity is measured by observing the time required for a certain volume of the liquid to flow, under certain conditions as to head, through a short tube of small bore.

The resistance caused by a fluid to the relative motion of its parts is called *dynamic* or *absolute viscosity*. It is the ratio of the shearing stress to the rate of shear deformation of a fluid. The coefficient of dynamic or absolute viscosity μ is the resistance caused by a layer of the fluid to the motion parallel to that layer, of another layer of the fluid at a unit distance from it, with a unit relative velocity.

The SI units of dynamic viscosity are N-s/m² and kg/m-s. The cgs unit of dynamic viscosity is the poise (P) (dyn-s/cm² or g/cm-s). The SI unit is ten times larger than the poise unit. The centipoise (cP) is 1/100 of poise. (Note that the dynamic viscosity of water at 20.2°C or 68.4°F is 1 centipoise.) The BES units of dynamic viscosity are lb_f-s/ft² and slug/ft-s. Note that

$$1 \text{ slug/ft-s} = 1 \text{ lb}_f\text{-s/ft}^2 = 47.9 \text{ kg/m-s} = 47.9 \text{ N-s/m}^2$$

$$1 \text{ P} = 100 \text{ cP} = 0.1 \text{ N-s/m}^2$$

The *kinematic viscosity* ν is the dynamic viscosity divided by mass density or

$$\nu = \frac{\mu}{\rho}$$

where ρ is the mass density of fluid. The SI unit of kinematic viscosity is m²/s, whereas the cgs unit of kinematic viscosity is the stoke (St) (cm²/s) and 1/100 stoke is called the centistoke (cSt). The BES unit of kinematic viscosity is ft²/s. In changing from the stoke to the poise, multiply by the mass density in g/cm³. Note that

$$1 \text{ m}^2\text{/s (SI unit of kinematic viscosity)}$$

$$= 10.764 \text{ ft}^2\text{/s (BES unit of kinematic viscosity)}$$

$$1 \text{ St} = 100 \text{ cSt} = 0.0001 \text{ m}^2\text{/s}$$

Table 4-2 summarizes the units used for dynamic and kinematic viscosities in various systems of units, and Table 4-3 shows the dynamic and kinematic viscosity of water. For hydraulic oils at normal operating conditions, the kinematic viscosity is about 5 to 100 centistokes (5×10^{-6} to 100×10^{-6} m²/s).

Petroleum oils tend to become thin as temperature increases and thicken as temperature decreases. If the system operates over a wide temperature range, fluid having a viscosity that is relatively less sensitive to temperature changes must be used.

Some additional remarks on hydraulic fluids. In concluding this section, some additional remarks are given below.

1. Although such fluids as water, crude oil, and vegetable or animal oils will transmit hydraulic power, they should not be used as hydraulic fluids

Table 4-2. UNITS FOR DYNAMIC AND KINEMATIC
            VISCOSITIES

System of units / Viscosity	Absolute systems			Gravitational systems	
	SI	mks	cgs	Metric engineering	British engineering
Dynamic viscosity μ	$\dfrac{\text{N-s}}{\text{m}^2}$ or $\dfrac{\text{kg}}{\text{m-s}}$	$\dfrac{\text{N-s}}{\text{m}^2}$ or $\dfrac{\text{kg}}{\text{m-s}}$	$\dfrac{\text{dyn-s}}{\text{cm}^2}$ or $\dfrac{\text{g}}{\text{cm-s}}$ (poise)	$\dfrac{\text{kg}_f\text{-s}}{\text{m}^2}$	$\dfrac{\text{lb}_f\text{-s}}{\text{ft}^2}$ or $\dfrac{\text{slug}}{\text{ft-s}}$
Kinematic viscosity ν	$\dfrac{\text{m}^2}{\text{s}}$	$\dfrac{\text{m}^2}{\text{s}}$	$\dfrac{\text{cm}^2}{\text{s}}$ (stoke)	$\dfrac{\text{m}^2}{\text{s}}$	$\dfrac{\text{ft}^2}{\text{s}}$

Table 4-3. DYNAMIC AND KINEMATIC VISCOSITY
            OF WATER

Temperature	Dynamic viscosity	Kinematic viscosity
°C	μ N-s/m^2	ν m^2/s
0	1.792×10^{-3}	1.792×10^{-6}
20	1.002×10^{-3}	1.004×10^{-6}
40	0.653×10^{-3}	0.658×10^{-6}
60	0.467×10^{-3}	0.475×10^{-6}
80	0.355×10^{-3}	0.365×10^{-6}
100	0.282×10^{-3}	0.295×10^{-6}
°F	μ lb$_f$-s/ft^2	ν ft^2/s
32	3.742×10^{-5}	1.929×10^{-5}
100	1.418×10^{-5}	0.736×10^{-5}
212	0.589×10^{-5}	0.318×10^{-5}

because of their lack of ability to lubricate properly and resist rust, corrosion, foaming, and so on.

2. The operating life of a hydraulic fluid depends on its oxidation resistance. Oxidation of hydraulic fluid is caused by air, heat, and contamination. Note that any hydraulic fluid combines with air to a certain extent, especially at high operating temperatures. Note also that the operating temperature of the hydraulic system should be kept between 30 and 60°C. For operating temperature above 70°C, oxidation is accelerated. Premium-grade fluids usually contain inhibitors to slow down oxidation.

3. When operating at high temperatures, fluid properties of significance are lubricity, viscosity, thermal stability, fluid stability, weight, and bulk modulus. (Note that these are not independent variables.)

4. For hydraulic systems located near high-temperature sources, fire-resistant fluids should be used. These fluids are available in several general types, such as water-glycol, synthetic oil, and water-oil emulsions. (In the water-oil emulsions the oil forms around the water molecules to provide lubricity.)

4-4 BASIC LAWS OF FLUID FLOW

Here we derive the basic equations that govern fluid flows, such as continuity equations, Euler's equation, and Bernoulli's equation. We begin with definitions of Reynolds number, laminar and turbulent flows, and other necessary terminology and then derive the equations.

Reynolds number. The forces that affect fluid flow are due to gravity, bouyancy, fluid inertia, viscosity, surface tension, and similiar factors. In many flow situations, the forces resulting from fluid inertia and viscosity are most significant. In fact, fluid flows in many important situations are dominated by either inertia or viscosity of the fluid. The dimensionless ratio of inertia force to viscous force is called the *Reynolds number*. Thus a large Reynolds number indicates the dominance of inertia force and a small number the dominance of viscosity. The Reynolds number R is given by

$$R = \frac{\rho v D}{\mu}$$

where ρ is fluid mass density, μ dynamic viscosity, v the average velocity of flow, and D a characteristic length. For flow in pipes, the characteristic length is the inside pipe diameter. Since the average velocity v for flow in a pipe is

$$v = \frac{Q}{A} = \frac{4Q}{\pi D^2}$$

where Q is the volumetric flow rate, A the pipe area, and D the inside pipe diameter, the Reynolds number for flow in pipes can be given by

$$R = \frac{\rho v D}{\mu} = \frac{4\rho Q}{\pi \mu D}$$

Laminar flow and turbulent flow. Flow dominated by viscosity forces is called *laminar flow*. It is characterized by a smooth, parallel line motion of the fluid. When inertia forces dominate, the flow is called *turbulent flow* and is characterized by an irregular and eddylike motion of fluid. For a Reynolds number below 2000 or $R < 2000$, the flow is always laminar. For a Reynolds number above 4000 or $R > 4000$, the flow is usually turbulent except in special cases.

In capillary tubes, flow is laminar. If velocities are kept very low or viscosities are very high, flow in pipes of relatively large diameter may also result in laminar flow. In general, flow in a pipe is laminar if the passage cross section is comparatively small and/or the pipe length is relatively long. Otherwise turbulent flow results. It should be noted that laminar flow is temperature sensitive, for it depends on viscosity.

For laminar flow, the velocity profile in a pipe becomes parabolic as in Fig. 4-17(a). Figure 4-17(b) shows a velocity profile in a pipe for turbulent flow.

Industrial processes often involve the flow of liquids through connecting pipes and tanks. In hydraulic control systems there are many cases of flow through small passages, such as flow between spool and bore and

(a)

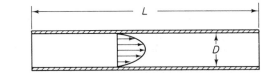

Laminar flow in pipe

(b)

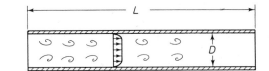

Turbulent flow in pipe

Fig. 4-17. (a) Velocity profile for laminar flow; (b) velocity profile for turbulent flow.

between piston and cylinder. The properties of such flow through small passages depend on the Reynolds number of flow involved in each situation.

Streamline. A *streamline* refers to a continuous line drawn through the fluid so that it has the direction of the velocity vector at every point (Fig. 4-18). Therefore no flow can cross a streamline.

Stream tube. A *stream tube* is the tube made by all the streamlines passing through a closed curve (Fig. 4-19). No flow can flow through its walls because the velocity vector has no component normal to the tube surface.

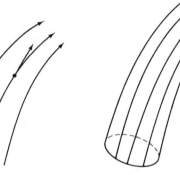

Fig. 4-18. Streamlines.   Fig. 4-19. Stream tube.

Steady flow. If pressure, velocity, density, temperature, and similiar factors at any point in the fluid do not change with time, the flow is said to be *steady*. That is, in steady flow at any point held constant in space

$$\frac{\partial p}{\partial t} = 0, \qquad \frac{\partial \mathbf{v}}{\partial t} = 0, \qquad \frac{\partial \rho}{\partial t} = 0, \qquad \frac{\partial T}{\partial t} = 0$$

where p, $\mathbf{v}$, ρ, and T are pressure, velocity vector, density, and temperature, respectively.

Flow is said to be *unsteady* if the condition at any point changes with time. The analysis of unsteady flow is much more complex than that of steady flow.

Control volume. A *control volume* refers to a region in space. Although entirely arbitrary, the size and shape of the control volume are frequently chosen in order to simplify analysis. The use of a control volume is convenient in the analysis of situations where flow occurs into and out of the space.

Continuity equations. The continuity equations are obtained by applying the principle of conservation of mass to the flow. This principle states that the mass within a system remains constant with time.

The continuity equation for a control volume states that the time rate of increase of mass within a control volume is equal to the net rate of mass inflow to the control volume.

Consider steady flow through the stream tube shown in Fig. 4-20(a), where the control volume constitutes the walls of the stream tube and the cross sections dA_1 and dA_2 that are normal to the stream tube. If we define ρ_1 and ρ_2 as the mass densities at the cross sections dA_1 and dA_2, respectively, and v_1 and v_2 as the velocities of the fluid at the cross sections dA_1 and dA_2, respectively, then by applying the principle of conservation of mass, we obtain

$$\rho_1 v_1 \, dA_1 = \rho_2 v_2 \, dA_2$$

This is the continuity equation applied to two cross sections along a stream tube in steady flow.

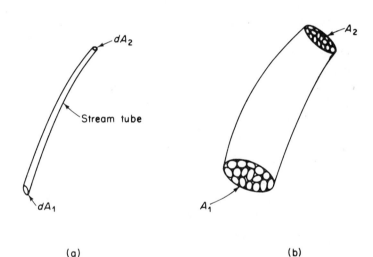

(a)   (b)

Fig. 4-20. (a) Stream tube; (b) collection of stream tubes.

For a collection of stream tubes as shown in Fig. 4-20(b), if the average densities are ρ_1 and ρ_2 over the cross sections A_1 and A_2, respectively, and the average velocities are V_1 and V_2 over the cross sections A_1 and A_2, respectively, then

$$\rho_1 V_1 A_1 = \rho_2 V_2 A_2$$

where

$$V_1 = \frac{1}{A_1} \int_{A_1} v_1 \, dA_1, \qquad V_2 = \frac{1}{A_2} \int_{A_2} v_2 \, dA_2$$

By defining the discharges Q_1 and Q_2 as

$$Q_1 = A_1 V_1, \qquad Q_2 = A_2 V_2$$

we can write the continuity equation as

$$\rho_1 Q_1 = \rho_2 Q_2$$

For incompressible steady flow, we have $\rho_1 = \rho_2$. Hence

$$Q_1 = Q_2$$

or

$$A_1 V_1 = A_2 V_2$$

This means that the flow rate of liquid in a pipe is constant at any cross section.

Euler's equation of motion. Consider an infinitesimal stream tube of length ds as shown in Fig. 4-21. Consider also the control volume composed of the wall of the stream tube between sections 1 and 2 plus the end areas of sections 1 and 2 that are normal to the stream tube. Let us fix this control volume in space and consider the flow through it. To simplify the analysis, we assume that the viscosity is zero or that the fluid is frictionless.

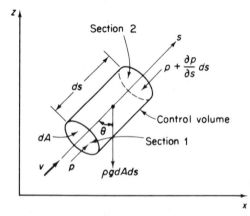

Fig. 4-21. Infinitesimal stream tube of length ds.

The mass of fluid in the control volume is $\rho \, dA \, ds$ and the acceleration of this mass is dv/dt. The pressure force acting on section 1 in the positive direction of s is $p \, dA$ and that acting on section 2 in the negative direction of s is $[p + (\partial p/\partial s) \, ds] \, dA$. The gravity force is $\rho g \, dA \, ds$. Any forces on the sides of the control volume are normal to s and do not enter the equation. Applying Newton's second law, we have the equation of motion

$$m \frac{dv}{dt} = p \, dA - \left(p + \frac{\partial p}{\partial s} \, ds \right) dA - \rho g \, dA \, ds \cos \theta$$

where $m = \rho\, dA\, ds$. So

$$\rho\, dA\, ds\, \frac{dv}{dt} = -\frac{\partial p}{\partial s}\, ds\, dA - \rho g\, dA\, ds\, \cos\theta$$

or

$$\frac{dv}{dt} = -\frac{1}{\rho}\frac{\partial p}{\partial s} - g\cos\theta \qquad (4\text{-}1)$$

In general, velocity v depends on s and t, or $v = v(s, t)$. Therefore

$$\frac{dv}{dt} = \frac{\partial v}{\partial s}\frac{ds}{dt} + \frac{\partial v}{\partial t} = v\frac{\partial v}{\partial s} + \frac{\partial v}{\partial t} \qquad (4\text{-}2)$$

By substituting Eq. (4-2) into Eq. (4-1), we find

$$v\frac{\partial v}{\partial s} + \frac{\partial v}{\partial t} = -\frac{1}{\rho}\frac{\partial p}{\partial s} - g\cos\theta$$

And noting that $\cos\theta = \partial z/\partial s$, where z is the vertical displacement, this last equation can be written

$$v\frac{\partial v}{\partial s} + \frac{\partial v}{\partial t} + \frac{1}{\rho}\frac{\partial p}{\partial s} + g\frac{\partial z}{\partial s} = 0 \qquad (4\text{-}3)$$

which is *Euler's equation of motion*.

For steady flow, $\partial v/\partial t = 0$, and Eq. (4-3) simplifies to

$$v\frac{\partial v}{\partial s} + \frac{1}{\rho}\frac{\partial p}{\partial s} + g\frac{\partial z}{\partial s} = 0$$

For steady flow, since v, p, and z are functions of s only, this last equation may be rewritten as

$$v\frac{dv}{ds} + \frac{1}{\rho}\frac{dp}{ds} + g\frac{dz}{ds} = 0$$

or

$$v\, dv + \frac{dp}{\rho} + g\, dz = 0 \qquad (4\text{-}4)$$

which is *Euler's equation of motion for steady flow*.

Bernoulli's equation. For steady, frictionless (meaning that the fluid has negligible viscosity), incompressible flow, Eq. (4-4) can be integrated to give

$$\frac{v^2}{2} + \frac{p}{\rho} + gz = \text{constant} \qquad (4\text{-}5)$$

This equation is the energy equation for steady flow through a control volume. By dividing both sides of Eq. (4-5) by g, we have

$$\frac{v^2}{2g} + \frac{p}{\gamma} + z = \text{constant} \qquad (4\text{-}6)$$

where $\gamma = \rho g$. This equation is called *Bernoulli's equation*. Each term of it has the dimension of length. Equation (4-6) shows that along a stream tube the sum of the velocity head $v^2/(2g)$, the pressure head p/γ, and the potential head z is constant (Fig. 4-22). If the velocity at a section increases, the pressure head plus the potential head must decrease and vice versa—that is, the total head at any section is constant.

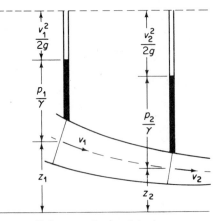

Fig. 4-22. Diagram illustrating that the sum of velocity head, pressure head, and potential head is constant—Bernoulli's equation.

For unsteady flow, Eq. (4-3) can be rewritten as

$$\frac{\partial v}{\partial t} + \frac{\partial}{\partial s}\left(\frac{v^2}{2} + \frac{p}{\rho} + gz\right) = 0$$

Integration of this last equation along the stream tube yields

$$\int_0^s \frac{\partial v}{\partial t}\,ds + \frac{v^2}{2} + \frac{p}{\rho} + gz = \text{constant}$$

At cross sections 1 and 2, we obtain

$$\int_0^{s_1} \frac{\partial v}{\partial t}\,ds + \frac{v_1^2}{2} + \frac{p_1}{\rho} + gz_1 = \int_0^{s_2} \frac{\partial v}{\partial t}\,ds + \frac{v_2^2}{2} + \frac{p_2}{\rho} + gz_2$$

or

$$\left(\frac{v_1^2}{2} + \frac{p_1}{\rho} + gz_1\right) - \left(\frac{v_2^2}{2} + \frac{p_2}{\rho} + gz_2\right) = \int_{s_1}^{s_2} \frac{\partial v}{\partial t}\,ds \qquad (4\text{-}7)$$

Equation (4-7) is called the *energy equation for unsteady flow* through a control volume.

Flow through orifice. An *orifice* is a sudden restriction of short length in a flow passage. Two types of flow regime exist, depending on whether viscous or inertia forces dominate [Fig. 4-23(a) and (b)]. Because of the con-

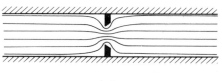

(a)

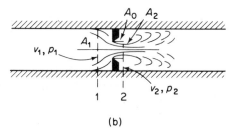

Fig. 4-23. (a) Flow through an orifice when Reynolds number is low; (b) flow through an orifice when Reynolds number is high.

(b)

tinuity law, the velocity of flow through an orifice must increase above that in the upstream region.

In Fig. 4-23(a) the pressure drop is caused by internal shear forces that result from viscosity. This situation occurs when Reynolds numbers are low. Figure 4-23(b) shows the case where the pressure drop across the orifice is caused by the acceleration of the fluid from the upstream velocity to the higher jet velocity. The situation here arises when Reynolds numbers are high. The downstream flow becomes turbulent. Since most important orifice flows occur as in Fig. 4-23(b), in the following we consider this case in detail.

Referring to Fig. 4-23(b), the fluid velocity is increased to the jet velocity between sections 1 and 2. The area of the issuing jet is smaller than the orifice area. The point along the jet where the jet area becomes a minimum is called the *vena contracta*. The ratio of stream area A_2 at the vena contracta to the orifice area A_0 is called the *contraction coefficient* C_c or

$$A_2 = C_c A_0$$

Since the flow between sections 1 and 2 is streamline, Bernoulli's equation can be applied. From Eq. (4-6) we obtain, at sections 1 and 2, the equation

$$\frac{v_1^2}{2g} + \frac{p_1}{\gamma} + z_1 = \frac{v_2^2}{2g} + \frac{p_2}{\gamma} + z_2 \qquad (4\text{-}8)$$

If we assume $z_1 = z_2$, then Eq. (4-8) becomes

$$v_2^2 - v_1^2 = \frac{2g}{\gamma}(p_1 - p_2) \qquad (4\text{-}9)$$

From the continuity equation, we have

$$v_1 A_1 = v_2 A_2 \qquad (4\text{-}10)$$

where A_1 and A_2 are the stream areas at sections 1 and 2, respectively.

Using Eqs. (4-9) and (4-10), we find

$$v_2 = \frac{1}{\sqrt{1 - (A_2/A_1)^2}} \sqrt{\frac{2g}{\gamma}(p_1 - p_2)}$$

Then the volumetric flow rate at the vena contracta is

$$v_2 A_2 = \frac{A_2}{\sqrt{1 - (A_2/A_1)^2}} \sqrt{\frac{2g}{\gamma}(p_1 - p_2)}$$

$$= \frac{C_c A_0}{\sqrt{1 - (C_c^2 A_0^2/A_1^2)}} \sqrt{\frac{2g}{\gamma}(p_1 - p_2)} \qquad (4\text{-}11)$$

Equation (4-11) gives the flow rate through the orifice. It is, however, approximate because viscous friction of the fluid was not considered. To account for the neglected viscous friction, an empirical factor called the *velocity coefficient* C_v is introduced to give the flow rate Q:

$$Q = C_v v_2 A_2$$

or

$$Q = \frac{C_v C_c A_0}{\sqrt{1 - (C_c^2 A_0^2/A_1^2)}} \sqrt{\frac{2g}{\gamma}(p_1 - p_2)} = c A_0 \sqrt{\frac{2g}{\gamma}(p_1 - p_2)} \qquad (4\text{-}12)$$

where c, the discharge coefficient, is

$$c = \frac{C_v C_c}{\sqrt{1 - (C_c^2 A_0^2/A_1^2)}}$$

The value of the discharge coefficient c is usually obtained experimentally.

In the case of hydraulic valves where the throttle area is adjusted to control the pressure and flow rate, Eq. (4-12) serves as a basic equation.

Comments. In conclusion, we would like to mention that excessive friction loss in hydraulic lines should be avoided. When a fluid flows in a hydraulic line, some of the energy being transferred is lost in the form of heat energy that results from friction. In designing hydraulic lines, the principal causes of excessive friction should be removed, such as excessive length of lines, excessive number of bends, fittings, and valves, excessive fluid velocity as a result of undersized lines, and excessive viscosity of fluid.

4-5   MATHEMATICAL MODELING OF HYDRAULIC SYSTEMS

Industrial processes often involve systems consisting of liquid-filled tanks connected by pipes having orifices, valves, or other flow-restricting devices. Dynamic characteristics of such systems can be analyzed by using the fundamental laws (Sec. 4-4) that govern the flow of liquids. In this section we are concerned with the mathematical modeling of such hydraulic systems. (Mathematical modeling of a hydraulic valve is presented in Sec. 4-6 and details of mathematical modeling of hydraulic controllers in Chapter 8.)

In Chapters 2 and 3 it was stated that three types of basic elements exist for mechanical and electrical systems: inertia elements, spring elements, and damper elements for mechanical systems; resistance elements, capacitance elements, and inductance elements for electrical systems. And as with mechanical and electrical systems, there are three types of basic elements for the hydraulic systems that concern us here—resistance elements, capacitance elements, and inertance elements. [Note that the terms *inertia, inductance,* and *inertance* represent inertia effects of systems. The term *inertia* is used with mechanical systems, the term *inductance* with electrical systems, and the term *inertance* with fluid (hydraulic and penumatic) systems.]

We begin this section with discussions on liquid flow from an orifice in the wall of a tank, followed by definitions of resistance, capacitance, and inertance for hydraulic systems. Then we obtain mathematical models of liquid-level systems in terms of resistance and capacitance. The section concludes with a simple response analysis of liquid-level systems.

Flow from orifice in tank wall. Consider the liquid flow through an orifice in the wall of a tank. Referring to the liquid-level system of Fig. 4-24, assume that liquid having small or negligible viscosity spouts through the orifice and that the flow is turbulent. The cross section of the jet is smaller than the orifice area. The cross section where the contraction is greatest is the vena contracta. The streamlines are parallel throughout the jet at this section and the pressure is atmospheric.

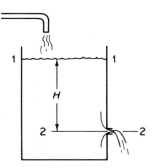

Let us denote the head at the level of the orifice by H. It is measured from the center of the orifice to the free surface and is assumed to be held constant.

If we apply Bernoulli's equation from the free surface (level 1–1) to the center of the vena contracta (level 2–2), then

Fig. 4-24. Liquid-level system.

$$\frac{v_1^2}{2g} + \frac{p_1}{\gamma} + z_1 = \frac{v_2^2}{2g} + \frac{p_2}{\gamma} + z_2$$

Let us choose the atmospheric pressure as pressure datum and level 2–2 as elevation datum. By substituting $v_1 = 0$, $p_1 = 0$, $z_1 = H$, $p_2 = 0$, and $z_2 = 0$ into this last equation, we have

$$H = \frac{v_2^2}{2g}$$

or

$$v_2 = \sqrt{2gH}$$

As a result of liquid friction (due to viscosity) at the orifice, the actual velocity is 1 to 2% less than that given by this last equation. To account for the friction loss, we generally introduce the velocity coefficient C_v. The actual discharge Q from the orifice is the product of the actual velocity at the vena contracta and the area of the jet. In terms of the contraction coefficient C_c or

$$C_c = \frac{A_2}{A_0}$$

where A_0 is the orifice area and A_2 is the jet area, the actual discharge can be given as

$$Q = C_v C_c A_0 \sqrt{2gH} = c A_0 \sqrt{2gH} \tag{4-13}$$

where $c = C_v C_c$ is the discharge coefficient.

The standard orifice for measuring or regulating purposes is the sharp-edged orifice or thin-plate orifice. The value of the discharge coefficient for these orifices is about 0.61.

Resistance. *Resistance* of a physical element (whether mechanical, electrical, hydraulic, or pneumatic) can be defined as the change in potential required to make a unit change in current, flow rate, or velocity, or

$$\text{Resistance} = \frac{\text{change in potential}}{\text{change in current, flow rate, or velocity}}$$

For liquid flow in pipes, orifices, valves, or any other flow-restricting devices, the potential may correspond to either differential pressure (N/m²) (pressure difference between upstream and downstream of a flow-restricting device) or differential head (m), and the flow rate may be the volumetric liquid flow rate (m³/s). By applying the preceding general definition of resistance to the liquid flow, we have

$$\text{Resistance } R = \frac{\text{change in differential pressure}}{\text{change in flow rate}} \quad \frac{\text{N/m}^2}{\text{m}^3\text{/s}} \text{ or } \frac{\text{N-s}}{\text{m}^5}$$

or

$$\text{Resistance } R = \frac{\text{change in differential head}}{\text{change in flow rate}} \quad \frac{\text{m}}{\text{m}^3\text{/s}} \text{ or } \frac{\text{s}}{\text{m}^2}$$

Example 4-1. Consider the systems shown in Fig. 4-25(a) and (b). In part (a) the orifice in a connecting pipe restricts flow. Similarly, in part (b), the valve in a pipe also restricts flow. The dynamic properties of such systems do not depend on the physical construction of the device that causes this restriction. Consequently, these two systems can be treated similarly by defining resistance of the flow through an orifice or valve in a pipe.

Referring to Fig. 4-25(b) [or (a)], let us choose head as a measure of potential.

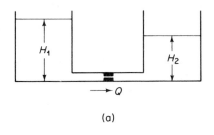

(a)

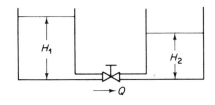

Fig. 4-25. Liquid-level systems.   (b)

Then the resistance can be defined as the change in the differential head required to cause a unit change in flow rate or

$$\text{Resistance } R = \frac{\text{change in differential head}}{\text{change in flow rate}} \quad \frac{\text{m}}{\text{m}^3/\text{s}}$$

$$= \frac{d(H_1 - H_2)}{dQ} \quad \text{s/m}^2$$

Liquid flow resistance depends on flow conditions—laminar flow and turbulent flow. Let us consider laminar flow resistance first.

For laminar flow, flow rate Q m³/s and differential head $(H_1 - H_2)$ m are proportional or

$$Q = K_l(H_1 - H_2)$$

where K_l is a proportionality constant. Therefore laminar flow resistance R_l can be given by

$$R_l = \frac{d(H_1 - H_2)}{dQ} = \frac{H_1 - H_2}{Q} = \frac{1}{K_l} \quad \text{s/m}^2$$

Note that laminar flow resistance is constant.

In considering laminar flow through a cylindrical pipe, the relationship between differential head $h \ (= H_1 - H_2)$ m and flow rate Q m³/s is given by the Hagen-Poiseuille formula

$$h = \frac{128\nu L}{g\pi D^4} Q$$

where

ν = kinematic viscosity, m²/s
L = length of pipe, m
D = diameter of pipe, m

So laminar flow resistance R_l for the liquid flow through cylindrical pipes is given by

$$R_l = \frac{dh}{dQ} = \frac{128\nu L}{g\pi D^4} \text{ s/m}^2$$

In practice, it should be noted that laminar flow through pipes seldom occurs in industrial processes.

Let us turn next to turbulent flow resistance R_t. For turbulent flow, referring to Eq. (4-12) or (4-13), the flow rate through the restriction can be given by

$$Q = K_t\sqrt{H_1 - H_2} \tag{4-14}$$

where K_t is a constant. Since Q and $(H_1 - H_2)$ are related by a nonlinear equation, turbulent flow resistance R_t is not constant. From Eq. (4-14) we have

$$\frac{d(H_1 - H_2)}{dQ} = \frac{2(H_1 - H_2)}{Q}$$

Thus turbulent flow resistance R_t is given by

$$R_t = \frac{d(H_1 - H_2)}{dQ} = \frac{2(H_1 - H_2)}{Q}$$

The fact that turbulent flow resistance R_t is not constant but depends on flow rate Q and differential head $(H_1 - H_2)$ means that we must define it at an operating condition (as to flow rate and differential head) and use this value of resistance in the neighborhood of the operating condition only.

<hr>

Capacitance. *Capacitance* of a physical element can be defined as the change in the quantity of material or distance required to make a unit change in potential or

$$\text{Capacitance} = \frac{\text{change in quantity of material or distance}}{\text{change in potential}}$$

For a liquid-filled tank system, the quantity of material may be the volume of liquid (m^3), and the potential may be either pressure (N/m^2) or head (m). If we apply the preceding general definition of capacitance to the liquid-filled tank system, the result is

$$\text{Capacitance } C = \frac{\text{change in quantity of liquid}}{\text{change in pressure}} \quad \frac{\text{m}^3}{\text{N/m}^2} \text{ or } \frac{\text{m}^5}{\text{N}}$$

or

$$\text{Capacitance } C = \frac{\text{change in quantity of liquid}}{\text{change in head}} \quad \frac{\text{m}^3}{\text{m}} \text{ or } \text{m}^2$$

In deriving mathematical models of liquid-level systems, it is convenient to choose head as a measure of potential, since with this choice the capacitance of a liquid-filled tank coincides with the cross-sectional area of the tank. If it is constant, the capacitance is constant for any head. It should be noted that capacitance (m^2) is different from capacity (m^3).

Inertance. The terms *inertance, inertia,* or *inductance* refer to the change in potential required to make a unit rate of change in flow rate, velocity, or current [change in flow rate per second, change in velocity per second (acceleration), or change in current per second], or

Inertance (inertia or inductance)

$$= \frac{\text{change in potential}}{\text{change in flow rate (velocity or current) per second}}$$

For the inertia effect of liquid flow in pipes, tubes, and similar devices, the potential may be either pressure (N/m^2) or head (m), and the change in flow rate per second may be the volumetric liquid flow acceleration (m^3/s^2). Application of the preceding general definition of inertance, inertia, or inductance to liquid flow gives

$$\text{Inertance } I = \frac{\text{change in pressure}}{\text{change in flow rate per second}} \quad \frac{N/m^2}{m^3/s^2} \text{ or } \frac{N\text{-}s^2}{m^5}$$

or

$$\text{Inertance } I = \frac{\text{change in head}}{\text{change in flow rate per second}} \quad \frac{m}{m^3/s^2} \text{ or } \frac{s^2}{m^2}$$

Example 4-2. Consider a liquid flow in a pipe. Liquid flow inertance is the potential difference (either pressure difference or head difference) between two sections in the pipe required to cause a unit rate of change in flow rate (a unit volumetric flow acceleration).

Suppose that the cross-sectional area of a pipe is constant and equal to A m^2 and that the pressure difference between two sections in the pipe is Δp N/m^2. Then the force $A \, \Delta p$ will accelerate the liquid between the two sections or

$$M \frac{dv}{dt} = A \, \Delta p$$

where M kg is the mass of liquid in the pipe between the two sections and v m/s is the velocity of liquid flow. Note that mass M is equal to $\rho A L$, where ρ kg/m^3 is the density, and L m is the distance between two sections considered. Therefore the last equation can be written

$$\rho A L \frac{dv}{dt} = A \, \Delta p$$

Noting that Av m^3/s is the volumetric flow rate and defining $Q = Av$ m^3/s, we can rewrite this last equation as

$$\frac{\rho L}{A} \frac{dQ}{dt} = \Delta p \tag{4-15}$$

If pressure (N/m^2) is chosen as a measure of potential, then liquid flow inertance I is obtained as

$$\text{Liquid flow inertance } I = \frac{\Delta p}{dQ/dt} = \frac{\rho L}{A} \quad \frac{N\text{-}s^2}{m^5}$$

If head (m) is chosen as a measure of potential, then noting that $\Delta p = \Delta h \rho g$, where Δh is the differential head, Eq. (4-15) becomes

$$\frac{\rho L}{A}\frac{dQ}{dt} = \Delta h \rho g$$

or

$$\frac{L}{Ag}\frac{dQ}{dt} = \Delta h$$

Consequently,

$$\text{Liquid flow inertance } I = \frac{\Delta h}{dQ/dt} = \frac{L}{Ag}\quad\frac{s^2}{m^2}$$

In order to illustrate the calculation of liquid flow inertance, consider water flow through a pipe whose cross-sectional area is constant and is 1×10^{-3} m² and where two sections are 15 m apart. Then

$$I = \frac{\rho L}{A} = \frac{1000 \times 15}{1 \times 10^{-3}}\frac{kg}{m^3}\frac{m}{m^2} = 1.5 \times 10^7 \text{ N-s}^2/\text{m}^5$$

or

$$I = \frac{L}{Ag} = \frac{15}{1 \times 10^{-3} \times 9.81}\frac{m}{m^2}\frac{s^2}{m} = 1529 \text{ s}^2/\text{m}^2$$

This means that if there is a differential head of 1 m between two sections that are 15 m apart, the volumetric water flow acceleration dQ/dt is

$$\frac{dQ}{dt} = \frac{\Delta h}{I} = \frac{\Delta h}{L/Ag} = \frac{1}{1529} = 0.000654 \text{ m}^3/\text{s}^2$$

Comments.

1. In deriving mathematical models of hydraulic systems in terms of resistance, capacitance, and inertance, these quantities should be expressed in terms of consistent units. For instance, if we choose pressure (N/m², kg_f/cm², lb_f/in.², etc.) or head (m, cm, in., etc.) as a measure of potential, the same unit of measure of potential must be used to express resistance, capacitance, and inertance. Similar comments apply to liquid flow rate (m³/s, cm³/s, in.³/s, etc.) and quantity of liquid (m³, cm³, in.³, etc.). As long as we use consistent units, the mathematical model remains the same.

2. Liquid capacitance and liquid flow inertance store energy as a result of the pressure and flow, respectively, and liquid flow resistance dissipates energy.

3. Inertia elements in mechanical systems and inductance elements in electrical systems are important elements in describing system dynamics. However, in deriving mathematical models of liquid-filled tanks connected by pipes with orifices, valves, and so on, only resistance and capacitance are important, and the effects of liquid flow inertance may be negligible. Such

liquid flow inertance becomes important only in special cases. For instance, it plays a dominant role in vibration transmitted through water, such as water hammer that results from the inertia effects of water flow in pipes and the elastic or capacitance effects of water flow in pipes. Note that this vibration or wave propagation results from inertance-capacitance effects of hydraulic circuits—comparable to free vibration in a mechanical spring-mass system or free oscillation in an electrical LC circuit.

Mathematical modeling of liquid-level systems. Turning to the liquid-level system shown in Fig. 4-26(a), let us obtain a mathematical model of it. If the operating condition as to the head and flow rate varies little for the time

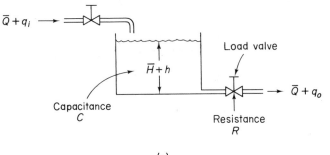

(a)

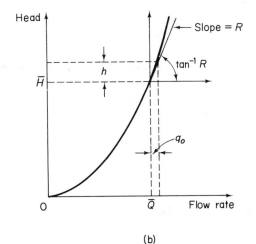

(b)

Fig. 4-26. (a) Liquid-level system; (b) head versus flow rate curve.

period considered, a mathematical model can easily be found in terms of resistance and capacitance. In the present analysis, we assume that the liquid outflow from the valve is turbulent.

Let us define

$\bar{H}$ = steady-state head (before any change has occurred), m
h = small deviation of head from its steady-state value, m
$\bar{Q}$ = steady-state flow rate (before any change has occurred), m³/s
q_i = small deviation of inflow rate from its steady-state value, m³/s
q_o = small deviation of outflow rate from its steady-state value, m³/s

The change in the liquid stored in the tank during dt seconds is equal to the net inflow to the tank during the same dt seconds, and so

$$C\,dh = (q_i - q_o)\,dt \qquad (4\text{-}16)$$

where C is the capacitance of the tank.

Resistance R of liquid flow through a valve is, by definition,

$$R = \frac{dH}{dQ}$$

where, for turbulent flow, Q is related to H by

$$Q = K\sqrt{H}$$

Since flow rate Q is proportional to the square root of head H, the value of resistance R is not constant. In practical situations, although the exact equation relating head and flow rate may not be known, an experimental curve relating head and flow rate may be available. Consider the head versus flow rate curve shown in Fig. 4-26(b), which may be either experimental or theoretical. Resistance R at the operating point ($H = \bar{H}, Q = \bar{Q}$) is equal to the slope of the curve at that point which is equal to $2\bar{H}/\bar{Q}$. (Clearly, as the operating point moves, the value of resistance R changes.)

Note that if the operating condition varies little, that is, if the changes in head and flow rate are small during the period of operation considered, then the value of resistance R may be considered constant during the entire period of operation and the system can be linearized using an average resistance value.

In the present system we defined h and q_o as small deviations from steady-state head and steady-state outflow rate, respectively. Thus

$$dH = h, \qquad dQ = q_o$$

And average resistance R may be written as

$$R = \frac{dH}{dQ} = \frac{h}{q_o}$$

By substituting $q_o = h/R$ into Eq. (4-16), we obtain

$$C\frac{dh}{dt} = q_i - \frac{h}{R}$$

or

$$RC\frac{dh}{dt} + h = Rq_i \tag{4-17}$$

Note that RC has the dimension of time and is the time constant of the system. Equation (4-17) is a linearized mathematical model for the system when h is considered the system output. Such a linearized mathematical model is valid, provided that changes in the head and flow rate from their respective steady-state values are small.

If q_o (the change in the outflow rate) rather than h (the change in head) is considered the system output, then another mathematical model may be obtained. Substituting $h = Rq_o$ into Eq. (4-17) gives

$$RC\frac{dq_o}{dt} + q_o = q_i \tag{4-18}$$

which is also a linearized mathematical model for the system.

Notice that this liquid-level system is analogous to the electrical system shown in Fig. 4-27. A mathematical model for the latter is

$$RC\frac{de_o}{dt} + e_o = e_i \tag{4-19}$$

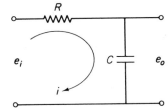

Fig. 4-27. Electrical system analogous to the liquid-level system shown in Fig. 4-26(a).

Comparing Eqs. (4-18) and (4-19), we see that they are of the same form and •
are thus analogous.

Example 4-3. Referring to the liquid-level system of Fig. 4-26(a), assume that the tank is a circular one with radius 1.7 m and that the steady-state operating condition corresponds to

$$\bar{H} = 2 \text{ m}, \qquad \bar{Q} = 0.5 \text{ m}^3/\text{min}$$

When the inflow rate is changed from 0.5 m³/min to 0.6 m³/min (or $q_i = 0.1$ m³/min), what is the change h in head as a function of time?

Since $\bar{Q}$ and $\bar{H}$ are related by

$$\bar{Q} = K\sqrt{\bar{H}}$$

the coefficient K is determined from

$$0.5 = K\sqrt{2}$$

as

$$K = 0.3536$$

The new steady-state head $\bar{H} + h(\infty)$ due to the change in the inflow rate can be found from

$$\bar{Q} + q_i = K\sqrt{\bar{H} + h(\infty)}$$

as

$$\bar{H} + h(\infty) = \left(\frac{\bar{Q} + q_i}{K}\right)^2 = \left(\frac{0.6}{0.3536}\right)^2 = 2.88$$

So average resistance R for the transient period is

$$R = \frac{d\bar{H}}{d\bar{Q}} = \frac{[\bar{H} + h(\infty)] - \bar{H}}{(\bar{Q} + q_i) - \bar{Q}} = \frac{2.88 - 2}{0.6 - 0.5} = 8.8 \text{ min/m}^2$$

Capacitance C is the same as the surface area of the tank, 9.08 m². The mathematical model of the system defined by Eq. (4-17) now becomes

$$8.8 \times 9.08 \frac{dh}{dt} + h = 8.8 \times 0.1$$

or

$$79.9 \frac{dh}{dt} + h = 0.88 \qquad\qquad (4\text{-}20)$$

Note that the initial condition is $h(0) = 0$.
    Let us define $x = h - 0.88$. Then Eq. (4-20) becomes

$$79.9 \frac{dx}{dt} + x = 0, \qquad x(0) = -0.88$$

By assuming the solution $x(t)$ as

$$x(t) = Ae^{\lambda t}$$

we obtain

$$79.9A\lambda e^{\lambda t} + Ae^{\lambda t} = 0$$

The characteristic equation is thus

$$79.9\lambda + 1 = 0$$

from which

$$\lambda = -\frac{1}{79.9}$$

Also, from the initial condition

$$x(0) = A = -0.88$$

and it follows that

$$x(t) = -0.88e^{-(1/79.9)t}$$

Consequently, $h(t)$ can be obtained as

$$h(t) = x(t) + 0.88 = 0.88[1 - e^{-(1/79.9)t}]$$

This equation gives the change in head as a function of time. As $t \longrightarrow \infty$, $h(t)$ approaches 0.88 m. (The total head $\bar{H} + h$ will approach 2.88 m.)

*4-6 LINEARIZATION OF NONLINEAR SYSTEMS

In this section we present a linearization technique that is applicable to many nonlinear systems. The process of linearizing nonlinear systems is important, for by linearizing nonlinear equations, it is possible to apply numerous linear analysis methods that will produce information on the behavior of nonlinear systems. The linearization procedure presented here is based on the expansion of the nonlinear function into a Taylor series about the operating point and the retention of only the linear term. Because we neglect higher-order terms of Taylor series expansion, these neglected terms must be small enough—that is, the variables deviate only slightly from the operating condition.

Linearization of $z = f(x)$ about a point $(\bar{x}, \bar{z})$. Consider a nonlinear system whose input is x and output is z. Thus the relationship between z and x may be written

$$z = f(x) \tag{4-21}$$

If the normal operating condition corresponds to a point $(\bar{x}, \bar{z})$, then Eq. (4-21) can be expanded into a Taylor series about this point as follows:

$$z = f(x) = f(\bar{x}) + \frac{df}{dx}(x - \bar{x}) + \frac{1}{2!}\frac{d^2f}{dx^2}(x - \bar{x})^2 + \cdots \tag{4-22}$$

where the derivatives df/dx, d^2f/dx^2, ... are evaluated at the operating point, $x = \bar{x}$, $z = \bar{z}$. If the variation $x - \bar{x}$ is small, we can neglect the higher-order terms in $x - \bar{x}$. Noting that $\bar{z} = f(\bar{x})$, Eq. (4-22) can be written

$$z - \bar{z} = a(x - \bar{x}) \tag{4-23}$$

where

$$a = \frac{df}{dx}\bigg|_{x=\bar{x}, z=\bar{z}}$$

---

*Starred sections deal with more challenging topics than the rest of the book. Depending on course objectives, these sections (although important) may be omitted from classroom discussions without losing the continuity of the main subject matter.

Equation (4-23) indicates that $z - \bar{z}$ is proportional to $x - \bar{x}$. It is a linear mathematical model for the nonlinear system given by Eq. (4-21) near the operating point $x = \bar{x}$, $z = \bar{z}$.

Linearization of $z = f(x, y)$ about a point $(\bar{x}, \bar{y}, \bar{z})$. Next, consider a nonlinear system whose output z is a function of two inputs x and y so that

$$z = f(x, y) \tag{4-24}$$

In order to obtain a linear mathematical model for this nonlinear system about an operating point $(\bar{x}, \bar{y}, \bar{z})$, we expand Eq. (4-24) into a Taylor series about this point. Then Eq. (4-24) becomes

$$z = f(\bar{x}, \bar{y}) + \left[\frac{\partial f}{\partial x}(x - \bar{x}) + \frac{\partial f}{\partial y}(y - \bar{y})\right]$$
$$+ \frac{1}{2!}\left[\frac{\partial^2 f}{\partial x^2}(x - \bar{x})^2 + 2\frac{\partial^2 f}{\partial x\,\partial y}(x - \bar{x})(y - \bar{y}) + \frac{\partial^2 f}{\partial y^2}(y - \bar{y})^2\right] + \cdots$$

where the partial derivatives are evaluated at the operating point, $x = \bar{x}$, $y = \bar{y}$, $z = \bar{z}$. Near this point, the higher-order terms may be neglected. Noting that $\bar{z} = f(\bar{x}, \bar{y})$, a linear mathematical model of this nonlinear system near the operating point $x = \bar{x}$, $y = \bar{y}$, $z = \bar{z}$ is

$$z - \bar{z} = a(x - \bar{x}) + b(y - \bar{y})$$

where

$$a = \frac{\partial f}{\partial x}\bigg|_{x=\bar{x}, y=\bar{y}, z=\bar{z}}$$

$$b = \frac{\partial f}{\partial y}\bigg|_{x=\bar{x}, y=\bar{y}, z=\bar{z}}$$

It is important to remember that in the present linearization procedure the deviations of the variables from the operating condition must be sufficiently small. Otherwise this procedure does not apply.

Linearization of valve characteristics. Figure 4-28(a) shows a hydraulic servo consisting of a four-way spool valve and a power cylinder and piston. We shall apply the linearization technique just presented to obtain a linearized mathematical model of the four-way spool valve. The valve, which is assumed underlapped and symmetrical, admits hydraulic fluid under high pressure into a power cylinder that contains a large piston, so that a large hydraulic force is established to move a load. We assume that the load inertia and friction are small compared to the large hydraulic force. In the present analysis, the hydraulic fluid is assumed to be incompressible and the inertia force of the power piston negligible. We also assume that, as is usually the case, the orifice area (the width of the slot in the valve sleeve) at each port is proportional to the valve displacement x.

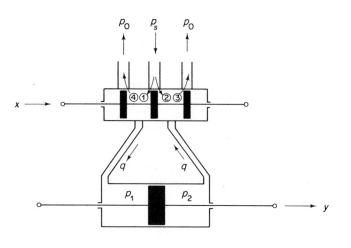

(a)

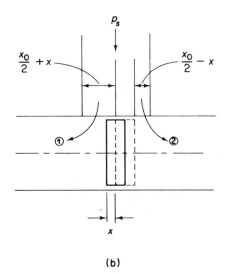

(b)

Fig. 4-28. (a) Hydraulic servo system; (b) enlarged diagram of the valve orifice area.

In Fig. 4-28(b) we have an enlarged diagram of the valve orifice area. Let us define the valve orifice areas of ports 1, 2, 3, 4, as A_1, A_2, A_3, A_4, respectively and also define the flow rates through ports 1, 2, 3, 4, as q_1, q_2, q_3, q_4, respectively. Note that since the valve is symmetrical, $A_1 = A_3$ and $A_2 = A_4$. Assuming the displacement x to be small, we obtain

$$A_1 = A_3 = k\left(\frac{x_0}{2} + x\right)$$

$$A_2 = A_4 = k\left(\frac{x_0}{2} - x\right)$$

where k is a constant.

Furthermore, we shall assume that the return pressure p_0 in the return line is small and thus can be neglected. Then referring to Fig. 4-28(a), flow rates through valve orifices are

$$q_1 = c_1 A_1 \sqrt{\frac{2g}{\gamma}(p_s - p_1)} = C_1\sqrt{p_s - p_1}\left(\frac{x_0}{2} + x\right)$$

$$q_2 = c_2 A_2 \sqrt{\frac{2g}{\gamma}(p_s - p_2)} = C_2\sqrt{p_s - p_2}\left(\frac{x_0}{2} - x\right)$$

$$q_3 = c_1 A_3 \sqrt{\frac{2g}{\gamma}(p_2 - p_0)} = C_1\sqrt{p_2 - p_0}\left(\frac{x_0}{2} + x\right) = C_1\sqrt{p_2}\left(\frac{x_0}{2} + x\right)$$

$$q_4 = c_2 A_4 \sqrt{\frac{2g}{\gamma}(p_1 - p_0)} = C_2\sqrt{p_1 - p_0}\left(\frac{x_0}{2} - x\right) = C_2\sqrt{p_1}\left(\frac{x_0}{2} - x\right)$$

where $C_1 = c_1 k\sqrt{2g/\gamma}$ and $C_2 = c_2 k\sqrt{2g/\gamma}$. Then the flow rate q to the left-hand side of the power piston is

$$q = q_1 - q_4 = C_1\sqrt{p_s - p_1}\left(\frac{x_0}{2} + x\right) - C_2\sqrt{p_1}\left(\frac{x_0}{2} - x\right) \qquad (4\text{-}25)$$

The flow rate from the right-hand side of the power piston to the drain is the same as this q and is given by

$$q = q_3 - q_2 = C_1\sqrt{p_2}\left(\frac{x_0}{2} + x\right) - C_2\sqrt{p_s - p_2}\left(\frac{x_0}{2} - x\right)$$

Note that the fluid is incompressible and that the valve is symmetrical. So we have $q_1 = q_3$ and $q_2 = q_4$. By equating q_1 and q_3, we obtain

$$p_s - p_1 = p_2$$

or

$$p_s = p_1 + p_2$$

If we define the pressure difference across the power piston as Δp or

$$\Delta p = p_1 - p_2$$

then

$$p_1 = \frac{p_s + \Delta p}{2}, \qquad p_2 = \frac{p_s - \Delta p}{2}$$

For the symmetrical valve shown in Fig. 4-28(a), the pressure in each side of the power piston is $\frac{1}{2}p_s$ when no load is applied, or $\Delta p = 0$. As the spool valve is displaced, the pressure in one line increases as the pressure in the other line descreases by the same amount.

In terms of p_s and Δp, we can rewrite the flow rate q given by Eq. (4-25) as

$$q = q_1 - q_4 = C_1\sqrt{\frac{p_s - \Delta p}{2}}\left(\frac{x_0}{2} + x\right) - C_2\sqrt{\frac{p_s + \Delta p}{2}}\left(\frac{x_0}{2} - x\right)$$

Noting that the supply pressure p_s is constant, the flow rate q can be written as a function of the valve displacement x and pressure difference Δp or

$$q = C_1\sqrt{\frac{p_s - \Delta p}{2}}\left(\frac{x_0}{2} + x\right) - C_2\sqrt{\frac{p_s + \Delta p}{2}}\left(\frac{x_0}{2} - x\right) = f(x, \Delta p)$$

By applying the linearization technique presented earlier in this section to this case, the linearized equation about point $x = \bar{x}$, $\Delta p = \Delta\bar{p}$, $q = \bar{q}$ is

$$q - \bar{q} = a(x - \bar{x}) + b(\Delta p - \Delta\bar{p}) \tag{4-26}$$

where

$$\bar{q} = f(\bar{x}, \Delta\bar{p})$$

$$a = \frac{\partial f}{\partial x}\bigg|_{x=\bar{x}, \Delta p=\Delta\bar{p}, q=\bar{q}} = C_1\sqrt{\frac{p_s - \Delta\bar{p}}{2}} + C_2\sqrt{\frac{p_s + \Delta\bar{p}}{2}}$$

$$b = \frac{\partial f}{\partial \Delta p}\bigg|_{x=\bar{x}, \Delta p=\Delta\bar{p}, q=\bar{q}} = -\left[\frac{C_1}{2\sqrt{2}\sqrt{p_s - \Delta\bar{p}}}\left(\frac{x_0}{2} + \bar{x}\right)\right.$$
$$\left. + \frac{C_2}{2\sqrt{2}\sqrt{p_s + \Delta\bar{p}}}\left(\frac{x_0}{2} - \bar{x}\right)\right] < 0$$

Coefficients a and b here are called *valve coefficients*. Equation (4-26) is a linearized mathematical model of the four-way spool valve near an operating point $x = \bar{x}$, $\Delta p = \Delta\bar{p}$, $q = \bar{q}$. The values of valve coefficients a and b vary with the operating point. Note that $\partial f/\partial \Delta p$ is negative and so b is negative.

Since the normal operating point is the point where $\bar{x} = 0$, $\Delta\bar{p} = 0$, $\bar{q} = 0$, near the normal operating point Eq. (4-26) becomes

$$q = K_1 x - K_2 \Delta p \tag{4-27}$$

where

$$K_1 = (C_1 + C_2)\sqrt{\frac{p_s}{2}} > 0$$

$$K_2 = (C_1 + C_2)\frac{x_0}{4\sqrt{2}\sqrt{p_s}} > 0$$

(Note that $C_1 = C_2$ when $x = 0$, $\Delta p = 0$.) Equation (4-27) is a linearized mathematical model of the four-way spool valve near the origin ($\bar{x} = 0$,

$\Delta \bar{p} = 0$, $\bar{q} = 0$). Note that the region near the origin is most important because system operation usually occurs near this point. Such a linearized mathematical model is useful in analyzing the performance of hydraulic control valves.

Concluding comments for the chapter. In this chapter we have briefly presented the basic materials of hydraulic systems and mathematical modeling techniques for such systems. Additional and more detailed discussions of hydraulic control valves are given in Chapter 8, where control systems and various types of automatic controllers are treated.

REFERENCES

4-1   BARNA, P. S., *Fluid Mechanics for Engineers*, 2nd ed., London, England: Butterworth & Company, Ltd., 1964.

4-2   BAYLEY, F. J., *An Introduction to Fluid Dynamics*, London, England: George Allen & Unwin, Ltd., 1958.

4-3   HOHMANN, C. J., "Pumps for Fluid Power, Part 2: for Aircraft," *Mechanical Engineering*, **90**, No. 10, October 1968, pp. 38–41.

4-4   MERRITT, H. E., *Hydraulic Control Systems*, New York: John Wiley & Sons, Inc., 1967.

4-5   MURRAY, J. F., "Pumps for Fluid Power, Part 3: for Extreme Environments," *Mechanical Engineering*, **90**, No. 11, November 1968, pp. 43–47.

4-6   OGATA, K., *Modern Control Engineering*, Englewood Cliffs, N.J.: Prentice-Hall, Inc., 1970.

4-7   STREETER, V. L., AND E. B. WYLIE, *Fluid Mechanics*, 6th ed., New York: McGraw-Hill Book Company, Inc., 1975.

4-8   THOMAS, G. M., AND R. W. HENKE, "Pumps for Fluid Power, Part 1: Basic Briefing," *Mechanical Engineering*, **90**, No. 9, September 1968, pp. 41–46.

EXAMPLE PROBLEMS AND SOLUTIONS

PROBLEM A-4-1. Water is compressed in a cylinder. If the volume of water is 1×10^{-3} m³ (1000 cm³) at pressure 1.7×10^5 N/m² abs (170 kPa abs), what is the volume of water when pressure of 8×10^5 N/m² abs (800 kPa abs) is applied? Assume that the bulk modulus K of water is 2.1×10^9 N/m².

Solution. Since bulk modulus K is given by

$$K = \frac{dp}{-dV/V}$$

by substituting the given numerical values, we obtain

$$2.1 \times 10^9 = \frac{(8 - 1.7) \times 10^5}{-dV/(1 \times 10^{-3})} = \frac{6.3 \times 10^2}{-dV}$$

or

$$-dV = \frac{6.3 \times 10^2}{2.1 \times 10^9} = 3 \times 10^{-7}$$

Hence the volume of water at pressure 8×10^5 N/m² abs is

$$(1 \times 10^{-3} - 3 \times 10^{-7}) \text{ m}^3 = 999.7 \times 10^{-6} \text{ m}^3 = 999.7 \text{ cm}^3$$

PROBLEM A-4-2. If the dynamic viscosity of a petroleum-based oil is 8 cP and the specific gravity is 0.83, determine the dynamic viscosity μ in SI and BES units. Also, determine the kinematic viscosity v in SI and BES units.

Solution. Since μ is 8 cP, $\mu = 0.08$ g/cm-s. Then

$$\mu = 0.008 \frac{\text{N-s}}{\text{m}^2} = 0.008 \frac{\text{kg}}{\text{m-s}} \qquad \text{(SI unit)}$$

$$= 0.000167 \frac{\text{lb}_f\text{-s}}{\text{ft}^2} = 0.000167 \frac{\text{slug}}{\text{ft-s}} \qquad \text{(BES unit)}$$

The kinematic viscosity v is obtained from $v = \mu/\rho$. Since $\rho = 830$ kg/m³ $= 1.610$ slug/ft³, we have

$$v = \frac{0.008}{830} = 9.64 \times 10^{-6} \frac{\text{m}^2}{\text{s}} \qquad \text{(SI unit)}$$

$$= \frac{0.000167}{1.610} = 1.037 \times 10^{-4} \frac{\text{ft}^2}{\text{s}} \qquad \text{(BES unit)}$$

PROBLEM A-4-3. Consider the rolling motion of the ship shown in Fig. 4-29. The force due to buoyancy is $-w$ and that due to gravity is w. These two forces produce a couple that causes rolling motion of the ship. The point where the vertical line through the center of buoyancy C' intersects the symmetrical line through the

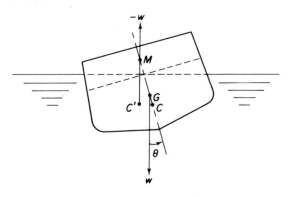

Fig. 4-29. Rolling motion of a ship.

center of gravity, which is in the ship's centerline plane, is called the *metacenter*. The metacenter is shown as point M. Define

R = distance of the metacenter to the center of gravity of the ship = $\overline{MG}$

J = moment of inertia of the ship about its longitudinal centroidal axis

Derive the equation of rolling motion of the ship when rolling angle θ is small.

Solution. From Fig. 4-29 we obtain

$$J\ddot{\theta} = -wR \sin \theta$$

or

$$J\ddot{\theta} + wR \sin \theta = 0$$

For small θ, we have $\sin \theta \doteqdot \theta$. Hence the equation of rolling motion of the ship is

$$J\ddot{\theta} + wR\theta = 0$$

The natural frequency of the rolling motion is $\sqrt{wR/J}$. Note that the distance $R (= \overline{MG})$ is considered positive when the couple of weight and buoyancy tends to rotate the ship toward the upright position. That is, R is positive if point M is above point G and negative if point M is below point G.

PROBLEM A-4-4. Assuming that the hydraulic power unit shown in Fig. 4-30 is used as a pump and that the left side of the piston is at the atmospheric pressure, show that

$$F_p v_p = pQ_p$$

where F_p is the force applied to the piston, v_p the velocity of the piston, p the gage pressure of the fluid in the discharge chamber, and Q_p the discharge rate.

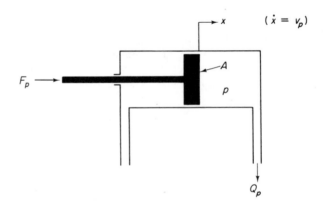

Fig. 4-30. Hydraulic power unit.

Solution. Define the piston area as A. Then the pressure p developed in the discharge chamber is

$$p = \frac{F_p}{A}$$

The discharge rate Q_p is

$$Q_p = Av_p$$

So

$$F_p v_p = pQ_p$$

Thus in the hydraulic pump the mechanical power $F_p v_p$ is transformed into the hydraulic power pQ_p in the absence of friction loss.

PROBLEM A-4-5. Actual spool valves are either overlapped or underlapped because of manufacturing tolerances. Consider the overlapped and underlapped spool valves shown in Fig. 4-31(a) and (b). Sketch curves relating the uncovered port area A versus displacement x.

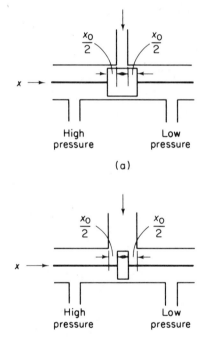

Fig. 4-31. (a) Overlapped spool valve; (b) underlapped spool valve.

Solution. For the overlapped valve, a dead zone exists between $-\frac{1}{2}x_0$ and $\frac{1}{2}x_0$, or $-\frac{1}{2}x_0 < x < \frac{1}{2}x_0$. The uncovered port area A versus displacement x curve is shown in Fig. 4-32(a). Such an overlapped valve is unfit as a control valve.

For the underlapped valve, the port area A versus displacement x curve is shown in Fig. 4-32(b). The effective curve for the underlapped region has a higher slope, meaning a higher sensitivity. Valves used for controls are usually underlapped.

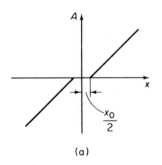

(a)

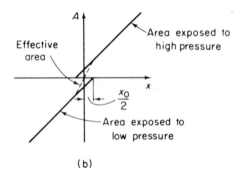

Effective
area

Area exposed to
high pressure

$\frac{x_0}{2}$

Area exposed to
low pressure

(b)

Fig. 4-32. (a) Uncovered port area A versus displacement x curve for the overlapped valve; (b) uncovered port area A versus displacement x curve for the underlapped valve.

PROBLEM A-4-6. In Fig. 4-33 the Venturi meter is used to determine flow rate in a horizontal pipe. Assume that water is flowing. Assume also that the diameter at section 1 is 0.15 m and that at section 2 is 0.1 m. Find the flow rate Q m³/s through the pipe when $p_1 - p_2 = 0.1373 \times 10^5$ N/m² ($= 13.73$ kPa).

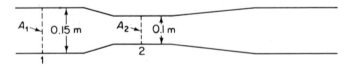

A_1 0.15 m A_2 0.1 m

1   2

Fig. 4-33. Venturi meter.

Solution. From Bernoulli's equation, Eq. (4-6),

$$\frac{v_1^2}{2g} + \frac{p_1}{\gamma} + z_1 = \frac{v_2^2}{2g} + \frac{p_2}{\gamma} + z_2$$

Since $z_1 = z_2$, we have

$$\frac{p_1 - p_2}{\gamma} = \frac{v_2^2}{2g} - \frac{v_1^2}{2g} \qquad (4\text{-}28)$$

From the continuity equation

$$Q = A_1v_1 = A_2v_2$$

where A_1 and A_2 are cross-sectional areas at sections 1 and 2, respectively.
From the problem statement

$$A_1 = \left(\frac{0.15}{2}\right)^2 \pi \text{ m}^2, \qquad A_2 = \left(\frac{0.1}{2}\right)^2 \pi \text{ m}^2$$

Consequently,

$$v_2 = \frac{A_1}{A_2}v_1 = 2.25v_1$$

From Eq. (4-28)

$$\frac{p_1 - p_2}{\gamma} = \frac{(2.25)^2v_1^2}{2g} - \frac{v_1^2}{2g} = 4.0625\frac{v_1^2}{2g}$$

or

$$v_1^2 = \frac{2g(p_1 - p_2)}{\gamma \times 4.0625} = \frac{2(p_1 - p_2)}{\rho \times 4.0625}$$

So noting that $\rho = \gamma/g = 1000 \text{ kg/m}^3$, we have

$$Q = A_1v_1 = \left(\frac{0.15}{2}\right)^2 \pi \sqrt{\frac{2(p_1 - p_2)}{\rho \times 4.0625}} = 1.767 \times 10^{-2} \sqrt{\frac{2 \times 0.1373 \times 10^5}{10^3 \times 4.0625}}$$

$$= 0.04594 \text{ m}^3/\text{s}$$

PROBLEM A-4-7. Consider the liquid-level system shown in Fig. 4-34. The flow rate Q through the orifice is equal to $cA_0\sqrt{2gH} = K\sqrt{H}$, where A_0 is the orifice area, c is the discharge coefficient, g is the gravitational acceleration constant, H is the head above the center of the orifice, and $K = cA_0\sqrt{2g}$. The capacitance of the tank is constant and is equal to C. Assume that at $t = 0$ the head is H_0. Find time t to lower the head from H_0 to H_1 ($0 < H_1 < H_0$), both heads measured from the center of the orifice.

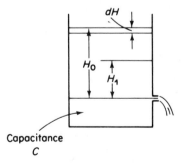

Fig. 4-34. Liquid-level system.

Solution. Assume that the flow rate Q is measured in cubic meters per second, the capacitance in square meters, and the head in meters. Then the liquid discharged from the orifice in dt seconds is $Q\,dt$, which is equal to the reduction in volume in the tank for the same dt seconds. Then

$$Q \, dt = -C \, dH$$

And so

$$dt = -\frac{C}{Q} \, dH = -\frac{C}{K\sqrt{H}} \, dH$$

Let us assume that $H = H_1$ at $t = t_1$. It follows that

$$t_1 = \int_0^{t_1} dt = \int_{H_0}^{H_1} \frac{-C}{K\sqrt{H}} \, dH = -\frac{C}{K} \int_{H_0}^{H_1} \frac{dH}{\sqrt{H}}$$

$$= -\frac{C}{K} 2\sqrt{H} \Big|_{H_0}^{H_1} = \frac{2C}{K}(\sqrt{H_0} - \sqrt{H_1})$$

Thus the time necessary to lower the head from H_0 meters to H_1 meters is $(2C/K)$ $\cdot(\sqrt{H_0} - \sqrt{H_1})$ seconds.

PROBLEM A-4-8. In the liquid-level system of Fig. 4-35 assume that the outflow rate Q m³/s through the outflow valve is related to the head H m by

$$Q = K\sqrt{H} = 0.01\sqrt{H}$$

Assume also that when the inflow rate Q_i is 0.015 m³/s, the head stays constant. At $t = 0$ the inflow valve is closed and so there is no inflow for $t \geq 0$. Find the time necessary to empty the tank to half the original head. The capacitance of the tank is 2 m².

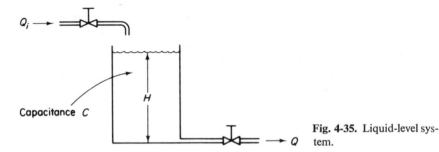

Fig. 4-35. Liquid-level system.

Solution. When the head is stationary, the inflow rate equals the outflow rate. Thus head H_0 at $t = 0$ is obtained from

$$0.015 = 0.01\sqrt{H_0}$$

or

$$H_0 = 2.25 \text{ m}$$

The equation for the system for $t > 0$ is

$$-C \, dH = Q \, dt$$

or

$$\frac{dH}{dt} = -\frac{Q}{C} = \frac{-0.01\sqrt{H}}{2}$$

Consequently,

$$\frac{dH}{\sqrt{H}} = -0.005 \, dt$$

Assume that $H = 1.125$ m at $t = t_1$. Integrating both sides of this last equation, we have

$$\int_{2.25}^{1.125} \frac{dH}{\sqrt{H}} = \int_0^{t_1} (-0.005)dt = -0.005t_1$$

So it follows that

$$2\sqrt{H}\,\Big|_{2.25}^{1.125} = 2\sqrt{1.125} - 2\sqrt{2.25} = -0.005t_1$$

or

$$t_1 = 176 \text{ s}$$

PROBLEM A-4-9. Assume that the liquid-level system shown in Fig. 4-35 is at steady state with inflow rate equal to $\bar{Q}$ m³/s and head equal to $\bar{H}$ m. At $t = 0$ the inflow rate is changed from $\bar{Q}$ to $\bar{Q} + 0.001$ m³/s. After a sufficient time has elapsed, a steady state is reached with a new head equal to $\bar{H} + 0.05$ m.

Assuming that the cross-sectional area of the tank is 2 m², determine the average resistance of the outflow valve. What is the time constant of the system?

Solution. Average resistance R of the outflow valve is given by

$$R = \frac{dH}{dQ} = \frac{0.05}{0.001} = 50 \text{ s/m}^2$$

Capacitance C of the tank is

$$C = 2 \text{ m}^2$$

Therefore the time constant T of the system is

$$T = RC = 50 \times 2 = 100 \text{ s}$$

PROBLEM A-4-10. Consider the flow of water through a capillary tube shown in Fig. 4-36. Assuming that the temperature of water is 20°C and that the flow is laminar, obtain resistance R of the capillary tube.

Solution. From the Hagen-Poiseuille formula we have

$$h = \frac{128vL}{g\pi D^4} Q$$

So resistance R is obtained as

$$R = \frac{dh}{dQ} = \frac{128vL}{g\pi D^4} \tag{4-29}$$

Noting that the kinematic viscosity v of water at temperature 20°C is 1.004×10^{-6} m²/s, we obtain, by substituting numerical values into Eq. (4-29),

$$R = \frac{128 \times 1.004 \times 10^{-6} \times 1}{9.81 \times 3.14 \times (3 \times 10^{-3})^4} = 5.15 \times 10^4 \text{ s/m}^2$$

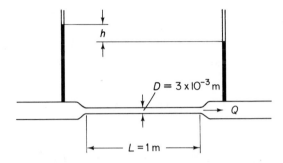

Fig. 4-36. Flow of water through a capillary tube.

PROBLEM A-4-11. Consider the liquid-level system of Fig. 4-37(a). The head versus flow rate curve is shown in Fig. 4-37(b). Assume that at steady state the liquid flow rate is 4×10^{-4} m³/s and the steady-state head is 1 m. At $t = 0$ the inflow valve is opened farther and the inflow rate is changed to 4.5×10^{-4} m³/s. Determine the average resistance R of the outflow valve. Also, determine the change in head as a function of time. The capacitance C of the tank is 0.02 m².

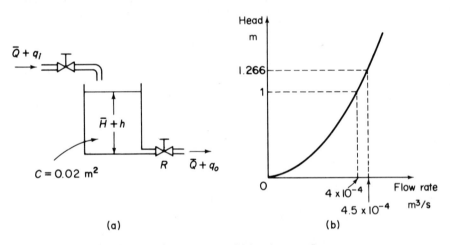

Fig. 4-37. (a) Liquid-level system; (b) head versus flow rate curve.

Solution. The flow rate through the outflow valve can be assumed as

$$Q = K\sqrt{H}$$

Next, from the curve given in Fig. 4-37(b) we see that

$$4 \times 10^{-4} = K\sqrt{1}$$

or

$$K = 4 \times 10^{-4}$$

So if the steady-state flow rate is changed to $4.5 \times 10^{-4}\,\text{m}^3/\text{s}$, then the new steady-state head can be obtained from

$$4.5 \times 10^{-4} = 4 \times 10^{-4}\,\sqrt{H}$$

or

$$H = 1.266\,\text{m}$$

This means that the change in head is $1.266 - 1 = 0.266\,\text{m}$. The average resistance R of the outflow valve is then

$$R = \frac{dH}{dQ} = \frac{1.266 - 1}{(4.5 - 4) \times 10^{-4}} = 0.532 \times 10^4\,\text{s/m}^2$$

Noting that the change in the liquid stored in the tank during dt seconds is equal to the net inflow to the tank during the same dt seconds, we have

$$C\,dh = (q_i - q_o)\,dt$$

where q_i and q_o are the changes in the inflow rate and outflow rate of the tank, respectively, and h is the change in the head. Thus

$$C\frac{dh}{dt} = q_i - q_o$$

Since

$$R = \frac{h}{q_o}$$

it follows that

$$C\frac{dh}{dt} = q_i - \frac{h}{R}$$

or

$$RC\frac{dh}{dt} + h = Rq_i$$

Substituting $R = 0.532 \times 10^4\,\text{s/m}^2$, $C = 0.02\,\text{m}^2$, and $q_i = 0.5 \times 10^{-4}\,\text{m}^3/\text{s}$ into this last equation yields

$$0.532 \times 10^4 \times 0.02\,\frac{dh}{dt} + h = 0.532 \times 10^4 \times 0.5 \times 10^{-4}$$

or

$$106.4\,\frac{dh}{dt} + h = 0.266$$

Finally, solving for h,

$$h(t) = 0.266(1 - e^{-t/106.4})\,\text{m}$$

This last equation gives the change in head as a function of time.

Problem A-4-12. For the liquid-level system shown in Fig. 4-38, the steady-state flow rate through the tanks is $\bar{Q}$ and steady-state heads of tank 1 and tank 2 are $\bar{H}_1$ and $\bar{H}_2$, respectively. At $t = 0$ the inflow rate is changed from $\bar{Q}$ to $\bar{Q} + q$, where q is a small change in the inflow rate. The corresponding changes in the heads

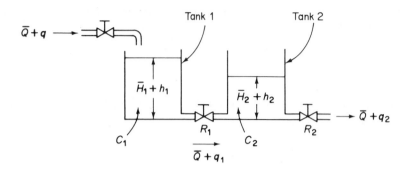

Fig. 4-38. Liquid-level system.

(h_1 and h_2) and changes in flow rates (q_1 and q_2) are assumed to be small. The capacitances of tank 1 and tank 2 are C_1 and C_2, respectively. The resistance of the valve between the tanks is R_1 and that of the outflow valve is R_2.

Assuming q as the input and q_2 as the output, derive a mathematical model (differential equation) for the system.

Solution. For tank 1, we have

$$C_1 \, dh_1 = (q - q_1) \, dt$$

where

$$q_1 = \frac{h_1 - h_2}{R_1}$$

Hence

$$C_1 \frac{dh_1}{dt} + \frac{h_1}{R_1} = q + \frac{h_2}{R_1} \tag{4-30}$$

For tank 2, we get

$$C_2 \, dh_2 = (q_1 - q_2) \, dt$$

where

$$q_2 = \frac{h_2}{R_2}$$

Therefore

$$C_2 \frac{dh_2}{dt} + \frac{h_2}{R_1} + \frac{h_2}{R_2} = \frac{h_1}{R_1} \tag{4-31}$$

By eliminating h_1 from Eqs. (4-30) and (4-31), the result is

$$R_1 C_1 R_2 C_2 \frac{d^2 h_2}{dt^2} + (R_1 C_1 + R_2 C_2 + R_2 C_1) \frac{dh_2}{dt} + h_2 = R_2 q$$

Noting that $h_2 = R_2 q_2$, we obtain

$$R_1 C_1 R_2 C_2 \frac{d^2 q_2}{dt^2} + (R_1 C_1 + R_2 C_2 + R_2 C_1) \frac{dq_2}{dt} + q_2 = q$$

This is the desired mathematical model or differential equation relating q_2 and q.

PROBLEM A-4-13. Referring to the liquid-level system of Fig. 4-39, the inflow rate and outflow rate at steady state are $\bar{Q}$, the flow rate between the tanks is zero, and the heads of tank 1 and tank 2 are both $\bar{H}$. At $t = 0$ the inflow rate is changed from $\bar{Q}$ to $\bar{Q} + q$, where q is a small change in the inflow rate. The resulting changes in the heads (h_1 and h_2) and flow rates (q_1 and q_2) are assumed to be small. The capacitances of tank 1 and tank 2 are C_1 and C_2, respectively. The resistance of the valve between the tanks is R_1 and that of the outflow valve is R_2.

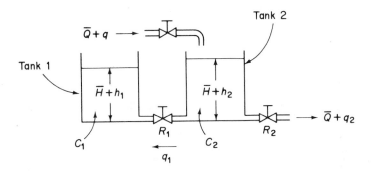

Fig. 4-39. Liquid-level system.

Derive mathematical models for the system when (a) q is the input and h_2 the output, (b) q is the input and q_2 the output, and (c) q is the input and h_1 the output.

Solution. For tank 1, we have

$$C_1 \, dh_1 = q_1 \, dt$$

where

$$q_1 = \frac{h_2 - h_1}{R_1}$$

Consequently,

$$R_1 C_1 \frac{dh_1}{dt} + h_1 = h_2 \tag{4-32}$$

For tank 2, we get

$$C_2 \, dh_2 = (q - q_1 - q_2) \, dt$$

where

$$q_1 = \frac{h_2 - h_1}{R_1}, \qquad q_2 = \frac{h_2}{R_2}$$

It follows that

$$R_2 C_2 \frac{dh_2}{dt} + \frac{R_2}{R_1} h_2 + h_2 = R_2 q + \frac{R_2}{R_1} h_1 \tag{4-33}$$

By eliminating h_1 from Eqs. (4-32) and (4-33), we have

$$R_1 C_1 R_2 C_2 \frac{d^2 h_2}{dt^2} + (R_1 C_1 + R_2 C_2 + R_2 C_1) \frac{dh_2}{dt} + h_2 = R_1 R_2 C_1 \frac{dq}{dt} + R_2 q \tag{4-34}$$

This is a desired mathematical model in which q is considered the input and h_2 the output.

Then the substitution of $h_2 = R_2 q_2$ into Eq. (4-34) gives

$$R_1 C_1 R_2 C_2 \frac{d^2 q_2}{dt^2} + (R_1 C_1 + R_2 C_2 + R_2 C_1) \frac{dq_2}{dt} + q_2 = R_1 C_1 \frac{dq}{dt} + q$$

This last equation is also a desired mathematical model in which q is considered the input and q_2 the output.

Finally, elimination of h_2 from Eqs. (4-32) and (4-33) yields

$$R_1 C_1 R_2 C_2 \frac{d^2 h_1}{dt^2} + (R_1 C_1 + R_2 C_2 + R_2 C_1) \frac{dh_1}{dt} + h_1 = R_2 q$$

which is a mathematical model of the system in which q is considered the input and h_1 the output.

PROBLEM A-4-14. The flow rate Q and head H in the liquid-level system of Fig. 4-35 are related by

$$Q = K\sqrt{H}$$

Assume that at steady state the head is $\bar{H}$ and the flow rate is $Q_i = Q = \bar{Q}$. Find a linearized mathematical model relating the flow rate to head near steady-state point $H = \bar{H}$, $Q = \bar{Q}$.

Solution. We shall demonstrate two approaches for obtaining a linearized mathematical model relating the flow rate and head near steady-state point $H = \bar{H}$, $Q = \bar{Q}$.

The first is to find resistance R of the outflow valve. Since resistance R is given by

$$R = \frac{dH}{dQ} = \frac{2H}{Q}$$

near steady-state point $H = \bar{H}$, $Q = \bar{Q}$,

$$\frac{dH}{dQ} = \frac{H - \bar{H}}{Q - \bar{Q}} = R = \frac{2\bar{H}}{\bar{Q}}$$

Hence

$$Q - \bar{Q} = \frac{1}{R}(H - \bar{H})$$

or

$$Q - \bar{Q} = \frac{\bar{Q}}{2\bar{H}}(H - \bar{H}) \qquad (4\text{-}35)$$

This equation is a linearized mathematical model relating flow rate Q and head H near steady-state point $H = \bar{H}$, $Q = \bar{Q}$.

The second approach is to expand the nonlinear equation

$$Q = K\sqrt{H} = f(H)$$

into a Taylor series about point $H = \bar{H}$, $Q = \bar{Q}$.

$$Q = f(H) = f(\bar{H}) + \frac{df}{dH}(H - \bar{H}) + \frac{1}{2!}\frac{d^2 f}{dH^2}(H - \bar{H})^2 + \cdots$$

By neglecting higher-order terms in $H - \bar{H}$, we have

$$Q - \bar{Q} = a(H - \bar{H})$$

where

$$\bar{Q} = f(\bar{H}) = K\sqrt{\bar{H}}$$

$$a = \frac{df}{dH}\bigg|_{H=\bar{H},\,Q=\bar{Q}} = \frac{K}{2\sqrt{H}}\bigg|_{H=\bar{H},\,Q=\bar{Q}} = \frac{K}{2\sqrt{\bar{H}}} = \frac{\bar{Q}}{2\bar{H}}$$

It follows that

$$Q - \bar{Q} = \frac{\bar{Q}}{2\bar{H}}(H - \bar{H})$$

This equation is identical to Eq. (4-35) and is a linearized mathematical model relating flow rate Q and head H near steady-state point $H = \bar{H}$, $Q = \bar{Q}$.

*Problem A-4-15. Find a linearized equation for

$$z = 0.4x^3 = f(x)$$

about a point $\bar{x} = 2$, $\bar{z} = 3.2$.

Solution. The Taylor series expansion of $f(x)$ about point $(2, 3.2)$ is

$$z - \bar{z} = a(x - \bar{x})$$

where

$$a = \frac{df}{dx}\bigg|_{x=2,\,z=3.2} = 1.2x^2\bigg|_{x=2,\,z=3.2} = 4.8$$

So the linear approximation of the given nonlinear equation is

$$z - 3.2 = 4.8(x - 2) \tag{4.36}$$

Figure 4-40 depicts a nonlinear curve $z = 0.4x^3$ and the linear equation given by Eq. (4-36). Note that the straight-line approximation of the cubic curve is valid near point $(2, 3.2)$.

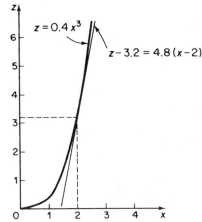

Fig. 4-40. Nonlinear curve $z = 0.4x^3$ and its linear approximation at point $\bar{x} = 2$ and $\bar{z} = 3.2$.

*PROBLEM A-4-16. Linearize the nonlinear equation

$$z = xy$$

in the region $5 \leq x \leq 7$, $10 \leq y \leq 12$. Find the error if the linearized equation is used to calculate the value of z when $x = 5$, $y = 10$.

Solution. Since the region considered is given by $5 \leq x \leq 7$, $10 \leq y \leq 12$, choose $\bar{x} = 6$, $\bar{y} = 11$. Then $\bar{z} = \bar{x}\bar{y} = 66$. Let us obtain a linearized equation for the nonlinear equation near a point $\bar{x} = 6$, $\bar{y} = 11$, $\bar{z} = 66$.

Expanding the nonlinear equation into a Taylor series about point $x = \bar{x}$, $y = \bar{y}$, $z = \bar{z}$ and neglecting the higher-order terms, we have

$$z - \bar{z} = a(x - \bar{x}) + b(y - \bar{y})$$

where

$$a = \frac{\partial(xy)}{\partial x}\bigg|_{x=\bar{x}, y=\bar{y}, z=\bar{z}} = \bar{y} = 11$$

$$b = \frac{\partial(xy)}{\partial y}\bigg|_{x=\bar{x}, y=\bar{y}, z=\bar{z}} = \bar{x} = 6$$

Hence the linearized equation is

$$z - 66 = 11(x - 6) + 6(y - 11)$$

or

$$z = 11x + 6y - 66$$

When $x = 5$, $y = 10$, the value of z given by the linearized equation is

$$z = 11x + 6y - 66 = 55 + 60 - 66 = 49$$

The exact value of z is $z = xy = 50$. The error is thus $50 - 49 = 1$. In terms of percentage, the error is 2%.

*PROBLEM A-4-17. Figure 4-41 shows a hydraulic servo consisting of a spool valve and a power cylinder and piston. Assume that the spool valve is symmetrical and

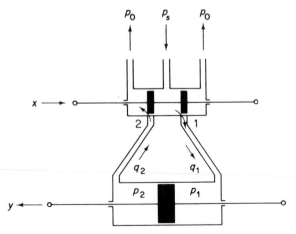

Fig. 4-41. Hydraulic servo system.

has no overlapping, that the valve orifice areas are proportional to the valve displacement x, and that the orifice coefficient and the pressure drop across the orifice are constant and independent of valve position. In addition, assume the following: the supply pressure is p_s, the return pressure p_0 in return line is small and can be neglected, the hydraulic fluid is incompressible, the inertia force of the power piston and load reactive forces are negligible compared with the hydraulic force developed by the power piston, and the leakage flow around the spool valve from the supply pressure side to the return pressure side is negligible.

Derive a linearized mathematical model of the spool valve near the origin.

Solution. Let us define the valve orifice areas of port 1 and port 2 as A_1 and A_2, respectively. Then $A_1 = A_2 = kx$, where k is a constant. Referring to Fig. 4-41, the flow rates through valve orifices are

$$q_1 = cA_1 \sqrt{\frac{2g}{\gamma}(p_s - p_1)} = C\sqrt{p_s - p_1}\, x$$

$$q_2 = cA_2 \sqrt{\frac{2g}{\gamma}(p_2 - p_0)} = C\sqrt{p_2 - p_0}\, x = C\sqrt{p_2}\, x$$

where $C = ck\sqrt{2g/\gamma}$. Since we assumed no leakage flow around the spool valve from the supply pressure side to the return pressure side, these two equations are the only flow rate equations that concern us in the present analysis.

Noting that $q_1 = q_2$, we have

$$p_s - p_1 = p_2$$

Let us define the pressure difference across the power piston as Δp or

$$\Delta p = p_1 - p_2$$

Then p_1 and p_2 can be written

$$p_1 = \frac{p_s + \Delta p}{2}, \qquad p_2 = \frac{p_s - \Delta p}{2}$$

The flow rate q_1 to the right-hand side of the power piston is

$$q_1 = C\sqrt{p_s - p_1}\, x = C\sqrt{\frac{p_s - \Delta p}{2}}\, x = f(x, \Delta p)$$

The linearized equation near an operating point $x = \bar{x}$, $\Delta p = \Delta \bar{p}$, $q_1 = \bar{q}_1$ is

$$q_1 - \bar{q}_1 = a(x - \bar{x}) + b(\Delta p - \Delta \bar{p}) \tag{4-37}$$

where

$$a = \frac{\partial f}{\partial x}\bigg|_{x = \bar{x},\, \Delta p = \Delta \bar{p},\, q_1 = \bar{q}_1} = C\sqrt{\frac{p_s - \Delta \bar{p}}{2}}$$

$$b = \frac{\partial f}{\partial \Delta p}\bigg|_{x = \bar{x},\, \Delta p = \Delta \bar{p},\, q_1 = \bar{q}_1} = -\frac{C}{2\sqrt{2}\sqrt{p_s - \Delta \bar{p}}}\, \bar{x} \leq 0$$

Near the origin ($\bar{x} = 0$, $\Delta \bar{p} = 0$, $\bar{q}_1 = 0$) Eq. (4-37) becomes

$$q_1 = K_1 x - K_2 \Delta p$$

where

$$K_1 = C\sqrt{\frac{p_s - \Delta\bar{p}}{2}}\bigg|_{\bar{x}=0,\,\Delta\bar{p}=0,\,\bar{q}_1=0} = C\sqrt{\frac{p_s}{2}}$$

$$K_2 = \frac{C}{2\sqrt{2}\sqrt{p_s - \Delta\bar{p}}}\,\bar{x}\bigg|_{\bar{x}=0,\,\Delta\bar{p}=0,\,\bar{q}_1=0} = 0$$

Hence

$$q_1 = K_1 x$$

which is a linear mathematical model near the origin of the spool valve shown in Fig. 4-41.

*PROBLEM A-4-18. Consider again the hydraulic servo shown in Fig. 4-41 and assume that the input to the servo is the displacement x of the spool valve and that the output is the displacement y of the power piston. The positive directions of x and y are indicated in the diagram. Assuming the hydraulic fluid as incompressible and the inertia force of the power piston and the load reactive forces as negligible compared with the hydraulic force developed by the power piston, derive a mathematical model of the system relating displacements x and y when x is small.

Solution. Since the hydraulic fluid is incompressible, we have

$$A\rho \, dy = q_1 \, dt$$

where A (m²) is the power piston area, ρ (kg/m³) is the mass density of the fluid, dy (m) is the displacement of the power piston during dt (s), and q_1 (kg/s) is the flow rate of fluid to the right-hand side of the power piston. This last equation can be written

$$A\rho \frac{dy}{dt} = q_1 \tag{4-38}$$

For small x, a linear mathematical model of the spool valve was obtained in Problem A-4-17 as

$$q_1 = K_1 x \tag{4-39}$$

So by eliminating q_1 from Eqs. (4-38) and (4-39), we have

$$\frac{dy}{dt} = \frac{K_1}{A\rho} x = Kx$$

where $K = K_1/(A\rho)$. If we integrate both sides of this last equation, the result is

$$y = K \int x \, dt$$

which is a mathematical model of the system relating displacements x and y. Note that the output displacement y is proportional to the integral of the valve displacement x.

*PROBLEM A-4-19. The hydraulic servo system of Fig. 4-42 consists of a spool valve, a power cylinder and piston, and load elements (mass, viscous friction, and

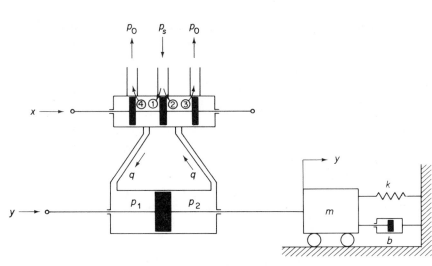

Fig. 4-42. Hydraulic servo system.

spring elements). Assuming the inertia force of the power piston as negligible and load reactive forces also as negligible, derive a mathematical model for the system. Assume also that the spool valve is symmetrical and that the valve orifice areas are proportional to the valve displacement x.

Solution. If the inertia force of the power piston and the load reactive forces are negligible, it can be assumed that the orifice coefficient and the pressure drop across the orifice are constant and independent of valve position.

Since valve orifice areas are assumed to be proportional to the valve displacement, the flow rate q (kg/s) may be written

$$q = K_1 x$$

where x (m) is the displacement of the valve and K_1 (kg/m-s) is a constant.

For the power piston

$$A\rho \, dy = q \, dt$$

where A (m²) is the power piston area and ρ (kg/m³) is the density of oil. Hence

$$\frac{dy}{dt} = \frac{q}{A\rho} = \frac{K_1}{A\rho} x = Kx$$

where $K = K_1/(A\rho)$. Integrating both sides of this last equation yields

$$y = K \int x \, dt$$

Thus the power piston displacement y is proportional to the integral of the valve displacement x. The dynamic characteristics of the servo system shown in Fig. 4-42 are the same as those of the servo system in Fig. 4-41.

It is important to point out that the present analysis applies only when the load reactive forces and power piston inertia force are negligible.

*PROBLEM A-4-20.** Referring again to Fig. 4-42 and assuming that the load reactive forces are not negligible, derive a mathematical model. Assume also that the mass of the power piston is included in the load mass m.

Solution. In deriving a mathematical model of the system when the load reactive forces are not negligible, such effects as the pressure drop across the orifice, the leakage of oil around the valve and around the piston, and the compressibility of the oil must be considered.

The pressure drop across the orifice is a function of the supply pressure p_s and the pressure difference $\Delta p = p_1 - p_2$. Thus the flow rate q is a nonlinear function of valve displacement x and pressure difference Δp or

$$q = f(x, \Delta p)$$

Linearizing this nonlinear equation about the origin ($x = 0, \Delta p = 0, q = 0$), we obtain, referring to Eq. (4-27),

$$q = K_1 x - K_2 \Delta p \qquad (4\text{-}40)$$

The flow rate q can be considered as consisting of three parts

$$q = q_0 + q_L + q_C \qquad (4\text{-}41)$$

where

$q_0 =$ useful flow rate to the power cylinder causing
   power piston to move, kg/s
$q_L =$ leakage flow rate, kg/s
$q_C =$ equivalent compressibility flow rate, kg/s

Let us obtain specific expressions for q_0, q_L, and q_C. The flow $q_0\,dt$ to the left-hand side of the power piston causes the piston to move to the right by dy. So we have

$$A\rho\,dy = q_0\,dt$$

where A (m²) is the power piston area, ρ (kg/m³) the density of oil, and dy (m) the displacement of the power piston. Then

$$q_0 = A\rho \frac{dy}{dt} \qquad (4\text{-}42)$$

The leakage component q_L can be written

$$q_L = L\,\Delta p \qquad (4\text{-}43)$$

where L is the leakage coefficient of the system.

The equivalent compressibility flow rate q_C can be expressed in terms of the effective bulk modulus K of oil (including the effects of entrapped air, expansion of pipes, etc.), where

$$K = \frac{d\,\Delta p}{-dV/V}$$

(Here dV is negative and so $-dV$ is positive.) Rewriting this last equation gives

$$-dV = \frac{V}{K} d\,\Delta p$$

or

$$\rho \frac{-dV}{dt} = \frac{\rho V}{K} \frac{d\,\Delta p}{dt}$$

Noting that $q_c = \rho(-dV)/dt$, we find

$$q_c = \frac{\rho V}{K} \frac{d\,\Delta p}{dt} \tag{4-44}$$

where V is the effective volume of oil under compression (that is, approximately half the total power cylinder volume).

Using Eqs. (4-40) through (4-44),

$$q = K_1 x - K_2\,\Delta p = A\rho \frac{dy}{dt} + L\,\Delta p + \frac{\rho V}{K} \frac{d\,\Delta p}{dt}$$

or

$$A\rho \frac{dy}{dt} + \frac{\rho V}{K} \frac{d\,\Delta p}{dt} + (L + K_2)\,\Delta p = K_1 x \tag{4-45}$$

The force developed by the power piston is $A\,\Delta p$, and this force is applied to the load elements. Thus

$$m \frac{d^2 y}{dt^2} + b \frac{dy}{dt} + ky = A\,\Delta p \tag{4-46}$$

Eliminating Δp from Eqs. (4-45) and (4-46) results in

$$\frac{\rho V m}{KA} \frac{d^3 y}{dt^3} + \left[\frac{\rho V b}{KA} + \frac{(L + K_2)m}{A} \right] \frac{d^2 y}{dt^2}$$

$$+ \left[A\rho + \frac{\rho V k}{KA} + \frac{(L + K_2)b}{A} \right] \frac{dy}{dt} + \frac{(L + K_2)k}{A} y = K_1 x$$

This is a mathematical model of the system relating the valve spool displacement x and the power piston displacement y when the load reactive forces are not negligible.

PROBLEMS

PROBLEM B-4-1. A liquid is compressed in a cylinder. If the volume of the liquid is 2×10^{-3} m³ at pressure 1×10^6 N/m² abs (1 MPa abs) and 1.9995×10^{-3} m³ at pressure 1.5×10^6 N/m² abs (1.5 MPa abs), find the bulk modulus of elasticity.

PROBLEM B-4-2. Pascal's law states that pressure at any point in a static liquid is the same in every direction and exerts equal force on equal areas. Referring to Fig. 4-43, if a force of P_1 is applied to the left-side piston, find the force P_2 acting on the right-side piston. Also, find the distance x_2 traveled by the piston on the right when the one on the left is moved by x_1.

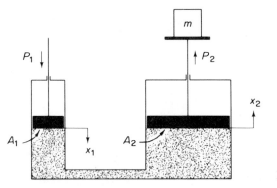

Fig. 4-43. Hydraulic system.

PROBLEM B-4-3. Figure 4-44 shows an accumulator that uses a spring. Obtain the maximum energy to be stored by this accumulator. Assume that pressure ranges from $p_{\min}$ to $p_{\max}$ as shown in the diagram and that the spring displacement is $x_{\max}$ ($x_{\min}$) when pressure p (gage pressure) is $p_{\max}$ ($p_{\min}$).

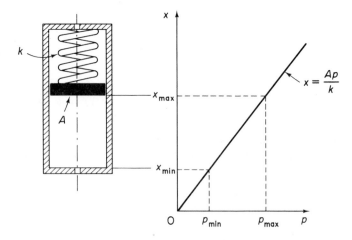

Fig. 4-44. Accumulator and its displacement versus pressure curve.

PROBLEM B-4-4. In Fig. 4-45 a water reservoir is connected by a long pipeline to a hydraulic generator system. The valve at the end of the pipeline is controlled by a turbine governor and may rapidly stop the water flow if the generator loses its load. Explain the role of the surge tank in such a system.

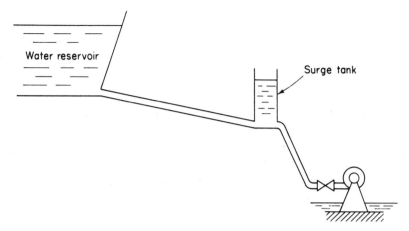

Fig. 4-45. Water reservoir and hydraulic generator system.

PROBLEM B-4-5. Consider the hydraulic power unit shown in Fig. 4-46. When it is used as a motor, show that

$$pQ_m = F_m v_m$$

where p is the gage pressure of the supply fluid, Q_m the flow rate to the cylinder, v_m the velocity of the piston, and F_m the force applied to the load. Assume that the left side of the piston is at the atmospheric pressure.

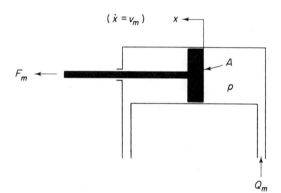

Fig. 4-46. Hydraulic power unit.

PROBLEM B-4-6. Referring to the Venturi meter shown in Fig. 4-33, find the flow rate Q through the pipe when $p_1 - p_2 = 1 \times 10^4$ N/m² (10 kPa). Assume that oil of mass density 820 kg/m³ is flowing.

PROBLEM B-4-7. For the liquid-level system shown in Fig. 4-47, assume that at $t = 0$ the head H is 5 m above the orifice. Find the velocity of the flow through the orifice at $t = 0$. If the flow rate at $t = 0$ is 0.04 m³/s, how long will it take to lower the head to 3 m above the orifice? Assume the capacitance of the tank to be 20 m².

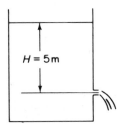

Fig. 4-47. Liquid-level system.

PROBLEM B-4-8. Consider a liquid-level system where the tank has a cross-sectional area of 4 m² at the level of the orifice. The cross-sectional area varies linearly with the level so that it is 2 m² at the level of 5 m above the center of the orifice. Assume that the orifice flow rate is

$$Q = cA_0\sqrt{2gH} = K\sqrt{H}$$

where $c = 0.62$, $A_0 = 0.01$ m², $g = 9.81$ m/s², H is the level above the orifice in m, and $K = cA_0\sqrt{2g}$. Find the time in seconds needed to lower the level from 5 m to 3 m above the orifice.

PROBLEM B-4-9. In the system shown in Fig. 4-48 the head is kept at 1 m for $t \leq 0$. The inflow valve opening is changed at $t = 0$ and the inflow rate is 0.05 m³/s for $t \geq 0$. Determine the time needed to fill the tank to a 2.5 m level. Assume that the outflow rate Q m³/s and head H m are related by

$$Q = 0.02\sqrt{H}$$

The capacitance of the tank is 2 m².

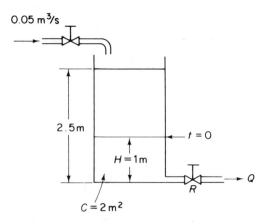

Fig. 4-48. Liquid-level system.

PROBLEM B-4-10. Referring to Fig. 4-49, assume that the outflow valve has been closed for $t < 0$ and that the heads of both tanks are equal, or $H_1 = H_2$. At $t = 0$ the outflow valve is opened. Assuming that the flows through the valves are laminar, derive a mathematical model relating the head of tank 2 and time t.

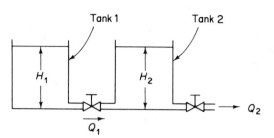

Fig. 4-49. Liquid-level system.

PROBLEM B-4-11. Derive a mechanical analog of the liquid-level system shown in Fig. 4-26(a) and given by Eq. (4-18).

PROBLEM B-4-12. At steady state the flow rate throughout the liquid-level system shown in Fig. 4-50 is $\bar{Q}$ and the heads of tank 1 and tank 2 are $\bar{H}_1$ and $\bar{H}_2$, respectively. At $t = 0$ the inflow rate is changed from $\bar{Q}$ to $\bar{Q} + q$, where q is a small change in the inflow rate. The resulting changes in the heads (h_1 and h_2) and flow rates (q_1 and q_2) are assumed to be small. The capacitances of tank 1 and tank 2 are C_1 and C_2, respectively. The resistance of the outflow valve of tank 1 is R_1 and that of tank 2 is R_2. Obtain a mathematical model of the system when q is the input and q_2 the output.

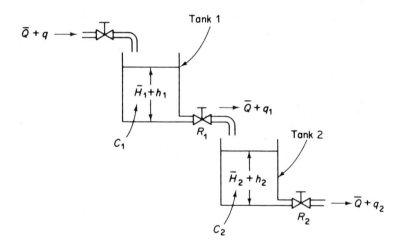

Fig. 4-50. Liquid-level system.

PROBLEM B-4-13. Find an electrical analog of the liquid-level system shown in Fig. 4-50.

PROBLEM B-4-14. Derive an electrical analog of the liquid-level system shown in Fig. 4-38 when q is the input and q_2 the output.

PROBLEM B-4-15. Find a mechanical analog of the liquid-level system shown in Fig. 4-38 when q is the input and q_2 the output.

PROBLEM B-4-16. Consider the hydraulic system shown in Fig. 4-51. Assuming that the piston displacement x is the input and the cylinder displacement y is the output, derive a mathematical model for the system.

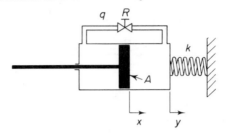

Fig. 4-51. Hydraulic system.

***PROBLEM B-4-17.** Obtain a linear approximation of

$$Q = 0.1\sqrt{H} = f(H)$$

about an operating point $H = 4$, $Q = 0.2$.

***PROBLEM B-4-18.** Find a linearized equation of

$$z = 5x^2$$

about a point $x = 2$, $z = 20$.

***PROBLEM B-4-19.** Linearize the nonlinear equation

$$z = \frac{x}{y}$$

in the region defined by $90 \leq x \leq 110$, $45 \leq y \leq 55$.

***PROBLEM B-4-20.** Linearize the nonlinear equation

$$z = x^2 + 2xy + 5y^2$$

in the region defined by $10 \leq x \leq 12$, $4 \leq y \leq 6$.

5

PNEUMATIC SYSTEMS

5-1 INTRODUCTION

Pneumatic systems are fluid systems that use air as the medium for transmitting signals and power. (Although the most common fluid in these systems is air, other gases can be used as well.)

Pneumatic systems are extensively used in the automation of production machinery and in the field of automatic controllers. For instance, pneumatic circuits that convert the energy of compressed air into mechanical energy enjoy wide usage, and various types of pneumatic controllers are found in industry. In addition, since early 1960s pneumatic devices called fluidic devices have been applied as decision elements or logic circuits in automatic warehousing, sequencing, and similar operations. And because pneumatic systems are so widely found in industry, engineers should be as familiar with the basic principles of pneumatic components and systems as with those of hydraulic systems.

Figures 5-1 to 5-3 illustrate three examples of air utilization. In Fig. 5-1 we have a schematic diagram of an air lift pump; Fig. 5-2 shows an air cushion on a wheel system and Fig. 5-3 a mechanical finger. (In Fig. 5-3 A and B are links and C is attached to the piston rod. When the piston rod moves upward, the mechanical finger catches a workpiece. When the piston rod moves downward, it will release the workpiece.)

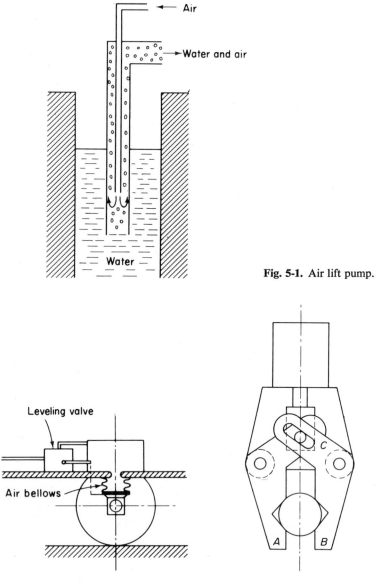

Fig. 5-1. Air lift pump.

Fig. 5-2. Air cushion.

Fig. 5-3. Mechanical finger.

The use of air in industries can be classified in the following way.

1. Oxygen in the air is utilized. (Combustion systems)
2. Relative flow of air is utilized. (Airplanes, parachutes, etc.)

3. Force due to wind is utilized. (Yachts, air lift pumps, etc.)
4. Energy of compressed air is utilized. (Air brakes, compressed-air tools, etc.)
5. Compressibility of air is utilized. (Air cushions)
6. Use of certain phenomena due to airflow. (Fluidics)

Comparison between pneumatic systems and hydraulic systems. As noted, the fluid generally found in pneumatic systems is air; in hydraulic systems it is oil. And it is primarily the different properties of the fluids involved that characterize the differences between the two systems. These differences can be listed as follows.

1. Air and gases are compressible, whereas oil is incompressible.
2. Air lacks lubricating property and always contains water vapor. Oil functions as a hydraulic fluid as well as a lubricator.
3. The normal operating pressure of penumatic systems is very much lower than that of hydraulic systems.
4. Output powers of pneumatic systems are considerably less than those of hydraulic systems.
5. Accuracy of penumatic actuators is poor at low velocities, whereas accuracy of hydraulic actuators may be made satisfactory at all velocities.
6. In pneumatic systems external leakage is permissible to a certain extent, but internal leakage must be avoided because the effective pressure difference is rather small. In hydraulic systems internal leakage is permissible to a certain extent, but external leakage must be avoided.
7. No return pipes are required in pneumatic systems when air is used, whereas they are always needed in hydraulic systems.
8. Normal operating temperature for pneumatic systems is 5 to 60°C. The pneumatic system, however, can be operated in the 0 to 200°C range. Pneumatic systems are insensitive to temperature changes, in contrast to hydraulic systems, where fluid friction due to viscosity depends greatly on temperature. Normal operating temperature for hydraulic systems is 20 to 70°C.

Outline of the chapter. Section 5-1 has presented a brief introduction to pneumatic systems. In Sec. 5-2 we discuss pneumatic components and systems, including pumps, actuators, and valves, followed by the physical and thermodynamic properties of air and other gases in Sec. 5-3. Section 5-4 describes the flow of gases through orifices, and Sec. 5-5 mathematical modeling of pneumatic systems. After considering introductory materials of fluidic devices in Sec. 5-6, we conclude the chapter with a description of digital fluidics and logic circuits in Sec. 5-7.

5-2   PNEUMATIC SYSTEMS

Pneumatic forces perform various functions—pushing, pulling, catching, for instance—as in pneumatic hoists, pneumatic tools, mechanical fingers, and similiar devices. In this section we present such pneumatic components as compressors that produce compressed air, pneumatic actuators that convert pneumatic energy into mechanical energy to perform useful mechanical work, and pneumatic valves that control pressure and/or flow. (Pneumatic devices, such as fluidic devices, are discussed in Sec. 5-6 and Sec. 5-7. Conventional pneumatic controllers are presented in Chapter 8.)

Figure 5-4 shows a functional diagram of a simple pneumatic circuit, the major components of which are a compressor, a filter, a lubricator, valves, and an actuator. In the following pages we describe each of these components briefly.

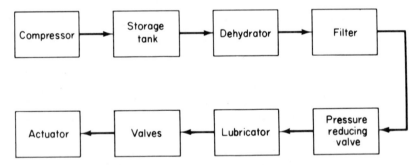

Fig. 5-4. Functional diagram of a simple pneumatic circuit.

Compressors. As the name implies, *compressors* are machines for compressing air or gas. They can be classified as to two types: positive displacement and centrifugal. The *positive displacement* type includes all machines that operate by taking a quantity of air or gas into a closed space where its volume is reduced and its pressure increased. Such compressors can be divided into reciprocating and rotary compressors. The second type, *centrifugal*, also includes axial compressors. Schematic diagrams of these compressors appear in Fig. 5-5.

Centrifugal compressors for pressure below 1×10^5 N/m² gage (0.1 MPa gage) are generally known as *blowers* or *fans*. When pressures are above 2×10^5 N/m² gage (0.2 MPa gage) in centrifugal compressors, the kinetic

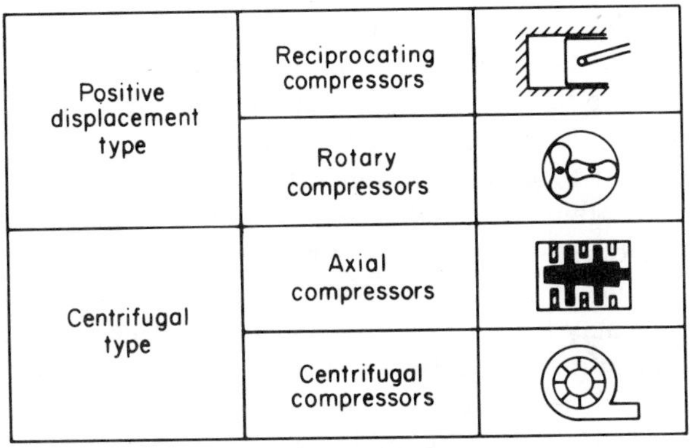

Positive displacement type	Reciprocating compressors	
	Rotary compressors	
Centrifugal type	Axial compressors	
	Centrifugal compressors	

Fig. 5-5. Compressors.

energy is normally recovered as pressure. In blowers or fans, however, the kinetic energy usually dissipates itself in eddies. Note that for pressure conversion from N/m^2 to kg_f/cm^2 or $lb_f/in.^2$,

$$1 \text{ MPa} = 10^6 \text{ N/m}^2 = 10.197 \text{ kg}_f/cm^2 = 145 \text{ lb}_f/in.^2 = 145 \text{ psi}$$

$$0 \text{ N/m}^2 \text{ gage} = 0 \text{ kg}_f/cm^2 \text{ gage} = 0 \text{ psig} = 1.0133 \times 10^5 \text{ N/m}^2 \text{ abs}$$

$$= 1.0332 \text{ kg}_f/cm^2 \text{ abs} = 14.70 \text{ lb}_f/in.^2 \text{ abs} = 14.7 \text{ psia}$$

The type of compressor known as reciprocating can produce high pressure. If pressures are from $5 \times 10^5 \text{ N/m}^2$ gage to $35 \times 10^5 \text{ N/m}^2$ gage (0.5 MPa gage to 3.5 MPa gage), two-stage compressors are used, and pressures up to $8 \times 10^6 \text{ N/m}^2$ (8 MPa) require three-stage compressors. When pressures range from $15 \times 10^6 \text{ N/m}^2$ to $35 \times 10^6 \text{ N/m}^2$ (15 MPa to 35 MPa) or even higher, then four-stage compressors are necessary. In order to obtain high pressures, air (or gas) must be cooled during its passage from stage to stage.

Because reciprocating compressors operate at constant speed, independent of the demand for compressed air, various types of unloaders have been used for economy. When the predetermined pressure is exceeded, the unloader prevents further compression of air until the pressure falls a predetermined amount, at which stage the compressor resumes compressing air again.

In the centrifugal compressor, large clearances exist between the rotor and the stationary parts. The only rubbing parts are the bearings. Since air and gases have low densities, the centrifugal compressors are run at high speed. In addition, they maintain a fairly constant pressure over a wide range of inlet volume. For every speed, however, there is a certain inlet volume below which the operation will become unstable. (At low load, a phenomenon known as *pressure surging* or *pulsations* may occur because of the compressibility of air or gas. In this situation, a slight adjustment of the operating condition may stop the surging.)

Pneumatic filters and lubricators. The greatest problem in pneumatic systems is the maintenance of a clean, dry air supply at constant pressure. Moisture, corrosive liquids, or foreign particles carried into the pneumatic system from the air supply will cause trouble. As air is compressed, temperature rises and relative humidity becomes low. When compressed air is cooled by a compressor aftercooler and relative humidity becomes high, moisture will condense in the storage tank. Most of the moisture in the air is removed as condensed water from the storage tank. Any remaining moisture and foreign particles may be removed by a *pneumatic filter*. To ensure that the pressure drop resulting from filtration is small, the capacity of the pneumatic filter should be sufficiently large.

For pneumatic controllers and fluidic devices, the air supply should be free from oil. In other applications, however, the supply of air should contain atomized oil to lubricate the pneumatic actuator. The *lubricator* is a devise that atomizes oil into the flow of air in order to lubricate the pneumatic actuator. A pneumatic filter, a pressure control valve, and a lubricator are usually combined into one unit, known as an *air pressure control unit*, as shown in Fig. 5-6.

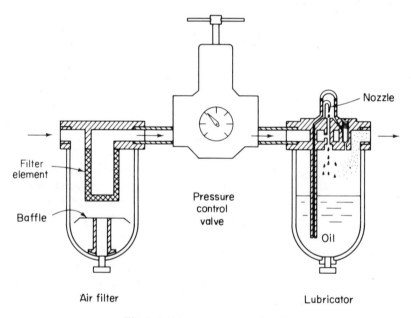

Fig. 5-6. Air pressure control unit.

Pneumatic actuators. *Pneumatic actuators,* which convert pneumatic energy into mechanical energy, can be divided into two types: the pneumatic cylinder (for linear motion) and the pneumatic motor (for continuous rotary motion). Most commonly used pneumatic actuators fall into the cylinder group. Various nonlinear motions (such as limited angle rotary motion) can be obtained by combining appropriate mechanisms with the linear motion of the cylinder-type actuator.

Pneumatic cylinders. *Pneumatic cylinders* can be classified as the piston type, the plunger type (ram), and the bellows type. Schematic diagrams of each are given in Fig. 5-7. The bellows-type cylinders have no rubbing parts, but they must be of short stroke with large diameter.

(a)

(b)

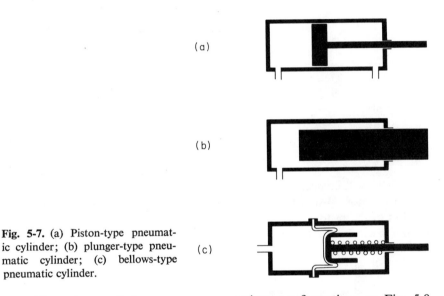

Fig. 5-7. (a) Piston-type pneumatic cylinder; (b) plunger-type pneumatic cylinder; (c) bellows-type pneumatic cylinder.

(c)

Piston-type cylinders can assume various configurations as Fig. 5-8 shows.

When using air as a medium for transmission of power, it is important to recognize the effect of compressibility on the performance of the system. Consider the pneumatic system shown in Fig. 5-9. When compressed air enters from port 1, the pressure in chamber A builds up until the pressure force exceeds the maximum static friction force that exists between the cylinder surface and the piston surface. This pressure buildup is rapid, since the volume of chamber A is small. As motion starts, the friction is rapidly reduced because the sliding friction is considerably less than the maximum static friction. Consequently, the piston will exhibit an impulsive motion and will hit the clamp almost instantaneously. Notice that immediately after the impulsive motion of the piston has started, the volume of chamber A increases quickly. This situation will cause a sudden pressure drop in the chamber because the air inflow to the chamber cannot catch up with the

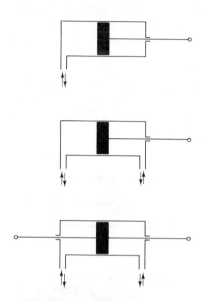

Fig. 5-8. Piston-type cylinders.

increase in the volume of the chamber. (In some cases, the clamping force may prove insufficient until the compressed air is filled to the increased volume of chamber *A*.) The compressed air supplied to chamber *A* once the piston has hit the clamp will do less useful work than it could, since the air will soon be purged into atmosphere as the valve is switched and the piston is moved to the left. (Such an energy loss lowers the efficiency of the pneumatic system.) As a result of this situation, a precise speed control of the piston is impossible.

In some applications, if the required stroke is very short, a speed control valve to control piston speed will be powerless because a short stroke will end before the speed becomes uniform. Then using a long cylinder and reducing the stroke by means of a link mechanism (Fig. 5-10) will be necessary.

For most piston-type cylinders, the accurate control of low-speed motion of the piston is difficult. In order to obtain such control, pneumatic cylinders can be combined with hydraulic cylinders as in Fig. 5-11. (In each

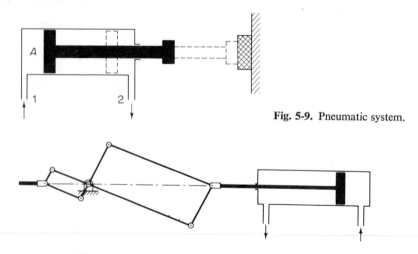

Fig. 5-9. Pneumatic system.

Fig. 5-10. Link mechanism to reduce the stroke.

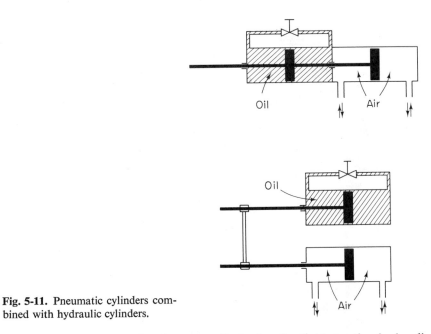

Fig. 5-11. Pneumatic cylinders combined with hydraulic cylinders.

diagram a flow control valve controls hydraulic fluid to the hydraulic cylinder. Since the pistons of both the hydraulic and the pneumatic cylinders are mechanically connected, the speeds of both pistons are similarly controlled.)

Comments on pneumatic cylinders. It is important that proper care (such as listed below) be taken if trouble-free operation of pneumatic cylinders is to be achieved.

1. The piston rod should be free from bending moments.
2. In dealing with a load of large inertia, a stopper must be provided, in addition to a cushion mechanism in the cylinder.
3. If a cylinder is to last, a sufficient amount of atomized oil should be added to the clean (dust and moisture-free) air of the cylinder.

Pneumatic motors. There are two kinds of *pneumatic motors*, piston and vane. Figure 5-12 shows an example of the former. By supplying compressed air to three cylinders in proper order, the crank can be rotated in the direction desired. Such a piston-type pneumatic motor rotates at a slow speed but has a large output torque.

A vane-type pneumatic motor is illustrated in Fig. 5-13. When compressed air is supplied to compartments, the rotor is rotated because of the

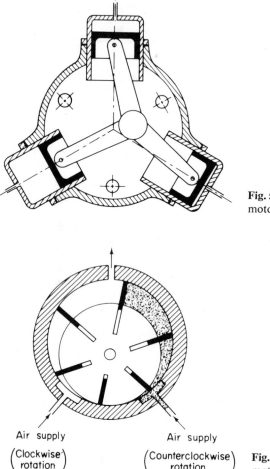

Fig. 5-12. Piston-type pneumatic-motor.

Air supply

$\begin{pmatrix} \text{Clockwise} \\ \text{rotation} \end{pmatrix}$

Air supply

$\begin{pmatrix} \text{Counterclockwise} \\ \text{rotation} \end{pmatrix}$

Fig. 5-13. Vane-type pneumatic motor.

unbalance of force applied to the vanes. This type rotates at a high speed, but the output power is rather limited.

Pneumatic motors find extensive use in such devices as pneumatic drills and pneumatic grinders as well as in many mining machines. Their widespread use is based on the following factors.

1. If the pneumatic motor is overloaded, the air pressure force and the load force balance each other and the motor simply stops without being damaged.
2. The pneumatic motor is fire and explosion proof.

3. The pneumatic motor has a large starting torque.

4. Quick starts and stops are possible.

5. Reversal of the direction of rotation is easy.

6. The pneumatic motor is lighter in weight compared to the electric motor of the same output capacity.

Example 5-1. For the pneumatic three-pulley hoist of Fig. 5-14, assume that piston area A of the pneumatic actuator is 15 in.² and that supply pressure p_1 of air is 70 psig. Find mass m of the maximum weight that can be lifted.

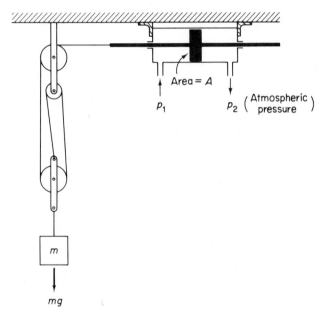

Fig. 5-14. Pneumatic three-pulley hoist.

Since tension F in the rope is the same on its entire length and three ropes support weight mg, we obtain

$$3F = mg$$

The lifting force is equal to tension F. Therefore

$$A(p_1 - p_2) = F = \frac{mg}{3}$$

or

$$mg = 3A(p_1 - p_2) = 3 \times 15 \times 70 = 3150 \text{ lb}_f$$

Note that mass m in slugs is

$$m = \frac{3150}{32.2} = 97.8 \text{ slugs}$$

Example 5-2. Solve the same problem as in Example 5-1 in terms of SI units.
Since 1 in.^{2} = 2.54^2 × 10^{-4} m^2, piston area A is

$$A = 96.77 \times 10^{-4} \text{ m}^2$$

Pressure difference $p_1 - p_2$ is

$$p_1 - p_2 = 70 \frac{\text{lb}_f}{\text{in.}^2} = \frac{70 \times 4.448}{2.54^2 \times 10^{-4}} \frac{\text{N}}{\text{m}^2} = 48.26 \times 10^4 \frac{\text{N}}{\text{m}^2}$$

Consequently,

$$mg = 3A(p_1 - p_2) = 3 \times 96.77 \times 10^{-4} \times 48.26 \times 10^4 = 1.401 \times 10^4 \text{ N}$$

Mass m of the maximum weight that can be lifted is

$$m = \frac{1.401 \times 10^4}{9.81} = 1428 \text{ kg}$$

Pressure control valves. In a pneumatic system a quantity of compressed air is stored in a tank. As need occurs, compressed air is taken from the tank and pressure is reduced by means of a pressure control valve to a desired value to ensure proper operation of pneumatic devices. Pressure control valves can be divided into pressure-reducing valves and relief valves.

Pressure-reducing valves. Figure 5-15 shows a nonrelief, direct-acting, pressure-reducing valve. When the large spring is lowered by rotating the handle, the valve rod is lowered, thereby allowing the air to flow from the primary to the secondary side. If pressure at the secondary side rises, the diaphragm will be pushed up, a step that tends to close the air passage. In this way, the air flow is controlled and pressure at the secondary side is kept constant.

A pilot-acting, pressure-reducing valve is depicted in Fig. 5-16. Here pressure control of the secondary side occurs through air pressure rather than through the spring as in the case of the direct-acting valve. The operating principle is essentially the same as for the direct-acting valve of Fig. 5-15.

Advantages of pilot-acting valves are as follows.

1. The flow characteristics of the pilot-acting valve are superior to those of the direct-acting valve.

2. Pressure control of large air flow occurs easily with pilot-acting valves.

3. Remote control is possible in the pilot-acting valve, whereas it is impossible in direct-acting valves.

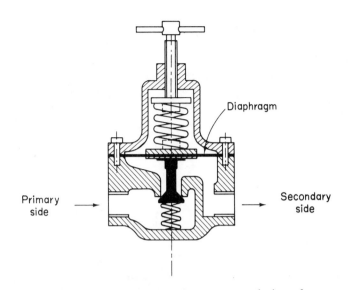

Fig. 5-15. Nonrelief, direct-acting, pressure-reducing valve.

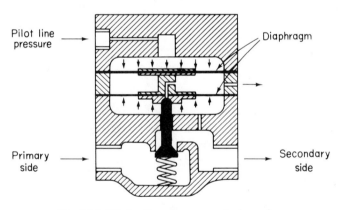

Fig. 5-16. Pilot-acting, pressure-reducing valve.

Relief valves. In pneumatic circuits, air pressure in pipes is controlled by means of pressure-reducing valves. Air pressure in the circuit, however, may rise abnormally high as a result of the mulfunctioning of some circuit components. In this case, a *relief valve* is used to release excess air to atmosphere. Relief valves are either the direct-acting type or the pilot-acting type.

An example of a direct-acting relief valve appears in Fig. 5-17. Such direct-acting relief valves are installed in most air tanks.

Figure 5-18 shows a pilot-acting relief valve. When the pressure in the circuit rises above a predetermined value, the auxiliary valve opens and the back pressure of the main valve is lowered, consequently, the main valve retreats and allows the air to escape to the atmosphere. This type of pilot-acting relief valve is suitable when the cracking pressure is 10^6 N/m² (1 MPa) gage (approximately 10 kg$_f$/cm² gage or 145 psig) or higher.

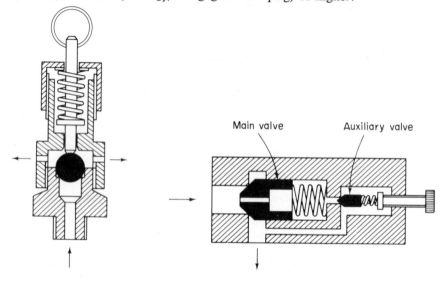

Fig. 5-17. Direct-acting relief valve.

Fig. 5-18. Pilot-acting relief valve.

Flow control valves. Flow rates can be controlled by *flow control valves*, which come in the form of poppet valves, needle valves, and so on. Figure 5-19 shows a poppet valve. It opens fully when the vertical spindle is lowered about one-quarter of the port diameter. In general, this type has good flow characteristics.

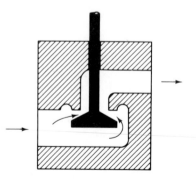

Fig. 5-19. Poppet valve.

Directional control valves. Valves that control the direction of flow are called *directional control valves*. They are necessary, for instance, in switching the direction of motion of the power piston. Directional control valves can be classified as slide valves and spool valves.

Examples of slide valves are shown in Fig. 5-20. This type has a long life and can be made in a small size, but it will require rather a large force to operate.

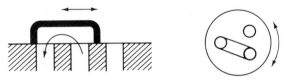

Fig. 5-20. Slide valves.

Figure 5-21 shows a spool valve, which is a balanced valve that requires a small force to operate. Since both the sleeve and spool are precision made, any dust in the air supply will cause difficulty in normal operation.

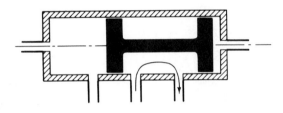

Fig. 5-21. Spool valve.

Magnetic valves. Extensively used to control flows in pneumatic systems, magnetic valves operate on the principle of on or off (open or closed).

Figure 5-22(a) shows a *two-port, direct-acting magnetic valve* in which the magnet is at the off position and the valve is at the closed position. Figure 5-22(b) shows the magnet at the on position and the valve at the open position. (The valve position is switched by means of a solenoid.)

A *three-port, two-position, direct-acting magnetic valve* with magnet off and on is illustrated in Fig. 5-23(a) and (b), respectively. These valves are used for sequential control, high-pressure, low-pressure switching, and similiar operations. For large-capacity valves, pilot-acting rather than direct-acting valves are normally preferred.

Three-port pneumatic pilot valves. Figure 5-24 shows a *three-port pneumatic pilot valve* that is used for switching flow paths. [Figure 5-24(a) corresponds to the closed position and Fig. 5-24(b) to the open position.] Differing from magnetic valves, this kind does not use electricity and so can be used to advantage when temperature or humidity is very high or when explosive gases are handled.

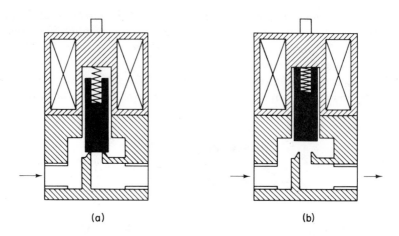

Fig. 5-22. Two-port, direct-acting magnetic valve; (a) valve-closed (magnet off) position; (b) valve-open (magnet on) position.

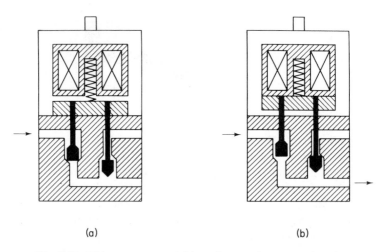

Fig. 5-23. Three-port, two-position, direct-acting magnetic valve; (a) valve-closed (magnet off) position; (b) valve-open (magnet on) position.

Interface valves. With the advent of fluidic devices pilot line pressure is becoming lower and lower. The output pressure of the fluidic devices is of the order of 1×10^4 N/m² gage (approximately 0.1 kg$_f$/cm² gage or 1.4 psig). Pneumatic pilot valves used in connection with fluidic devices are called *interface valves*. Pilot line pressure of the interface valve is generally very low, somewhere around 7×10^2 to 1×10^4 N/m² gage (approximately 0.007 to 0.1 kg$_f$/cm² gage or 0.1 to 1.4 psig).

In Fig. 5-25(a), which shows an interface valve, pressure applied to the piston is atmospheric and the main valve is at the closed position. When pilot line pressure is applied to the piston as shown in Fig. 5-25(b), the plunger closes the breed nozzle. This step causes the pressure of the chamber above

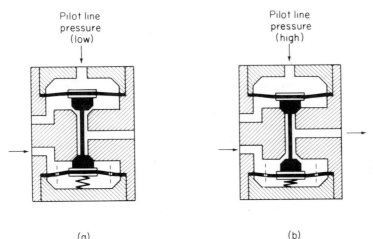

Fig. 5-24. Three-port pneumatic pilot valve; (a) valve-closed position; (b) valve-open position.

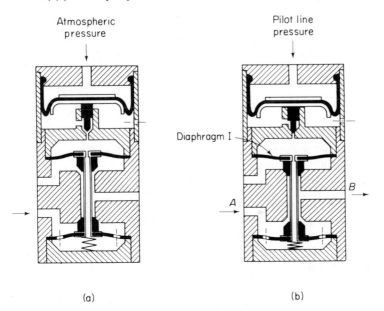

Fig. 5-25. Interface valve; (a) valve-closed position; (b) valve-open position.

diaphragm I to rise, with the result that the diaphragms are pushed down and the main valve opens and the flow from port *A* to port *B* results.

Check valves. A *check valve* (Fig. 5-26) allows air or gas to flow in one direction only by means of a spring and valve.

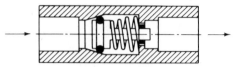

Fig. 5-26. Check valve.

Shuttle valves. Figure 5-27 shows a *shuttle valve*, a device that is essentially a combination of two check valves. The direction of flow can be made from *A* to *C* or from *B* to *C* but not from *A* to *B* or from *B* to *A*.

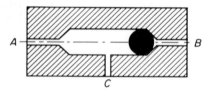

Fig. 5-27. Shuttle valve.

Summary of pneumatic systems. Either alone or combined with hydraulic or electric systems, pneumatic systems are widely used in industry. In particular, hybrid pneumatic and electric systems are often employed in sequential control. Such hybrid systems permit the advantages of pneumatic and electric or hydraulic systems to be utilized and the disadvantages to be compensated.

The advantages of pneumatic systems over other systems are listed below.

1. In the pneumatic system output power can be easily controlled.
2. Actuator speed may vary extensively, although exact speed control is difficult to achieve.
3. Overloading will not harm pneumatic systems.
4. Since compressed air can be stored in a tank, the pneumatic system can respond to occasional heavy demand even if the air compressor of the system is of small size.
5. The pneumatic system can be operated at a broad temperature range and is fire and explosion proof.

Some disadvantages of pneumatic systems, in general, are as follows.

1. Air has no ability to lubricate moving parts.

2. Moisture and foreign particles in air may cause difficulty in the normal operation of pneumatic systems.
3. Efficiency of pneumatic systems is generally low (20 to 30%).
4. Compressibility of air causes delay in response.

5-3 PHYSICAL AND THERMODYNAMIC PROPERTIES OF GASES

In this section we first review properties of air and then briefly discuss thermodynamic properties of gases.

Physical properties of air. Air that does not contain moisture is called *dry air*. Volumetric composition of dry air at sea level is approximately

$$N_2: \quad 78\%$$
$$O_2: \quad 21\%$$
$$Ar, CO_2, etc.: \quad 1\%$$

Physical properties of air and other gases at standard pressure and temperature are shown in Table 5-1. Standard pressure p and temperature t are defined as

$$p = 1.0133 \times 10^5 \text{ N/m}^2 \text{ abs} = 1.0332 \text{ kg}_f/\text{cm}^2 \text{ abs}$$
$$= 14.7 \text{ lb}_f/\text{in.}^2 \text{ abs} = 14.7 \text{ psia}$$
$$t = 0°C = 273 \text{ K} = 32°F = 492°R$$

Table 5-1. PROPERTIES OF GASES

Gas	Molecular weight	Gas constant R_{gas}		Specific heat kcal/kg K or Btu/lb °R		Specific heat ratio c_p/c_v
		N-m/kg K	ft-lb$_f$/lb °R	c_p	c_v	
Air	29.0	287	53.3	0.240	0.171	1.40
Hydrogen (H$_2$)	2.02	4121	766	3.40	2.42	1.41
Nitrogen (N$_2$)	28.0	297	55.2	0.248	0.177	1.40
Oxygen (O$_2$)	32.0	260	48.3	0.218	0.156	1.40
Water vapor (H$_2$O)	18.0	462	85.8	0.444	0.334	1.33

The density ρ, specific volume v, and specific weight γ of air at standard pressure and temperature are

$$\rho = 1.293 \text{ kg/m}^3$$
$$v = 0.7733 \text{ m}^3/\text{kg}$$
$$\gamma = 12.68 \text{ N/m}^3$$

Units of heat. *Heat* is energy transferred from one body to another because of a temperature difference. The SI unit of heat is the joule (J). Other units of heat commonly used in engineering calculations are kilocalorie (kcal) and Btu (British thermal unit).

$$1 \text{ J} = 1 \text{ N-m} = 2.389 \times 10^{-4} \text{ kcal} = 9.480 \times 10^{-4} \text{ Btu}$$
$$1 \text{ kcal} = 4186 \text{ J} = \frac{1}{0.860} \text{ Wh} = 1.163 \text{ Wh}$$
$$1 \text{ Btu} = 1055 \text{ J} = 778 \text{ ft-lb}_f$$

From the engineering point of view, the kilocalorie can be considered that energy needed to raise the temperature of one kilogram of water from 14.5 to 15.5°C. The Btu can be considered as the energy required to raise one pound of water one degree Fahrenheit at some arbitrarily chosen temperature level. (These units give roughly the same values as those defined above.)

Perfect gas law. Consider a perfect gas changing from a state represented by p_1, V_1, T_1 to a state represented by p_2, V_2, T_2. If the temperature is kept constant at T but the pressure (absolute pressure) is changed from p_1 to p_2, then the volume of gas will change from V_1 to V' such that

$$p_1 V_1 = p_2 V' \tag{5-1}$$

If pressure is kept constant but temperature is increased from T_1 to T_2, then the volume of gas reaches V_2. Thus

$$\frac{V'}{T_1} = \frac{V_2}{T_2} \tag{5-2}$$

By eliminating V' between Eqs. (5-1) and (5-2), we have

$$\frac{p_1 V_1}{T_1} = \frac{p_2 V_2}{T_2}$$

This means that for a fixed quantity of a perfect gas, no matter what physical changes occur, pV/T will be constant. Consequently, we can write

$$\frac{pV}{T} = \text{constant}$$

At low pressures and high enough temperatures all gases approach a condition such that

$$pV = mRT \tag{5-3}$$

where p (N/m²) is the absolute pressure of gas, V (m³) the volume of gas, m (kg) the mass of gas, T (K) the absolute temperature of gas, and R (N-m/kg K) a gas constant that depends on gas.

If the volume of gas corresponds to one molecular weight, the gas constant is the same for all gases. So if we define the volume occupied by one mole of gas as $\bar{v}$ (m³/kg mole), the perfect gas law becomes

$$p\bar{v} = \bar{R}T \qquad (5\text{-}4)$$

where $\bar{R}$ is called the *universal gas constant*. The value of the universal gas constant is

$$\bar{R} = 8314 \text{ N-m/kg-mole K} = 1545 \text{ ft-lb}_f/\text{lb-mole } °R$$

The gas that satisfies Eq. (5-4) is defined as the *perfect gas*. Real gases below critical pressure and above critical temperature tend to obey the perfect gas law.

Example 5-3. Find the value of R_{air}, the gas constant for air.

From Eq. (5-3)

$$R = \frac{pV}{mT} = \frac{pv}{T}$$

where $v = V/m = $ specific volume. The specific volume of air at standard pressure and temperature is

$$v = 0.7733 = \frac{1}{1.293} \text{ m}^3/\text{kg}$$

Standard pressure and temperature are

$$p = 1.0133 \times 10^5 \text{ N/m}^2 \text{ abs}$$

$$T = 273 \text{ K}$$

Then it follows that

$$R_{air} = \frac{1.0133 \times 10^5}{1.293 \times 273} = 287 \text{ N-m/kg K} = 29.27 \text{ kg}_f\text{-m/kg K}$$

In terms of BES units,

$$v = 12.39 \text{ ft}^3/\text{lb}$$

$$p = 14.7 \text{ lb}_f/\text{in.}^2 \text{ abs}$$

$$T = 492°R$$

And so

$$R_{air} = \frac{14.7 \times 144 \times 12.39}{492} = 53.3 \text{ ft-lb}_f/\text{lb } °R$$

Thermodynamic properties of gases. If a gas acquires heat from the surroundings, a portion of the energy is used as an addition to the internal energy (as a temperature rise) and the rest as an external work (as an expansion of volume). Thus heat can be converted into work and vice versa.

Even though energy is transformed from one form to another, it can be neither created nor destroyed. This fact is referred to as the *first law of thermodynamics.*

Between mechanical work L (N-m) and heat energy Q (kcal), the following relationship exists:

$$L = JQ \quad \text{or} \quad Q = AL$$

where

$J =$ mechanical equivalent of heat $= 4186$ N-m/kcal
$\quad = 426.9$ kg$_f$-m/kcal $= 778$ ft-lb$_f$/Btu $= 4.186$ J/cal
$A =$ thermal equivalent of work
$\quad = \dfrac{1}{J} =$ reciprocal of mechanical equivalent of heat

As noted, if heat Q is added to a system from the surroundings, then a portion of heat is stored as an addition of the internal energy as a temperature rise and the rest is transformed into an external work. So

$$Q = U_2 - U_1 + AL$$

where

$U_1 =$ internal energy at the initial state
$U_2 =$ internal energy at the final state
$AL =$ heat transformed into mechanical work

Specific heats. A *specific heat* of a gas is defined as the ratio of the amount of heat required to raise the unit mass of the gas one degree to that required to raise the unit mass of water one degree at some specified temperature, using the same system of units. Two specific heats are generally used for gases—that at constant pressure (c_p) and that at constant volume (c_v).

Changes of state for perfect gas. A process is called a *reversible* process if both the system and the surroundings can be returned to their original states; otherwise it is defined as *irreversible*. All actual processes are irreversible.

Let us briefly discuss changes of state for a perfect gas. Figure 5-28 shows pressure-volume curves of a perfect gas. In the following analysis, subscripts 1 and 2 refer to the initial and final states, respectively.

1. Change of state at constant volume $(p_2/p_1 = T_2/T_1)$. This corresponds to the change of state when the volume is kept constant as shown by curve 1 in Fig. 5-28. The heat Q_v kcal added to the system of m kg gas from the surroundings becomes an addition to the internal energy, for since the volume stays constant, no external work is done. Therefore

$$L = 0$$

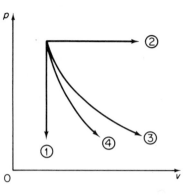

Fig. 5-28. Pressure-volume curves of a perfect gas.

and

$$U_2 - U_1 = Q_v = mc_v(T_2 - T_1) \quad \text{kcal}$$

2. Change of state at constant pressure $(V_2/V_1 = T_2/T_1)$. It corresponds to the change of state when the pressure is kept constant as shown by curve 2 in Fig. 5-28. If heat Q_p kcal is added to the system of m kg gas from the surroundings, a portion of it is used to expand the volume and the rest is kept as an addition of the internal energy. Referring to Eq. (5-3),

$$L = p(V_2 - V_1) = mR(T_2 - T_1) \quad \text{N-m}$$
$$U_2 - U_1 = Q_p - AL = mc_p(T_2 - T_1) - AmR(T_2 - T_1)$$
$$= m(c_p - AR)(T_2 - T_1) \quad \text{kcal}$$

3. Change of state at constant temperature (Isothermal) $(p_2/p_1 = V_1/V_2)$. This situation corresponds to the change of state when the temperature is kept constant as shown by curve 3 in Fig. 5-28. Here the heat Q_T kcal added to the system from the surroundings is used as an external work—there is no increase in the internal energy because of no change in termperature. Hence $U_2 - U_1 = 0$ and the work done is

$$L = \int_{V_1}^{V_2} p \, dV = mRT \ln \frac{V_2}{V_1} \quad \text{N-m}$$

and

$$AL = Q_T \quad \text{kcal}$$

4. Reversible adiabatic change of state (Isentropic) $(p_1 V_1^k = p_2 V_2^k)$. Adiabatic change of state refers to the state in which no heat is transferred to or from the system. Reversible adiabatic (frictionless adiabatic) change of state is called *isentropic* change of state. The adiabatic change is shown by curve 4 in Fig. 5-28. The relationship between pressure p and volume V is

$$pV^k = \text{constant}$$

where k is called the *adiabatic exponent*. For a perfect gas, it is the same as

c_p/c_v or

$$k = \frac{c_p}{c_v} = 1.40$$

In the adiabatic change of state heat Q kcal transferred to or from the system is zero. Thus

$$Q = 0$$

And so the work done L by the gas is equal to the change in the internal energy or

$$AL = U_1 - U_2 = mc_v(T_1 - T_2) = \frac{c_v}{R}(p_1 V_1 - p_2 V_2) \quad \text{kcal}$$

Notice that compression or expansion of air in the pneumatic cylinder is nearly adiabatic.

5. *Polytropic change of state* ($p_1 V_1^n = p_2 V_2^n$). The actual change of state for any real gas does not fall exactly into any of the four listed above. It may be represented by properly choosing the value of n of the equation

$$pV^n = \text{constant}$$

The change of state given by this last equation is called *polytropic*, and the exponent n is called the *polytropic exponent*. The polytropic change of state is quite general in that it can cover all four changes of state mentioned above by properly choosing the value of n. In fact, by giving the polytropic exponent different values, the preceding changes may be made special cases of the polytropic change; that is, for $n = 1$, $n = 0$, $n = \infty$, and $n = k$, the change of state is isothermal, constant pressure, constant volume, and isentropic, respectively.

*5-4 FLOW OF GASES THROUGH ORIFICES

Since gas is compressible, the flow of gas through pipes and orifices is more complicated than the flow of liquid. This section presents an analysis of gas flow through an orifice. In industrial pneumatic pressure systems, laminar flow seldom occurs; consequently, we are concerned here with turbulent flow through pipes, orifices, and valves.

We begin by deriving equations for the flow of gas through an orifice and show that under certain conditions the speed of gas through it becomes equal to the speed of sound. Then we obtain equations for the mass flow rate for gas flow through an orifice. Finally, mass flow rate equations for pneumatic circuits are derived.

*Starred sections deal with more challenging topics than the rest of the book. Depending on course objectives, these sections (although important) may be omitted from classroom discussions without losing the continuity of the main subject matter.

Flow of perfect gas through an orifice. The flow of a real gas through orifices and nozzles can be approximated by isentropic (frictionless adiabatic) flow if the friction effects and heat transfer are negligible.

Let us consider the steady flow of a perfect gas through an orifice as in Fig. 5-29. Here cross section 1 is taken at upstream from the orifice. The cross section at the vena contracta (where the issuing jet area becomes a minimum) is denoted by cross section 2.

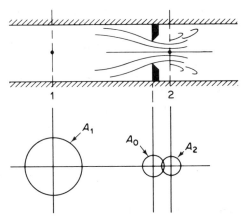

Fig. 5-29. Steady flow of a perfect gas through an orifice.

The stream area A_2 of the issuing jet at the vena contracta is smaller than the orifice area A_0. The ratio of A_2 and A_0 is the contraction coefficient C_c or

$$A_2 = C_c A_0$$

Referring to Fig. 5-29, the state of gas at section 1 is p_1, v_1, T_1 and that at section 2 is p_2, v_2, T_2. The velocities at section 1 and section 2 are denoted by w_1 and w_2, respectively. Pressures p_1 and p_2 are absolute pressures.

For isentropic change of state, we have

$$p_1 v_1^k = p_2 v_2^k = \text{constant}$$

Noting that the specific volume v is the reciprocal of the density,

$$v = \frac{1}{\rho}$$

Hence

$$p_1 = p_1^k p_2 p_2^{-k} \tag{5-5}$$

In Sec. 4-4 we derived Euler's equation, Eq. (4-4), rewritten thus

$$w\, dw + \frac{dp}{\rho} + g\, dz = 0$$

where w is used for velocity. Neglecting elevation change, Euler's equation becomes

$$w \, dw + \frac{dp}{\rho} = 0$$

Then by differentiating Eq. (5-5) with respect to p_1 and noting that $p_2 \rho_2^{-k} = $ constant, we have

$$dp_1 = k \rho_1^{k-1} \, dp_1 p_2 \rho_2^{-k} \qquad (5\text{-}6)$$

Euler's equation at section 1 is

$$w_1 \, dw_1 + \frac{dp_1}{\rho_1} = 0$$

Substituting Eq. (5-6) into this last equation gives

$$w_1 \, dw_1 + p_2 \rho_2^{-k} k \rho_1^{k-2} \, dp_1 = 0$$

By noting again that $p_2 \rho_2^{-k} = p_1 \rho_1^{-k} = $ constant and integrating this last equation, we find

$$\frac{w_1^2}{2} + p_2 \rho_2^{-k} k \frac{\rho_1^{k-1}}{k-1} = \text{constant}$$

or

$$\frac{w_1^2}{2} + \frac{k}{k-1} \frac{p_1}{\rho_1} = \text{constant}$$

Therefore we obtain

$$\frac{w_1^2}{2} + \frac{k}{k-1} \frac{p_1}{\rho_1} = \frac{w_2^2}{2} + \frac{k}{k-1} \frac{p_2}{\rho_2} \qquad (5\text{-}7)$$

From the continuity equation

$$\rho_1 w_1 A_1 = \rho_2 w_2 A_2$$

it follows that

$$w_1 = \frac{\rho_2}{\rho_1} \frac{A_2}{A_1} w_2 = \left(\frac{p_2}{p_1} \right)^{1/k} \frac{A_2}{A_1} w_2 \qquad (5\text{-}8)$$

If we substitute Eq. (5-8) into Eq. (5-7) and simplify, the result is

$$\frac{w_2^2}{2} = \frac{\dfrac{k}{k-1} \left(\dfrac{p_1}{\rho_1} - \dfrac{p_2}{\rho_2} \right)}{1 - \left(\dfrac{p_2}{p_1} \right)^{2/k} \left(\dfrac{A_2}{A_1} \right)^2} \qquad (5\text{-}9)$$

If area A_2 is sufficiently small compared with area A_1, then noting that $p_2/p_1 < 1$, we may assume

$$1 - \left(\frac{p_2}{p_1} \right)^{2/k} \left(\frac{A_2}{A_1} \right)^2 \doteq 1$$

With this assumption Eq. (5-9) simplifies to

$$\frac{w_2^2}{2} = \frac{k}{k-1} \left(\frac{p_1}{\rho_1} - \frac{p_2}{\rho_2} \right)$$

Substitution of $\rho_2 = (p_2/p_1)^{1/k}\rho_1$ into this last equation gives

$$\frac{w_2^2}{2} = \frac{k}{k-1}\frac{p_1}{\rho_1}\left[1 - \left(\frac{p_2}{p_1}\right)^{(k-1)/k}\right]$$

or

$$w_2 = \sqrt{\frac{2k}{k-1}\frac{p_1}{\rho_1}\left[1 - \left(\frac{p_2}{p_1}\right)^{(k-1)/k}\right]} \tag{5-10}$$

The mass flow rate G is

$$G = \rho_2 w_2 A_2 = \rho_2 A_2 \sqrt{\frac{2k}{k-1}\frac{p_1}{\rho_1}\left[1 - \left(\frac{p_2}{p_1}\right)^{(k-1)/k}\right]}$$

Since $\rho_2 = (p_2/p_1)^{1/k}\rho_1$ and $A_2 = C_c A_0$, this last equation may be written

$$G = C_c A_0 \sqrt{\frac{2k}{k-1}p_1\rho_1\left[\left(\frac{p_2}{p_1}\right)^{2/k} - \left(\frac{p_2}{p_1}\right)^{(k+1)/k}\right]}$$

In deriving it, friction effects due to viscosity of gas were not considered. Including both the effects of the neglected friction and the contraction coefficient C_c, we can introduce a discharge coefficient c (the exact value of which may be determined experimentally) and write the mass flow rate as

$$G = c A_0 \sqrt{\frac{2k}{k-1}p_1\rho_1\left[\left(\frac{p_2}{p_1}\right)^{2/k} - \left(\frac{p_2}{p_1}\right)^{(k+1)/k}\right]} \tag{5-11}$$

Noting that $p_1 = \rho_1 R T_1$, Eq. (5-11) may be modified into the following form.

$$G = c A_0 \rho_1 \sqrt{\frac{2k}{k-1}R T_1\left[\left(\frac{p_2}{p_1}\right)^{2/k} - \left(\frac{p_2}{p_1}\right)^{(k+1)/k}\right]}$$

$$= c A_0 \frac{p_1}{\sqrt{T_1}}\sqrt{\frac{2k}{k-1}\frac{1}{R}\left[\left(\frac{p_2}{p_1}\right)^{2/k} - \left(\frac{p_2}{p_1}\right)^{(k+1)/k}\right]} \tag{5-12}$$

Critical pressure, critical velocity, and maximum mass flow rate. For given values of p_1, ρ_1, A_0, and c, the mass flow rate G becomes a function only of p_2. A curve relating G and p_2 is shown in Fig. 5-30. The mass flow rate becomes maximum at point B. The particular value of pressure p_2 that corresponds to point B can be obtained as that pressure p_2 for which

$$\frac{\partial G}{\partial p_2} = 0$$

Referring to Eq. (5-12), this condition can be modified to

$$\frac{\partial\left[\left(\frac{p_2}{p_1}\right)^{2/k} - \left(\frac{p_2}{p_1}\right)^{(k+1)/k}\right]}{\partial p_2} = 0$$

which results in

$$\frac{2}{k}\left(\frac{p_2}{p_1}\right)^{(2/k)-1}\left(\frac{1}{p_1}\right) - \frac{k+1}{k}\left(\frac{p_2}{p_1}\right)^{1/k}\left(\frac{1}{p_1}\right) = 0$$

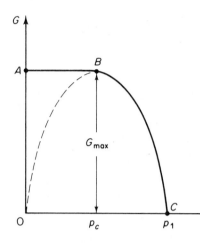

Fig. 5-30. Curve relating mass flow rate G and pressure p_2.

By denoting this particular value of p_2 as p_c, we have

$$\left(\frac{p_c}{p_1}\right)^{(k-1)/k} = \frac{2}{k+1}$$

or

$$p_c = \left(\frac{2}{k+1}\right)^{k/(k-1)} p_1 \tag{5-13}$$

The pressure p_c given by Eq. (5-13) is called the *critical pressure*.

The maximum mass flow rate $G_{\max}$ that occurs when $p_2 = p_c$ is obtained by substituting $p_2 = p_c$ into Eq. (5-11).

$$
\begin{aligned}
G_{\max} &= cA_0\sqrt{\frac{2k}{k-1}p_1\rho_1\left[\left(\frac{p_c}{p_1}\right)^{2/k} - \left(\frac{p_c}{p_1}\right)^{(k+1)/k}\right]} \\
&= cA_0\sqrt{\frac{2k}{k-1}p_1\rho_1\left[\left(\frac{2}{k+1}\right)^{2/(k-1)} - \left(\frac{2}{k+1}\right)^{(k+1)/(k-1)}\right]} \\
&= cA_0\sqrt{\frac{2k}{k+1}\left(\frac{2}{k+1}\right)^{2/(k-1)} p_1\rho_1} \tag{5-14}
\end{aligned}
$$

Since the value of cA_0 is a constant and the value of k for a given gas is also a constant, the maximum mass flow rate $G_{\max}$ depends only on the condition at section 1.

Critical velocity. Substituting $p_2 = p_c$ into Eq. (5-10) gives the critical velocity w_c.

$$
\begin{aligned}
w_c &= \sqrt{\frac{2k}{k-1}\frac{p_1}{\rho_1}\left[1 - \left(\frac{p_c}{p_1}\right)^{(k-1)/k}\right]} \\
&= \sqrt{\frac{2k}{k-1}\frac{p_1}{\rho_1}\left(1 - \frac{2}{k+1}\right)} \\
&= \sqrt{\frac{2k}{k+1}\frac{p_1}{\rho_1}} \tag{5-15}
\end{aligned}
$$

Note that for isentropic change of state

$$\frac{p_1}{\rho_1^k} = \frac{p_c}{\rho_c^k} \tag{5-16}$$

From Eq. (5-13) we have

$$p_1 = p_c \left(\frac{2}{k+1}\right)^{-k/(k-1)} \tag{5-17}$$

And so, using Eqs. (5-16) and (5-17), we find

$$\frac{p_1^k}{\rho_1^k} = \frac{p_1}{\rho_1^k} p_1^{k-1} = \frac{p_c}{\rho_c^k} p_c^{k-1} \left(\frac{2}{k+1}\right)^{-k} = \frac{p_c^k}{\rho_c^k} \left(\frac{2}{k+1}\right)^{-k}$$

or

$$\frac{p_1}{\rho_1} = \left(\frac{2}{k+1}\right)^{-1} \frac{p_c}{\rho_c}$$

By substituting this last equation into Eq. (5-15), the critical velocity w_c is given as

$$w_c = \sqrt{\frac{2k}{k+1}\left(\frac{2}{k+1}\right)^{-1}\frac{p_c}{\rho_c}} = \sqrt{k\frac{p_c}{\rho_c}} = \sqrt{kRT_c}$$

Note that $c = \sqrt{kRT}$ is the speed of sound. (Refer to Problem A-5-9.) Hence the speed of w_c is equal to the speed of sound. (The speed of sound in a gas depends on the nature of gas and its absolute temperature.)

Summary of gas flow through an orifice. The relationship between the mass flow rate of gas through an orifice and pressure drop cannot be expressed by a single equation because there are two distinct types of flow conditions, sonic and subsonic. Under the sonic flow condition pressure p_2 at the vena contracta remains at critical pressure p_c, independent of the pressure in the downstream. Thus no effects in the downstream will be reflected back into pressure p_2.

Referring to Fig. 5-30, if the value of p_2 is varied from a very small value (compared to p_1) to the value equal to p_1, then the mass flow rate G follows curve ABC. The mass flow rate is constant at G_{max} until p_2 increases and becomes equal to p_c, and thereafter the mass flow rate decreases and finally becomes equal to zero when p_2 reaches p_1. The gas flow is sonic between points A and B and subsonic between points B and C.

Alternate forms of mass flow rate equations. The mass flow rate for the subsonic flow condition is given by Eq. (5-11) or Eq. (5-12). Let us introduce the expansion factor ϵ:

$$\epsilon = \sqrt{\frac{k}{k-1}\frac{p_1^2}{p_2(p_1-p_2)}\left[\left(\frac{p_2}{p_1}\right)^{2/k} - \left(\frac{p_2}{p_1}\right)^{(k+1)/k}\right]} \tag{5-18}$$

which can also be written

$$\epsilon = \sqrt{\frac{k}{k-1}\frac{1}{(p_1/p_2)-1}\left(\frac{p_1}{p_2}\right)^{(k-1)/k}\left[\left(\frac{p_1}{p_2}\right)^{(k-1)/k}-1\right]}$$

The value of the expansion factor ϵ depends on the values of k and p_2/p_1. It is, however, approximately constant for $p_1 \geq p_2 \geq p_c$.

In terms of this expansion factor, Eq. (5-12) may be written

$$G = cA_0\epsilon\sqrt{\frac{2}{RT_1}}\sqrt{p_2(p_1-p_2)} \tag{5-19}$$

The advantage of introducing the expansion factor ϵ is that for $A_2^2 \ll A_1^2$, where areas A_1 and A_2 are defined in Fig. 5-29, exact equations for the mass flow rate can be approximated by simpler expressions at the expense of introducing an error of only a few percent.

The mass flow rate for the sonic flow condition is given by Eq. (5-14). Two alternate forms for this equation appear below. By substituting $p_1 = \rho_1 RT_1$ into Eq. (5-14), we find

$$G_{\max} = cA_0\rho_1\left(\frac{2}{k+1}\right)^{1/(k-1)}\sqrt{\frac{2kRT_1}{k+1}}$$

Substituting $\rho_1 = p_1/(RT_1)$ into Eq. (5-14) gives

$$G_{\max} = cA_0\frac{p_1}{\sqrt{RT_1}}\sqrt{k\left(\frac{2}{k+1}\right)^{(k+1)/(k-1)}} \tag{5-20}$$

Equations (5-19) and (5-20) serve as basic equations for the calculation of the mass flow rate for gas flow through an orifice.

Airflow through an orifice. The preceding analysis can be applied to the analysis of airflow in the pneumatic controller, airflow from the supply to the power cylinder through the directional control valve, and so forth. In the following material we derive mass flow rate equations for air by substituting appropriate constant values for k and R into Eqs. (5-19) and (5-20).

Consider airflow through an orifice. (Refer to Fig. 5-29.) Note first that from Eqs. (5-16) and (5-17) we have

$$\frac{\rho_c}{\rho_1} = \left(\frac{2}{k+1}\right)^{1/(k-1)} \tag{5-21}$$

Also

$$\frac{T_c}{T_1} = \frac{p_c}{p_1}\frac{\rho_1}{\rho_c} = \frac{2}{k+1} \tag{5-22}$$

For air, $k = 1.40$. So by using Eqs. (5-13), (5-21), and (5-22), we obtain

$$\frac{p_c}{p_1} = \left(\frac{2}{k+1}\right)^{k/(k-1)} = 0.528 \qquad \text{for } k = 1.40$$

$$\frac{\rho_c}{\rho_1} = \left(\frac{2}{k+1}\right)^{1/(k-1)} = 0.634 \qquad \text{for } k = 1.40$$

$$\frac{T_c}{T_1} = \frac{2}{k+1} = 0.833 \qquad \text{for } k = 1.40$$

Referring to Fig. 5-29, the last three equations show that the critical pressure for airflow is 52.8% of the pressure at section 1, the density is reduced by 37%, and the absolute temperature drops about 17% from section 1 to section 2.

Mass flow rate equation for air when $p_2 > 0.528p_1$. When the pressure condition across the orifice is such that $p_2 > 0.528p_1$, the speed of airflow through the orifice is subsonic and the mass flow rate can be obtained by substituting $R = R_{air}$ into Eq. (5-19) as follows:

$$G_{air} = cA_0\epsilon\sqrt{\frac{2}{R_{air}T_1}}\sqrt{p_2(p_1 - p_2)} \tag{5-23}$$

By assuming that A_0 is measured in m², p_1 and p_2 in N/m² abs, and T_1 in K, and by substituting $R_{air} = 287$ N-m/kg K into this last equation, we have

$$G_{air} = 0.0835cA_0\epsilon\frac{1}{\sqrt{T_1}}\sqrt{p_2(p_1 - p_2)} \text{ kg/s} \tag{5-24}$$

where the value of the expansion factor ϵ given by Eq. (5-18) can be approximated for $k = 1.40$ and $1 \geq p_2/p_1 \geq 0.528$ by the linear equation

$$\epsilon = 0.97 + 0.0636\left(\frac{p_2}{p_1} - 0.528\right) \qquad \left(1 \geq \frac{p_2}{p_1} \geq 0.528\right) \tag{5-25}$$

The value of ϵ varies linearly from $\epsilon = 1$ at $p_2/p_1 = 1$ to $\epsilon = 0.97$ at $p_2/p_1 = 0.528$.

Since the value of ϵ is nearly constant for $1 \geq p_2/p_1 \geq 0.528$, by using an average value of 0.985 for ϵ, the mass flow rate G_{air} is given by

$$G_{air} = 0.0822cA_0\frac{1}{\sqrt{T_1}}\sqrt{p_2(p_1 - p_2)} \text{ kg/s} \qquad \left(1 \geq \frac{p_2}{p_1} > 0.528\right) \tag{5-26}$$

where A_0 is measured in m², p_1 and p_2 in N/m² abs, and T_1 in K. This is an approximate mass flow rate equation for airflow through an orifice when $p_2 > 0.528p_1$ and $A_2^2 \ll A_1^2$. The mass flow rate G_{air} depends on the value of c, A_0, upstream temperature T_1, upstream absolute pressure p_1, and absolute pressure p_2. As long as $p_2 > 0.528p_1$, the airflow speed is subsonic and Eq. (5-26) gives the mass flow rate.

Mass flow rate equation for air when $p_2 \leq 0.528p_1$. When the pressure condition across the orifice is such that $p_2 \leq 0.528p_1$, airflow through the orifice is at the speed of sound and the mass flow rate is not influenced by the orifice back pressure. The mass flow rate for this condition is obtained by substituting $k = 1.40$ and $R = 287$ N-m/kg K into Eq. (5-20)—that is,

$$G_{air, max} = 0.0404cA_0\frac{p_1}{\sqrt{T_1}} \text{ kg/s} \qquad \left(\frac{p_2}{p_1} \leq 0.528\right) \tag{5-27}$$

where A_0 is measured in m², p_1 in N/m² abs, and T_1 in K. The maximum mass flow rate $G_{\text{air, max}}$ depends on the values of c, A_0, upstream temperature T_1, and upstream absolute pressure p_1, but it is independent of absolute pressure p_2. As long as $p_2 \leq 0.528p_1$, the flow speed is sonic and the mass flow rate stays constant at $G_{\text{air, max}}$.

Note that, as a matter of course, Eqs. (5-24) and (5-27) give the same mass flow rate when $p_2 = 0.528p_1$. To verify this statement, notice first that from Eq. (5-25) we have $\epsilon = 0.97$ for $p_2/p_1 = 0.528$. Then substituting $\epsilon = 0.97$ and $p_2 = 0.528p_1$ into Eq. (5-24) gives

$$G_{\text{air}} = 0.0835 \times 0.97cA_0 \frac{1}{\sqrt{T_1}} \sqrt{0.528p_1(1 - 0.528)p_1} = 0.0404cA_0 \frac{p_1}{\sqrt{T_1}}$$

which is the same as Eq. (5-27).

Example 5-4. Referring to Fig. 5-29, assume that $A_0 = 3 \times 10^{-4}$ m², $c = 0.68$, $p_1 = 2.5 \times 10^5$ N/m² abs, and $T_1 = 273$ K. Assuming also that air is flowing, calculate the mass flow rates for the following two cases.

(a) Pressure p_2 is 2×10^5 N/m² abs and the flow condition is subsonic.

(b) The flow condition is sonic—that is, $p_2 \leq 0.528p_1$, where p_1 and p_2 are absolute pressures.

For subsonic flow conditions, Eq. (5-26) applies; for sonic flow conditions, Eq. (5-27) may be used.

(a) The mass flow rate is obtained by substituting the given numerical values into Eq. (5-26) as

$$G_{\text{air}} = 0.0822 \times 0.68 \times 3 \times 10^{-4} \frac{1}{\sqrt{273}} \sqrt{2 \times 10^5(2.5 \times 10^5 - 2 \times 10^5)}$$

$$= 0.101 \text{ kg/s}$$

(b) The mass flow rate is obtained by substituting the given numerical values into Eq. (5-27) as

$$G_{\text{air, max}} = 0.0404 \times 0.68 \times 3 \times 10^{-4} \frac{2.5 \times 10^5}{\sqrt{273}} = 0.125 \text{ kg/s}$$

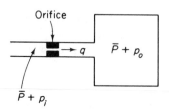

Fig. 5-31. Pneumatic pressure system.

Example 5-5. Consider the pneumatic pressure system shown in Fig. 5-31 and assume that, for $t < 0$, the absolute pressure of air in the system is $\bar{P}$. At $t = 0$ the pressure at the left side of the orifice is changed from $\bar{P}$ to $\bar{P} + p_i$, where p_i may be positive or negative. (In the present analysis, we assume that p_i is positive. If p_i is negative, the direction of flow is reversed.) Then the pressure of air in the vessel will change from $\bar{P}$ to $\bar{P} + p_o$. We assume that changes in pressure (both p_i and p_o) are sufficiently small compared with absolute pressure $\bar{P}$. Assuming that the temperature of the system remains constant, show that the mass flow rate q is approximately proportional to the square root of the pressure difference $\Delta p = p_i - p_o$. Plot a typical Δp versus q curve.

Since both p_i and p_0 are sufficiently small compared with $\bar{P}$, the flow condition is subsonic [$\bar{P} + p_o > 0.528(\bar{P} + p_i)$] and Eq. (5-23) applies to this case. By substituting $p_1 = \bar{P} + p_i$ and $p_2 = \bar{P} + p_o$ into Eq. (5-23), we have

$$q = G_{air} = cA_0\epsilon\sqrt{\frac{2}{R_{air}T_1}}\,\sqrt{(\bar{P} + p_o)(p_i - p_o)}$$

$$= cA_0\epsilon\sqrt{\frac{2\bar{P}}{R_{air}T_1}}\,\sqrt{1 + \frac{p_o}{\bar{P}}}\,\sqrt{p_i - p_o} \qquad (5\text{-}28)$$

Since p_o is assumed to be sufficiently small compared with $\bar{P}$, as a first approximation we have

$$\sqrt{1 + \frac{p_o}{\bar{P}}} \doteq 1 + \frac{1}{2}\frac{p_o}{\bar{P}} \doteq 1$$

Then Eq. (5-28) may be simplified to

$$q = K\sqrt{p_i - p_o} = K\sqrt{\Delta p} \qquad (5\text{-}29)$$

where

$$K = cA_0\epsilon\sqrt{\frac{2\bar{P}}{R_{air}T_1}} = 0.985cA_0\sqrt{\frac{2\bar{P}}{R_{air}T_1}} = \text{constant}$$

$$\Delta p = p_i - p_o$$

Thus the mass flow rate q is proportional to the square root of the pressure difference $\Delta p = p_i - p_o$ if p_o is sufficiently small compared with $\bar{P}$, where $\bar{P}$ is the absolute pressure of air at steady state. Figure 5-32 depicts a typical Δp versus q curve based on Eq. (5-29).

Observe that if the mass flow rate q of the pneumatic pressure system shown in Fig. 5-31 is experimentally measured and the pressure difference Δp is plotted against the mass flow rate q, then a nonlinear curve similar to that shown in Fig. 5-32 can be obtained. Such a curve plays an important role in determining a mathematical model of the pneumatic system shown in Fig. 5-31. (For details, see Sec. 5-5.)

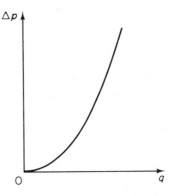

Fig. 5-32. Pressure difference versus mass flow rate curve.

Effective cross-sectional areas of pneumatic components. In deriving Eqs. (5-26) and (5-27), it was assumed that air flows through an orifice. These two equations, however, can serve as basic equations for calculating mass flow rate for airflow through valves or other flow-restricting devices in pneumatic circuits if the flow through a pneumatic component involving a valve or other flow-restricting device can be considered equivalent to flow through an orifice.

Although the area A_0 may not be defined for a valve or other flow-restricting device, it may be possible to define an equivalent orifice area for a given pneumatic component involving the same devices. Such an equivalent orifice area may be determined as follows. Apply pressure p_1 at the inlet of the pneumatic component. Then pressure p_2 will appear at the output of the component. For a given pressure p_1, if we measure temperature T_1, mass flow rate G, and pressure p_2, then an equivalent orifice area cA_0 can be estimated by using Eq. (5-26) or (5-27). (Pressures p_1 and p_2 are measured in N/m² abs and temperature T_1 in K.) This type of equivalent orifice area cA_0 is often called the *effective cross-sectional area* of a pneumatic component. The concept of the effective cross-sectional area is useful in calculating the mass flow rate in pneumatic components.

Airflow through a valve in a pneumatic circuit. If a valve is opened in a pneumatic circuit, the pressure ratio p_2/p_1 (where p_1 and p_2 are the absolute pressures at upstream and downstream of the valve) may be small ($p_2/p_1 \leq 0.528$) initially so that the airflow through the valve becomes a sonic flow and the mass flow rate can be given by Eq. (5-27), where the effective cross-sectional area of the valve is used for cA_0. As time elapses and the pressure ratio p_2/p_1 increases and becomes $p_2/p_1 > 0.528$, airflow through the valve becomes subsonic and the mass flow rate is determined according to Eq. (5-26), where the effective cross-sectional area of the valve is used for cA_0.

5-5   MATHEMATICAL MODELING
OF PNEUMATIC SYSTEMS

As in mechanical, electrical, and hydraulic systems, three types of elements—resistance, capacitance, and inertance—constitute the basic elements of pneumatic systems. Mathematical models of these systems can be written in terms of the three.

We begin this section by deriving specific expressions of the basic elements of pneumatic systems based on the general definitions of resistance, capacitance, and inertance presented in Section 4-5, followed by the derivation of a mathematical model of a pneumatic pressure system consisting of a vessel and a pipe with an orifice.

Resistance. Airflow *resistance* in pipes, orifices, valves, and any other flow-restricting devices can be defined as the change in differential pressure (existing between upstream and downstream of a flow-restricting device)

(N/m²) required to make a unit change in mass flow rate (kg/s) or

$$\text{Resistance } R = \frac{\text{change in differential pressure}}{\text{change in mass flow rate}} \quad \frac{\text{N/m}^2}{\text{kg/s}} \text{ or } \frac{\text{N-s}}{\text{kg-m}^2}$$

Therefore resistance R can be expressed as

$$R = \frac{d(\Delta p)}{dq} \qquad (5\text{-}30)$$

where $d(\Delta p)$ is a change in the differential pressure and dq is a change in the mass flow rate.

For steady flow, where $\Delta p = \text{constant} = \Delta \bar{p}$ and $q = \text{constant} = \bar{q}$, resistance R at this operating condition can be obtained easily if an experimental curve relating Δp and q is available. Referring to Fig. 5-33, consider a small change $d(\Delta p)$ near the operating condition $\Delta \bar{p}$. The corresponding small change dq about the operating condition $\bar{q}$ can be found from the curve Δp versus q. Resistance R is then $d(\Delta p)/dq$ and is equal to the slope of the line approximating the curve near the operating condition point $\Delta p = \Delta \bar{p}, q = \bar{q}$ as shown in the figure. Notice that the value of airflow resistance R is not constant but varies with the change in the operating condition.

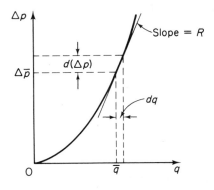

Fig. 5-33. Diagram showing resistance R as the slope of a pressure difference versus mass flow rate curve at the operating point.

<hr>

Example 5-6. Consider airflow through a valve and assume that upstream absolute pressure is p_1 and downstream absolute pressure is p_2. Assume also that differential pressure $\Delta p = p_1 - p_2$ is sufficiently small compared with p_1 and that the mass flow rate through the valve is q. Determine resistance R of the valve.

When the pressure drop at a flow-restricting device (such as an orifice and a valve) is sufficiently small, the mass flow rate q is proportional to the square root of the pressure drop $\Delta p = p_1 - p_2$. [Refer to Eq. (5-29).] Thus

$$q = K\sqrt{\Delta p} \qquad (5\text{-}31)$$

where K is a constant. Then referring to Eq. (5-30) resistance R at any operating point $\Delta p = \Delta \bar{p}, q = \bar{q}$ is found as

$$R = \frac{d(\Delta p)}{dq}\bigg|_{\Delta p = \Delta \bar{p},\, q = \bar{q}} = \frac{2\,\Delta p}{q}\bigg|_{\Delta p = \Delta \bar{p},\, q = \bar{q}} = \frac{2\,\Delta \bar{p}}{\bar{q}} \qquad (5\text{-}32)$$

Note that Eq. (5-32) is an approximate equation, since it is based on Eq. (5-31), which is true when Δp is small compared with absolute pressure p_1.

Pneumatic systems may operate in such a way that the average or steady-state flow through the valve is zero; that is, the normal operating condition is $\Delta \bar{p} = 0$, $\bar{q} = 0$. If the normal operating condition is $\Delta \bar{p} = 0$, $\bar{q} = 0$ and the ranges of Δp and q are $-\Delta p_1 < \Delta p < \Delta p_1$ and $-q_1 < q < q_1$, respectively, then, for practical purposes, the average resistance R may be approximated by the slope of the straight line connecting point $\Delta p = \Delta p_1, q = q_1$ and point $\Delta p = -\Delta p_1, q = -q_1$ as shown in Fig. 5-34.

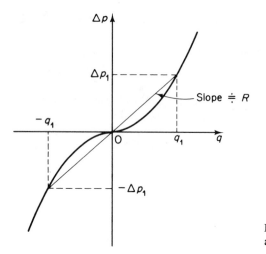

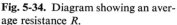

Fig. 5-34. Diagram showing an average resistance R.

In modeling a pneumatic system mathematically, it is always desirable to have an experimental curve, relating Δp and q for the entire ranges of operation, available so that resistance R can be determined graphically with reasonable accuracy.

Capacitance. For a pneumatic pressure vessel, *capacitance* can be defined as the change in the mass of air (kg) in the vessel required to make a unit change in pressure (N/m²) or

$$\text{Capacitance } C = \frac{\text{change in mass of air}}{\text{change in pressure}} \quad \frac{\text{kg}}{\text{N/m}^2} \text{ or } \frac{\text{kg-m}^2}{\text{N}}$$

which may be expressed as

$$C = \frac{dm}{dp} = V \frac{d\rho}{dp} \quad \frac{\text{kg}}{\text{N/m}^2} \tag{5-33}$$

where
m = mass of air in vessel, kg
p = absolute pressure of air, N/m²
V = volume of vessel, m³
ρ = mass density of air, kg/m³

Such a capacitance C may be calculated by use of the perfect gas law.
As given in Section 5-3, for air we have

$$pv = \frac{p}{\rho} = \frac{\bar{R}}{M}T = R_{air}T \tag{5-34}$$

where

$\quad p =$ absolute pressure of air, N/m^2
$\quad v =$ specific volume of air, m^3/kg
$\quad M =$ molecular weight of air per mole, kg/kg-mole
$\quad \bar{R} =$ universal gas constant, N-m/kg-mole K
$R_{air} =$ gas constant of air, N-m/kg K
$\quad T =$ absolute temperature of air, K

If the change of state of air is between isothermal and adiabatic, then the
expansion process can be expressed as polytropic and can be given by

$$\frac{p}{\rho^n} = \text{constant} \tag{5-35}$$

where

$$n = \text{polytropic exponent}$$

Since dp/dp can be obtained from Eq. (5-35) as

$$\frac{d\rho}{dp} = \frac{\rho}{np}$$

by substituting Eq. (5-34) into this last equation, we have

$$\frac{d\rho}{dp} = \frac{1}{nR_{air}T} \tag{5-36}$$

Then capacitance C of a vessel is found from Eqs. (5-33) and (5-36) as

$$C = \frac{V}{nR_{air}T} \quad \frac{\text{kg}}{\text{N}/\text{m}^2} \tag{5-37}$$

Note that if gas other than air is filled in a pressure vessel, capacitance C is
given by

$$C = \frac{V}{nR_{gas}T} \quad \frac{\text{kg}}{\text{N}/\text{m}^2} \tag{5-38}$$

where R_{gas} is the gas constant for the particular gas involved.
From the preceding analysis it is clear that capacitance C of a pressure
vessel is not constant but depends on the expansion process involved, the
nature of gas (air, N_2, H_2, etc.), and the temperature of the gas in the vessel.
The value of polytropic exponent n is approximately constant ($n = 1.0$ to 1.2)
for gases in uninsulated metal vessels.

Example 5-7. Find the capacitance C of a 2 m^3 pressure vessel that contains air at
50°C. Assume that the expansion and compression of air occur slowly and that

there is sufficient time for heat to transfer to and from the vessel so that the expansion process may be considered isothermal, or $n = 1$.

Capacitance C is found by substituting $V = 2$ m³, $R_{air} = 287$ N-m/kg K, $T = 273 + 50 = 323$ K, and $n = 1$ into Eq. (5-37) as follows:

$$C = \frac{V}{nR_{air}T} = \frac{2}{1 \times 287 \times 323} = 2.16 \times 10^{-5} \text{ kg-m}^2/\text{N}$$

Example 5-8. Referring to Example 5-7, if hydrogen (H_2) rather than air is filled in the same pressure vessel, what is the capacitance? Assume that the temperature of gas is 50°C and that the expansion process is isothermal, or $n = 1$.

The gas constant for hydrogen is

$$R_{H_2} = 4121 \text{ N-m/kg K}$$

By substituting $V = 2$ m³, $R_{H_2} = 4121$ N-m/kg K, $T = 273 + 50 = 323$ K, and $n = 1$ into Eq. (5-38), we have

$$C = \frac{V}{nR_{H_2}T} = \frac{2}{1 \times 4121 \times 323} = 1.50 \times 10^{-6} \text{ kg-m}^2/\text{N}$$

Inertance. *Inertance* in a pneumatic system refers to the change in pressure (N/m²) required to make a unit rate of change in mass flow rate (that is, the change in mass flow rate per second) (kg/s²) or

$$\text{Inertance } I = \frac{\text{change in pressure}}{\text{change in mass flow rate per second}} \quad \frac{\text{N/m}^2}{\text{kg/s}^2} \text{ or } \frac{1}{\text{m}}$$

Air (or gas) in pipes may exhibit sustained vibrations (acoustic resonance) because the air (or gas) has inertia and, moreover, it is elastic. Note that an inertance–capacitance combination in a pneumatic system acts like a spring–mass combination in a mechanical system, thus causing vibrations.

Example 5-9. Consider airflow in a pipe and derive airflow inertance.

Airflow inertance can be obtained as the pressure difference between two sections in the pipe required to cause a unit rate of change in flow rate. It is similar to the liquid flow inertance presented in Example 4-2.

Suppose that the cross-sectional area of a pipe is constant and equal to A m² and that the pressure difference between two sections in the pipe is Δp N/m². Then force $A \Delta p$ will accelerate the air between the two sections according to Newton's second law or

$$M\frac{dv}{dt} = A \Delta p \tag{5-39}$$

where M kg is the mass of air in the pipe between two sections and v m/s is the air velocity. Noting that

$$M = \rho AL$$

where ρ kg/m³ is the density of air and L m is the distance between two sections,

Eq. (5-39) can be written

$$\rho A L \frac{dv}{dt} = A\,\Delta p$$

In terms of the mass flow rate $Q = \rho A v$ kg/s, this last equation may be written

$$L\frac{dQ}{dt} = A\,\Delta p$$

Then airflow inertance I is obtained as

$$\text{Airflow inertance } I = \frac{\Delta p}{dQ/dt} = \frac{L}{A}\ \frac{\text{N/m}^2}{\text{kg/s}^2}\ \text{or}\ \frac{1}{\text{m}}$$

Mathematical modeling of a pneumatic system. The pneumatic pressure system shown in Fig. 5-35(a) consists of a pressure vessel and connecting pipe with a valve. In the figure

$\bar{P} = $ steady-state pressure of the system, N/m²
$p_i = $ small change in inflow pressure, N/m²
$p_o = $ small change in air pressure in vessel, N/m²
$V = $ volume of vessel, m³
$m = $ mass of air in vessel, kg
$q = $ mass flow rate, kg/s

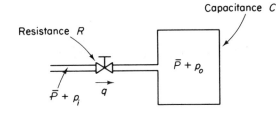

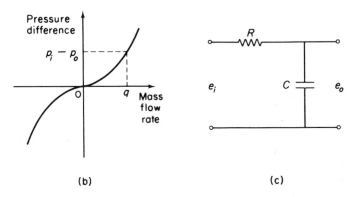

Fig. 5-35. (a) Pneumatic pressure system; (b) pressure difference versus mass flow rate curve; (c) analogous electrical system.

Let us obtain a mathematical model of this pneumatic pressure system. Assume that the system operates in such a way that the average flow through the valve is zero, or the normal operating condition corresponds to $p_i - p_o = 0$, $q = 0$, and that the flow condition is subsonic for the entire range of operation.

Referring to Fig. 5-35(b), average resistance R of the valve can be written

$$R = \frac{p_i - p_o}{q}$$

And referring to Eq. (5-33), capacitance C of the pressure vessel can be written

$$C = \frac{dm}{dp_o}$$

or

$$C \, dp_o = dm$$

This last equation states that capacitance C times pressure change dp_o (during dt seconds) is equal to dm, the change in the mass of air in the vessel (during dt seconds). The change in mass dm is equal to the mass flow during dt seconds, or $q \, dt$. Hence

$$C \, dp_o = q \, dt$$

By substituting $q = (p_i - p_o)/R$ into this last equation, we have

$$C \, dp_o = \frac{p_i - p_o}{R} \, dt$$

Rewriting,

$$RC \frac{dp_o}{dt} + p_o = p_i \tag{5-40}$$

Equation (5-40) is a mathematical model for the system shown in Fig. 5-35(a).

Note that the pneumatic pressure system considered here is analogous to the electrical system shown in Fig. 5-35(c), where its mathematical model is

$$RC \frac{de_o}{dt} + e_o = e_i$$

Note also that in the mathematical models for the two systems, RC has the dimension of time and is the time constant of the respective system.

Comments. Industrial pneumatic controllers may consist of bellows, nozzles, orifices, valves, and pipes connecting these elements. The materials presented in this section are applicable to the mathematical modeling and analysis of pneumatic controllers. Since pneumatic controllers are described in detail in Chapter 8, further discussion of pneumatic pressure systems is postponed until then.

Fluid devices through which air (or gases or liquids) flows in intricate and precise channels and performs sensing, logic, amplification, control, information processing, and so on, are called *fluidic devices*. These devices have no moving parts and are usually made of glass, plastic, aluminum, brass, or stainless steel. *Fluidics* is the general study of fluidic devices and systems.

A fluidic control system is one in which fluidic devices (fluid amplifiers, interface devices, etc.) are used to control a system. Fluidic control can be digital or analog.

Fluid amplifiers. The basic element controlling the flow of fluid is the *fluid amplifier*, which enables a flow or pressure to be controlled by one or more input signals that are of a lower pressure or flow value than the fluid being controlled. Fluid amplifiers utilize a variety of basic principles in their operation and may be classified as

1. Wall-attachment amplifiers
2. Turbulence amplifiers
3. Jet interaction amplifiers
4. Vortex amplifiers

Most commercially available fluid amplifiers fall into these four categories or combinations thereof.

Definitions. Before discussing fluidic devices, let us define certain terminologies.

Input signal. An input signal in fluidic devices is a pressure or flow that is directed into an input port to control an element or logic function.

Output signal. An output signal in fluidic devices is the pressure or flow leaving the output port of an element or logic function.

Vent. A vent is a port that allows fluid to exhaust from an element to a reference or ambient pressure.

Fan-in ratio. Fan-in ratio refers to the number of separate inputs available on an element.

Fan-out ratio. Fan-out ratio means the number of similar elements that can be operated in parallel from a single fluidic element. Here "similar" refers to the impedance and not to devices performing the same function.

---

*Starred sections deal with more challenging topics than the rest of the book. Depending on course objectives, these sections (although important) may be omitted from classroom discussions without losing the continuity of the main subject matter.

[As in the case of electrical systems, the total restriction to circuit flow, represented by resistance, capacitance, and inertance, combined into a resultant, is called *fluidic impedance*. (Impedance is a change in pressure divided by change in flow.)]

Fluidic transducer. A fluidic transducer is a device that uses a fluid dynamic phenomenon for its operation to convert a signal from one medium (i.e., air pressure) to a signal in other medium (i.e., electric voltage).

Wall-attachment effect. Liquid flowing from a faucet will attract the air around it and make it move in the same direction. In Fig. 5-36 an adequate

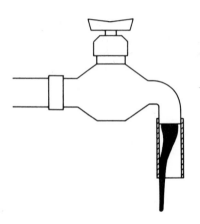

supply of air can be drawn into the larger space to the right of the stream to replace that extracted by the action of the stream. However, the left side has a limited space, and it is difficult for air to replace that which is drawn out. A partial vacuum forms at the left, and the stream moves toward the wall to make up the loss. Assisted by atmospheric pressure, the stream gets closer to the wall and eventually attaches to it. It will remain attached until disturbed. This phenomenon is called the *wall effect*.

A fluid jet can be diverted from its normal path of flow by introducing another jet (control jet) perpendicular to the first as shown in Fig. 5-37. If the fluid jet enters a relatively narrow chamber and if the jet touches one wall, it attaches itself to the wall as shown in Fig. 5-38. It is possible to break such

Fig. 5-36. Liquid flow from a faucet.

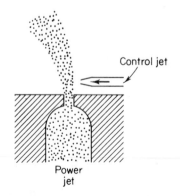

Fig. 5-37. Fluid jet diverted from its normal path by control jet.

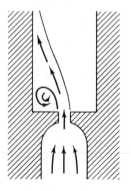

Fig. 5-38. Wall-attachment phenomenon.

wall effects if a stream or jet is applied to the low-pressure region below the point where the jet touches the wall. This fact makes it possible to design a bistable device, or flip-flop, by providing control jets on either side of the main jet. A bistable device is one that has two output possibilities and that will alternate from one output to another on receiving correctly phased input signals. Such a fluidic bistable device, called a *fluidic digital amplifier*, is suitable for logic operations that use binary signals. (A digital amplifier is an element that operates on the "all or nothing" principle. It will give a full output or no output, depending on the input or control signal applied.)

Wall-attachment amplifiers. Fluidic amplifiers whose operating principle is based on wall-attachment phenomenon are called *wall-attachment amplifiers*. An example of one appears in Fig. 5-39. It is a bistable fluidic amplifier or flip-flop and operates as follows.

1. A control signal at "*a*" provides output flow in port *A*.
2. Removal of control pressure does not change flow (clings to wall *C*).
3. A signal at "*b*" switches flow to port *B*.

Thus this device has a memory. Such a bistable amplifier or flip-flop serves as a memory bank in a fluidic system.

In fluidic devices a percentile representation of output capture as related to supply, such as output pressure versus input pressure, is commonly used to indicate the degree of recovery. For the bistable amplifier considered here, the maximum output pressure is about 35% of the supply pressure, whereas the maximum flow out is about 50% of the supply flow. The remaining 50% of the supply flow will exit through the exhaust vents. (The power recovery is normally around 15% of the power supplied.) The minimum control pressure needed to cause switching is about 10% of the supply pressure. The fan-out ratio of wall-attachment amplifiers is about four.

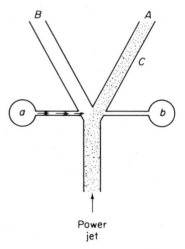

Fig. 5-39. Wall-attachment amplifier.

Turbulence amplifiers. Another type of fluidic devices is the *turbulence amplifier*, which depends on change in flow conditions that result in a change from laminar to turbulent flow in a fluid stream. Figure 5-40 shows a schematic diagram of the turbulence amplifier. A fluid can be made to flow in a straight laminar stream across a gap to produce an output, or it can be

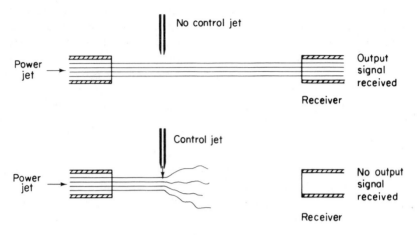

Fig. 5-40. Turbulence amplifier.

prevented from doing so by a small jet that makes the stream turbulent. If a long smooth pipe of small bore is supplied with low-pressure air, it is possible to emit a laminar stream. This stream will remain laminar for a considerable distance before becoming turbulent. (Laminar stream leaving a capillary tube of diameter d may remain laminar for approximately $100d$.) If, therefore, an output tube is placed in line with the supply tube before the point of turbulence, the fluid stream will cross the gap to flow into it, and an output will result. An input control signal placed at right angle to the main jet moves the point of turbulence, decreases flow into the output pipe, and, in fact, prevents an output signal. The power of the input signal necessary to interrupt the laminar jet and make it turbulent is extremely small. So such a device yields a large-pressure gain. The fan-out ratio of turbulence amplifiers is about eight, which is about twice as high as that of wall-attachment amplifiers. Note that the higher the fan-out ratio, the easier it is to interconnect and build into a system. In the turbulence amplifier several input jets can be positioned around the stream, thereby making it possible to combine many input signals into one element. (Thus the fan-in ratio can be made reasonably high—for instance, five or six.)

Jet interaction amplifiers. Although wall-attachment amplifiers are basically bistable devices, they can be changed into proportional devices by widening the passages following the nozzle where wall attachment occurs, as shown in Fig. 5-41. In this amplifier the main jet flow is distributed between the two outlet passages according to the balance of control jets. This type of fluidic amplifier is called a *jet interaction amplifier*. It is a proportional amplifier.

The operating principle of jet interaction amplifiers does not depend on wall-attachment phenomenon but on a direct interaction between control jets and the main power jet. Referring to Fig. 5-41, the input signal is the differential pressure existing across ports "*a*" and "*b*". The output is the differential pressure existing across channels "*A*" and "*B*". So it is an analog unit. (If the control supply is made large enough, however, the output switches completely and the amplifier becomes a digital logic element.)

To acquire some idea of the gain of such an amplifier, note that a pressure gain of 10 and a power gain of 100 per stage are possible. If two such devices are cascaded (see Fig. 5-42), then the gains are multiplied.

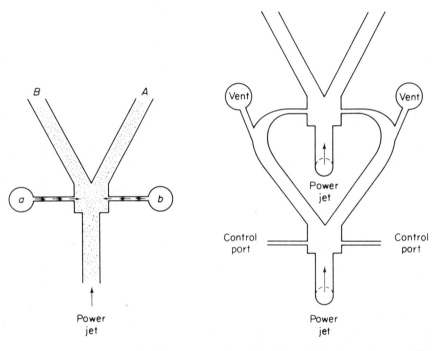

Fig. 5-41. Jet interaction amplifier.   **Fig. 5-42.** Cascaded jet interaction amplifier.

Vortex amplifiers. Figure 5-43 is a schematic diagram of a *vortex amplifier* consisting of a cylindrical vortex chamber, supply port, control port, and outlet port connected to a receiver tube. As explained below, a vortex amplifier acts as a valve that controls flow by means of a constrained vortex. (A control jet is used to generate a vortex or spiral flow.) As the working medium, air, gases, or liquids may be used.

Referring to Fig. 5-43, the power jet (main power flow) is admitted through the supply port. If no control jet is present, there is a steady-state uniform jet that leaves the outlet port and enters the receiver tube. Any excess flow passes into the vent and is discharged to the atmosphere if air is

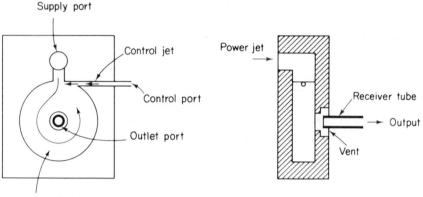

Fig. 5-43. Vortex amplifier.

the fluid medium, or to a low-pressure reservoir if gases (other than air) or liquids are used.

The output flow rate is determined by the area of the outlet port and the supply pressure. Since the supply port is made larger than the outlet port, if no control jet is present, the pressure in the vortex chamber is constant and is equal to the supply pressure. When the control jet is admitted through the tangential control port, it mixes with the power jet and generates a vortex in the chamber. Because of the conservation of angular momentum, as the radius decreases, the tangential velocity increases and the pressure in the vortex decreases. Because of the radial pressure gradient in the vortex, the outlet flow rate is decreased as the vortex is made stronger. In fact, as the vortex becomes stronger, most of the flow fans out through the vent and little flows through the receiver tube. Therefore a vortex amplifier acts essentially as a fluidic nonmoving-parts valve that offers a simple and reliable means of controlling fluid flows. In most applications, vortex amplifiers will control flow from full output to about 10% of the full output.

Advantages of the vortex amplifier used as a fluid control valve over conventional control valves are (a) reliability is good because no moving parts are involved and (b) there is no hysteresis in the valve characteristic curves.

The disadvantages can be stated as: (a) the output flow rate cannot be reduced to exactly zero, (b) the control pressure must be 30 to 70% greater than the supply pressure, and (c) the power consumption may be high, although vortex amplifiers may recover up to 40% of the power supplied to it.

If air is used as a fluid medium, supply pressure is normally 1.4×10^5 to 7×10^5 N/m^2 gage (approximately 20.3 to 101.5 psig, which corresponds approximately to 1.43 to 7.14 kg$_f$/cm^2 gage), but in some special cases the supply pressure may be made much lower or much higher.

Speed of response of fluidic devices. As far as the speed of response of fluidic devices is concerned, they are comparable to conventional pneumatic and hydraulic devices or electromechanical relays. This means that the speed of response of fluidic devices is very much slower than that of electronic devices. Electronic responses are generally expressed in terms of microseconds (10^{-6} second) or nanoseconds (10^{-9} second), whereas fluidic responses are expressed in terms of milliseconds (10^{-3} second).

Advantages of fluidic devices. The advantages of fluidic devices over other types of devices performing similar functions can be summarized as follows.

1. Because of the absence of mechanical moving parts, fluidic devices are rugged.
2. The devices are simple; of small size; highly reliable; long lasting; may be operated with air, gas, or any low-viscosity liquid; and require very low maintenance.
3. They are fire and explosion proof if air (or any other noncombustible fluid) is used as the fluid medium.
4. Connections can be changed while the fluidic system is under power without the shock hazard inherent in electrical and electronic systems.
5. Fluidic devices are unaffected by high or low temperature (if air or gas is used as the fluid medium), vibration, G-force (acceleration), electric or magnetic field, and nuclear radiation, all of which void electrical components. (Thus they can perform control and computational functions in adverse environments and are ideal for operations in hazardous locations, such as mines, refineries, and chemical plants, in which other devices may fail.)
6. If mass produced, they are inexpensive.

Disadvantages of fluidic devices. The disadvantages of fluidic devices are listed below.

1. Fluidic devices require very clean fluid, especially in amplifier circuits.
2. Their response is much slower than corresponding electronic devices.
3. They consume relatively high power. (While a fluidic device is operating, there is a constant flow of fluid through it. High-power

consumption can be a distinct disadvantage in certain applications where available power is quite limited, such as in a satellite system.)

4. Design of suitable interface devices has lagged.

*5-7 DIGITAL FLUIDICS AND LOGIC CIRCUITS

Many of the electronic circuits used in computing equipment function as high-speed switches. The operation that they perform can also be handled by fluidic devices but with much lower speeds. Because fluidic switching circuits produce output signals only when certain input conditions exist, they are referred to as *logic circuits* or *decision elements*.

After first reviewing mathematical logic, we describe basic fluidic logic circuits and then discuss general procedures for building desired logic circuits by using NOR elements.

Digital fluidics. *Digital fluidic devices* are those fluidic components that perform logic functions. (A logic function is an assembly of logic elements that can produce a desired effect when certain conditions are satisfied.)

A *gate* is a device or circuit that allows passage of a signal only if certain control requirements have been satisfied. As we shall see, fluidic logic gates can be built into familiar digital circuits.

In addition to performing the same logic functions as their electronic counterparts, digital fluidics offer high reliability. Although their operating speed is much lower than that of electronic devices, when reliability in extreme environments (for instance, high temperature or high radiation) is more important than speed of operation, they are far superior and, in some cases, may be the only solution. Today fluidic components that are capable of providing logic functions are commercially available. Some present-day applications of digital fluidics occur frequently in automatic warehousing, machine feeding, sequencing, handling, and similiar operations.

Mathematical logic. Logic functions are frequently written in the function notation of boolean algebra, a type of algebra particularly suited to the description and design of switching circuits.

Logic is essentially two-valued—that is, it is either true or not true. The convention is that 1 (ONE) stands for the "yes" and 0 (ZERO) stands for the "no" of a proposition. The same convention applies in this book; and the symbol "1" denotes the presence of a signal (positive signal or true statement) and the symbol "0" denotes the absence of a signal (zero signal or a false statement).

---

*Starred sections deal with more challenging topics than the rest of the book. Depending on course objectives, these sections (although important) may be omitted from classroom discussions without losing the continuity of the main subject matter.

In any control system it is the function of the logic circuitry to accept input information (signals), make decisions based on this information, and initiate signals for control action. When making these decisions, basic logic operators define the desired relationships between the information received and thus determine the decision that is made.

In the following material input signals are denoted by letters A, B, C, and so on and output signals by letters occurring at the end of the alphabet, X, Y, Z. Signs between letters show their relationship.

$\cdot$ denotes "and"

$+$ denotes "or"

Thus $A \cdot B$ is read A and B. As in other types of algebraic notation, the dot sign can be omitted and may be written AB. Similarly, $A + B$ is read A or B. If a bar is placed above a letter, it means its complement or negative logic value. So if $A = 1$, then $\bar{A} = 0$.

Basic logic functions. Three of the basic logic functions are AND, OR, and NOT. Their symbols are repeated below.

AND: A and B and $C = A \cdot B \cdot C = ABC$

OR: A or B or $C = A + B + C$

NOT: Not $A = \bar{A}$

Notice that there are two OR relationships—the INCLUSIVE OR and the EXCLUSIVE OR. The difference between the two can be seen from the following interpretations of A or B.

INCLUSIVE OR: A or B or both A and B

EXCLUSIVE OR: A or B, but not both A and B

The EXCLUSIVE OR function (frequently called the *half adder*) delivers an output signal only when its two input signals are different. The half adder circuit is often used as a comparator, and it provides a convenient way of performing the binary addition of two input signals. In this book OR refers to INCLUSIVE OR unless specifically stated otherwise.

There are other logic functions besides AND, OR, and NOT. Such functions, however, may be made up of combinations of the three. For instance, consider EXCLUSIVE OR. The mathematical expression for EXCLUSIVE OR is

$$(A + B)\overline{AB}$$

which is interpreted as A or B but not both A and B. Consider another example, INHIBITOR. "A is inhibited by B" means that A occurs if B is not present. The mathematical logic expression for the INHIBITOR is

$$A\bar{B}$$

Basic identities and laws for mathematical logic. First, we present basic identities.

$$A0 = 0$$
$$1A = A$$
$$A\bar{A} = 0$$
$$AA = A$$
$$A + A = A$$
$$0 + A = A$$
$$1 + A = 1$$
$$A + \bar{A} = 1$$
$$\bar{\bar{A}} = A$$

The basic laws that all logic expressions follow are

Commutative law:

$$A + B = B + A$$
$$AB = BA$$

Associative law:

$$A + (B + C) = (A + B) + C$$
$$A(BC) = (AB)C$$

Distributive law:

$$A(B + C) = AB + AC$$
$$(A + B)(C + D) = AC + BC + AD + BD$$

Table 5-2 shows some useful identities for logic equations. (In Table 5-2, identities 15, 16, 19, 20, 21, and 22 are proved in Problem A-5-15.)

Basic logic circuits. The most fundamental operators that form logic functions are

OR, AND, NOT, NOR, NAND, FLIP-FLOP

These operators are described below, together with their respective truth tables. (A *truth table* is a tabular correlation of input and output relationships for logic elements.) It will be seen that digital fluidic devices can gate, or inhibit, signal transmission by the application or removal of input signals.

OR circuit. Figure 5-44(a) shows a two-input *OR circuit* and its truth table. This circuit has two inputs and produces an output when the input signal is applied to one or both of the input terminals. If the circuit has more than two input terminals, the application of any one or more of a number of input signals yields a positive output signal. That is, the OR circuit performs

Table 5-2. SOME USEFUL IDENTITIES FOR LOGIC EQUATIONS

1	$A0 = 0$
2	$1A = A$
3	$A\bar{A} = 0$
4	$AA = A$
5	$A + A = A$
6	$0 + A = A$
7	$1 + A = 1$
8	$A + \bar{A} = 1$
9	$\bar{\bar{A}} = A$
10	$A + B = B + A$
11	$AB = BA$
12	$A + (B + C) = (A + B) + C = A + B + C$
13	$A(BC) = (AB)C = ABC$
14	$A(B + C) = AB + AC$
15	$A(A + B + C) = A$
16	$A + BC = (A + B)(A + C)$
17	$\overline{A + B} = \bar{A}\bar{B}$
18	$\overline{AB} = \bar{A} + \bar{B}$
19	$A + AB = A$
20	$A + \bar{A}B = A + B$
21	$A\bar{B} + \bar{A}B = (A + B)(\bar{A} + \bar{B})$
22	$(A + \bar{B})(\bar{A} + B) = AB + \bar{A}\bar{B}$

the logic operation of producing a true ("1") output when any one or more of its inputs are true ("1"), and a false ("0") output when none of its inputs is true. In Fig. 5-44(b) an electrical analog of the two-input OR circuit is shown. Closing either one or both of the switches causes a signal to be delivered at X. Figure 5-44(c) illustrates a three-input OR circuit. The output X is $A + B + C$. This is read: X is equal to A or B or C.

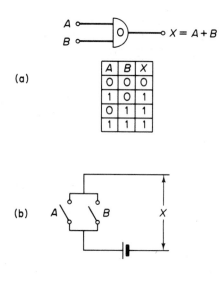

(a)

(b)

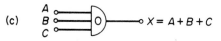

(c)

Fig. 5-44. (a) Two-input OR circuit and its truth table; (b) electrical analog; (c) three-input OR circuit.

AND circuit. The *AND circuit* has two or more inputs and one output. One characteristic of this circuit is that it produces an output only when all inputs are applied simultaneously. In other words, the AND function requires that all input signals be simultaneously present before an output signal is delivered. Figure 5-45(a) shows a two-input AND circuit (with the inputs designated as A and B) and its truth table. A positive signal appears at output X only when signals are present at both inputs A and B. The output is $X = AB$. This is read: X is equal to A and B. In Fig. 5-45(b) we have an electrical analog of the two-input AND circuit. Both switches must be closed before a positive signal appears at X. Figure 5-45(c) shows a three-input AND circuit.

NOT circuit. The *NOT circuit* [Fig. 5-46(a)] performs the function of negation. It is called an "inverter" because the application of an input signal destroys the output. This circuit produces no output when the input is

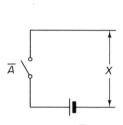

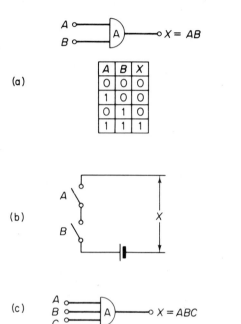

(a)

(b)

(c)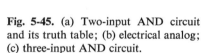

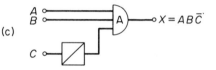

(a)

(b)

Switch open: $\bar{A} = 0$

Switch closed: $\bar{A} = 1$

(c)

Fig. 5-45. (a) Two-input AND circuit and its truth table; (b) electrical analog; (c) three-input AND circuit.

Fig. 5-46. (a) NOT circuit; (b) electrical analog; (c) combination of AND and NOT circuits.

present. If the input signal is removed, the output will be positive. If the input terminal is designated A and the output X, then $X = \bar{A}$. This is read: X is not equal to A. Figure 5-46(b) depicts an electrical analog of the NOT circuit. The application of signal A breaks the circuit and output X will cease. (Note that closing the switch corresponds to $\bar{A} = 1$, which implies $A = 0$. An open switch corresponds to $\bar{A} = 0$, or $A = 1$.)

A combination of an AND and a NOT circuit is shown in Fig. 5-46(c). The output is $X = AB\bar{C}$. If an input is applied to terminal C ($C = 1$), then there will be no output from the NOT circuit and hence no input to one of the AND terminals and no output at X ($X = 0$). The presence of an input to the NOR circuit inhibits the circuit and so no output is produced. (Output from this circuit occurs only when input C is absent and A and B occur simultaneously.) Such a circuit is called an *inhibitor* circuit. It utilizes a supplementary input signal to inhibit or turn off the output. Thus an output from the circuit will occur only when the inhibiting signal is absent. This feature is useful in digital circuits when we wish to obtain an output from a circuit only when certain conditions have been satisfied in other system components.

NOR circuit. The *NOR circuit* is a combination of an OR and a NOT circuit. Figure 5-47(a) illustrates the NOR circuit and its truth table. The NOR function requires that all input signals be removed before an output is possible. Thus in Fig. 5-47(a) the output X is equal to $\overline{A + B}$. Neither A nor B can be present if output X is required.

Figure 5-47(b) shows an electrical analog of the NOR circuit. The operation of one switch will open the circuit and prevent the positive output at X.

(a)

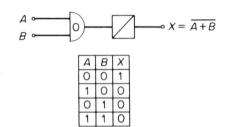

A	B	X
0	0	1
1	0	0
0	1	0
1	1	0

(b)

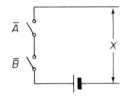

Switch open: $\overline{A} = 0$, $\overline{B} = 0$
Switch closed: $\overline{A} = 1$, $\overline{B} = 1$

(c)

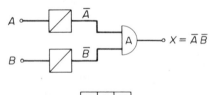

$X = \overline{A}\,\overline{B}$

A	B	X
0	0	1
1	0	0
0	1	0
1	1	0

(d)

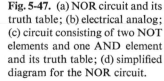

$X = \overline{A + B} = \overline{A}\,\overline{B}$

Fig. 5-47. (a) NOR circuit and its truth table; (b) electrical analog; (c) circuit consisting of two **NOT** elements and one **AND** element and its truth table; (d) simplified diagram for the NOR circuit.

Note that the truth table for the circuit shown in Fig. 5-47(c), which consists of two NOT elements and one AND element, is the same as that for the circuit shown in Fig. 5-47(a). The output X of the circuit shown in Fig. 5-47(c) is $X = \bar{A}\bar{B}$. Since the truth tables for the two circuits are the same, we obtain

$$\overline{A + B} = \bar{A}\bar{B}$$

This relationship holds true for circuits with many inputs, and the general expression is

$$\overline{A + B + C + \cdots} = \bar{A}\bar{B}\bar{C}\cdots$$

This relationship is called *De Morgan's summation law.*

The NOR element is important in logic circuits. Although it can be built of other logic elements (such as OR and NOT elements), this element may be considered a primitive logic function and as such can be used as a basic building block from which any other logic function is generated. We shall discuss this subject in detail later on. The simplified diagram of the NOR circuit consisting of a triangle, input terminals, and output terminal, as shown in Fig. 5-47(d), can be used with advantage to draw logic diagrams.

NAND circuit. The NAND function is the inverse of the AND function. The *NAND circuit* is a combination of an AND and a NOT circuit. Figure 5-48(a) shows a NAND circuit and its truth table. The output X is $\overline{AB}$. In the NAND circuit all input signals must be present before the output ceases. So both A and B must be applied to stop the output.

Figure 5-48(b) shows an electrical analog of the NAND circuit. Here the normally closed switches must all be open before the output signal at X ceases.

Note that the circuit shown in Fig. 5-48(c), which consists of two NOT elements and one OR element, has the same truth table as the one in Fig. 5-48(a). Since the output of the circuit in Fig. 5-48(a) is $X = \overline{AB}$ and that in Fig. 5-48(c) is $X = \bar{A} + \bar{B}$, we see that

$$\overline{AB} = \bar{A} + \bar{B}$$

This relationship holds true for the case of many inputs, and the general expression is

$$\overline{ABC\cdots} = \bar{A} + \bar{B} + \bar{C} + \cdots$$

This relationship is called *De Morgan's multiplication law.*

FLIP-FLOP. The *flip-flop* is a memory function. In Fig. 5-49(a) a flip-flop circuit that consists of two NOR elements is shown. The output of one NOR element is fed to the input of the other NOR element.

If a brief input signal (such as a pulse signal) is applied to one of the

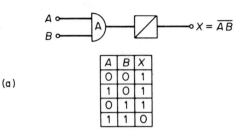

(a)

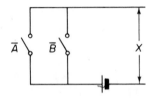

(b)

Switch open: $\bar{A} = 0$, $\bar{B} = 0$
Switch closed: $\bar{A} = 1$, $\bar{B} = 1$

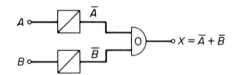

(c)

A	B	X
0	0	1
1	0	1
0	1	1
1	1	0

Fig. 5-48. (a) NAND circuit and its truth table; (b) electrical analog; (c) circuit consisting of two NOT elements and one OR element and its truth table.

input terminals, this circuit converts the brief input signal to a maintained signal. This maintained signal can be removed by applying a second brief input signal at the other input terminal.

An electrical analog of the flip-flop circuit appears in Fig. 5-49(b). In this circuit a momentary operation of normally open contact A causes the solenoid coil to energize, which, in turn, closes contact C. So current will continue to flow through contact C to the solenoid, thereby maintaining this condition and providing a signal at X. If normally closed contact B is momentarily opened, the current flow to the solenoid will be interrupted, contact C will open, and output X will cease.

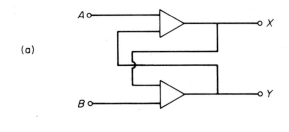

(a)

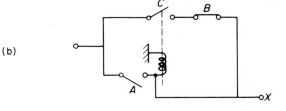

(b)

Fig. 5-49. (a) Flip-flop circuit;
(b) electrical analog.

Example 5-10. In order to illustrate the decision-making properties of logic circuits, two examples are given here. Consider the circuits shown in Fig. 5-50(a) and (b). Each circuit recognizes certain conditions of input and produces an output when these input conditions exist.

The circuit shown in Fig. 5-50(a) produces an output at X if either C is present or A and B occur simultaneously or both. Thus $X = AB + C$.

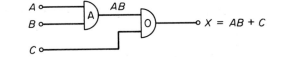

(a)

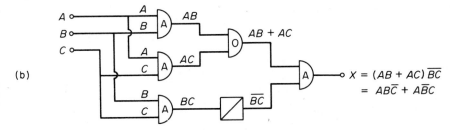

(b)

Fig. 5-50. Logic circuits.

In the circuit of Fig. 5-50(b) three two-input AND circuits detect the three combinations of two inputs: A and B, A and C, and B and C. If the combination of (A and B), or (A and C), but not (B and C) occurs, the circuit produces the output. Thus the output X is equal to $(AB + AC)\overline{BC} = AB\bar{C} + A\bar{B}C$, as can be seen from the fact that

$$(AB + AC)\overline{BC} = (AB + AC)(\bar{B} + \bar{C})$$
$$= AB\bar{B} + AC\bar{B} + AB\bar{C} + AC\bar{C}$$

Since $B\bar{B} = 0$, $C\bar{C} = 0$, we have

$$(AB + AC)\overline{BC} = AB\bar{C} + A\bar{B}C$$

In this logic circuit the combination of B and C is detected by one of the AND circuits. This signal, acting through the inverter, inhibits the output. (If B and C occur simultaneously, there is no output.)

Building logic circuits by use of NOR elements. It will be shown below that NOR elements can be used to build any logic circuits. Schematic diagrams of one type of fluidic NOR element appear in Fig. 5-51. This type of NOR element is called a *flowboard amplifier*. It is a turbulence amplifier and its operation is illustrated in the figure. Twenty or more of such NOR

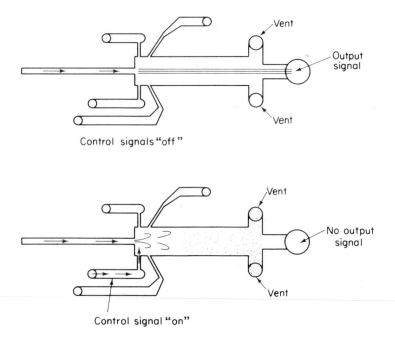

Control signals "off"

Control signal "on"

Fig. 5-51. Flowboard amplifier.

elements can be assembled in one board, and these elements can be connected to produce desired logic functions.

The flowboard amplifier has several advantages. It allows high-speed switching while, at the same time, being insensitive to vibration; it can be combined in circuits without tuning; and it is load insensitive.

Building OR, AND, NAND circuits by use of NOR elements only. A diagram of a NOR element and its output is shown in Fig. 5-52(a). In order to build an OR circuit, we need two NOR elements as in Fig. 5-52(b). An AND circuit can be built by use of three NOR elements as in Fig. 5-52(c). Similarly, a NAND circuit can be built by using four NOR elements as in Fig. 5-52(d).

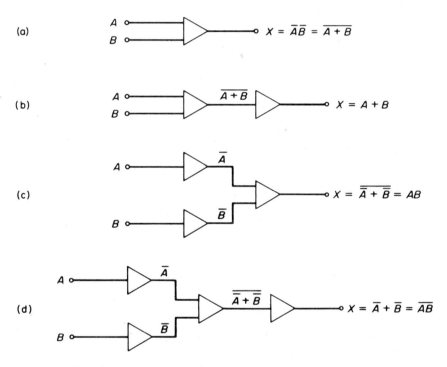

Fig. 5-52. Building logic circuits by use of NOR elements only; (a) NOR circuit; (b) OR circuit; (c) AND circuit; (d) NAND circuit.

General procedures for building desired logic circuits by use of NOR elements. Suppose that we wish to obtain a logic circuit for the system whose truth table is given in Table 5-3. The condition on A, B, and C under which X becomes "1" is $A = 1$, $B = 0$, $C = 1$. Hence

$$X = A\bar{B}C = \overline{\bar{A} + B + \bar{C}}$$

The logic circuit for this logic expression appears in Fig. 5-53.

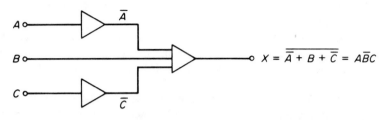

Fig. 5-53. Logic circuit.

Next, let us consider another problem. Assume that we wish to build a logic circuit that will give the truth table shown in Table 5-4. In order to find

Table 5-3. TRUTH TABLE

A	B	C	X
1	1	1	0
1	1	0	0
1	0	1	1
1	0	0	0
0	1	1	0
0	1	0	0
0	0	1	0
0	0	0	0

Table 5-4. TRUTH TABLE

A	B	C	X
1	1	1	0
1	1	0	1
1	0	1	1
1	0	0	1
0	1	1	1
0	1	0	0
0	0	1	1
0	0	0	1

the logic expression for X that is equivalent to the information given in this table, we determine the conditions in the left-hand columns under which a "1" appears in each element of the X column. In this problem six logic expressions are involved, and the mathematical logic expression for the output X is

$$X = AB\bar{C} + A\bar{B}C + A\bar{B}\bar{C} + \bar{A}BC + \bar{A}\bar{B}C + \bar{A}\bar{B}\bar{C}$$

This expression can be simplified by using the formulas given in Table 5-2.

$$X = A\bar{C}(B + \bar{B}) + A\bar{B}C + \bar{A}BC + \bar{A}\bar{B}(C + \bar{C})$$
$$= A\bar{C} + A\bar{B}C + \bar{A}BC + \bar{A}\bar{B}$$
$$= A(\bar{C} + \bar{B}C) + \bar{A}(BC + \bar{B})$$

$$= A(\bar{C} + \bar{B}) + \bar{A}(\bar{B} + C)$$
$$= (A + \bar{A})\bar{B} + A\bar{C} + \bar{A}C$$
$$= \bar{B} + A\bar{C} + \bar{A}C$$
$$= \bar{B} + \overline{\bar{A} + C} + \overline{A + \bar{C}}$$

Figure 5-54 shows this logic circuit. Similarly, any logic circuit can be constructed by use of NOR elements only.

Notice that the same result can be obtained much more easily if we write the logic expression for $\bar{X}$. (Since the number of "0" elements in the X column is much smaller than the number of "1" elements in the same column, the logic expression for $\bar{X}$ is much simpler than that for X.) The conditions in the left-hand columns under which a "0" appears in each element of the X column are ABC and $\bar{A}B\bar{C}$. Thus we obtain

$$\bar{X} = ABC + \bar{A}B\bar{C}$$
$$= B(AC + \bar{A}\bar{C})$$
$$= B(\bar{A} + C)(A + \bar{C})$$

And so

$$X = \overline{B(\bar{A} + C)(A + \bar{C})} = \bar{B} + \overline{\bar{A} + C} + \overline{A + \bar{C}}$$

The logic circuit for this logic expression is the same as the one shown in Fig. 5-54.

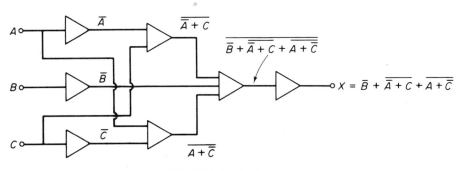

Fig. 5-54. Logic circuit.

Example 5-11. Construct a logic circuit that will give the outputs in the truth table of Table 5-5.

Since the numbers of "0" elements in the X and Y columns are much smaller than the numbers of "1" elements, we shall write logic expressions for $\bar{X}$ and $\bar{Y}$.

$$\bar{X} = \bar{A}\bar{B}\bar{C}$$
$$\bar{Y} = \bar{A}\bar{B}\bar{C} + \bar{A}B\bar{C}$$

Table 5-5. TRUTH TABLE

A	B	C	X	Y
1	1	1	1	1
1	1	0	1	1
1	0	1	1	1
1	0	0	1	1
0	1	1	1	1
0	1	0	1	0
0	0	1	1	1
0	0	0	0	0

These two expressions can be rewritten as

$$X = \overline{\overline{A}\overline{B}\overline{C}} = \overline{\overline{A + B + C}} = A + B + C \tag{5-41}$$

$$Y = \overline{\overline{Y}} = \overline{\overline{A}\overline{B}\overline{C} + \overline{A}\overline{B}C} = \overline{\overline{A}\overline{C}(\overline{B} + B)} = \overline{\overline{A}\overline{C}} = \overline{\overline{A + C}} = A + C \tag{5-42}$$

The logic circuit that will give these logic expressions is shown in Fig. 5-55.

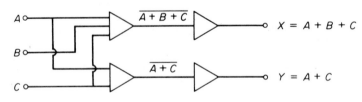

Fig. 5-55. Logic circuit.

Note that the same result can, of course, be obtained by writing logic expressions for X and Y and simplifying them.

$$X = ABC + AB\bar{C} + A\bar{B}C + A\bar{B}\bar{C} + \bar{A}BC + \bar{A}B\bar{C} + \bar{A}\bar{B}C \tag{5-43}$$

$$Y = ABC + AB\bar{C} + A\bar{B}C + A\bar{B}\bar{C} + \bar{A}BC + \bar{A}\bar{B}C \tag{5-44}$$

Equations (5-43) and (5-44) can be reduced to Eqs. (5-41) and (5-42), respectively. The reduction steps are shown below.

$$X = AB(C + \bar{C}) + A\bar{B}(C + \bar{C}) + \bar{A}B(C + \bar{C}) + \bar{A}\bar{B}C$$

$$= AB + A\bar{B} + \bar{A}B + \bar{A}\bar{B}C$$

$$= A(B + \bar{B}) + \bar{A}(B + \bar{B}C)$$

$$= A + \bar{A}(B + C) = (A + \bar{A}B) + (A + \bar{A}C)$$
$$= A + B + A + C = A + B + C$$
$$Y = AB(C + \bar{C}) + A\bar{B}(C + \bar{C}) + \bar{A}C(B + \bar{B})$$
$$= AB + A\bar{B} + \bar{A}C = A(B + \bar{B}) + \bar{A}C$$
$$= A + \bar{A}C = A + C$$

REFERENCES

5-1   ANDERSEN, B. W., *The Analysis and Design of Pneumatic Systems*, New York: John Wiley & Sons, Inc., 1967.

5-2   CHAPMAN, W. P., "Fluidics: Progress and Growing Pains," *Mechanical Engineering*, **89**, No. 10, October 1967, pp. 48–51.

5-3   HENKE, R. W., "Digital Fluidics Works Now," *Control Engineering*, **14**, No. 1, January 1967, pp. 100–104.

5-4   JASKOLSKI, E. P. AND D. T. CAMP, "Fluidic Threshold Logic," *Mechanical Engineering*, **94**, No. 9, September 1972, pp. 24-28.

5-5   WALKER, J. H., "Inspection and Sorting with Fluidics," *Mechanical Engineering*, **92**, No. 3, March 1970, pp. 14–19.

EXAMPLE PROBLEMS AND SOLUTIONS

PROBLEM A-5-1. Consider the pneumatic system shown in Fig. 5-56. The load consists of a mass m and friction. The frictional force is assumed to be $\mu N = \mu mg$. If $m = 1000$ kg, $\mu = 0.3$, and $p_1 - p_2 = 5 \times 10^5$ N/m^2, find the minimum area of the piston needed if the load is to be moved.

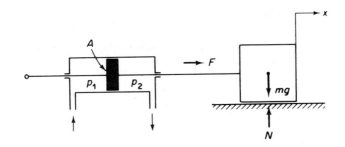

Fig. 5-56. Pneumatic system.

Solution. Assume that the minimum area of the piston is A m^2. Then the minimum force needed to move the load mass is

$$F = A(p_1 - p_2) = \mu mg$$

Hence

$$A = \frac{\mu mg}{p_1 - p_2} = \frac{0.3 \times 1000 \times 9.81}{5 \times 10^5} = 0.00589 \text{ m}^2$$

Thus the minimum area of the piston is 58.9 cm².

PROBLEM A-5-2. In the system shown in Fig. 5-57 assume that the friction force acting on the mass is μN. Find the minimum pressure difference $\Delta p = p_1 - p_2$ necessary to move the load m. The piston area is A.

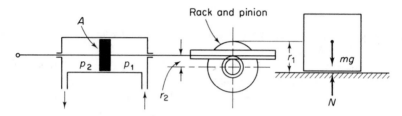

Fig. 5-57. Pneumatic system.

Solution. The minimum force F necessary to move the load is

$$F = A(p_1 - p_2) = \frac{r_1}{r_2} \mu N = \frac{r_1}{r_2} \mu mg$$

Consequently,

$$\Delta p = p_1 - p_2 = \frac{r_1 \mu mg}{r_2 A}$$

PROBLEM A-5-3. A six-pulley hoist is shown in Fig. 5-58. If the piston area A is 30×10^{-4} m² and the pressure difference $p_1 - p_2$ is 5×10^5 N/m², find the mass m of the maximum load that can be pulled up.

Solution. The pneumatic force on the piston is

$$A(p_1 - p_2) = 30 \times 10^{-4} \times 5 \times 10^5 = 1500 \text{ N}$$

Note that in this system the piston pulls six ropes. Since tension in the rope is the same on its entire length, we obtain

$$6F = 1500 \text{ N}$$

where F is the tension in the rope and also the lifting force. This force should be equal to mg. Thus

$$F = mg$$

or

$$m = \frac{1500}{9.81 \times 6} = 25.5 \text{ kg}$$

PROBLEM A-5-4. If the pressure is -30 mm Hg gage and the atmospheric pressure is 755 mm Hg, what is the absolute pressure in N/m², kg$_f$/cm², and lb$_f$/in.² ?

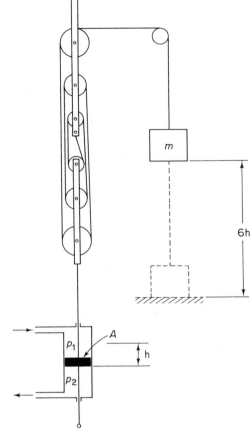

Fig. 5-58. Six-pulley hoist.

Solution. The absolute pressure p is

$$p = 755 - 30 = 725 \text{ mm Hg}$$

Since

$$1 \text{ N/m}^2 = 7.501 \times 10^{-3} \text{ mm Hg}$$

$$1 \text{ kg}_f/\text{cm}^2 = 735.6 \text{ mm Hg}$$

$$1 \text{ lb}_f/\text{in.}^2 = 51.71 \text{ mm Hg}$$

we obtain

$$p = \frac{725}{7.501 \times 10^{-3}} = 9.665 \times 10^4 \text{ N/m}^2 \text{ abs}$$

$$p = \frac{725}{735.6} = 0.986 \text{ kg}_f/\text{cm}^2 \text{ abs}$$

$$p = \frac{725}{51.71} = 14.02 \text{ lb}_f/\text{in.}^2 \text{ abs}$$

PROBLEM A-5-5. A body of mass 50 kg is raised by 30 m. Find the work done in terms of heat Q in kcal.

Solution. The work done L is

$$L = 50 \times 9.81 \times 30 = 1.47 \times 10^4 \text{ N-m}$$

Since the mechanical equivalent of heat is $J = 4186$ N-m/kcal, the heat Q in kcal is

$$Q = \frac{L}{J} = \frac{1.47 \times 10^4}{4186} = 3.51 \text{ kcal}$$

PROBLEM A-5-6. Figure 5-59 shows a safety valve for a boiler. Mass m of the weight is 20 kg. Neglecting the weight of the valve and lever, determine the distance $\overline{OC}$ so that the cracking pressure is 6×10^5 N/m^2 gage (which is equal to 6.12 kg$_f$/cm^2 gage or 87 psig). Area A of the valve is 15×10^{-4} m^2.

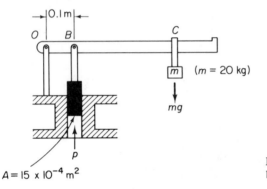

Fig. 5-59. Safety valve for a boiler.

Solution. The torque balance equation is

$$A \, \Delta p \times \overline{OB} = mg \times \overline{OC}$$

where Δp is the difference between the pressure inside the tank and the atmospheric pressure. Thus $\Delta p = 6 \times 10^5$ N/m^2. It follows that

$$\overline{OC} = \frac{A \, \Delta p \times \overline{OB}}{mg} = \frac{15 \times 10^{-4} \times 6 \times 10^5 \times 0.1}{20 \times 9.81} = 0.459 \text{ m}$$

PROBLEM A-5-7. Assume that a cylinder contains 0.1 kg of air at pressure 2×10^5 N/m^2 abs and temperature 20°C. If the air is compressed isentropically to 4×10^5 N/m^2 abs, find the final temperature and the work done on the gas.

Solution. For isentropic change of state,

$$p_1 V_1^k = p_2 V_2^k$$

So noting that $p_1 V_1 / T_1 = p_2 V_2 / T_2$, we have

$$\frac{T_2}{T_1} = \left(\frac{V_1}{V_2}\right)^{k-1} = \left(\frac{p_2}{p_1}\right)^{(k-1)/k}$$

or

$$T_2 = T_1 \left(\frac{p_2}{p_1}\right)^{(k-1)/k} = (273 + 20)\left(\frac{4 \times 10^5}{2 \times 10^5}\right)^{(1.4-1)/1.4}$$

$$= 293 \times 2^{0.2857} = 293 \times 1.22 = 357 \text{ K} = 84°\text{C}$$

Since the change of state is isentropic, the work done on the gas must equal the negative of its increase in internal energy. Thus the work done AL per m kg mass of air is

$$AL = U_1 - U_2 = mc_v(T_1 - T_2)$$

Therefore

$$AL = 0.1 \times 0.171 \times (293 - 357) = -1.10 \text{ kcal}$$

PROBLEM A-5-8. Air is compressed into a tank of volume 2 m³. The compressed air pressure is 5×10^5 N/m² gage and the temperature is 20°C. Find the mass of air in the tank. Also, find the specific volume and specific weight of the compressed air.

Solution. The pressure and temperature are

$$p = (5 + 1.0133) \times 10^5 \text{ N/m}^2 \text{ abs}$$

$$T = 273 + 20 = 293 \text{ K}$$

Referring to Example 5-3, the gas constant of air is $R_{air} = 287$ N-m/kg K. Therefore the mass of the compressed air is

$$m = \frac{pV}{R_{air}T} = \frac{6.0133 \times 10^5 \times 2}{287 \times 293} = 14.3 \text{ kg}$$

The specific volume v is

$$v = \frac{V}{m} = \frac{2}{14.3} = 0.140 \text{ m}^3/\text{kg}$$

The specific weight γ is

$$\gamma = \frac{mg}{V} = \frac{14.3 \times 9.81}{2} = 70.1 \text{ N/m}^3$$

PROBLEM A-5-9. Sound is a longitudinal wave phenomenon representing the propagation of compressional waves in an elastic medium. The speed c of propagation of sound waves is the square root of the ratio of the elastic modulus E to the density ρ of the medium or

$$c = \sqrt{\frac{E}{\rho}}$$

For gases

$$c = \sqrt{\frac{dp}{d\rho}}$$

Show that the speed c of sound can also be given by

$$c = \sqrt{kRT}$$

where

k = ratio of specific heats, c_p/c_v
R = gas constant
T = absolute temperature

Solution. Since the pressure and temperature changes due to passage of a sound wave are negligible, the process can be considered isentropic. Then

$$\frac{p}{\rho^k} = \text{constant}$$

Therefore

$$\frac{dp}{d\rho} = \frac{kp}{\rho}$$

Since $p = \rho RT$, we obtain

$$c = \sqrt{\frac{dp}{d\rho}} = \sqrt{\frac{kp}{\rho}} = \sqrt{kRT}$$

For a given gas, the values of k and R are constant. So the speed of sound in a gas is a function of its absolute temperature only.

PROBLEM A-5-10. Find the speed of sound in air when the temperature is 293 K.

Solution. Noting that for air

$$k = 1.40$$

$$R_{\text{air}} = 287 \text{ N-m/kg K}$$

we have

$$c = \sqrt{kR_{\text{air}}T} = \sqrt{1.40 \times 287 \times 293} = 343.1 \text{ m/s}$$
$$= 1235 \text{ km/h} = 1126 \text{ ft/s} = 768 \text{ mi/h}$$

PROBLEM A-5-11. In dealing with gas systems, we find it convenient to work in molar quantities because one mole of any gas contains the same number of molecules. Thus one mole occupies the same volume if measured under the same conditions of pressure and temperature.

At standard pressure and temperature (1.0133×10^5 N/m² abs and 273 K, or 14.7 psia and 492°R), one kg mole of any gas is found to occupy 22.4 m³ (or 1 lb mole of any gas is found to occupy 359 ft³). For instance, at standard pressure and temperature, the volume occupied by 2 kg of hydrogen, 32 kg of oxygen, or 28 kg of nitrogen is the same, 22.4 m³. This volume is called the *modal* volume and is denoted by $\bar{v}$.

If we consider one mole of gas, then

$$p\bar{v} = \bar{R}T \tag{5-45}$$

The value of $\bar{R}$ is the same for all gases under all conditions. The constant $\bar{R}$ is the universal gas constant.

Find the value of the universal gas constant in SI and BES units.

Solution. By substituting $p = 1.0133 \times 10^5$ N/m² abs, $\bar{v} = 22.4$ m³/kg-mole, and $T = 273$ K into Eq. (5-45), we obtain

$$\bar{R} = \frac{p\bar{v}}{T} = \frac{1.0133 \times 10^5 \times 22.4}{273} = 8314 \text{ N-m/kg-mole K}$$

This is the universal gas constant in SI units.

To obtain the universal gas constant in BES units, substitute $p = 14.7$ psia $= 14.7 \times 144 \text{ lb}_f/\text{ft}^2$ abs, $\bar{v} = 359$ ft³/lb-mole, and $T = 492°$R into Eq. (5-45).

$$\bar{R} = \frac{p\bar{v}}{T} = \frac{14.7 \times 144 \times 359}{492} = 1545 \text{ ft-lb}_f/\text{lb-mole }°\text{R}$$

$$= 1.985 \text{ Btu/lb-mole }°\text{R}$$

PROBLEM A-5-12. The molecular weight of a pure substance is the weight of one molecule of the substance compared to the weight of one oxygen atom, which is taken to be 16. That is, the molecular weight of carbon dioxide (CO_2) is $12 + (16 \times 2) = 44$. The molecular weights of oxygen (molecular) and water vapor are 32 and 18, respectively.

Determine the specific volume v of a mixture that consists of 100 m³ of oxygen, 5 m³ of carbon dioxide, and 20 m³ of water vapor when the pressure and temperature are 1.0133×10^5 N/m² abs and 294 K, respectively.

Solution. The mean molecular weight of the mixture is

$$M = \left(32 \times \frac{100}{125}\right) + \left(44 \times \frac{5}{125}\right) + \left(18 \times \frac{20}{125}\right) = 30.24$$

Thus

$$v = \frac{\bar{R}T}{Mp} = \frac{8314 \times 294}{30.24 \times 1.0133 \times 10^5} = 0.798 \text{ m}^3/\text{kg}$$

PROBLEM A-5-13. Referring to the pneumatic pressure system shown in Fig. 5-35(a), assume that the system is at steady state for $t < 0$ and that the steady-state pressure of the system is $\bar{P} = 5 \times 10^5$ N/m² abs. At $t = 0$ the inlet pressure is suddenly changed from $\bar{P}$ to $\bar{P} + p_i$, where p_i is a step change with a magnitude equal to 2×10^4 N/m². This step causes the air to flow into the vessel until pressure equalizes. Assume that the initial flow rate is $q(0) = 1 \times 10^{-4}$ kg/s. As air flows into the vessel, the pressure of air in the vessel rises from $\bar{P}$ to $\bar{P} + p_0$. Determine p_0 as a function of time. Assume that the expansion process is isothermal ($n = 1$), that the temperature of the entire system is constant at $T = 293$ K, and that the vessel has a capacity of 0.1 m³.

Solution. The average resistance of the valve is

$$R = \frac{\Delta p}{q} = \frac{2 \times 10^4}{1 \times 10^{-4}} = 2 \times 10^8 \text{ N-s/kg-m}^2$$

The capacitance of the vessel is

$$C = \frac{V}{nR_{\text{air}}T} = \frac{0.1}{1 \times 287 \times 293} = 1.19 \times 10^{-6} \text{ kg-m}^2/\text{N}$$

A mathematical model for this system is obtained from

$$C \, dp_0 = q \, dt$$

where

$$q = \frac{\Delta p}{R} = \frac{p_i - p_0}{R}$$

Thus

$$RC \frac{dp_0}{dt} + p_0 = p_i$$

By substituting the values of R, C, and p_i into this last equation, we have

$$2 \times 10^8 \times 1.19 \times 10^{-6} \frac{dp_0}{dt} + p_0 = 2 \times 10^4$$

or

$$238 \frac{dp_0}{dt} + p_0 = 2 \times 10^4 \tag{5-46}$$

Let us define

$$x(t) = p_0(t) - 2 \times 10^4 \tag{5-47}$$

Then by substituting Eq. (5-47) into Eq. (5-46), we obtain a differential equation in x as follows:

$$238 \frac{dx}{dt} + x = 0 \tag{5-48}$$

Noting that $p_0(0) = 0$, the initial condition for $x(t)$ is

$$x(0) = p_0(0) - 2 \times 10^4 = -2 \times 10^4$$

By assuming the exponential solution $x = Ke^{\lambda t}$ and substituting it into Eq. (5-48), we find the characteristic equation

$$238\lambda + 1 = 0$$

from which

$$\lambda = -0.0042$$

Hence $x(t)$ can be written

$$x(t) = Ke^{-0.0042t}$$

where K is a constant that can be determined from the initial condition.

$$x(0) = K = -2 \times 10^4$$

Thus

$$x(t) = -2 \times 10^4 e^{-0.0042t}$$

Substitution of this last equation into Eq. (5-47) yields

$$p_0(t) = x(t) + 2 \times 10^4 = 2 \times 10^4 (1 - e^{-0.0042t})$$

Since the time constant of the system is $RC = 238$ seconds, it takes approximately 950 seconds before the response settles within 2% of the total change.

*PROBLEM A-5-14. Figure 5-60 is a schematic diagram of a fluidic device. It is a slightly modified version of the wall-attachment amplifier shown in Fig. 5-39. By venting one side, a higher pressure exists on that side. The main jet is in port X unless an input signal exists at "A" or "B". (The output is at port Y if an input signal is present at "A" or "B." When the input signal is off, the main jet switches from port Y to port X.)

       Construct a truth table for this device. (A and B correspond to inputs and X and Y correspond to outputs.) What logic function does this device perform?

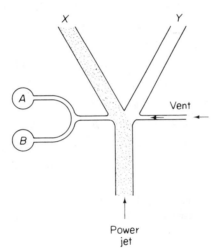

Fig. 5-60. Fluidic device.

Solution. The truth table for this device is shown in Table 5-6. From the table it can be seen that port X acts as a NOR device and port Y as an OR device.

Table 5-6. TRUTH TABLE

A	B	X	Y
1	1	0	1
1	0	0	1
0	1	0	1
0	0	1	0

*PROBLEM A-5-15.** Prove the following identities of mathematical logic.

1. $$A(A + B + C) = A$$

2. $$A + BC = (A + B)(A + C)$$

3. $$A + AB = A$$

4. $$A + \bar{A}B = A + B$$

5. $$A\bar{B} + \bar{A}B = (A + B)(\bar{A} + \bar{B})$$

6. $$(A + \bar{B})(\bar{A} + B) = AB + \bar{A}\bar{B}$$

Solution.

1. If A is 1, then $A(A + B + C) = 1$ regardless of B and C; and if A is 0, then $A(A + B + C) = 0$ regardless of B and C. Therefore

$$A(A + B + C) = A$$

2.
$$\begin{aligned}(A + B)(A + C) &= AA + BA + AC + BC \\ &= A(A + B + C) + BC = A + BC\end{aligned}$$

3.
$$A + AB = A(1 + B) = A\,1 = A$$

4.
$$\begin{aligned}A + B &= (A + \bar{A})(A + B) = AA + \bar{A}A + AB + \bar{A}B \\ &= A(A + B) + \bar{A}B = A + \bar{A}B\end{aligned}$$

5.
$$(A + B)(\bar{A} + \bar{B}) = A\bar{A} + B\bar{A} + A\bar{B} + B\bar{B} = A\bar{B} + \bar{A}B$$

6.
$$(A + \bar{B})(\bar{A} + B) = A\bar{A} + \bar{B}\bar{A} + AB + \bar{B}B = AB + \bar{A}\bar{B}$$

*PROBLEM A-5-16.** Consider the logic circuit shown in Fig. 5-61 and find the logic expression for X. Construct a truth table for this circuit. Show that the input signal C acts as an override signal.

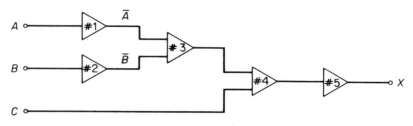

Fig. 5-61. Logic circuit.

Solution. The output of #3 NOR element is AB and the output of #4 NOR element is $\overline{AB + C}$. Hence the output X is equal to $AB + C$, or $X = AB + C$. From this expression we see that whenever $C = 1$, the output is $X = 1$. So the signal C is an override. It will activate the circuit no matter what values A and B have. The truth table for this circuit is shown in Table 5-7.

Table 5-7. TRUTH TABLE

A	B	C	X
1	1	1	1
1	1	0	1
1	0	1	1
1	0	0	0
0	1	1	1
0	1	0	0
0	0	1	1
0	0	0	0

Table 5-8. TRUTH TABLE

A	B	C	X
1	1	1	0
1	1	0	0
1	0	1	1
1	0	0	1
0	1	1	0
0	1	0	1
0	0	1	0
0	0	0	0

*PROBLEM A-5-17. Obtain a logic circuit that performs the logic function shown in Table 5-8.

Solution. The logic expression for X is

$$X = A\bar{B}C + A\bar{B}\bar{C} + \bar{A}B\bar{C}$$
$$= A\bar{B} + \bar{A}B\bar{C}$$
$$= \overline{\bar{A} + B} + \overline{A + \bar{B} + C}$$

The logic circuit for this expression is shown in Fig. 5-62.

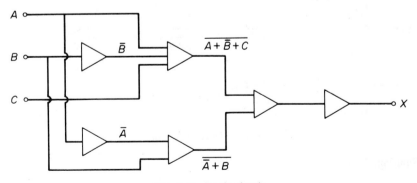

Fig. 5-62. Logic circuit.

*PROBLEM A-5-18. Obtain a logic circuit for the following logic expression. Use only NOR elements.

$$X = (A + BC)(\bar{A} + \bar{B}) + A(A + \bar{B} + C)$$

Solution. This logic expression can be simplified as follows:

$$\begin{aligned} X &= A\bar{A} + BC\bar{A} + A\bar{B} + BC\bar{B} + A \\ &= (A + \bar{A}BC) + A\bar{B} \\ &= A + BC + A\bar{B} \\ &= A(1 + \bar{B}) + BC \\ &= A + BC \\ &= A + \overline{\bar{B} + \bar{C}} \end{aligned}$$

A logic circuit for this simplified expression is given in Fig. 5-63.

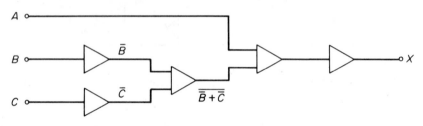

Fig. 5-63. Logic circuit.

*PROBLEM A-5-19. Derive the logic expression for the circuit shown in Fig. 5-64. Construct a truth table for this circuit.

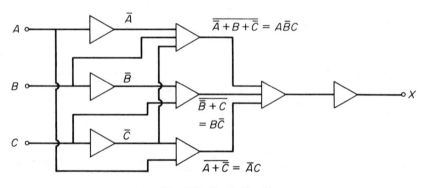

Fig. 5-64. Logic circuit.

Solution. From the diagram we obtain

$$\begin{aligned} X &= A\bar{B}C + B\bar{C} + \bar{A}C \\ &= (A\bar{B} + \bar{A})C + B\bar{C} \end{aligned}$$

$$= (\bar{A} + \bar{B})C + B\bar{C}$$
$$= \bar{A}C + \bar{B}C + B\bar{C}$$

A truth table for this logic circuit is shown in Table 5-9.

Table 5-9. TRUTH TABLE

A	B	C	X
1	1	1	0
1	1	0	1
1	0	1	1
1	0	0	0
0	1	1	1
0	1	0	1
0	0	1	1
0	0	0	0

*PROBLEM A-5-20. An automated production line performs a series of four tests on a manufactured product. Design a logic circuit that will simultaneously examine the results of all four tests and decide into which of three chutes the piece will drop. If it passes two or three tests, chute #2 is to open. If it passes one or none of the tests, it is to go to chute #3. Chute #1 accepts perfect units only.

Solution. Here the four tests can be considered as inputs to the system and chutes #1, #2, and #3 considered as outputs. By defining four tests as inputs A, B, C, and D, and chutes #1, #2, and #3 as outputs X, Y, and Z, respectively, we can construct a truth table for this problem as shown in Table 5-10. The logic expressions for X, Y, and Z can be obtained from this table. Since the number of zeros is smaller than the number of ones in column Y, it is simpler to obtain the desired logic expression for Y if we begin with $\bar{Y}$. From Table 5-10

$$X = ABCD$$
$$\bar{Y} = ABCD + A\bar{B}\bar{C}\bar{D} + \bar{A}B\bar{C}\bar{D} + \bar{A}\bar{B}C\bar{D} + \bar{A}\bar{B}\bar{C}D + \bar{A}\bar{B}\bar{C}\bar{D}$$
$$Z = A\bar{B}\bar{C}\bar{D} + \bar{A}B\bar{C}\bar{D} + \bar{A}\bar{B}C\bar{D} + \bar{A}\bar{B}\bar{C}D + \bar{A}\bar{B}\bar{C}\bar{D}$$

Hence

$$\bar{X} = \overline{ABCD} = \bar{A} + \bar{B} + \bar{C} + \bar{D}$$
$$X = \bar{\bar{X}} = \overline{\bar{A} + \bar{B} + \bar{C} + \bar{D}}$$

Table 5-10. TRUTH TABLE

A	B	C	D	X	Y	Z
1	1	1	1	1	0	0
1	1	1	0	0	1	0
1	1	0	1	0	1	0
1	0	1	1	0	1	0
0	1	1	1	0	1	0
1	1	0	0	0	1	0
1	0	1	0	0	1	0
1	0	0	1	0	1	0
0	1	1	0	0	1	0
0	1	0	1	0	1	0
0	0	1	1	0	1	0
1	0	0	0	0	0	1
0	1	0	0	0	0	1
0	0	1	0	0	0	1
0	0	0	1	0	0	1
0	0	0	0	0	0	1

Next, we simplify the logic expression for Z.

$$Z = A\bar{B}\bar{C}\bar{D} + \bar{A}B\bar{C}\bar{D} + \bar{A}\bar{B}C\bar{D} + \bar{A}\bar{B}\bar{C}D + \bar{A}\bar{B}\bar{C}\bar{D}$$
$$= (\bar{A}\bar{B}\bar{C}D + \bar{A}\bar{B}\bar{C}\bar{D}) + (A\bar{B}\bar{C}\bar{D} + \bar{A}\bar{B}\bar{C}\bar{D})$$
$$+ (\bar{A}B\bar{C}\bar{D} + \bar{A}\bar{B}\bar{C}\bar{D}) + (\bar{A}\bar{B}C\bar{D} + \bar{A}\bar{B}\bar{C}\bar{D})$$
$$= \bar{A}\bar{B}\bar{C}(D + \bar{D}) + \bar{B}\bar{C}\bar{D}(A + \bar{A})$$
$$+ \bar{A}\bar{B}\bar{D}(C + \bar{C}) + \bar{A}\bar{C}\bar{D}(B + \bar{B})$$
$$= \bar{A}\bar{B}\bar{C} + \bar{B}\bar{C}\bar{D} + \bar{A}\bar{B}\bar{D} + \bar{A}\bar{C}\bar{D}$$
$$= \overline{A + B + C} + \overline{B + C + D} + \overline{A + B + D} + \overline{A + C + D}$$

Noting that

$$\bar{Y} = ABCD + Z$$

we obtain

$$Y = \overline{ABCD + Z} = \overline{\bar{A} + \bar{B} + \bar{C} + \bar{D}} + Z$$

$$= \overline{\bar{A} + \bar{B} + \bar{C} + \bar{D}} + \overline{A + B + C} + \overline{B + C + D} + \overline{A + B + D} + \overline{A + C + D}$$

A logic circuit for this system is shown in Fig. 5-65.

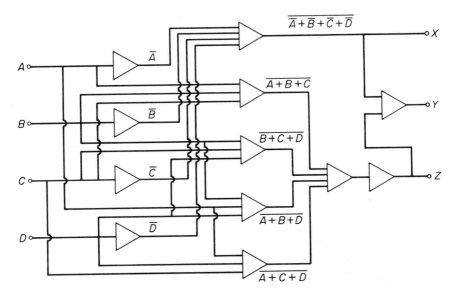

Fig. 5-65. Logic circuit.

PROBLEMS

PROBLEM B-5-1. Find the pressure difference $p_1 - p_2$ necessary to keep massless bar AB horizontal in the system shown in Fig. 5-66. Assume that $mg = 1000$ N and $A = 5 \times 10^{-3}$ m².

PROBLEM B-5-2. In the system of Fig. 5-67 mass m is to be pushed upward along the inclined plane by the pneumatic cylinder. Friction force μN is acting opposite to the direction of motion or intended motion. If the load is to be moved, show that the area A of the piston must not be smaller than

$$\frac{mg \sin (\theta + \alpha)}{(p_1 - p_2) \cos \theta}$$

where $\theta = \tan^{-1} \mu$ and α is the angle of inclination of the plane.

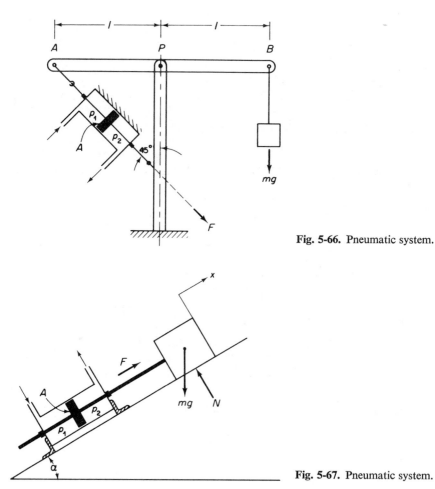

Fig. 5-66. Pneumatic system.

Fig. 5-67. Pneumatic system.

PROBLEM B-5-3. Figure 5-68 shows a toggle-joint. Prove that

$$F = 2\frac{l_1}{l_2}R$$

PROBLEM B-5-4. The system shown in Fig. 5-69 consists of a power cylinder and a rack-and-pinion mechanism to drive the load. Power piston D moves rack C, which, in turn, causes pinion B to rotate on rack A. Find displacement y of the output when the displacement of the power piston is x.

PROBLEM B-5-5. If the atmospheric pressure is 758 mm Hg and the measured pressure is 25 mm Hg gage, what is the absolute pressure in N/m², kg$_f$/cm², and lb$_f$/in.² ?

PROBLEM B-5-6. A mass of 100 kg is raised vertically by 10 m. Find the work done in terms of heat in kcal.

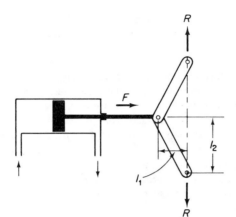

Fig. 5-68. Toggle-joint.

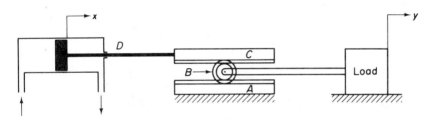

Fig. 5-69. Pneumatic system.

PROBLEM B-5-7. Air is compressed into a tank of volume 10 m³. The pressure is 7×10^5 N/m² gage and the temperature is 20°C. Find the mass of air in the tank. If the temperature of the compressed air is raised to 40°C, what is the gage pressure in N/m², kg_f/cm^2, and $lb_f/in.^2$?

PROBLEM B-5-8. Express R_{air}, the gas constant of air, in terms of kilocalories per kilogram kelvin.

PROBLEM B-5-9. A cylinder contains 0.1 kg of air. Assume that when the air is compressed with the work done of 1.2×10^4 N-m, a heat of 2 kcal is withdrawn to the surroundings. Find the increase in the internal energy of the air per kilogram.

PROBLEM B-5-10. The equation for the speed of sound in a gas is given by

$$c = \sqrt{\frac{dp}{d\rho}}$$

Show that it may also be given by

$$c = \sqrt{\frac{K}{\rho}}$$

where K is the bulk modulus of elasticity of the gas.

PROBLEM B-5-11. Obtain the capacitance C of a pneumatic pressure vessel that contains 10 m³ of air at temperature 20°C. Assume that the expansion process is isothermal.

PROBLEM B-5-12. The pneumatic pressure system shown in Fig. 5-70(a) consists of a pressure vessel and a pipe with an orifice. Assume that the system is at steady state for $t < 0$ and that the steady-state pressure is $\bar{P}$, where $\bar{P} = 2 \times 10^5$ N/m² abs. At $t = 0$ the input pressure is changed from $\bar{P}$ to $\bar{P} + p_i$, a step that will cause the pressure in the vessel to change from $\bar{P}$ to $\bar{P} + p_o$. Assume also that the operating range for pressure difference $\Delta p = p_i - p_o$ is between -3×10^4 N/m² and 3×10^4 N/m². The capacity of the vessel is 1×10^{-4} m³, and the Δp versus q (mass flow rate) curve is given in Fig. 5-70(b). The temperature of the entire system is 30°C, and the expansion process is assumed to be isothermal. Derive a mathematical model for the system.

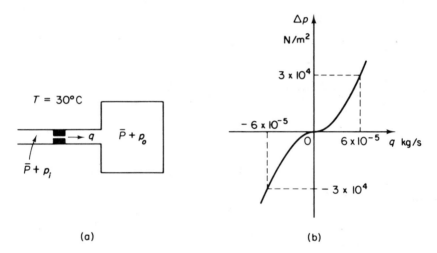

Fig. 5-70. (a) Pneumatic pressure system; (b) pressure difference versus mass flow rate curve.

PROBLEM B-5-13. Consider the pneumatic pressure system shown in Fig. 5-71. For $t < 0$, the inlet valve is closed, the outlet valve is fully opened to the atmosphere, and pressure p_2 in the vessel is at the atmospheric pressure. At $t = 0$ the inlet valve is fully opened. The inlet pipe is connected to a pressure source that supplies air at a constant pressure p_1, where $p_1 = 0.5 \times 10^5$ N/m² gage. Assume that the expansion process is isothermal ($n = 1$) and that the temperature of the entire system stays constant.

Determine steady-state pressure p_2 in the vessel after the inlet valve is fully opened, assuming that the inlet and outlet valves are identical ones—that is, both valves have identical flow characteristics.

*PROBLEM B-5-14. Explain how the circuit shown in Fig. 5-72 acts as a memory circuit.

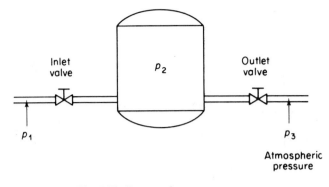

Fig. 5-71. Pneumatic pressure system.

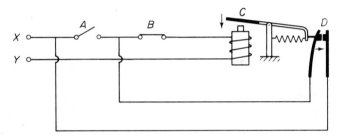

Fig. 5-72. Electrical memory circuit.

*PROBLEM B-5-15. Figure 5-73 is a schematic diagram of a fluidic device. What function does this device perform?

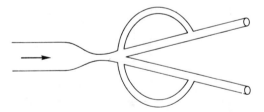

Fig. 5-73. Fluidic device.

*PROBLEM B-5-16. Obtain a logic circuit for the following logic expression by using only NOR elements.

$$X = A(B + \bar{C})(\bar{A} + C)$$

*PROBLEM B-5-17. Draw a logic circuit that will perform the logic function shown in Table 5-11.

*PROBLEM B-5-18. Construct a logic circuit that will give the truth table shown in Table 5-12.

Table 5-11. TRUTH TABLE

A	B	C	X
1	1	1	0
1	1	0	1
1	0	1	1
1	0	0	1
0	1	1	0
0	1	0	0
0	0	1	1
0	0	0	1

Table 5-12. TRUTH TABLE

A	B	C	X
1	1	1	1
1	1	0	1
1	0	1	1
1	0	0	0
0	1	1	1
0	1	0	0
0	0	1	0
0	0	0	0

*PROBLEM B-5-19. For the logic circuit shown in Fig. 5-74, derive the logic expression for X and Y. Construct a truth table for this circuit.

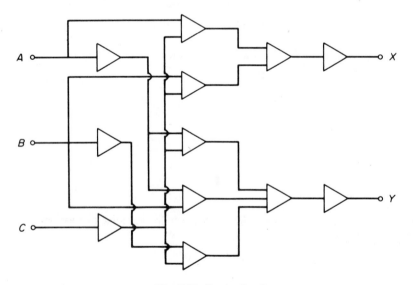

Fig. 5-74. Logic circuit.

*PROBLEM B-5-20. Using NOR elements only, construct a logic circuit that represents the following statement. A gate is normally in closed position. The gate is to open if switches A and B are both opened and/or emergency switch C is opened.

6

LAPLACE TRANSFORMATION

6-1 INTRODUCTION

The Laplace transform method is an operational method that can be used advantageously in solving linear, time-invariant, differential equations. Its main advantage is that differentiation of the time function corresponds to multiplication of the transform by a complex variable s, and thus the differential equations in time become algebraic equations in s. The solution of the differential equation can then be found by using a Laplace transform table or the partial-fraction expansion technique. Another advantage of the Laplace transform method is that, in solving the differential equation, the initial conditions are automatically taken care of, and both the particular solution and the complementary solution can be obtained simultaneously.

Emphasis throughout this chapter is not on mathematical rigor but on the methods of application in problems associated with the analysis and design of linear systems.

The outline of the rest of the chapter is as follows. Section 6-2 reviews complex numbers, complex variables, and complex functions, which are normally included in the mathematics courses required by sophomore-level engineering students. Section 6-3 defines Laplace transformation and derives Laplace transforms of simple time functions, and in Sec. 6-4 useful theorems on Laplace transformation are given. The inverse Laplace transformation, the mathematical process of obtaining time functions from Laplace-transformed complex variable expressions, is covered in Sec. 6-5. The final section, Sec. 6-6, presents the solution of linear, time-invariant, differential equations through the Laplace transform method.

6-2   COMPLEX NUMBERS, COMPLEX VARIABLES, AND COMPLEX FUNCTIONS

Since this section is a review of complex numbers, complex algebra, complex variables, and complex functions and since most of the material covered is generally included in the basic mathematics courses required of engineering students, it can be omitted entirely or used simply for personal reference.

Complex numbers. Using notation $j = \sqrt{-1}$, all numbers in engineering calculations can be expressed as

$$z = x + jy$$

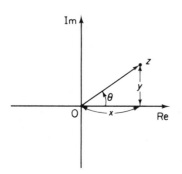

where z is called a *complex number* and x and y its *real* and *imaginary parts*, respectively. Note that both x and y are real and that the j is the only imaginary quantity in the expression. The complex plane representation of z is shown in Fig. 6-1. (Note also that the real axis and the imaginary axis define the complex plane and that the combination of a real number and an imaginary number defines a point in the complex plane.) A complex number z can be considered a point in the complex plane or a directed line segment to the point; both interpretations are useful.

Fig. 6-1. Complex plane representation of a complex number z.

The magnitude, or absolute value, of z is defined as the length of the directed line segment shown in Fig. 6-1. The angle of z is the angle that the directed line segment makes with the positive real axis. A counterclockwise rotation is defined as the positive direction for the measurement of angles.

$$\text{Magnitude of } z = |z| = \sqrt{x^2 + y^2}, \qquad \text{Angle of } z = \theta = \tan^{-1}\frac{y}{x}$$

In terms of magnitude and angle, z can be expressed as $z = |z|\underline{/\theta}$, which is called the *polar form* of z, as opposed to the form $z = x + jy$, which is called the *rectangular form* of z. Thus z can be written

$$z = x + jy = |z|\underline{/\theta}$$

where $x = |z| \cos \theta$ and $y = |z| \sin \theta$.

Complex conjugate. The *complex conjugate* of $z = x + jy$ is defined as

$$\bar{z} = x - jy$$

It has ths same real part as z and an imaginary part that is the negative of the imaginary part of z. Figure 6-2 shows both z and $\bar{z}$. Note that

$$z = x + jy = |z|\underline{/\theta} = |z|(\cos\theta + j\sin\theta)$$
$$\bar{z} = x - jy = |z|\underline{/-\theta} = |z|(\cos\theta - j\sin\theta)$$

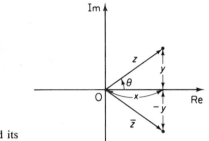

Fig. 6-2. Complex number z and its complex conjugate $\bar{z}$.

Euler's theorem. The power series expansions of $\cos\theta$ and $\sin\theta$ are, respectively,

$$\cos\theta = 1 - \frac{\theta^2}{2!} + \frac{\theta^4}{4!} - \frac{\theta^6}{6!} + \cdots$$

$$\sin\theta = \theta - \frac{\theta^3}{3!} + \frac{\theta^5}{5!} - \frac{\theta^7}{7!} + \cdots$$

And so

$$\cos\theta + j\sin\theta = 1 + (j\theta) + \frac{(j\theta)^2}{2!} + \frac{(j\theta)^3}{3!} + \frac{(j\theta)^4}{4!} + \cdots$$

Since

$$e^x = 1 + x + \frac{x^2}{2!} + \frac{x^3}{3!} + \cdots$$

we see that

$$\cos\theta + j\sin\theta = e^{j\theta}$$

This is known as *Euler's theorem.*

By using Euler's theorem, we can express sine and cosine in complex form. Noting that $e^{-j\theta}$ is the complex conjugate of $e^{j\theta}$ and that

$$e^{j\theta} = \cos\theta + j\sin\theta$$
$$e^{-j\theta} = \cos\theta - j\sin\theta$$

we find, after adding and subtracting these two equations,

$$\cos\theta = \frac{e^{j\theta} + e^{-j\theta}}{2}$$

$$\sin\theta = \frac{e^{j\theta} - e^{-j\theta}}{2j}$$

Various forms for representing complex numbers. A complex number can be written in various forms.

$$z = x + jy$$
$$z = |z|(\cos\theta + j\sin\theta) \qquad \text{(rectangular forms)}$$

$$z = |z|\underline{/\theta}$$
$$z = |z|e^{j\theta} \qquad \text{(polar forms)}$$

In converting complex numbers to polar form from rectangular,

$$|z| = \sqrt{x^2 + y^2}, \qquad \theta = \tan^{-1}\frac{y}{x}$$

To convert complex numbers to rectangular form from polar,

$$x = |z|\cos\theta, \qquad y = |z|\sin\theta$$

Complex algebra. If the complex numbers are written in a suitable form, operations like addition, subtraction, multiplication, and division can be performed easily.

Equality of complex numbers. Two complex numbers z and w are said to be equal if and only if their real parts are equal and their imaginary parts are equal. So if two complex numbers are written

$$z = x + jy, \qquad w = u + jv$$

then $z = w$ if and only if $x = u$ and $y = v$.

Addition. Two complex numbers in rectangular form are added by adding the real parts and the imaginary parts separately.

$$z + w = (x + jy) + (u + jv) = (x + u) + j(y + v)$$

Subtraction. Subtraction can be considered as addition of the negative.

$$z - w = (x + jy) - (u + jv) = (x - u) + j(y - v)$$

Note that addition and subtraction can be done easily on the rectangular plane.

Multiplication. If a complex number is multiplied by a real number, then it is a complex number whose real and imaginary parts are multiplied by that real number.

$$az = a(x + jy) = ax + jay$$

If two complex numbers appear in rectangular form and we want the product in rectangular form, multiplication is accomplished by using the fact that $j^2 = -1$. Thus if two complex numbers are written

$$z = x + jy, \qquad w = u + jv$$

then

$$zw = (x + jy)(u + jv) = xu + jyu + jxv + j^2yv$$
$$= (xu - yv) + j(xv + yu)$$

In the polar form, multiplication of two complex numbers can be done easily. The magnitude of the product is the product of the two magnitudes, and the angle of the product is the sum of the two angles. So if two complex numbers are written

$$z = |z| \underline{/\theta}, \qquad w = |w| \underline{/\phi}$$

then

$$zw = |z||w| \underline{/\theta + \phi}$$

Multiplication by j. It is important to note that multiplication by j is equivalent to counterclockwise rotation by 90°. For example, if

$$z = x + jy$$

then

$$jz = j(x + jy) = jx + j^2y = -y + jx$$

or noting that $j = 1\underline{/90°}$, if

$$z = |z| \underline{/\theta}$$

then

$$jz = 1\underline{/90°} \cdot |z| \underline{/\theta} = |z| \underline{/\theta + 90°}$$

Figure 6-3 illustrates multiplication of a complex number z by j.

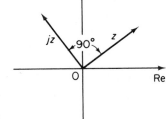

Fig. 6-3. Multiplication of a complex number z by j.

Division. If a complex number $z = |z| \underline{/\theta}$ is divided by another complex number $w = |w| \underline{/\phi}$, then

$$\frac{z}{w} = \frac{|z| \underline{/\theta}}{|w| \underline{/\phi}} = \frac{|z|}{|w|} \underline{/\theta - \phi}$$

Therefore the result consists of the quotient of the magnitudes and the difference of the angles.

Division in rectangular form is inconvenient but can be done by multiplying the denominator and numerator by the complex conjugate of the denominator. This procedure converts the denominator to a real number and thus simplifies division. For instance,

$$\frac{z}{w} = \frac{x + jy}{u + jv} = \frac{(x + jy)(u - jv)}{(u + jv)(u - jv)} = \frac{(xu + yv) + j(yu - xv)}{u^2 + v^2}$$

$$= \frac{xu + yv}{u^2 + v^2} + j\frac{yu - xv}{u^2 + v^2}$$

Division by j. Note that division by j is equivalent to clockwise rotation by 90°. For example, if $z = x + jy$, then

$$\frac{z}{j} = \frac{x + jy}{j} = \frac{(x + jy)j}{jj} = \frac{jx - y}{-1} = y - jx$$

or

$$\frac{z}{j} = \frac{|z|\underline{/\theta}}{1\underline{/90°}} = |z|\underline{/\theta - 90°}$$

Figure 6-4 illustrates division of a complex number z by j.

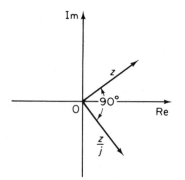

Fig. 6-4. Division of a complex number z by j.

Powers and roots. Multiplying z by n times, we obtain

$$z^n = (|z|\underline{/\theta})^n = |z|^n\underline{/n\theta}$$

Extracting the nth root of a complex number is equivalent to raising the number to the $1/n$th power.

$$z^{1/n} = (|z|\underline{/\theta})^{1/n} = |z|^{1/n}\underline{\left/\frac{\theta}{n}\right.}$$

For instance,

$$(8.66 - j5)^3 = (10\underline{/-30°})^3 = 1000\underline{/-90°} = 0 - j1000 = -j1000$$
$$(2.12 - j2.12)^{1/2} = (9\underline{/-45°})^{1/2} = 3\underline{/-22.5°}$$

Comments. It is important to note that

$$|zw| = |z||w|$$
$$|z + w| \neq |z| + |w|$$

Complex variable. A complex number has a real part and an imaginary part, both of which are constant. If the real part and/or imaginary part are variables, a complex number is called a *complex variable.* In the Laplace transformation we use the notation s as a complex variable—that is,

$$s = \sigma + j\omega$$

where σ is the real part and ω is the imaginary part.

Complex function. A complex function $F(s)$, a function of s, has a real part and an imaginary part or

$$F(s) = F_x + jF_y$$

where F_x and F_y are real quantities. The magnitude of $F(s)$ is $\sqrt{F_x^2 + F_y^2}$, and the angle θ of $F(s)$ is $\tan^{-1} (F_y/F_x)$. The angle is measured counterclockwise from the positive real axis. The complex conjugate of $F(s)$ is $\bar{F}(s) = F_x - jF_y$.

Complex functions commonly encountered in linear systems analysis are single-valued functions of s and are uniquely determined for a given value of s. Typical such functions have the form

$$F(s) = \frac{K(s + z_1)(s + z_2) \cdots (s + z_m)}{(s + p_1)(s + p_2) \cdots (s + p_n)}$$

Points at which $F(s)$ equals zero are called *zeros*. That is, $s = -z_1, s = -z_2,$ $\ldots, s = -z_m$ are zeros of $F(s)$. Points at which $F(s)$ equals infinity are called *poles*. That is, $s = -p_1, s = -p_2, \ldots, s = -p_n$ are poles of $F(s)$. If the denominator of $F(s)$ involves k-multiple factors $(s + p)^k$, then $s = -p$ is called a *multiple pole* of order k. If $k = 1$, the pole is called a *simple pole.*

To illustrate, consider the complex function

$$G(s) = \frac{K(s + 2)(s + 10)}{s(s + 1)(s + 5)(s + 15)^2}$$

$G(s)$ has zeros at $s = -2, s = -10$, simple poles at $s = 0, s = -1, s = -5$, and a double pole (multiple pole of order 2) at $s = -15$. Note that $G(s)$ becomes zero at $s = \infty$. Since for large values of s

$$G(s) \doteq \frac{K}{s^3}$$

$G(s)$ possesses a triple zero (multiple zero of order 3) at $s = \infty$. If points at infinity are included, $G(s)$ has the same number of poles as zeros. To summarize, $G(s)$ has five zeros ($s = -2, s = -10, s = \infty, s = \infty, s = \infty$) and five poles ($s = 0, s = -1, s = -5, s = -15, s = -15$).

6-3 LAPLACE TRANSFORMATION

In addition to a definition of the Laplace transformation, this section includes examples on how to obtain Laplace transforms of common functions.

Laplace transformation. Let us define

$f(t) =$ a time function such that $f(t) = 0$ for $t < 0$

$s =$ a complex variable

$\mathcal{L} =$ an operational symbol indicating that the quantity that it prefixes is to be transformed by the Laplace integral

$$\int_0^\infty e^{-st}\,dt$$

$F(s) =$ Laplace transform of $f(t)$

Then the Laplace transform of $f(t)$ is given by

$$\mathcal{L}[f(t)] = F(s) = \int_0^\infty e^{-st}\,dt[f(t)] = \int_0^\infty f(t)e^{-st}\,dt$$

The reverse process of finding the time function $f(t)$ from the Laplace transform $F(s)$ is called the *inverse Laplace transformation*. The notation for the inverse Laplace transformation is $\mathcal{L}^{-1}$. Thus

$$\mathcal{L}^{-1}[F(s)] = f(t)$$

Existence of Laplace transform. The Laplace transform of a function $f(t)$ exists if the Laplace integral converges. The integral will converge if $f(t)$ is sectionally continuous in every finite interval in the range $t > 0$ and if it is of exponential order as t approaches infinity. A function $f(t)$ is said to be of exponential order if a real, positive constant σ exists such that the function

$$e^{-\sigma t}\,|f(t)|$$

approaches zero as t approaches infinity. If the limit of the function $e^{-\sigma t}\,|f(t)|$ approaches zero for σ greater than σ_c and the limit approaches infinity for σ less than σ_c, the value σ_c is called the *abscissa of convergence*.

It can be seen that, for such functions as t, $\sin \omega t$, and $t \sin \omega t$, the abscissa of convergence is equal to zero. For functions like e^{-ct}, te^{-ct}, and $e^{-ct}\sin \omega t$, the abscissa of convergence is equal to $-c$. In the case of functions that increase faster than the exponential function, however, it is impossible to find suitable values of the abscissa of convergence. Consequently, such functions as e^{t^2} and te^{t^2} do not possess Laplace transforms.

Nevertheless, it should be noted that although e^{t^2} for $0 \leq t \leq \infty$ does not possess a Laplace transform, the time function defined by

$$f(t) = e^{t^2} \quad \text{for } 0 \leq t \leq T < \infty$$
$$= 0 \quad \text{for } t < 0, T < t$$

does, since $f(t) = e^{t^2}$ for only a limited time interval $0 \leq t \leq T$ and not for

$0 \leq t \leq \infty$. Such a signal can be physically generated. Note that the signals that can be physically generated always have corresponding Laplace transforms.

If functions $f_1(t)$ and $f_2(t)$ are both Laplace transformable, then the Laplace transform of $f_1(t) + f_2(t)$ is given by

$$\mathcal{L}[f_1(t) + f_2(t)] = \mathcal{L}[f_1(t)] + \mathcal{L}[f_2(t)]$$

Exponential function. Consider the exponential function

$$
\begin{aligned}
f(t) &= 0 && \text{for } t < 0 \\
&= Ae^{-\alpha t} && \text{for } t \geq 0
\end{aligned}
$$

where A and α are constants. The Laplace transform of this exponential function can be obtained as follows:

$$\mathcal{L}[Ae^{-\alpha t}] = \int_0^\infty Ae^{-\alpha t}e^{-st}\, dt = A \int_0^\infty e^{-(\alpha+s)t}\, dt = \frac{A}{s + \alpha}$$

In performing this integration, we assumed that the real part of s is greater than $-\alpha$ (the abscissa of convergence) so that the integral converges. The Laplace transform $F(s)$ of any Laplace transformable function $f(t)$ obtained in this way is valid throughout the entire s plane except at the poles of $F(s)$. (Although we do not present a proof of this statement, it can be proved by use of the theory of complex variables.)

Step function. Consider the step function

$$
\begin{aligned}
f(t) &= 0 && \text{for } t < 0 \\
&= A && \text{for } t > 0
\end{aligned}
$$

where A is a constant. Note that it is a special case of the exponential function $Ae^{-\alpha t}$, where $\alpha = 0$. The step function is undefined at $t = 0$. Its Laplace transform is given by

$$\mathcal{L}[A] = \int_0^\infty Ae^{-st}\, dt = \frac{A}{s}$$

The step function whose height is unity is called a *unit-step function*. The unit-step function that occurs at $t = t_0$ is frequently written $1(t - t_0)$, a notation that will be used in this book. The preceding step function whose height is A can be written $A1(t)$.

The Laplace transform of the unit-step function that is defined by

$$
\begin{aligned}
1(t) &= 0 && \text{for } t < 0 \\
&= 1 && \text{for } t > 0
\end{aligned}
$$

is $1/s$ or

$$\mathcal{L}[1(t)] = \frac{1}{s}$$

Physically, a step function occurring at $t = t_0$ corresponds to a constant signal suddenly applied to the system at time t equals t_0.

Ramp function. Consider the ramp function

$$f(t) = 0 \qquad \text{for } t < 0$$
$$= At \qquad \text{for } t \geq 0$$

where A is a constant. The Laplace transform of this ramp function is obtained as

$$\mathcal{L}[At] = \int_0^\infty At e^{-st}\, dt = At \frac{e^{-st}}{-s}\Big|_0^\infty - \int_0^\infty \frac{A e^{-st}}{-s}\, dt$$

$$= \frac{A}{s} \int_0^\infty e^{-st}\, dt = \frac{A}{s^2}$$

Sinusoidal function. The Laplace transform of the sinusoidal function

$$f(t) = 0 \qquad\qquad \text{for } t < 0$$
$$= A \sin \omega t \qquad \text{for } t \geq 0$$

where A and ω are constants, is obtained as follows. Noting that

$$e^{j\omega t} = \cos \omega t + j \sin \omega t$$
$$e^{-j\omega t} = \cos \omega t - j \sin \omega t$$

$\sin \omega t$ can be written

$$\sin \omega t = \frac{1}{2j}(e^{j\omega t} - e^{-j\omega t})$$

Hence

$$\mathcal{L}[A \sin \omega t] = \frac{A}{2j} \int_0^\infty (e^{j\omega t} - e^{-j\omega t}) e^{-st}\, dt$$

$$= \frac{A}{2j} \frac{1}{s - j\omega} - \frac{A}{2j} \frac{1}{s + j\omega} = \frac{A\omega}{s^2 + \omega^2}$$

Similarly, the Laplace transform of $A \cos \omega t$ can be derived as follows:

$$\mathcal{L}[A \cos \omega t] = \frac{As}{s^2 + \omega^2}$$

Comments. The Laplace transform of any Laplace transformable function $f(t)$ can be found by multiplying $f(t)$ by e^{-st} and then integrating the product from $t = 0$ to $t = \infty$. Once we know the method of obtaining the Laplace transform, however, it is not necessary to derive the Laplace transform of $f(t)$ each time. Laplace transform tables can conveniently be used to find the transform of a given function $f(t)$. Table 6-1 shows Laplace transforms of time functions that will frequently appear in linear systems analysis. In Table 6-2 the properties of Laplace transforms are given. Most are derived or proved in Section 6-4.

6-4 LAPLACE TRANSFORM THEOREMS

In the following discussion we present Laplace transforms of several functions as well as theorems on the Laplace transformation that are useful in the study of linear systems.

Translated function. Let us obtain the Laplace transform of the translated function $f(t - \alpha)1(t - \alpha)$, where $\alpha \geq 0$. This function is zero for $t < \alpha$. The functions $f(t)1(t)$ and $f(t - \alpha)1(t - \alpha)$ are shown in Fig. 6-5.

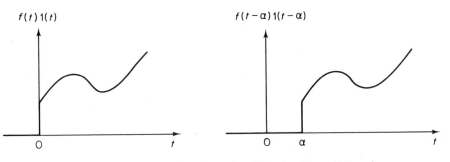

$f(t)1(t)$ $f(t - \alpha)1(t - \alpha)$

Fig. 6-5. Function $f(t)1(t)$ and translated function $f(t - \alpha)1(t - \alpha)$.

By definition, the Laplace transform of $f(t - \alpha)1(t - \alpha)$ is

$$\mathcal{L}[f(t - \alpha)1(t - \alpha)] = \int_0^\infty f(t - \alpha)1(t - \alpha)e^{-st}\, dt$$

By changing the independent variable from t to τ, where $\tau = t - \alpha$, we obtain

$$\int_0^\infty f(t - \alpha)1(t - \alpha)e^{-st}\, dt = \int_{-\alpha}^\infty f(\tau)1(\tau)e^{-s(\tau+\alpha)}\, d\tau$$

Noting that $f(\tau)1(\tau) = 0$ for $\tau < 0$, we can change the lower limit of integration from $-\alpha$ to 0. Thus

$$\int_{-\alpha}^\infty f(\tau)1(\tau)e^{-s(\tau+\alpha)}\, d\tau = \int_0^\infty f(\tau)1(\tau)e^{-s(\tau+\alpha)}\, d\tau$$

$$= \int_0^\infty f(\tau)e^{-s\tau}e^{-\alpha s}\, d\tau$$

$$= e^{-\alpha s}\int_0^\infty f(\tau)e^{-s\tau}\, d\tau = e^{-\alpha s}F(s)$$

where

$$F(s) = \mathcal{L}[f(t)] = \int_0^\infty f(t)e^{-st}\, dt$$

And so

$$\mathcal{L}[f(t - \alpha)1(t - \alpha)] = e^{-\alpha s}F(s) \qquad \alpha \geq 0$$

317

Table 6-1. LAPLACE TRANSFORM PAIRS

	$f(t)$	$F(s)$
1	Unit impulse $\delta(t)$	1
2	Unit step $1(t)$	$\dfrac{1}{s}$
3	t	$\dfrac{1}{s^2}$
4	$\dfrac{t^{n-1}}{(n-1)!}$ $(n = 1, 2, 3, \ldots)$	$\dfrac{1}{s^n}$
5	t^n $(n = 1, 2, 3, \ldots)$	$\dfrac{n!}{s^{n+1}}$
6	e^{-at}	$\dfrac{1}{s+a}$
7	te^{-at}	$\dfrac{1}{(s+a)^2}$
8	$\dfrac{1}{(n-1)!}t^{n-1}e^{-at}$ $(n = 1, 2, 3, \ldots)$	$\dfrac{1}{(s+a)^n}$
9	$t^n e^{-at}$ $(n = 1, 2, 3, \ldots)$	$\dfrac{n!}{(s+a)^{n+1}}$
10	$\sin \omega t$	$\dfrac{\omega}{s^2 + \omega^2}$
11	$\cos \omega t$	$\dfrac{s}{s^2 + \omega^2}$
12	$\sinh \omega t$	$\dfrac{\omega}{s^2 - \omega^2}$
13	$\cosh \omega t$	$\dfrac{s}{s^2 - \omega^2}$
14	$\dfrac{1}{a}(1 - e^{-at})$	$\dfrac{1}{s(s+a)}$
15	$\dfrac{1}{b-a}(e^{-at} - e^{-bt})$	$\dfrac{1}{(s+a)(s+b)}$
16	$\dfrac{1}{b-a}(be^{-bt} - ae^{-at})$	$\dfrac{s}{(s+a)(s+b)}$

Table 6-1. (CONTINUED)

	$f(t)$	$F(s)$
17	$\dfrac{1}{ab}\left[1 + \dfrac{1}{a-b}(be^{-at} - ae^{-bt})\right]$	$\dfrac{1}{s(s+a)(s+b)}$
18	$\dfrac{1}{a^2}(1 - e^{-at} - ate^{-at})$	$\dfrac{1}{s(s+a)^2}$
19	$\dfrac{1}{a^2}(at - 1 + e^{-at})$	$\dfrac{1}{s^2(s+a)}$
20	$e^{-at}\sin \omega t$	$\dfrac{\omega}{(s+a)^2 + \omega^2}$
21	$e^{-at}\cos \omega t$	$\dfrac{s+a}{(s+a)^2 + \omega^2}$
22	$\dfrac{\omega_n}{\sqrt{1-\zeta^2}}e^{-\zeta\omega_n t}\sin \omega_n\sqrt{1-\zeta^2}\,t$	$\dfrac{\omega_n^2}{s^2 + 2\zeta\omega_n s + \omega_n^2}$
23	$-\dfrac{1}{\sqrt{1-\zeta^2}}e^{-\zeta\omega_n t}\sin (\omega_n\sqrt{1-\zeta^2}\,t - \phi)$ $\phi = \tan^{-1}\dfrac{\sqrt{1-\zeta^2}}{\zeta}$	$\dfrac{s}{s^2 + 2\zeta\omega_n s + \omega_n^2}$
24	$1 - \dfrac{1}{\sqrt{1-\zeta^2}}e^{-\zeta\omega_n t}\sin (\omega_n\sqrt{1-\zeta^2}\,t + \phi)$ $\phi = \tan^{-1}\dfrac{\sqrt{1-\zeta^2}}{\zeta}$	$\dfrac{\omega_n^2}{s(s^2 + 2\zeta\omega_n s + \omega_n^2)}$
25	$1 - \cos \omega t$	$\dfrac{\omega^2}{s(s^2 + \omega^2)}$
26	$\omega t - \sin \omega t$	$\dfrac{\omega^3}{s^2(s^2 + \omega^2)}$
27	$\sin \omega t - \omega t \cos \omega t$	$\dfrac{2\omega^3}{(s^2 + \omega^2)^2}$
28	$\dfrac{1}{2\omega}t\sin \omega t$	$\dfrac{s}{(s^2 + \omega^2)^2}$
29	$t\cos \omega t$	$\dfrac{s^2 - \omega^2}{(s^2 + \omega^2)^2}$
30	$\dfrac{1}{\omega_2^2 - \omega_1^2}(\cos \omega_1 t - \cos \omega_2 t)\quad (\omega_1^2 \neq \omega_2^2)$	$\dfrac{s}{(s^2 + \omega_1^2)(s^2 + \omega_2^2)}$
31	$\dfrac{1}{2\omega}(\sin \omega t + \omega t \cos \omega t)$	$\dfrac{s^2}{(s^2 + \omega^2)^2}$

Table 6-2. PROPERTIES OF LAPLACE TRANSFORMS

1	$\mathcal{L}[Af(t)] = AF(s)$
2	$\mathcal{L}[f_1(t) \pm f_2(t)] = F_1(s) \pm F_2(s)$
3	$\mathcal{L}_{\pm}\left[\dfrac{d}{dt}f(t)\right] = sF(s) - f(0\pm)$
4	$\mathcal{L}_{\pm}\left[\dfrac{d^2}{dt^2}f(t)\right] = s^2 F(s) - sf(0\pm) - \dot{f}(0\pm)$
5	$\mathcal{L}_{\pm}\left[\dfrac{d^n}{dt^n}f(t)\right] = s^n F(s) - \displaystyle\sum_{k=1}^{n} s^{n-k}\overset{(k-1)}{f}(0\pm)$ where $\overset{(k-1)}{f}(t) = \dfrac{d^{k-1}}{dt^{k-1}}f(t)$
6	$\mathcal{L}_{\pm}\left[\displaystyle\int f(t)\,dt\right] = \dfrac{F(s)}{s} + \dfrac{\left[\int f(t)\,dt\right]_{t=0\pm}}{s}$
7	$\mathcal{L}_{\pm}\left[\displaystyle\iint f(t)\,dt\,dt\right] = \dfrac{F(s)}{s^2} + \dfrac{\left[\int f(t)\,dt\right]_{t=0\pm}}{s^2} + \dfrac{\left[\iint f(t)\,dt\,dt\right]_{t=0\pm}}{s}$
8	$\mathcal{L}_{\pm}\left[\displaystyle\int \cdots \int f(t)\,(dt)^n\right] = \dfrac{F(s)}{s^n} + \displaystyle\sum_{k=1}^{n} \dfrac{1}{s^{n-k+1}}\left[\int \cdots \int f(t)\,(dt)^k\right]_{t=0\pm}$
9	$\mathcal{L}\left[\displaystyle\int_0^t f(t)\,dt\right] = \dfrac{F(s)}{s}$
10	$\displaystyle\int_0^{\infty} f(t)\,dt = \lim_{s\to 0} F(s) \quad$ if $\displaystyle\int_0^{\infty} f(t)\,dt$ exists
11	$\mathcal{L}[e^{-at}f(t)] = F(s+a)$
12	$\mathcal{L}[f(t-\alpha)1(t-\alpha)] = e^{-\alpha s}F(s) \qquad \alpha \geq 0$
13	$\mathcal{L}[tf(t)] = -\dfrac{dF(s)}{ds}$
14	$\mathcal{L}[t^2 f(t)] = \dfrac{d^2}{ds^2}F(s)$
15	$\mathcal{L}[t^n f(t)] = (-1)^n \dfrac{d^n}{ds^n}F(s) \qquad n = 1, 2, 3, \ldots$
16	$\mathcal{L}\left[\dfrac{1}{t}f(t)\right] = \displaystyle\int_s^{\infty} F(s)\,ds$
17	$\mathcal{L}\left[f\!\left(\dfrac{t}{a}\right)\right] = aF(as)$

This last equation states that the translation of the time function $f(t)1(t)$ by α (where $\alpha \geq 0$) corresponds to the multiplication of the transform $F(s)$ by $e^{-\alpha s}$.

Pulse function. Consider the pulse function

$$f(t) = \frac{A}{t_0} \qquad \text{for } 0 < t < t_0$$
$$= 0 \qquad \text{for } t < 0, t_0 < t$$

where A and t_0 are constants.

The pulse function here may be considered a step function of height A/t_0 that begins at $t = 0$ and that is superimposed by a negative step function of height A/t_0 beginning at $t = t_0$, that is,

$$f(t) = \frac{A}{t_0} 1(t) - \frac{A}{t_0} 1(t - t_0)$$

Then the Laplace transform of $f(t)$ is obtained as

$$\mathcal{L}[f(t)] = \mathcal{L}\left[\frac{A}{t_0} 1(t)\right] - \mathcal{L}\left[\frac{A}{t_0} 1(t - t_0)\right]$$
$$= \frac{A}{t_0 s} - \frac{A}{t_0 s} e^{-st_0}$$
$$= \frac{A}{t_0 s}(1 - e^{-st_0}) \qquad\qquad (6\text{-}1)$$

Impulse function. The impulse function is a special limiting case of the pulse function. Consider the impulse function

$$f(t) = \lim_{t_0 \to 0} \frac{A}{t_0} \qquad \text{for } 0 < t < t_0$$
$$= 0 \qquad \text{for } t < 0, t_0 < t$$

Since the height of the impulse function is A/t_0 and the duration is t_0, the area under the impulse is equal to A. As the duration t_0 approaches zero, the height A/t_0 approaches infinity, but the area under the impulse remains equal to A. Note that the magnitude of the impulse is measured by its area.

Referring to Eq. (6-1), the Laplace transform of this impulse function is shown to be:

$$\mathcal{L}[f(t)] = \lim_{t_0 \to 0}\left[\frac{A}{t_0 s}(1 - e^{-st_0})\right]$$
$$= \lim_{t_0 \to 0} \frac{\dfrac{d}{dt_0}[A(1 - e^{-st_0})]}{\dfrac{d}{dt_0}(t_0 s)} = \frac{As}{s} = A$$

Thus the Laplace transform of the impulse function is equal to the area under the impulse.

The impulse function whose area is equal to unity is called the *unit-impulse function* or the *Dirac delta function*. The unit-impulse function occurring at $t = t_0$ is usually denoted by $\delta(t - t_0)$. $\delta(t - t_0)$ satisfies the following:

$$\delta(t - t_0) = 0 \qquad \text{for } t \neq t_0$$

$$\delta(t - t_0) = \infty \qquad \text{for } t = t_0$$

$$\int_{-\infty}^{\infty} \delta(t - t_0)\, dt = 1$$

It should be mentioned that an impulse that has an infinite magnitude and zero duration is mathematical fiction and does not occur in physical systems. If, however, the magnitude of a pulse input to a system is very large and its duration is very short compared to the system time constants, then we can approximate the pulse input by an impulse function. For instance, if a force or torque input $f(t)$ is applied to a system for a very short time duration $0 < t < t_0$, where the magnitude of $f(t)$ being sufficiently large so that the integral $\int_0^{t_0} f(t)\, dt$ is not negligible, then this input can be considered an impulse input. (Note that when we describe the impulse input, the area or magnitude of the impulse is most important, but the exact shape of the impulse is usually immaterial.) The impulse input supplies energy to the system in an infinitesimal time.

The concept of the impulse function is quite useful in differentiating discontinuous functions. The unit-impulse function $\delta(t - t_0)$ can be considered the derivative of the unit-step function $1(t - t_0)$ at the point of discontinuity $t = t_0$ or

$$\delta(t - t_0) = \frac{d}{dt} 1(t - t_0)$$

Conversely, if the unit-impulse function $\delta(t - t_0)$ is integrated, the result is the unit-step function $1(t - t_0)$. With the concept of the impulse function we can differentiate a function containing discontinuities, giving impulses, the magnitudes of which are equal to the magnitude of each corresponding discontinuity.

Multiplication of $f(t)$ by $e^{-\alpha t}$. If $f(t)$ is Laplace transformable, its Laplace transform being $F(s)$, then the Laplace transform of $e^{-\alpha t}f(t)$ is obtained as

$$\mathcal{L}[e^{-\alpha t}f(t)] = \int_0^{\infty} e^{-\alpha t}f(t)e^{-st}\, dt = F(s + \alpha) \tag{6-2}$$

We see that the multiplication of $f(t)$ by $e^{-\alpha t}$ has the effect of replacing s by $(s + \alpha)$ in the Laplace transform. Conversely, changing s to $(s + \alpha)$ is equivalent to multiplying $f(t)$ by $e^{-\alpha t}$. (Note that α may be real or complex.)

The relationship given by Eq. (6-2) is useful in finding the Laplace transforms of such functions as $e^{-\alpha t} \sin \omega t$ and $e^{-\alpha t} \cos \omega t$. For instance, since

$$\mathcal{L}[\sin \omega t] = \frac{\omega}{s^2 + \omega^2} = F(s), \qquad \mathcal{L}[\cos \omega t] = \frac{s}{s^2 + \omega^2} = G(s)$$

it follows from Eq. (6-2) that the Laplace transforms of $e^{-\alpha t} \sin \omega t$ and $e^{-\alpha t} \cos \omega t$ are given, respectively, by

$$\mathcal{L}[e^{-\alpha t} \sin \omega t] = F(s + \alpha) = \frac{\omega}{(s + \alpha)^2 + \omega^2}$$

$$\mathcal{L}[e^{-\alpha t} \cos \omega t] = G(s + \alpha) = \frac{s + \alpha}{(s + \alpha)^2 + \omega^2}$$

Comments on the lower limit of the Laplace integral. In some cases, $f(t)$ possesses an impulse function at $t = 0$. Then the lower limit of the Laplace integral must be clearly specified as to whether it is $0-$ or $0+$, since the Laplace transforms of $f(t)$ differ for these two lower limits. If such a distinction of the lower limit of the Laplace integral is necessary, we use the notations

$$\mathcal{L}_+[f(t)] = \int_{0+}^{\infty} f(t)e^{-st} \, dt$$

$$\mathcal{L}_-[f(t)] = \int_{0-}^{\infty} f(t)e^{-st} \, dt = \mathcal{L}_+[f(t)] + \int_{0-}^{0+} f(t)e^{-st} \, dt$$

If $f(t)$ involves an impulse function at $t = 0$, then

$$\mathcal{L}_+[f(t)] \neq \mathcal{L}_-[f(t)]$$

since

$$\int_{0-}^{0+} f(t)e^{-st} \, dt \neq 0$$

for such a case. Obviously, if $f(t)$ does not possess an impulse function at $t = 0$ (that is, if the function to be transformed is finite between $t = 0-$ and $t = 0+$), then

$$\mathcal{L}_+[f(t)] = \mathcal{L}_-[f(t)]$$

Differentiation theorem. The Laplace transform of the derivative of a function $f(t)$ is given by

$$\mathcal{L}\left[\frac{d}{dt}f(t)\right] = sF(s) - f(0) \qquad (6\text{-}3)$$

where $f(0)$ is the initial value of $f(t)$ evaluated at $t = 0$.

For a given function $f(t)$, the values of $f(0+)$ and $f(0-)$ may be the same or different as illustrated in Fig. 6-6. The distinction between $f(0+)$ and $f(0-)$ is important when $f(t)$ has a discontinuity at $t = 0$ because in such a

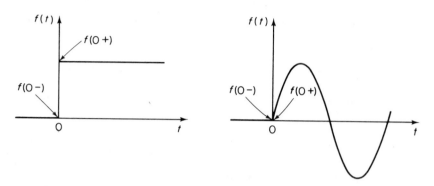

Fig. 6-6. Step function and sine function indicating initial values at $t = 0-$ and $t = 0+$.

case $df(t)/dt$ will involve an impulse function at $t = 0$. If $f(0+) \neq f(0-)$, Eq. (6-3) must be modified to

$$\mathcal{L}_+\left[\frac{d}{dt}f(t)\right] = sF(s) - f(0+)$$

$$\mathcal{L}_-\left[\frac{d}{dt}f(t)\right] = sF(s) - f(0-)$$

To prove the differentiation theorem, Eq. (6-3), we proceed as follows. Integrating the Laplace integral by parts gives

$$\int_0^\infty f(t)e^{-st}\,dt = f(t)\frac{e^{-st}}{-s}\Big|_0^\infty - \int_0^\infty \left[\frac{d}{dt}f(t)\right]\frac{e^{-st}}{-s}\,dt$$

Hence

$$F(s) = \frac{f(0)}{s} + \frac{1}{s}\mathcal{L}\left[\frac{d}{dt}f(t)\right]$$

It follows that

$$\mathcal{L}\left[\frac{d}{dt}f(t)\right] = sF(s) - f(0)$$

Similarly, we obtain the following relationship for the second derivative of $f(t)$:

$$\mathcal{L}\left[\frac{d^2}{dt^2}f(t)\right] = s^2F(s) - sf(0) - \dot{f}(0)$$

where $\dot{f}(0)$ is the value of $df(t)/dt$ evaluated at $t = 0$. To derive this equation, define

$$\frac{d}{dt}f(t) = g(t)$$

Then

$$\mathcal{L}\left[\frac{d^2}{dt^2}f(t)\right] = \mathcal{L}\left[\frac{d}{dt}g(t)\right] = s\mathcal{L}[g(t)] - g(0)$$

$$= s\mathcal{L}\left[\frac{d}{dt}f(t)\right] - \dot{f}(0)$$

$$= s^2F(s) - sf(0) - \dot{f}(0)$$

Similarly, for the nth derivative of $f(t)$, we obtain

$$\mathcal{L}\left[\frac{d^n}{dt^n}f(t)\right] = s^nF(s) - s^{n-1}f(0) - s^{n-2}\dot{f}(0) - \cdots - \overset{(n-1)}{f}(0)$$

where $f(0), \dot{f}(0), \ldots, \overset{(n-1)}{f}(0)$ represent the values of $f(t), df(t)/dt, \ldots,$ $d^{n-1}f(t)/dt^{n-1}$, respectively, evaluated at $t = 0$. If the distinction between $\mathcal{L}_+$ and $\mathcal{L}_-$ is necessary, we substitute $t = 0+$ or $t = 0-$ into $f(t), df(t)/dt,$ $\ldots, d^{n-1}f(t)/dt^{n-1}$, depending on whether we take $\mathcal{L}_+$ or $\mathcal{L}_-$.

Note that in order for Laplace transforms of derivatives of $f(t)$ to exist, $d^nf(t)/dt^n$ $(n = 1, 2, 3, \ldots)$ must be Laplace transformable.

Note also that if all the initial values of $f(t)$ and its derivatives are equal to zero, then the Laplace transform of the nth derivative of $f(t)$ is given by $s^nF(s)$.

Example 6-1. Consider the cosine function.

$$g(t) = 0 \qquad \text{for } t < 0$$
$$= A \cos \omega t \qquad \text{for } t \geq 0$$

The Laplace transform of this cosine function can be obtained directly as in the case of the sinusoidal function (see Sec. 6-3). The use of the differentiation theorem, however, will be demonstrated here by deriving the Laplace transform of the cosine function from the Laplace transform of the sine function. If we define

$$f(t) = 0 \qquad \text{for } t < 0$$
$$= \sin \omega t \qquad \text{for } t \geq 0$$

then

$$\mathcal{L}[\sin \omega t] = F(s) = \frac{\omega}{s^2 + \omega^2}$$

The Laplace transform of the cosine function is obtained as

$$\mathcal{L}[A \cos \omega t] = \mathcal{L}\left[\frac{d}{dt}\left(\frac{A}{\omega}\sin \omega t\right)\right] = \frac{A}{\omega}[sF(s) - f(0)]$$

$$= \frac{A}{\omega}\left[\frac{s\omega}{s^2 + \omega^2} - 0\right] = \frac{As}{s^2 + \omega^2}$$

Final value theorem. The final value theorem relates the steady-state behavior of $f(t)$ to the behavior of $sF(s)$ in the neighborhood of $s = 0$. This theorem, however, applies if and only if $\lim_{t \to \infty} f(t)$ exists [which means that $f(t)$ settles down to a definite value for $t \to \infty$.] If all poles of $sF(s)$ lie in the left half s plane, $\lim_{t \to \infty} f(t)$ exists. Yet if $sF(s)$ has poles on the imaginary axis or in the right half s plane, $f(t)$ will contain oscillating or exponentially increasing time functions, respectively, and $\lim_{t \to \infty} f(t)$ will not exist. The final value theorem does not apply in such cases. For instance, if $f(t)$ is the sinusoidal function $\sin \omega t$, $sF(s)$ has poles at $s = \pm j\omega$ and $\lim_{t \to \infty} f(t)$ does not exist. Therefore this theorem is not applicable to such a function.

The final value theorem may be stated as follows. If $f(t)$ and $df(t)/dt$ are Laplace transformable, if $F(s)$ is the Laplace transform of $f(t)$, and if $\lim_{t \to \infty} f(t)$ exists, then

$$\lim_{t \to \infty} f(t) = \lim_{s \to 0} sF(s)$$

To prove the theorem, we let s approach zero in the equation for the Laplace transform of the derivative of $f(t)$ or

$$\lim_{s \to 0} \int_0^\infty \left[\frac{d}{dt} f(t) \right] e^{-st}\, dt = \lim_{s \to 0} \left[sF(s) - f(0) \right]$$

Since $\lim_{s \to 0} e^{-st} = 1$, we obtain

$$\int_0^\infty \left[\frac{d}{dt} f(t) \right] dt = f(t) \Big|_0^\infty = f(\infty) - f(0)$$

$$= \lim_{s \to 0} sF(s) - f(0)$$

from which

$$f(\infty) = \lim_{t \to \infty} f(t) = \lim_{s \to 0} sF(s)$$

Example 6-2. Given

$$\mathscr{L}[f(t)] = F(s) = \frac{1}{s(s + 1)}$$

what is $\lim_{t \to \infty} f(t)$?

Since the pole of $sF(s) = 1/(s + 1)$ lies in the left half s plane, $\lim_{t \to \infty} f(t)$ exists. So the final value theorem is applicable in this case.

$$\lim_{t \to \infty} f(t) = f(\infty) = \lim_{s \to 0} sF(s) = \lim_{s \to 0} \frac{s}{s(s + 1)} = \lim_{s \to 0} \frac{1}{s + 1} = 1$$

In fact, this result can easily be verified, since

$$f(t) = 1 - e^{-t} \qquad \text{for } t \geq 0$$

Comments. The final value theorem does not apply if $sF(s)$ possesses poles on the $j\omega$ axis or in the right half s plane. It should be noted that $sF(s)$ may formally possess a single pole at the origin, provided that it does not possess any other poles in the right half s plane, including the $j\omega$ axis, and provided that $\lim_{s \to 0} sF(s)$ exists.

Consider the function

$$f(t) = 0 \qquad t < 0$$
$$= t \qquad 0 \leq t \leq T$$
$$= T \qquad T < t$$

This function is plotted in Fig. 6-7. The function $f(t)$ can be considered as

$$f(t) = t1(t) - (t - T)1(t - T)$$

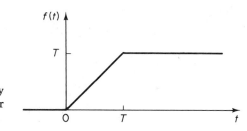

Fig. 6-7. Function $f(t)$ defined by $f(t) = 0$ for $t < 0$, $f(t) = t$ for $0 \leq t \leq T$, and $f(t) = T$ for $T < t$.

The Laplace transform of $f(t)$ is

$$\mathcal{L}[f(t)] = F(s) = \mathcal{L}[t1(t)] - \mathcal{L}[(t - T)1(t - T)]$$
$$= \frac{1}{s^2} - \frac{e^{-Ts}}{s^2} = \frac{1 - e^{-Ts}}{s^2}$$

Consequently,

$$sF(s) = \frac{1 - e^{-Ts}}{s}$$

Thus $sF(s)$ formally possesses a single pole at the origin but no other poles in the right half s plane, including the $j\omega$ axis. Since

$$\lim_{s \to 0} sF(s) = \lim_{s \to 0} \frac{1 - e^{-Ts}}{s}$$
$$= \lim_{s \to 0} \frac{\frac{d}{ds}(1 - e^{-Ts})}{\frac{d}{ds}s} = \lim_{s \to 0} \frac{Te^{-Ts}}{1} = T$$

we see that the limit of $sF(s)$ exists as $s \to 0$. Hence the final value theorem applies to this case or

$$\lim_{t \to \infty} f(t) = \lim_{s \to 0} sF(s) = T$$

Initial value theorem. The initial value theorem discussed below is the counterpart of the final value theorem. By using this theorem, we are able to find the value of $f(t)$ at $t = 0+$ directly from the Laplace transform of $f(t)$. The initial value theorem does not give the value of $f(t)$ at exactly $t = 0$ but at a time slightly greater than zero.

The initial value theorem may be stated as follows. If $f(t)$ and $df(t)/dt$ are both Laplace transformable and if $\lim_{s\to\infty} sF(s)$ exists, then

$$f(0+) = \lim_{s\to\infty} sF(s)$$

To prove this theorem, we use the equation for the $\mathcal{L}_+$ transform of $df(t)/dt$.

$$\mathcal{L}_+\left[\frac{d}{dt}f(t)\right] = sF(s) - f(0+)$$

For the time interval $0+ \leq t \leq \infty$, as s approaches infinity, e^{-st} approaches zero. (Note that we must use $\mathcal{L}_+$ rather than $\mathcal{L}_-$ for this condition.) And so

$$\lim_{s\to\infty} \int_{0+}^{\infty} \left[\frac{d}{dt}f(t)\right]e^{-st}\,dt = \lim_{s\to\infty}[sF(s) - f(0+)] = 0$$

or

$$f(0+) = \lim_{s\to\infty} sF(s)$$

In applying the initial value theorem, we are not limited as to the locations of the poles of $sF(s)$. Thus the initial value theorem is valid for the sinusoidal function.

It should be noted that the initial value theorem and the final value theorem provide a convenient check on the solution, since they enable us to predict the system behavior in the time domain without actually transforming functions in s back to time functions.

Integration theorem. If $f(t)$ is of exponential order, then the Laplace transform of $\int f(t)\,dt$ exists and is given by

$$\mathcal{L}\left[\int f(t)\,dt\right] = \frac{F(s)}{s} + \frac{f^{-1}(0)}{s} \tag{6-4}$$

where $F(s) = \mathcal{L}[f(t)]$ and $f^{-1}(0) = \int f(t)\,dt$, evaluated at $t = 0$.

Note that if $f(t)$ involves an impulse function at $t = 0$, then $f^{-1}(0+) \neq f^{-1}(0-)$. So if $f(t)$ involves an impulse function at $t = 0$, we must modify Eq. (6-4).

$$\mathcal{L}_+\left[\int f(t)\,dt\right] = \frac{F(s)}{s} + \frac{f^{-1}(0+)}{s}$$

$$\mathcal{L}_-\left[\int f(t)\,dt\right] = \frac{F(s)}{s} + \frac{f^{-1}(0-)}{s}$$

The integration theorem given by Eq. (6-4) can be proved in the following way. Integration by parts yields

$$\mathcal{L}\left[\int f(t)\, dt\right] = \int_0^\infty \left[\int f(t)\, dt\right] e^{-st}\, dt$$

$$= \left[\int f(t)\, dt\right] \frac{e^{-st}}{-s}\bigg|_0^\infty - \int_0^\infty f(t)\, \frac{e^{-st}}{-s}\, dt$$

$$= \frac{1}{s}\int f(t)\, dt\bigg|_{t=0} + \frac{1}{s}\int_0^\infty f(t) e^{-st}\, dt$$

$$= \frac{f^{-1}(0)}{s} + \frac{F(s)}{s}$$

and the theorem is proved.

We see that integration in the time domain is converted into division in the s domain. If the initial value of the integral is zero, the Laplace transform of the integral of $f(t)$ is given by $F(s)/s$.

The preceding integration theorem can be modified slightly to deal with the definite integral of $f(t)$. If $f(t)$ is of exponential order, the Laplace transform of the definite integral $\int_0^t f(t)\, dt$ can be given by

$$\mathcal{L}\left[\int_0^t f(t)\, dt\right] = \frac{F(s)}{s} \tag{6-5}$$

Note that if $f(t)$ involves an impulse function at $t = 0$, then $\int_{0+}^t f(t)\, dt \neq \int_{0-}^t f(t)\, dt$, and the following distinction must be observed:

$$\mathcal{L}_+\left[\int_{0+}^t f(t)\, dt\right] = \frac{\mathcal{L}_+[f(t)]}{s}$$

$$\mathcal{L}_-\left[\int_{0-}^t f(t)\, dt\right] = \frac{\mathcal{L}_-[f(t)]}{s}$$

To prove Eq. (6-5), first note that

$$\int_0^t f(t)\, dt = \int f(t)\, dt - f^{-1}(0)$$

where $f^{-1}(0)$ is equal to $\int f(t)\, dt$ evaluated at $t = 0$ and is a constant. Hence

$$\mathcal{L}\left[\int_0^t f(t)\, dt\right] = \mathcal{L}\left[\int f(t)\, dt - f^{-1}(0)\right]$$

$$= \mathcal{L}\left[\int f(t)\, dt\right] - \mathcal{L}[f^{-1}(0)]$$

Referring to Eq. (6-4) and noting that $f^{-1}(0)$ is a constant so that

$$\mathcal{L}[f^{-1}(0)] = \frac{f^{-1}(0)}{s}$$

we obtain

$$\mathscr{L}\left[\int_0^t f(t)\,dt\right] = \frac{F(s)}{s} + \frac{f^{-1}(0)}{s} - \frac{f^{-1}(0)}{s} = \frac{F(s)}{s}$$

6-5   INVERSE LAPLACE TRANSFORMATION

The inverse Laplace transformation refers to the process of finding the time function $f(t)$ from the corresponding Laplace transform $F(s)$. Several methods are available for finding the inverse Laplace transforms, the simplest of which are (a) to use tables of Laplace transforms to find the time function $f(t)$ corresponding to a given Laplace transform $F(s)$ and (b) to use the partial-fraction expansion method. In this section we present the latter. (We shall not discuss additional methods, such as using the inversion integral. For this method, see, for instance, Reference 6-4.)

Partial-fraction expansion method for finding inverse Laplace transforms. If $F(s)$, the Laplace transform of $f(t)$, is broken up into components

$$F(s) = F_1(s) + F_2(s) + \cdots + F_n(s)$$

and if the inverse Laplace transforms of $F_1(s)$, $F_2(s)$, ..., $F_n(s)$ are readily available, then

$$\mathscr{L}^{-1}[F(s)] = \mathscr{L}^{-1}[F_1(s)] + \mathscr{L}^{-1}[F_2(s)] + \cdots + \mathscr{L}^{-1}[F_n(s)]$$
$$= f_1(t) + f_2(t) + \cdots + f_n(t)$$

where $f_1(t), f_2(t), \ldots, f_n(t)$ are the inverse Laplace transforms of $F_1(s)$, $F_2(s)$, ..., $F_n(s)$, respectively. The inverse Laplace transform of $F(s)$ thus obtained is unique except possibly at points where the time function is discontinuous. Whenever the time function is continuous, the time function $f(t)$ and its Laplace transform $F(s)$ have a one-to-one correspondence.

For problems in systems analysis, $F(s)$ frequently occurs in the form

$$F(s) = \frac{A(s)}{B(s)}$$

where $A(s)$ and $B(s)$ are polynomials in s and the degree of $A(s)$ is not higher than that of $B(s)$.

The advantage of the partial-fraction expansion approach is that the individual terms of $F(s)$, resulting from the expansion into partial-fraction form, are very simple functions of s; consequently, it is not necessary to refer to a Laplace transform table if we memorize several simple Laplace transform pairs. It should be noted, however, that in applying the partial-fraction expansion technique in the search for the inverse Laplace transform of $F(s)$ = $A(s)/B(s)$, the roots of the denominator polynomial $B(s)$ must be known in advance. That is, this method does not apply until the denominator polynomial has been factored.

Consider $F(s)$ written in the factored form

$$F(s) = \frac{A(s)}{B(s)} = \frac{K(s + z_1)(s + z_2) \cdots (s + z_m)}{(s + p_1)(s + p_2) \cdots (s + p_n)}$$

where $p_1, p_2, \ldots, p_n$ and $z_1, z_2, \ldots, z_m$ are either real or complex quantities, but for each complex p_i or z_i there will occur the complex conjugate of p_i or z_i, respectively. Here the highest power of s in $B(s)$ is assumed to be higher than that in $A(s)$.

In the expansion of $A(s)/B(s)$ into partial-fraction form, it is important that the highest power of s in $B(s)$ be greater than the highest power of s in $A(s)$. If such is not the case, the numerator $A(s)$ must be divided by the denominator $B(s)$ in order to produce a polynomial in s plus a remainder (a ratio of polynomials in s whose numerator is of lower degree than the denominator.) (For details, see Example 6-4.)

Partial-fraction expansion when $F(s)$ involves distinct poles only. In this case, $F(s)$ can always be expanded into a sum of simple partial fractions

$$F(s) = \frac{A(s)}{B(s)} = \frac{a_1}{s + p_1} + \frac{a_2}{s + p_2} + \cdots + \frac{a_n}{s + p_n}$$

where a_k $(k = 1, 2, \ldots, n)$ are constants. The coefficient a_k is called the *residue* at the pole at $s = -p_k$. The value of a_k can be found by multiplying both sides of this last equation by $(s + p_k)$ and letting $s = -p_k$, which gives

$$\left[(s + p_k)\frac{A(s)}{B(s)} \right]_{s = -p_k} = \left[\frac{a_1}{s + p_1}(s + p_k) + \frac{a_2}{s + p_2}(s + p_k) \right.$$

$$\left. + \cdots + \frac{a_k}{s + p_k}(s + p_k) + \cdots + \frac{a_n}{s + p_n}(s + p_k) \right]_{s = -p_k}$$

$$= a_k$$

We see that all the expanded terms drop out with the exception of a_k. Thus the residue a_k is found from

$$a_k = \left[(s + p_k)\frac{A(s)}{B(s)} \right]_{s = -p_k} \tag{6-6}$$

Note that since $f(t)$ is a real function of time, if p_1 and p_2 are complex conjugates, then the residues a_1 and a_2 are also complex conjugates. Only one of the conjugates, a_1 or a_2, need be evaluated because the other is known automatically.

Since

$$\mathcal{L}^{-1}\left[\frac{a_k}{s + p_k} \right] = a_k e^{-p_k t}$$

$f(t)$ is obtained as

$$f(t) = \mathcal{L}^{-1}[F(s)] = a_1 e^{-p_1 t} + a_2 e^{-p_2 t} + \cdots + a_n e^{-p_n t} \qquad (t \geq 0)$$

===

Example 6-3. Find the inverse Laplace transform of

$$F(s) = \frac{s + 3}{(s + 1)(s + 2)}$$

The partial-fraction expansion of $F(s)$ is

$$F(s) = \frac{s + 3}{(s + 1)(s + 2)} = \frac{a_1}{s + 1} + \frac{a_2}{s + 2}$$

where a_1 and a_2 are found by using Eq. (6-6).

$$a_1 = \left[(s + 1)\frac{s + 3}{(s + 1)(s + 2)}\right]_{s=-1} = \left[\frac{s + 3}{s + 2}\right]_{s=-1} = 2$$

$$a_2 = \left[(s + 2)\frac{s + 3}{(s + 1)(s + 2)}\right]_{s=-2} = \left[\frac{s + 3}{s + 1}\right]_{s=-2} = -1$$

Thus

$$f(t) = \mathcal{L}^{-1}[F(s)]$$

$$= \mathcal{L}^{-1}\left[\frac{2}{s + 1}\right] + \mathcal{L}^{-1}\left[\frac{-1}{s + 2}\right]$$

$$= 2e^{-t} - e^{-2t} \qquad (t \geq 0)$$

Example 6-4. Obtain the inverse Laplace transform of

$$G(s) = \frac{s^3 + 5s^2 + 9s + 7}{(s + 1)(s + 2)}$$

Here, since the degree of the numerator polynomial is higher than that of the denominator polynomial, we must divide the numerator by the denominator.

$$G(s) = s + 2 + \frac{s + 3}{(s + 1)(s + 2)}$$

Note that the Laplace transform of the unit impulse function $\delta(t)$ is 1 and that the Laplace transform of $d\delta(t)/dt$ is s. The third term on the right-hand side of this last equation is $F(s)$ in Example 6-3. So the inverse Laplace transform of $G(s)$ is given as

$$g(t) = \frac{d}{dt}\delta(t) + 2\delta(t) + 2e^{-t} - e^{-2t} \qquad (t \geq 0-)$$

Example 6-5. Find the inverse Laplace transform of

$$F(s) = \frac{2s + 12}{s^2 + 2s + 5}$$

Notice that the denominator polynomial can be factored as

$$s^2 + 2s + 5 = (s + 1 + j2)(s + 1 - j2)$$

If the function $F(s)$ involves a pair of complex conjugate poles, it is convenient not to expand $F(s)$ into the usual partial fractions but to expand it into the sum of a damped sine and a damped cosine function.

Noting that $s^2 + 2s + 5 = (s + 1)^2 + 2^2$ and referring to the Laplace transforms of $e^{-\alpha t} \sin \omega t$ and $e^{-\alpha t} \cos \omega t$, rewritten thus,

$$\mathcal{L}[e^{-\alpha t} \sin \omega t] = \frac{\omega}{(s + \alpha)^2 + \omega^2}$$

$$\mathcal{L}[e^{-\alpha t} \cos \omega t] = \frac{s + \alpha}{(s + \alpha)^2 + \omega^2}$$

the given $F(s)$ can be written as a sum of a damped sine and a damped cosine function.

$$F(s) = \frac{2s + 12}{s^2 + 2s + 5} = \frac{10 + 2(s + 1)}{(s + 1)^2 + 2^2}$$

$$= 5 \frac{2}{(s + 1)^2 + 2^2} + 2 \frac{s + 1}{(s + 1)^2 + 2^2}$$

It follows that

$$f(t) = \mathcal{L}^{-1}[F(s)]$$

$$= 5\mathcal{L}^{-1} \left[\frac{2}{(s + 1)^2 + 2^2} \right] + 2\mathcal{L}^{-1} \left[\frac{s + 1}{(s + 1)^2 + 2^2} \right]$$

$$= 5e^{-t} \sin 2t + 2e^{-t} \cos 2t \qquad (t \geq 0)$$

Partial-fraction expansion when $F(s)$ involves multiple poles. Instead of discussing the general case, we shall use an example to show how to obtain the partial-fraction expansion of $F(s)$. (See also Problem A-6-14.)

Consider the following $F(s)$:

$$F(s) = \frac{s^2 + 2s + 3}{(s + 1)^3}$$

The partial-fraction expansion of this $F(s)$ involves three terms

$$F(s) = \frac{A(s)}{B(s)} = \frac{b_3}{(s + 1)^3} + \frac{b_2}{(s + 1)^2} + \frac{b_1}{s + 1}$$

where b_3, b_2, and b_1 are determined as follows. By multiplying both sides of this last equation by $(s + 1)^3$, we have

$$(s + 1)^3 \frac{A(s)}{B(s)} = b_3 + b_2(s + 1) + b_1(s + 1)^2 \qquad (6\text{-}7)$$

Then letting $s = -1$, Eq. (6-7) gives

$$\left[(s + 1)^3 \frac{A(s)}{B(s)} \right]_{s=-1} = b_3$$

Also, differentiation of both sides of Eq. (6-7) with respect to s yields

$$\frac{d}{ds} \left[(s + 1)^3 \frac{A(s)}{B(s)} \right] = b_2 + 2b_1(s + 1) \qquad (6\text{-}8)$$

if we let $s = -1$ in Eq. (6-8), then

$$\frac{d}{ds}\left[(s + 1)^3 \frac{A(s)}{B(s)}\right]_{s=-1} = b_2$$

By differentiating both sides of Eq. (6-8) with respect to s, the result is

$$\frac{d^2}{ds^2}\left[(s + 1)^3 \frac{A(s)}{B(s)}\right] = 2b_1$$

From the preceding analysis it can be seen that the values of b_1, b_2, and b_3 are found systematically as follows:

$$b_3 = \left[(s + 1)^3 \frac{A(s)}{B(s)}\right]_{s=-1}$$
$$= (s^2 + 2s + 3)_{s=-1}$$
$$= 2$$

$$b_2 = \left\{\frac{d}{ds}\left[(s + 1)^3 \frac{A(s)}{B(s)}\right]\right\}_{s=-1}$$
$$= \left[\frac{d}{ds}(s^2 + 2s + 3)\right]_{s=-1}$$
$$= (2s + 2)_{s=-1}$$
$$= 0$$

$$b_1 = \frac{1}{2!}\left\{\frac{d^2}{ds^2}\left[(s + 1)^3 \frac{A(s)}{B(s)}\right]\right\}_{s=-1}$$
$$= \frac{1}{2!}\left[\frac{d^2}{ds^2}(s^2 + 2s + 3)\right]_{s=-1}$$
$$= \frac{1}{2}(2) = 1$$

We thus obtain

$$f(t) = \mathcal{L}^{-1}[F(s)]$$
$$= \mathcal{L}^{-1}\left[\frac{2}{(s + 1)^3}\right] + \mathcal{L}^{-1}\left[\frac{0}{(s + 1)^2}\right] + \mathcal{L}^{-1}\left[\frac{1}{s + 1}\right]$$
$$= t^2 e^{-t} + 0 + e^{-t}$$
$$= (t^2 + 1)e^{-t} \qquad (t \geq 0)$$

6-6   SOLVING LINEAR, TIME-INVARIANT, DIFFERENTIAL EQUATIONS

In this section we are concerned with the use of the Laplace transform method in solving linear, time-invariant, differential equations.

The Laplace transform method yields the complete solution (complementary solution and particular solution) of linear, time-invariant, differential equations. Classical methods for finding the complete solution of a differential equation require the evaluation of the integration constants from the initial conditions. In the case of the Laplace transform method, however, this requirement is unnecessary because the initial conditions are automatically included in the Laplace transform of the differential equation.

If all initial conditions are zero, then the Laplace transform of the differential equation is obtained simply by replacing d/dt with s, d^2/dt^2 with s^2, and so on.

In solving linear, time-invariant, differential equations by the Laplace transform method, two steps are needed.

1. By taking the Laplace transform of each term in the given differential equation, convert the differential equation into an algebraic equation in s and obtain the expression for the Laplace transform of the dependent variable by rearranging the algebraic equation.
2. The time solution of the differential equation is obtained by finding the inverse Laplace transform of the dependent variable.

In the following discussion a few examples are used to demonstrate the solution of linear, time-invariant, differential equations by the Laplace transform method.

Example 6-6. Find the solution $x(t)$ of the differential equation

$$\ddot{x} + 3\dot{x} + 2x = 0, \qquad x(0) = a, \qquad \dot{x}(0) = b$$

where a and b are constants.

By writing the Laplace transform of $x(t)$ as $X(s)$ or

$$\mathcal{L}[x(t)] = X(s)$$

we obtain

$$\mathcal{L}[\dot{x}] = sX(s) - x(0)$$

$$\mathcal{L}[\ddot{x}] = s^2 X(s) - sx(0) - \dot{x}(0)$$

And so the given differential equation becomes

$$[s^2 X(s) - sx(0) - \dot{x}(0)] + 3[sX(s) - x(0)] + 2X(s) = 0$$

By substituting the given initial conditions into this last equation,

$$[s^2 X(s) - as - b] + 3[sX(s) - a] + 2X(s) = 0$$

or

$$(s^2 + 3s + 2)X(s) = as + b + 3a$$

Solving for $X(s)$, we have

$$X(s) = \frac{as + b + 3a}{s^2 + 3s + 2} = \frac{as + b + 3a}{(s+1)(s+2)} = \frac{2a+b}{s+1} - \frac{a+b}{s+2}$$

The inverse Laplace transform of $X(s)$ gives

$$x(t) = \mathcal{L}^{-1}[X(s)] = \mathcal{L}^{-1}\left[\frac{2a+b}{s+1}\right] - \mathcal{L}^{-1}\left[\frac{a+b}{s+2}\right]$$

$$= (2a+b)e^{-t} - (a+b)e^{-2t} \qquad (t \geq 0)$$

which is the solution of the given differential equation. Notice that the initial conditions a and b appear in the solution. Thus $x(t)$ has no undetermined constants.

Example 6-7. Find the solution $x(t)$ of the differential equation

$$\ddot{x} + 2\dot{x} + 5x = 3, \qquad x(0) = 0, \qquad \dot{x}(0) = 0$$

Noting that $\mathcal{L}[3] = 3/s$, $x(0) = 0$, and $\dot{x}(0) = 0$, the Laplace transform of the differential equation becomes

$$s^2 X(s) + 2s X(s) + 5 X(s) = \frac{3}{s}$$

Solving for $X(s)$, we find

$$X(s) = \frac{3}{s(s^2 + 2s + 5)} = \frac{3}{5}\frac{1}{s} - \frac{3}{5}\frac{s+2}{s^2 + 2s + 5}$$

$$= \frac{3}{5}\frac{1}{s} - \frac{3}{10}\frac{2}{(s+1)^2 + 2^2} - \frac{3}{5}\frac{s+1}{(s+1)^2 + 2^2}$$

Hence the inverse Laplace transform becomes

$$x(t) = \mathcal{L}^{-1}[X(s)]$$

$$= \frac{3}{5}\mathcal{L}^{-1}\left[\frac{1}{s}\right] - \frac{3}{10}\mathcal{L}^{-1}\left[\frac{2}{(s+1)^2 + 2^2}\right] - \frac{3}{5}\mathcal{L}^{-1}\left[\frac{s+1}{(s+1)^2 + 2^2}\right]$$

$$= \frac{3}{5} - \frac{3}{10}e^{-t}\sin 2t - \frac{3}{5}e^{-t}\cos 2t \qquad (t \geq 0)$$

which is the solution of the given differential equation.

REFERENCES

6-1   BOHN, E. V., *The Transform Analysis of Linear Systems*, Reading, Mass.: Addison-Wesley Publishing Company, Inc., 1963.

6-2   CHURCHILL, R. V., *Operational Mathematics*, New York: McGraw-Hill Book Company, Inc., 1958.

6-3   GOLDMAN, S., *Transformation Calculus and Electrical Transients*, Englewood Cliffs, N.J.: Prentice-Hall, Inc., 1949.

6-4   KAPLAN, W., *Operational Methods for Linear Systems*, Reading, Mass.: Addison-Wesley Publishing Company, Inc., 1962.

6-5   LEPAGE, W. R., *Complex Variables and the Laplace Transform for Engineers*, New York: McGraw-Hill Book Company, Inc., 1961.

6-6   OGATA, K., *Modern Control Engineering*, Englewood Cliffs, N.J.: Prentice-Hall, Inc., 1970.

EXAMPLE PROBLEMS AND SOLUTIONS

PROBLEM A-6-1. Obtain the real and imaginary parts of

$$\frac{2+j1}{3+j4}$$

Also, obtain the magnitude and angle of this complex quantity.

Solution.

$$\frac{2+j1}{3+j4} = \frac{(2+j1)(3-j4)}{(3+j4)(3-j4)} = \frac{6+j3-j8+4}{9+16} = \frac{10-j5}{25}$$

$$= \frac{2}{5} - j\frac{1}{5}$$

Hence

$$\text{Real part} = \frac{2}{5}, \qquad \text{Imaginary part} = -\frac{1}{5}$$

The magnitude and angle of this complex quantity are obtained as follows:

$$\text{Magnitude} = \sqrt{\left(\frac{2}{5}\right)^2 + \left(\frac{-1}{5}\right)^2} = \sqrt{\frac{5}{25}} = \frac{1}{\sqrt{5}} = 0.447$$

$$\text{Angle} = \tan^{-1}\frac{(-1/5)}{(2/5)} = \tan^{-1}\left(\frac{-1}{2}\right) = -26.6°$$

PROBLEM A-6-2. Find the Laplace transform of $f(t)$ defined by

$$f(t) = 0 \qquad (t < 0)$$
$$= te^{-3t} \qquad (t \geq 0)$$

Solution. Since

$$\mathcal{L}[t] = G(s) = \frac{1}{s^2}$$

referring to Eq. (6-2), we obtain

$$F(s) = \mathcal{L}[te^{-3t}] = G(s+3) = \frac{1}{(s+3)^2}$$

PROBLEM A-6-3. What is the Laplace transform of

$$f(t) = 0 \qquad (t < 0)$$
$$= \sin(\omega t + \theta) \qquad (t \geq 0)$$

where θ is a constant?

Solution. Noting that

$$\sin(\omega t + \theta) = \sin \omega t \cos \theta + \cos \omega t \sin \theta$$

we have

$$\mathcal{L}[\sin(\omega t + \theta)] = \cos \theta \, \mathcal{L}[\sin \omega t] + \sin \theta \, \mathcal{L}[\cos \omega t]$$

$$= \cos \theta \frac{\omega}{s^2 + \omega^2} + \sin \theta \frac{s}{s^2 + \omega^2}$$

$$= \frac{\omega \cos \theta + s \sin \theta}{s^2 + \omega^2}$$

PROBLEM A-6-4. In addition to the Laplace transform $F(s)$ of the function $f(t)$ shown in Fig. 6-8, find the limiting value of $F(s)$ as a approaches zero.

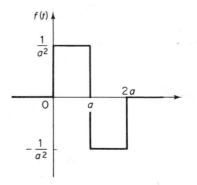

Fig. 6-8. Function $f(t)$.

Solution. The function $f(t)$ can be written

$$f(t) = \frac{1}{a^2}1(t) - \frac{2}{a^2}1(t-a) + \frac{1}{a^2}1(t-2a)$$

Then

$$F(s) = \mathcal{L}[f(t)]$$

$$= \frac{1}{a^2}\mathcal{L}[1(t)] - \frac{2}{a^2}\mathcal{L}[1(t-a)] + \frac{1}{a^2}\mathcal{L}[1(t-2a)]$$

$$= \frac{1}{a^2}\frac{1}{s} - \frac{2}{a^2}\frac{1}{s}e^{-as} + \frac{1}{a^2}\frac{1}{s}e^{-2as}$$

$$= \frac{1}{a^2 s}(1 - 2e^{-as} + e^{-2as})$$

As a approaches zero, we have

$$\lim_{a \to 0} F(s) = \lim_{a \to 0} \frac{1 - 2e^{-as} + e^{-2as}}{a^2 s} = \lim_{a \to 0} \frac{\frac{d}{da}(1 - 2e^{-as} + e^{-2as})}{\frac{d}{da}(a^2 s)}$$

$$= \lim_{a \to 0} \frac{2se^{-as} - 2se^{-2as}}{2as} = \lim_{a \to 0} \frac{e^{-as} - e^{-2as}}{a}$$

$$= \lim_{a \to 0} \frac{\frac{d}{da}(e^{-as} - e^{-2as})}{\frac{d}{da}(a)} = \lim_{a \to 0} \frac{-se^{-as} + 2se^{-2as}}{1}$$

$$= -s + 2s = s$$

PROBLEM A-6-5. Prove that if the Laplace transform of $f(t)$ is $F(s)$, then except at poles of $F(s)$

$$\mathcal{L}[tf(t)] = -\frac{d}{ds}F(s)$$

$$\mathcal{L}[t^2 f(t)] = \frac{d^2}{ds^2}F(s)$$

In general,

$$\mathcal{L}[t^n f(t)] = (-1)^n \frac{d^n}{ds^n} F(s) \qquad (n = 1, 2, 3, \ldots)$$

Solution.

$$\mathcal{L}[tf(t)] = \int_0^\infty tf(t)e^{-st}\, dt = -\int_0^\infty f(t) \frac{d}{ds}(e^{-st})\, dt$$

$$= -\frac{d}{ds} \int_0^\infty f(t)e^{-st}\, dt = -\frac{d}{ds} F(s)$$

Similarly, by defining $tf(t) = g(t)$, the result is

$$\mathcal{L}[t^2 f(t)] = \mathcal{L}[tg(t)] = -\frac{d}{ds} G(s) = -\frac{d}{ds} \left[-\frac{d}{ds} F(s) \right]$$

$$= (-1)^2 \frac{d^2}{ds^2} F(s) = \frac{d^2}{ds^2} F(s)$$

Repeating the same process, we obtain

$$\mathcal{L}[t^n f(t)] = (-1)^n \frac{d^n}{ds^n} F(s) \qquad (n = 1, 2, 3, \ldots)$$

PROBLEM A-6-6. Find the Laplace transform of $f(t)$ defined by

$$f(t) = 0 \qquad\qquad (t < 0)$$
$$= t^2 \sin \omega t \qquad (t \geq 0)$$

Solution. Since

$$\mathcal{L}[\sin \omega t] = \frac{\omega}{s^2 + \omega^2}$$

referring to Problem A-6-5, we have

$$\mathcal{L}[f(t)] = \mathcal{L}[t^2 \sin \omega t] = \frac{d^2}{ds^2} \left[\frac{\omega}{s^2 + \omega^2} \right] = \frac{-2\omega^3 + 6\omega s^2}{(s^2 + \omega^2)^3}$$

PROBLEM A-6-7. Prove that if the Laplace transform of $f(t)$ is $F(s)$, then

$$\mathcal{L}\left[f\!\left(\frac{t}{a} \right) \right] = aF(as) \qquad (a > 0)$$

Solution. If we define $t/a = \tau$ and $as = s_1$, then

$$\mathcal{L}\left[f\!\left(\frac{t}{a} \right) \right] = \int_0^\infty f\!\left(\frac{t}{a} \right) e^{-st}\, dt = \int_0^\infty f(\tau)e^{-as\tau} a\, d\tau$$

$$= a \int_0^\infty f(\tau)e^{-s_1\tau}\, d\tau = aF(s_1) = aF(as)$$

PROBLEM A-6-8. Prove that if $f(t)$ is of exponential order and if $\int_0^\infty f(t)\, dt$ exists

[which means that $\int_0^\infty f(t)\, dt$ assumes a definite value], then

$$\int_0^\infty f(t)\, dt = \lim_{s \to 0} F(s)$$

where $F(s) = \mathcal{L}[f(t)]$.

Solution. Note that

$$\int_0^\infty f(t)\, dt = \lim_{t\to\infty} \int_0^t f(t)\, dt$$

Referring to Eq. (6-5),

$$\mathcal{L}\left[\int_0^t f(t)\, dt\right] = \frac{F(s)}{s}$$

Since $\int_0^\infty f(t)\, dt$ exists, by applying the final value theorem to this case,

$$\lim_{t\to\infty} \int_0^t f(t)\, dt = \lim_{s\to 0} s\frac{F(s)}{s}$$

or

$$\int_0^\infty f(t)\, dt = \lim_{s\to 0} F(s)$$

PROBLEM A-6-9. The convolution of two time functions is defined by

$$\int_0^t f_1(\tau) f_2(t-\tau)\, d\tau$$

A commonly used notation for the convolution is $f_1(t)*f_2(t)$, which means that

$$f_1(t)*f_2(t) = \int_0^t f_1(\tau) f_2(t-\tau)\, d\tau = \int_0^t f_1(t-\tau) f_2(\tau)\, d\tau$$

If $f_1(t)$ and $f_2(t)$ are both Laplace transformable, then show that

$$\mathcal{L}\left[\int_0^t f_1(\tau) f_2(t-\tau)\, d\tau\right] = F_1(s)F_2(s)$$

where $F_1(s) = \mathcal{L}[f_1(t)]$ and $F_2(s) = \mathcal{L}[f_2(t)]$.

Solution. Noting that $1(t-\tau) = 0$ for $t < \tau$, we have

$$\mathcal{L}\left[\int_0^t f_1(\tau) f_2(t-\tau)\, d\tau\right] = \mathcal{L}\left[\int_0^\infty f_1(\tau) f_2(t-\tau) 1(t-\tau)\, d\tau\right]$$

$$= \int_0^\infty e^{-st}\left[\int_0^\infty f_1(\tau) f_2(t-\tau) 1(t-\tau)\, d\tau\right] dt$$

$$= \int_0^\infty f_1(\tau)\, d\tau \int_0^\infty f_2(t-\tau) 1(t-\tau) e^{-st}\, dt$$

$$= \int_0^\infty f_1(\tau)\, d\tau \int_\tau^\infty f_2(t-\tau) e^{-st}\, dt$$

Changing the order of integration is valid here, since $f_1(t)$ and $f_2(t)$ are both Laplace transformable, giving convergent integrals. If we substitute $\lambda = t - \tau$ into this last equation, the result is

$$\mathcal{L}\left[\int_0^t f_1(\tau) f_2(t-\tau)\, d\tau\right] = \int_0^\infty f_1(\tau) e^{-s\tau}\, d\tau \int_0^\infty f_2(\lambda) e^{-s\lambda}\, d\lambda$$

$$= F_1(s)F_2(s)$$

or

$$\mathcal{L}[f_1(t)*f_2(t)] = F_1(s)F_2(s)$$

Thus the Laplace transform of the convolution of two time functions is the product of their Laplace transforms.

PROBLEM A-6-10. Determine the Laplace transform of $f_1(t) * f_2(t)$, where

$$f_1(t) = f_2(t) = 0 \qquad \text{for } t < 0$$
$$f_1(t) = t \qquad \text{for } t \geq 0$$
$$f_2(t) = 1 - e^{-t} \qquad \text{for } t \geq 0$$

Solution. Note that

$$\mathcal{L}[t] = F_1(s) = \frac{1}{s^2}$$

$$\mathcal{L}[1 - e^{-t}] = F_2(s) = \frac{1}{s} - \frac{1}{s+1}$$

The Laplace transform of the convolution integral is given by

$$\mathcal{L}[f_1(t) * f_2(t)] = F_1(s)F_2(s) = \frac{1}{s^2}\left[\frac{1}{s} - \frac{1}{s+1}\right]$$

$$= \frac{1}{s^3} - \frac{1}{s^2(s+1)} = \frac{1}{s^3} - \frac{1}{s^2} + \frac{1}{s} - \frac{1}{s+1}$$

To verify that it is indeed the Laplace transform of the convolution integral, let us first perform integration of the convolution integral and then take its Laplace transform.

$$f_1(t) * f_2(t) = \int_0^t \tau[1 - e^{-(t-\tau)}]\, d\tau = \int_0^t (t - \tau)(1 - e^{-\tau})\, d\tau$$

$$= \frac{t^2}{2} - t + 1 - e^{-t}$$

And so

$$\mathcal{L}\left[\frac{t^2}{2} - t + 1 - e^{-t}\right] = \frac{1}{s^3} - \frac{1}{s^2} + \frac{1}{s} - \frac{1}{s+1}$$

PROBLEM A-6-11. Prove that if $f(t)$ is a periodic function with period T, then

$$\mathcal{L}[f(t)] = \frac{\displaystyle\int_0^T f(t)e^{-st}\, dt}{1 - e^{-Ts}}$$

Solution.

$$\mathcal{L}[f(t)] = \int_0^\infty f(t)e^{-st}\, dt = \sum_{n=0}^\infty \int_{nT}^{(n+1)T} f(t)\, e^{-st}\, dt$$

By changing the independent variable from t to τ, where $\tau = t - nT$

$$\mathcal{L}[f(t)] = \sum_{n=0}^\infty e^{-nTs} \int_0^T f(\tau)e^{-s\tau}\, d\tau$$

Noting that

$$\sum_{n=0}^\infty e^{-nTs} = 1 + e^{-Ts} + e^{-2Ts} + \cdots$$

$$= 1 + e^{-Ts}(1 + e^{-Ts} + e^{-2Ts} + \cdots)$$

$$= 1 + e^{-Ts}\left(\sum_{n=0}^\infty e^{-nTs}\right)$$

we obtain

$$\sum_{n=0}^{\infty} e^{-nTs} = \frac{1}{1 - e^{-Ts}}$$

It follows that

$$\mathcal{L}[f(t)] = \frac{\int_0^T f(t)e^{-st}\, dt}{1 - e^{-Ts}}$$

PROBLEM A-6-12. What is the Laplace transform of the periodic function shown in Fig. 6-9?

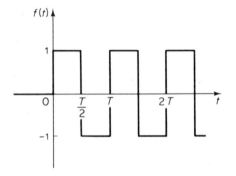

Fig. 6-9. Periodic function (square wave).

Solution. Note that

$$\int_0^T f(t)e^{-st}\, dt = \int_0^{T/2} e^{-st}\, dt + \int_{T/2}^{T} (-1)e^{-st}\, dt$$

$$= \frac{e^{-st}}{-s}\Big|_0^{T/2} - \frac{e^{-st}}{-s}\Big|_{T/2}^{T}$$

$$= \frac{e^{-(1/2)Ts} - 1}{-s} + \frac{e^{-Ts} - e^{-(1/2)Ts}}{s}$$

$$= \frac{1}{s}[e^{-Ts} - 2e^{-(1/2)Ts} + 1]$$

$$= \frac{1}{s}[1 - e^{-(1/2)Ts}]^2$$

Consequently,

$$F(s) = \frac{\int_0^T f(t)e^{-st}\, dt}{1 - e^{-Ts}} = \frac{(1/s)[1 - e^{-(1/2)Ts}]^2}{1 - e^{-Ts}}$$

$$= \frac{1 - e^{-(1/2)Ts}}{s[1 + e^{-(1/2)Ts}]} = \frac{1}{s} \tanh \frac{Ts}{4}$$

PROBLEM A-6-13. Find the inverse Laplace transform of $F(s)$, where

$$F(s) = \frac{1}{s(s^2 + 2s + 2)}$$

Solution. Since

$$s^2 + 2s + 2 = (s + 1 + j1)(s + 1 - j1)$$

we notice that $F(s)$ involves a pair of complex conjugate poles, and so we expand $F(s)$ into the form

$$F(s) = \frac{1}{s(s^2 + 2s + 2)} = \frac{a_1}{s} + \frac{a_2 s + a_3}{s^2 + 2s + 2}$$

where a_1, a_2, and a_3 are determined from

$$1 = a_1(s^2 + 2s + 2) + (a_2 s + a_3)s$$

By comparing coefficients of s^2, s, and s^0 terms on both sides of this last equation, respectively, we obtain

$$a_1 + a_2 = 0, \qquad 2a_1 + a_3 = 0, \qquad 2a_1 = 1$$

from which

$$a_1 = \frac{1}{2}, \qquad a_2 = -\frac{1}{2}, \qquad a_3 = -1$$

Therefore

$$F(s) = \frac{1}{2}\frac{1}{s} - \frac{1}{2}\frac{s + 2}{s^2 + 2s + 2}$$

$$= \frac{1}{2}\frac{1}{s} - \frac{1}{2}\frac{1}{(s + 1)^2 + 1^2} - \frac{1}{2}\frac{s + 1}{(s + 1)^2 + 1^2}$$

The inverse Laplace transform of $F(s)$ gives

$$f(t) = \frac{1}{2} - \frac{1}{2}e^{-t}\sin t - \frac{1}{2}e^{-t}\cos t \qquad (t \geq 0)$$

PROBLEM A-6-14. Derive the inverse Laplace transform of

$$F(s) = \frac{5(s + 2)}{s^2(s + 1)(s + 3)}$$

Solution.

$$F(s) = \frac{5(s + 2)}{s^2(s + 1)(s + 3)} = \frac{b_2}{s^2} + \frac{b_1}{s} + \frac{a_1}{s + 1} + \frac{a_2}{s + 3}$$

where

$$a_1 = \frac{5(s + 2)}{s^2(s + 3)}\Big|_{s=-1} = \frac{5}{2}$$

$$a_2 = \frac{5(s + 2)}{s^2(s + 1)}\Big|_{s=-3} = \frac{5}{18}$$

$$b_2 = \frac{5(s + 2)}{(s + 1)(s + 3)}\Big|_{s=0} = \frac{10}{3}$$

$$b_1 = \frac{d}{ds}\left[\frac{5(s + 2)}{(s + 1)(s + 3)}\right]_{s=0}$$

$$= \frac{5(s + 1)(s + 3) - 5(s + 2)(2s + 4)}{(s + 1)^2(s + 3)^2}\Big|_{s=0} = -\frac{25}{9}$$

Thus

$$F(s) = \frac{10}{3}\frac{1}{s^2} - \frac{25}{9}\frac{1}{s} + \frac{5}{2}\frac{1}{s + 1} + \frac{5}{18}\frac{1}{s + 3}$$

The inverse Laplace transform of $F(s)$ is

$$f(t) = \frac{10}{3}t - \frac{25}{9} + \frac{5}{2}e^{-t} + \frac{5}{18}e^{-3t} \qquad (t \geq 0)$$

PROBLEM A-6-15. Find the inverse Laplace transform of

$$F(s) = \frac{s^4 + 2s^3 + 3s^2 + 4s + 5}{s(s + 1)}$$

Solution. Since the numerator polynomial is of higher degree than the denominator polynomial, by dividing the numerator by the denominator until the remainder is a fraction, we have

$$F(s) = s^2 + s + 2 + \frac{2s + 5}{s(s + 1)} = s^2 + s + 2 + \frac{a_1}{s} + \frac{a_2}{s + 1}$$

where

$$a_1 = \frac{2s + 5}{s + 1}\bigg|_{s=0} = 5$$

$$a_2 = \frac{2s + 5}{s}\bigg|_{s=-1} = -3$$

It follows that

$$F(s) = s^2 + s + 2 + \frac{5}{s} - \frac{3}{s + 1}$$

The inverse Laplace transform of $F(s)$ is

$$f(t) = \mathcal{L}^{-1}[F(s)] = \frac{d^2}{dt^2}\,\delta(t) + \frac{d}{dt}\,\delta(t) + 2\,\delta(t) + 5 - 3e^{-t} \qquad (t \geq 0-)$$

PROBLEM A-6-16. Derive the inverse Laplace transform of

$$F(s) = \frac{1}{s(s^2 + \omega^2)}$$

Solution.

$$F(s) = \frac{1}{s(s^2 + \omega^2)} = \frac{1}{\omega^2}\left(\frac{1}{s} - \frac{s}{s^2 + \omega^2}\right)$$

$$= \frac{1}{\omega^2}\frac{1}{s} - \frac{1}{\omega^2}\frac{s}{s^2 + \omega^2}$$

Hence the inverse Laplace transform of $F(s)$ is obtained as

$$f(t) = \mathcal{L}^{-1}[F(s)] = \frac{1}{\omega^2}(1 - \cos \omega t) \qquad (t \geq 0)$$

PROBLEM A-6-17. Find the initial value of $df(t)/dt$ when the Laplace transform of $f(t)$ is given by

$$F(s) = \mathcal{L}[f(t)] = \frac{2s + 1}{s^2 + s + 1}$$

Solution. Using the initial value theorem,

$$\lim_{t \to 0+} f(t) = \lim_{s \to \infty} sF(s) = \lim_{s \to \infty} \frac{s(2s + 1)}{s^2 + s + 1} = 2$$

Since the $\mathcal{L}_+$ transform of $df(t)/dt = g(t)$ is given by

$$\mathcal{L}_+[g(t)] = sF(s) - f(0+)$$

$$= \frac{s(2s+1)}{s^2+s+1} - 2 = \frac{-s-2}{s^2+s+1}$$

the initial value of $df(t)/dt$ is obtained as

$$\lim_{t \to 0+} \frac{df(t)}{dt} = g(0+) = \lim_{s \to \infty} s[sF(s) - f(0+)]$$

$$= \lim_{s \to \infty} \frac{-s^2 - 2s}{s^2 + s + 1} = -1$$

PROBLEM A-6-18. Solve the differential equation

$$\dot{x} + 2x = \delta(t), \qquad x(0-) = 0$$

Solution. By Laplace transforming this differential equation,

$$[sX(s) - x(0-)] + 2X(s) = 1$$

or

$$(s+2)X(s) = 1$$

Then solving for $X(s)$ results in

$$X(s) = \frac{1}{s+2}$$

The inverse Laplace transform of $X(s)$ gives

$$x(t) = \mathcal{L}^{-1}[X(s)] = \mathcal{L}^{-1}\left[\frac{1}{s+2}\right] = e^{-2t} \qquad (t > 0)$$

$$= 0 \qquad (t < 0)$$

Thus $x(t)$ may be written as

$$x(t) = e^{-2t}1(t)$$

To verify this result, notice that

$$\dot{x}(t) = -2e^{-2t}1(t) + e^{-2t}\delta(t)$$

Since $\delta(t) = 0$ for $t \neq 0$, we have

$$e^{-2t}\delta(t) = \delta(t)$$

Therefore

$$\dot{x} + 2x = -2e^{-2t}1(t) + \delta(t) + 2e^{-2t}1(t) = \delta(t)$$

and

$$x(0-) = 0$$

PROBLEM A-6-19. What is the solution of the following differential equation?

$$2\ddot{x} + 7\dot{x} + 3x = 0, \qquad x(0) = 3, \qquad \dot{x}(0) = 0$$

Solution. Taking the Laplace transform of this differential equation, we have

$$2[s^2 X(s) - sx(0) - \dot{x}(0)] + 7[sX(s) - x(0)] + 3X(s) = 0$$

By substituting the given initial conditions into this last equation,

$$2[s^2 X(s) - 3s] + 7[sX(s) - 3] + 3X(s) = 0$$

or

$$(2s^2 + 7s + 3)X(s) = 6s + 21$$

Then solving for $X(s)$ yields

$$X(s) = \frac{6s + 21}{2s^2 + 7s + 3} = \frac{6s + 21}{(2s + 1)(s + 3)}$$

$$= \frac{7.2}{2s + 1} - \frac{0.6}{s + 3} = \frac{3.6}{s + 0.5} - \frac{0.6}{s + 3}$$

Finally, taking the inverse Laplace transform of $X(s)$, we find

$$x(t) = 3.6e^{-0.5t} - 0.6e^{-3t} \qquad (t \geq 0)$$

PROBLEM A-6-20. Obtain the solution of the differential equation

$$\dot{x} + ax = A \sin \omega t, \qquad x(0) = b$$

Solution. Laplace transforming both sides of this differential equation, we have

$$[sX(s) - x(0)] + aX(s) = A \frac{\omega}{s^2 + \omega^2}$$

or

$$(s + a)X(s) = \frac{A\omega}{s^2 + \omega^2} + b$$

By solving for $X(s)$, the result is

$$X(s) = \frac{A\omega}{(s + a)(s^2 + \omega^2)} + \frac{b}{s + a}$$

$$= \frac{A\omega}{a^2 + \omega^2}\left(\frac{1}{s + a} - \frac{s - a}{s^2 + \omega^2}\right) + \frac{b}{s + a}$$

$$= \left(b + \frac{A\omega}{a^2 + \omega^2}\right)\frac{1}{s + a} + \frac{Aa}{a^2 + \omega^2}\frac{\omega}{s^2 + \omega^2} - \frac{A\omega}{a^2 + \omega^2}\frac{s}{s^2 + \omega^2}$$

And so the inverse Laplace transform of $X(s)$ gives

$$x(t) = \mathcal{L}^{-1}[X(s)]$$

$$= \left(b + \frac{A\omega}{a^2 + \omega^2}\right)e^{-at} + \frac{Aa}{a^2 + \omega^2} \sin \omega t - \frac{A\omega}{a^2 + \omega^2} \cos \omega t \qquad (t \geq 0)$$

PROBLEMS

PROBLEM B-6-1. Derive the Laplace transform of the function

$$f(t) = 0 \qquad (t < 0)$$

$$= te^{-2t} \qquad (t \geq 0)$$

PROBLEM B-6-2. Find the Laplace transforms of the functions shown.

1. $\qquad$ $f_1(t) = 0$ $\qquad\qquad\qquad\qquad$ $(t < 0)$
    $\qquad\qquad\quad = 3 \sin (5t + 45°)$ $\qquad$ $(t \geq 0)$
2. $\qquad$ $f_2(t) = 0$ $\qquad\qquad\qquad\qquad$ $(t < 0)$
    $\qquad\qquad\quad = 0.03(1 - \cos 2t)$ $\qquad$ $(t \geq 0)$

PROBLEM B-6-3. Obtain the Laplace transform of the function defined by

$$f(t) = 0 \qquad\qquad (t < 0)$$
$$\quad = t^2 e^{-at} \qquad (t \geq 0)$$

PROBLEM B-6-4. Obtain the Laplace transform of the function

$$f(t) = 0 \qquad\qquad\qquad (t < 0)$$
$$\quad = \cos 2\omega t \cdot \cos 3\omega t \qquad (t \geq 0)$$

PROBLEM B-6-5. What is the Laplace transform of the function $f(t)$ shown in Fig. 6-10?

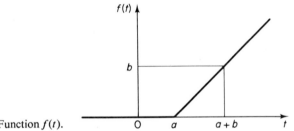

Fig. 6-10. Function $f(t)$.

PROBLEM B-6-6. What is the Laplace transform of the function $f(t)$ shown in Fig. 6-11? Also, what is the limiting value of $\mathcal{L}[f(t)]$ as a approaches zero?

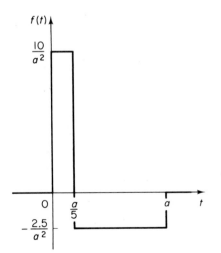

Fig. 6-11. Function $f(t)$.

PROBLEM B-6-7. Find the Laplace transform of the function $f(t)$ shown in Fig. 6-12; also find the limiting value of $\mathcal{L}[f(t)]$ as a approaches zero.

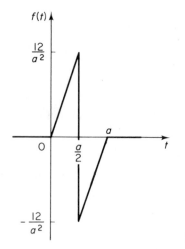

Fig. 6-12. Function $f(t)$.

PROBLEM B-6-8. Given

$$F(s) = \frac{5(s + 2)}{s(s + 1)}$$

obtain $f(\infty)$. Use the final value theorem.

PROBLEM B-6-9. Given

$$F(s) = \frac{2(s + 2)}{s(s + 1)(s + 3)}$$

obtain $f(0+)$. Use the initial value theorem.

PROBLEM B-6-10. Show that if $f(t)/t$ is Laplace transformable, then

$$\mathcal{L}\left[\frac{f(t)}{t}\right] = \int_{s}^{\infty} F(s) \, ds$$

PROBLEM B-6-11. Derive the Laplace transform of the periodic function shown in Fig. 6-13. (In the figure, dots imply the repetition of the pulse.)

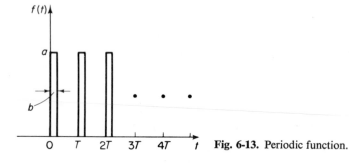

Fig. 6-13. Periodic function.

PROBLEM B-6-12. Obtain the Laplace transform of the periodic function shown in Fig. 6-14. (In the figure, dots imply the repetition of the impulse.)

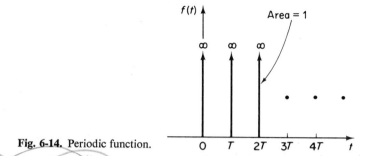

Fig. 6-14. Periodic function.

PROBLEM B-6-13. What are the inverse Laplace transforms of the functions shown?

1.
$$F_1(s) = \frac{s + 5}{(s + 1)(s + 3)}$$

2.
$$F_2(s) = \frac{3(s + 4)}{s(s + 1)(s + 2)}$$

PROBLEM B-6-14. Find the inverse Laplace transforms of the following functions.

1.
$$F_1(s) = \frac{6s + 3}{s^2}$$

2.
$$F_2(s) = \frac{5s + 2}{(s + 1)(s + 2)^2}$$

PROBLEM B-6-15. Find the inverse Laplace transform of
$$F(s) = \frac{2s^2 + 4s + 5}{s(s + 1)}$$

PROBLEM B-6-16. Derive the inverse Laplace transform of
$$F(s) = \frac{1}{s^2(s^2 + \omega^2)}$$

PROBLEM B-6-17. Find the solution $x(t)$ of the differential equation
$$\ddot{x} + 4x = 0, \qquad x(0) = 5, \qquad \dot{x}(0) = 0$$

PROBLEM B-6-18. Obtain the solution $x(t)$ of the differential equation
$$\ddot{x} + \omega_n^2 x = t, \qquad x(0) = 0, \qquad \dot{x}(0) = 0$$

PROBLEM B-6-19. Determine the solution $x(t)$ of the differential equation
$$2\ddot{x} + 2\dot{x} + x = 1, \qquad x(0) = 0, \qquad \dot{x}(0) = 2$$

PROBLEM B-6-20. Obtain the solution $x(t)$ of the differential equation
$$\ddot{x} + x = \sin 3t, \qquad x(0) = 0, \qquad \dot{x}(0) = 0$$

7

LINEAR SYSTEMS ANALYSIS

7-1 INTRODUCTION

In practice, the input signal to a dynamic system is not known in advance but is probably random in nature, and the instantaneous input cannot be expressed analytically. Only in certain special cases is the opposite true.

From the design point of view, it is convenient to have a basis for comparing the performances of the various systems. This basis may be established by specifying particular test signals and then comparing system responses to these input signals.

Many design criteria are based on such signals or on the response of systems to changes in initial conditions (without any test signals). The use of test signals can be justified because of a correlation between the way that a particular system responds to a typical test input signal and the ability of that system to cope with actual input signals.

Typical test signals. Today the test input signals in common use have the following forms: step function, ramp function, impulse function, and sinusoidal function. Through these test signals, which are very simple functions of time, mathematical and experimental analyses of systems can be done easily.

Which input signal to use in analyzing the dynamic characteristics of a system can be determined by the form of input that the system will most frequently be subjected to under normal operation. If it is subjected to sudden disturbances, a step function of time may be a good test signal; for a system subjected to shock inputs, an impulse function may be best. Once a system is designed on the basis of test signals, its performance in response to actual inputs is generally satisfactory. In addition, the use of such test signals enables us to compare the performance of all systems on the same basis.

Natural and forced responses. Consider a system defined by a differential equation—for instance,

$$\overset{(n)}{x} + a_1 \overset{(n-1)}{x} \cdots + a_{n-1}\dot{x} + a_n x = p(t) \tag{7-1}$$

where the coefficients $a_1, a_2, \ldots, a_n$ are constants, $x(t)$ is the dependent variable, t is the independent variable, and $p(t)$ is the input function.

The differential equation (7-1) has a complete solution $x(t)$ composed of two parts—the complementary solution $x_c(t)$ and the particular solution $x_p(t)$. The complementary solution $x_c(t)$ is found by equating the right-hand side of Eq. (7-1) to zero and solving the associated homogeneous differential equation. The particular solution $x_p(t)$ depends on the functional form of the input function $p(t)$.

If the complementary solution $x_c(t)$ approaches zero as time t approaches infinity so that

$$\lim_{t \to \infty} x(t) = \lim_{t \to \infty} x_p(t)$$

and if $\lim_{t \to \infty} x_p(t)$ is a bounded function of time, the system is said to be in a *steady state*.

Customarily, engineers call the complementary solution $x_c(t)$ and particular solution $x_p(t)$ the *natural* and *forced response*, respectively. Although the natural behavior of a system is not itself a response to any external or input function, a study of this type of behavior will reveal characteristics that will be useful in predicting the forced response as well.

Transient response and steady-state response. Both natural and forced responses of a dynamic system consist of two parts—the transient and the steady-state response. *Transient response* refers to the process generated in going from the initial state to the final state. By *steady-state response* we mean the way in which the system output behaves as t approaches infinity. The transient response of a dynamic system often exhibits damped vibrations before reaching a steady state.

Outline of the chapter. This chapter deals primarily with the transient and steady-state response analyses of systems. Laplace transform methods are used as a major tool in the analyses.

In Sec. 7-2 we deal with the transient response analysis of first-order systems subjected to step and ramp inputs. Section 7-3 begins with the transient response analysis of second-order systems subjected to initial conditions only, followed by the transient response of such systems to step and impulse inputs. Section 7-4 defines the transfer function and derives transfer functions for a few physical systems. System response to sinusoidal inputs is the subject of Sec. 7-5. Here the sinusoidal transfer function is defined and its use in the steady-state sinusoidal response (frequency response) is explained. Section 7-6 treats vibration isolation problems as an application of a steady-state response analysis of dynamic systems to sinusoidal inputs. Finally, analog computers that solve differential equations and simulate physical systems are discussed in Sec. 7-7.

7-2 TRANSIENT RESPONSE ANALYSIS OF FIRST-ORDER SYSTEMS

From time to time in Chapters 2 through 5 we analyzed the transient response of several first-order systems. Essentially, this section is a systematic review of the transient response analysis of first-order systems.

Here we consider a thermal system (a thin, glass-wall, mercury thermometer system) as an example of first-order systems. After first deriving a mathematical model of a thermometer system, we find the system's response to step and ramp inputs. Then we point out that the mathematical results obtained can be applied to any physical or nonphysical systems having the same mathematical model.

Mathematical modeling of a thermal system—thermometer system. Consider the thin, glass-wall, mercury thermometer system shown in Fig. 7-1. Assume that the thermometer is at a uniform temperature $\bar{\Theta}$ K (ambient temperature) and that at $t = 0$ it is immersed in a bath of temperature $\bar{\Theta} + \theta_b$ K, where θ_b is the bath temperature (which may be constant or changing) measured from the ambient temperature $\bar{\Theta}$. Let us denote the instantaneous thermometer temperature by $\bar{\Theta} + \theta$ K so that θ is the change in the thermometer temperature satisfying the condition that $\theta(0) = 0$.

Thermal systems like the present one can be described in terms of resistance and capacitance. Unlike mechanical, electrical, hydraulic, and pneumatic systems, however, these systems do not

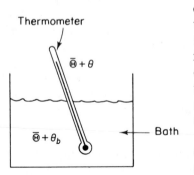

Fig. 7-1. Thin, glass-wall, mercury thermometer system.

have inertance or inertia. Let us characterize this thermometer system in terms of a thermal resistance R that resists the heat flow and a thermal capacitance C that stores heat.

The thermal resistance R for conduction or convection heat transfer is defined as the change in temperature difference (K) needed to cause a unit change in heat flow rate (J/s)—that is,

$$\text{Resistance } R = \frac{\text{change in temperature difference}}{\text{change in heat flow rate}} \quad \frac{\text{K}}{\text{J/s}}$$

The thermal capacitance C is defined as the change in quantity of heat (J) required to make a unit change in temperature (K) or

$$\text{Capacitance } C = \frac{\text{change in quantity of heat}}{\text{change in temperature}} \quad \frac{\text{J}}{\text{K}}$$

Note that thermal capacitance C is the product of the specific heat and mass of the material. The specific heat of a material is the ratio of the amount of heat required to raise the temperature of one kilogram of the material by one kelvin to that required to raise the temperature of one kilogram of water by one kelvin.

A mathematical model for this thermal system can be derived by considering heat balance as follows. The heat entering the thermometer during dt seconds is $q\,dt$, where q is the heat flow rate to the thermometer. This heat is stored in the thermal capacitance C of the thermometer, thereby raising its temperature by $d\theta$. Thus the heat balance equation is

$$C\,d\theta = q\,dt \tag{7-2}$$

Since thermal resistance R may be written

$$R = \frac{d(\Delta\theta)}{dq} = \frac{\Delta\theta}{q}$$

heat flow rate q may be given, in terms of thermal resistance R, as

$$q = \frac{(\bar{\Theta} + \theta_b) - (\bar{\Theta} + \theta)}{R} = \frac{\theta_b - \theta}{R}$$

where $\bar{\Theta} + \theta_b$ is the bath temperature and $\bar{\Theta} + \theta$ the thermometer temperature. Consequently, we can rewrite Eq. (7-2) as

$$C\frac{d\theta}{dt} = \frac{\theta_b - \theta}{R}$$

or

$$RC\frac{d\theta}{dt} + \theta = \theta_b \tag{7-3}$$

This is a mathematical model of the thermometer system. This system is analogous to the electrical system shown in Fig. 7-2, the mathematical model of which is

$$RC\frac{de_o}{dt} + e_o = e_i$$

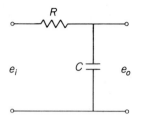

Fig. 7-2. Electrical system analogous to the system shown in Fig. 7-1.

By comparing the mathematical models of the thermal and electrical systems, the analogy is apparent. Therefore the two are analogous systems. Table 7-1 gives analogous electrical-thermal quantities.

Table 7-1. ELECTRICAL-THERMAL ANALOGY

Electrical systems		Thermal systems	
Voltage e	V	Temperature θ	K
Current i	A	Heat flow rate q	J/s
Charge q	C	Heat h	J
Resistance R	Ω	Thermal resistance R	K-s/J
Capacitance C	F	Thermal capacitance C	J/K

Step response of first-order system. For the thin, glass-wall, mercury thermometer system shown in Fig. 7-1, assume that the thermometer is at the ambient temperature $\bar{\Theta}$ K and that at $t = 0$ it is immersed in a water bath of temperature $\bar{\Theta} + \theta_b$ K. (θ_b is the difference between the bath temperature and the ambient temperature.) Let us define the instantaneous thermometer temperature as $\bar{\Theta} + \theta$ K. [Note that θ is the change in the thermometer temperature satisfying the condition that $\theta(0) = 0$.] We wish to find the response $\theta(t)$ when the bath temperature is constant, or θ_b is constant.

In the mathematical model for the present system, Eq. (7-3), RC is the time constant. Let us write $RC = T$. Then the mathematical model can be written

$$T\frac{d\theta}{dt} + \theta = \theta_b \tag{7-4}$$

The Laplace transform of Eq. (7-4) becomes

$$T[s\Theta(s) - \theta(0)] + \Theta(s) = \Theta_b(s)$$

Noting that $\theta(0) = 0$, this last equation simplifies to

$$\Theta(s) = \frac{1}{Ts + 1}\Theta_b(s) \tag{7-5}$$

Note that for $\theta_b =$ constant, we have

$$\Theta_b(s) = \frac{\theta_b}{s}$$

Hence Eq. (7-5) becomes

$$\Theta(s) = \frac{1}{Ts+1}\frac{\theta_b}{s} = \left[\frac{1}{s} - \frac{1}{s+(1/T)}\right]\theta_b$$

The inverse Laplace transform of this last equation gives

$$\theta(t) = (1 - e^{-t/T})\theta_b \qquad (7\text{-}6)$$

The response curve $\theta(t)$ is shown in Fig. 7-3. Equation (7-6) states that initially the response $\theta(t)$ is zero and that it finally becomes θ_b. (There is no steady-state error.) One of the important characteristics of such an exponential response curve is that at $t = T$ the value of $\theta(t)$ is $0.632\theta_b$, or the response $\theta(t)$ has reached 63.2% of its total change. This fact can readily be seen by substituting $e^{-1} = 0.368$ in Eq. (7-6).

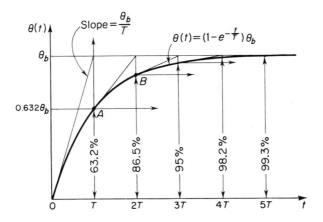

Fig. 7-3. Step response curve for the first-order system.

Another important property of the exponential response curve is that the slope of the tangent line at $t = 0$ is θ_b/T, since

$$\frac{d\theta}{dt}\Big|_{t=0} = \frac{\theta_b}{T}e^{-t/T}\Big|_{t=0} = \frac{\theta_b}{T}$$

The response would reach the final value at $t = T$ if it maintained its initial speed of response. The slope of the response curve $\theta(t)$ decreases monotonically from θ_b/T at $t = 0$ to zero at $t = \infty$.

Referring to Fig. 7-3, in one time constant the exponential response curve has gone from zero to 63.2% of the total change. In two time constants the response reaches 86.5% of the total change. At $t = 3T$, $4T$, and $5T$ the response $\theta(t)$ reaches 95, 98.2, and 99.3% of the total change, respectively.

So, for $t \geq 4T$, the response remains within 2% of the final value. As can be seen from Eq. (7-6), the steady state is reached mathematically only after an infinite time. In practice, however, a reasonable estimate of the response time is the length of time that the response curve needs to reach the 2% line of the final value, or four time constants.

Ramp response of first-order system. Consider again the thermometer system shown in Fig. 7-1. Assume that, for $t < 0$, both bath temperature and thermometer temperature are at steady state at the ambient temperature $\bar{\Theta}$ K and that, for $t \geq 0$, heat is added to the bath and the bath temperature changes linearly at the rate of r K/s—that is,

$$\theta_b(t) = rt$$

Let us derive the ramp response $\theta(t)$.

First, note that

$$\Theta_b(s) = \mathcal{L}[\theta_b(t)] = \mathcal{L}[rt] = \frac{r}{s^2}$$

By substituting this equation into Eq. (7-5), we find

$$\Theta(s) = \frac{1}{Ts + 1} \frac{r}{s^2} = r \left[\frac{1}{s^2} - \frac{T}{s} + \frac{T}{s + (1/T)} \right]$$

The inverse Laplace transform of this last equation gives

$$\theta(t) = r(t - T + Te^{-t/T}) \qquad \text{for } t \geq 0 \qquad (7\text{-}7)$$

The error $e(t)$ between the actual bath temperature and the indicated thermometer temperature is

$$e(t) = rt - \theta(t) = rT(1 - e^{-t/T})$$

As t approaches infinity, $e^{-t/T}$ approaches zero. Thus the error $e(t)$ approaches rT or

$$e(\infty) = rT$$

Ramp input rt and response $\theta(t)$ are shown in Fig. 7-4. The error in following the ramp input is equal to rT for sufficiently large t. The smaller the time constant T, the smaller the steady-state error in following the ramp input.

Comments. Since the mathematical analysis does not depend on the physical structure of the system, the preceding results for step and ramp responses can be applied to any systems having this mathematical model:

$$T \frac{dx_o}{dt} + x_o = x_i \qquad (7\text{-}8)$$

where

$T =$ time constant of the system
$x_i =$ input or forcing function
$x_o =$ output or response function

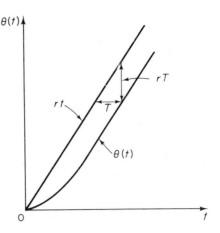

Fig. 7-4. Ramp response curve for the first-order system.

Referring to Eq. (7-6), for a step input $x_i(t) = r \cdot 1(t)$, any system described by Eq. (7-8) will exhibit the following response.

$$x_o(t) = (1 - e^{-t/T})r$$

Similarly, for a ramp input $x_i(t) = rt \cdot 1(t)$, any system described by Eq. (7-8) will exhibit the following response. [Refer to Eq. (7-7).]

$$x_o(t) = r(t - T + Te^{-t/T})$$

Many physical systems have the mathematical model of Eq. (7-8), and Table 7-2 shows several. (The systems are analogous.) All analogous systems will exhibit the same response to the same input function.

7-3   TRANSIENT RESPONSE ANALYSIS OF SECOND-ORDER SYSTEMS

Turning next to the transient response analysis of second-order systems, let us look at some typical second-order systems—spring-mass-damper systems and *LRC* circuits. Here we employ such systems as examples. The results obtained can be applied to the response of any analogous systems.

In the following pages we first discuss free vibrations of mechanical systems and then treat the step response of an electrical and a mechanical system, after which the impulse response of mechanical systems is described.

Free vibration without damping. Consider the spring-mass system shown in Fig. 7-5. We shall obtain the response of the system when the mass is displaced downward by a distance $x(0) = a$ and released with an initial velocity $\dot{x}(0) = b$.

Table 7-2. EXAMPLES OF PHYSICAL SYSTEMS HAVING THE MATHEMATICAL
MODEL OF THE FORM $T(dx_o/dt) + x_o = x_i$

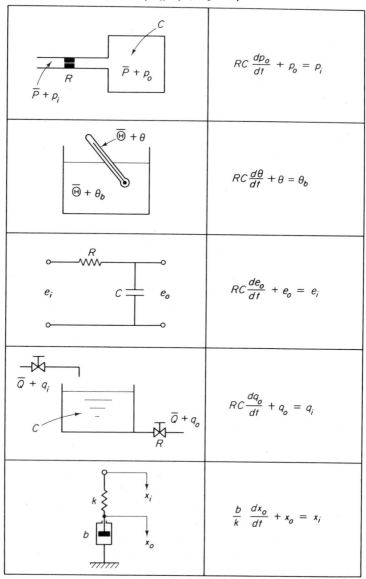

	$RC\dfrac{dp_o}{dt} + p_o = p_i$
	$RC\dfrac{d\theta}{dt} + \theta = \theta_b$
	$RC\dfrac{de_o}{dt} + e_o = e_i$
	$RC\dfrac{dq_o}{dt} + q_o = q_i$
	$\dfrac{b}{k}\dfrac{dx_o}{dt} + x_o = x_i$

The mathematical model of the system is

$$m\ddot{x} + kx = 0, \qquad x(0) = a, \qquad \dot{x}(0) = b$$

where displacement x is measured from the equilibrium position.

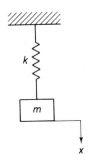

Fig. 7-5. Spring-mass system.

The solution of this last equation gives the response $x(t)$. In order to solve this differential equation, let us take the Laplace transform of both sides.

$$m[s^2 X(s) - sx(0) - \dot{x}(0)] + kX(s) = 0$$

which can be rewritten as

$$(ms^2 + k)X(s) = msa + mb$$

Solving for $X(s)$,

$$X(s) = \frac{sa + b}{s^2 + (k/m)}$$

$$= \frac{as}{s^2 + (\sqrt{k/m})^2} + \frac{b}{\sqrt{k/m}} \frac{\sqrt{k/m}}{s^2 + (\sqrt{k/m})^2}$$

The inverse Laplace transform of this last equation gives

$$x(t) = a \cos \sqrt{\frac{k}{m}} t + b\sqrt{\frac{m}{k}} \sin \sqrt{\frac{k}{m}} t$$

The response $x(t)$ consists of a sine and a cosine function and depends on the values of the initial conditions a and b. If, for example, the mass is released with zero velocity so that $b = 0$, then the motion $x(t)$ is a simple cosine function:

$$x(t) = a \cos \sqrt{\frac{k}{m}} t$$

Free vibration with viscous damping. Damping is always present in actual mechanical systems, although in some cases it may be negligibly small.

The mechanical system in Fig. 7-6 consists of a mass, spring, and damper. If the mass is pulled downward and released, it will vibrate freely. The amplitude of the resulting motion will decrease with each cycle at a rate that depends on the amount of viscous damping. (Since the damping force opposes motion, there is a continual loss of energy in the system.)

The mathematical model of this system is

$$m\ddot{x} + b\dot{x} + kx = 0 \tag{7-9}$$

where displacement x is measured from the equilibrium position.

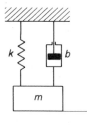

x **Fig. 7-6.** Spring-mass-damper system.

The character of the natural response of a second-order system like this one is determined by the roots of the characteristic equation

$$ms^2 + bs + k = 0$$

The two roots of this equation are

$$s = \frac{-b \pm \sqrt{b^2 - 4mk}}{2m}$$

If the damping coefficient b is small so that $b^2 < 4mk$, the roots of the characteristic equation are complex conjugates. The natural response is an exponentially decaying sinusoid, and the system is said to be *underdamped*.

If the damping coefficient b is increased, a point will be reached at which $b^2 = 4mk$. When the damping has reached this value ($b = 2\sqrt{mk}$), the two roots of the characteristic equation are equal. The system is said to be *critically damped*.

If the damping coefficient b is increased further so that $b^2 > 4mk$, the two roots become real. The response is the sum of two decaying exponentials and the system is said to be *overdamped*.

In solving Eq. (7-9) for the response $x(t)$, it is convenient to define

$$\omega_n = \sqrt{\frac{k}{m}} = \text{undamped natural frequency, rad/s}$$

$$\zeta = \text{damping ratio} = \frac{\text{actual damping value}}{\text{critical damping value}} = \frac{b}{2\sqrt{km}}$$

and rewrite Eq. (7-9) as follows:

$$\ddot{x} + 2\zeta\omega_n\dot{x} + \omega_n^2 x = 0 \qquad (7\text{-}10)$$

In the following discussion we shall use Eq. (7-10) as the system equation and derive the response $x(t)$ for three cases—underdamped case ($0 < \zeta < 1$), overdamped case ($\zeta > 1$), and critically damped case ($\zeta = 1$).

Case 1.   Underdamped case $(0 < \zeta < 1)$

The Laplace transform of Eq. (7-10) gives

$$[s^2 X(s) - sx(0) - \dot{x}(0)] + 2\zeta\omega_n[sX(s) - x(0)] + \omega_n^2 X(s) = 0$$

Solving for $X(s)$, we have

$$X(s) = \frac{(s + 2\zeta\omega_n)x(0) + \dot{x}(0)}{s^2 + 2\zeta\omega_n s + \omega_n^2} \tag{7-11}$$

or

$$X(s) = \frac{(s + \zeta\omega_n)x(0)}{(s + \zeta\omega_n)^2 + (\omega_n\sqrt{1 - \zeta^2})^2}$$
$$+ \frac{\zeta\omega_n x(0) + \dot{x}(0)}{\omega_n\sqrt{1 - \zeta^2}} \frac{\omega_n\sqrt{1 - \zeta^2}}{(s + \zeta\omega_n)^2 + (\omega_n\sqrt{1 - \zeta^2})^2}$$

The inverse Laplace transform of this last equation gives

$$x(t) = x(0)e^{-\zeta\omega_n t} \cos \omega_n\sqrt{1 - \zeta^2}\, t$$
$$+ \frac{\zeta\omega_n x(0) + \dot{x}(0)}{\omega_n\sqrt{1 - \zeta^2}} e^{-\zeta\omega_n t} \sin \omega_n\sqrt{1 - \zeta^2}\, t$$

Next, define

$$\omega_d = \omega_n\sqrt{1 - \zeta^2} = \text{damped natural frequency, rad/s}$$

Then the response $x(t)$ can be given by

$$x(t) = e^{-\zeta\omega_n t}\left\{ x(0) \cos \omega_d t + \left[\frac{\zeta}{\sqrt{1 - \zeta^2}}x(0) + \frac{1}{\omega_d}\dot{x}(0)\right] \sin \omega_d t \right\} \tag{7-12}$$

If the initial velocity is zero or $\dot{x}(0) = 0$, Eq. (7-12) simplifies to

$$x(t) = x(0)e^{-\zeta\omega_n t}\left(\cos \omega_d t + \frac{\zeta}{\sqrt{1 - \zeta^2}} \sin \omega_d t \right) \tag{7-13}$$

or

$$x(t) = \frac{x(0)}{\sqrt{1 - \zeta^2}} e^{-\zeta\omega_n t} \sin\left(\omega_d t + \tan^{-1} \frac{\sqrt{1 - \zeta^2}}{\zeta} \right)$$
$$= \frac{x(0)}{\sqrt{1 - \zeta^2}} e^{-\zeta\omega_n t} \cos\left(\omega_d t - \tan^{-1} \frac{\zeta}{\sqrt{1 - \zeta^2}} \right) \tag{7-14}$$

Notice that, in the present case, the damping introduces the term $e^{-\zeta\omega_n t}$ as a multiplying factor. This factor is a decreasing exponential and becomes smaller and smaller as time increases, thus causing the amplitude of the harmonic motion to decrease with time.

Case 2.   Overdamped case ($\zeta > 1$)

Here the two roots of the characteristic equation are real, and so Eq. (7-11) can be written

$$X(s) = \frac{(s + 2\zeta\omega_n)x(0) + \dot{x}(0)}{(s + \zeta\omega_n + \omega_n\sqrt{\zeta^2 - 1})(s + \zeta\omega_n - \omega_n\sqrt{\zeta^2 - 1})}$$
$$= \frac{a}{s + \zeta\omega_n + \omega_n\sqrt{\zeta^2 - 1}} + \frac{b}{s + \zeta\omega_n - \omega_n\sqrt{\zeta^2 - 1}}$$

where a and b are obtained as

$$a = \frac{(-\zeta + \sqrt{\zeta^2 - 1})x(0)}{2\sqrt{\zeta^2 - 1}} - \frac{\dot{x}(0)}{2\omega_n\sqrt{\zeta^2 - 1}}$$

$$b = \frac{(\zeta + \sqrt{\zeta^2 - 1})x(0)}{2\sqrt{\zeta^2 - 1}} + \frac{\dot{x}(0)}{2\omega_n\sqrt{\zeta^2 - 1}}$$

The inverse Laplace transform of $X(s)$ gives the response $x(t)$.

$$x(t) = ae^{-(\zeta\omega_n + \omega_n\sqrt{\zeta^2-1})t} + be^{-(\zeta\omega_n - \omega_n\sqrt{\zeta^2-1})t}$$

$$= \left[\frac{(-\zeta + \sqrt{\zeta^2 - 1})x(0)}{2\sqrt{\zeta^2 - 1}} - \frac{\dot{x}(0)}{2\omega_n\sqrt{\zeta^2 - 1}}\right]e^{-(\zeta\omega_n + \omega_n\sqrt{\zeta^2-1})t}$$

$$+ \left[\frac{(\zeta + \sqrt{\zeta^2 - 1})x(0)}{2\sqrt{\zeta^2 - 1}} + \frac{\dot{x}(0)}{2\omega_n\sqrt{\zeta^2 - 1}}\right]e^{-(\zeta\omega_n - \omega_n\sqrt{\zeta^2-1})t}$$

Notice that both terms on the right-hand side of this last equation decrease exponentially. The motion of the mass in this case is a gradual creeping back to the equilibrium position.

Case 3. Critically damped case ($\zeta = 1$)

In reality, all systems have a damping ratio greater or less than unity, and $\zeta = 1$ rarely occurs in practice. Nevertheless, there is a mathematical usefulness for the case $\zeta = 1$ as a reference. (The response does not exhibit any vibration, but it is the fastest among such nonvibratory motions.)

In the critically damped case, the damping ratio ζ is equal to unity. So the two roots of the characteristic equation are the same and are equal to the natural frequency ω_n. Equation (7-11) can, therefore, be written

$$X(s) = \frac{(s + 2\omega_n)x(0) + \dot{x}(0)}{s^2 + 2\omega_n s + \omega_n^2}$$

$$= \frac{(s + \omega_n)x(0) + \omega_n x(0) + \dot{x}(0)}{(s + \omega_n)^2}$$

$$= \frac{x(0)}{s + \omega_n} + \frac{\omega_n x(0) + \dot{x}(0)}{(s + \omega_n)^2}$$

The inverse Laplace transform of this last equation gives

$$x(t) = x(0)e^{-\omega_n t} + [\omega_n x(0) + \dot{x}(0)]te^{-\omega_n t}$$

The response $x(t)$ is similar to that found for the overdamped case. The mass, when displaced and released, will return to the equilibrium position without vibration.

Figure 7-7 shows the response $x(t)$ for the three cases (underdamped, critically damped, and overdamped) with initial conditions $x(0) \neq 0$ and $\dot{x}(0) = 0$.

Experimental determination of damping ratio. It is sometimes necessary to determine the damping ratios and damped natural frequencies of recorders

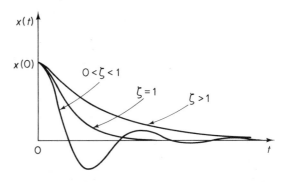

Fig. 7-7. Typical response curves of the spring-mass-damper system.

and other instruments. In order to ascertain the damping ratio and damped natural frequency of a system experimentally, a record of decaying or damped oscillations like that shown in Fig. 7-8 is needed. (Such an oscillation may be recorded by giving the system any convenient initial conditions.)

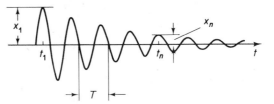

Fig. 7-8. Decaying oscillation.

The period of oscillation T can be measured directly from crossing points on the zero axis as shown in the figure.

To determine the damping ratio ζ from the rate of decay of the oscillation, we measure amplitudes—that is, at time $t = t_1$ we measure the amplitude x_1 and at time $t = t_1 + (n-1)T$ we measure the amplitude x_n. Note that it is necessary to choose n large enough so that x_n/x_1 is not near unity. Since the decay in amplitude from one cycle to the next may be represented as the ratio of the exponential multiplying factors at times t_1 and $t_1 + T$, referring to Eq. (7-12), we obtain

$$\frac{x_1}{x_2} = \frac{e^{-\zeta\omega_n t_1}}{e^{-\zeta\omega_n(t_1+T)}} = \frac{1}{e^{-\zeta\omega_n T}} = e^{\zeta\omega_n T}$$

Similarly,

$$\frac{x_1}{x_n} = \frac{1}{e^{-\zeta\omega_n(n-1)T}} = e^{(n-1)\zeta\omega_n T}$$

The logarithm of the ratio of succeeding amplitudes is called the *logarithmic decrement*. Thus

$$\text{Logarithmic decrement} = \ln\frac{x_1}{x_2} = \frac{1}{n-1}\left(\ln\frac{x_1}{x_n}\right) = \zeta\omega_n T$$

$$= \zeta\omega_n\frac{2\pi}{\omega_d} = \frac{2\pi\zeta}{\sqrt{1-\zeta^2}}$$

Once the amplitudes x_1 and x_n are measured and the logarithmic decrement is calculated, the damping ratio ζ is found from

$$\frac{1}{n-1}\left(\ln \frac{x_1}{x_n}\right) = \frac{2\pi\zeta}{\sqrt{1-\zeta^2}}$$

or

$$\zeta = \frac{\frac{1}{n-1}\left(\ln \frac{x_1}{x_n}\right)}{\sqrt{4\pi^2 + \left[\frac{1}{n-1}\left(\ln \frac{x_1}{x_n}\right)\right]^2}}$$

Example 7-1. In the system shown in Fig. 7-6 the numerical values of m, b, and k are given as $m = 1$ kg, $b = 2$ N-s/m, and $k = 100$ N/m. The mass is displaced 0.05 m and released without initial velocity. Find the frequency observed in the vibration. In addition, find the amplitude four cycles later.

The equation of motion for the system is

$$m\ddot{x} + b\dot{x} + kx = 0$$

Substituting the numerical values for m, b, and k into this equation gives

$$\ddot{x} + 2\dot{x} + 100x = 0$$

where the initial conditions are $x(0) = 0.05$ and $\dot{x}(0) = 0$. From this last equation the undamped natural frequency ω_n and the damping ratio ζ are found to be

$$\omega_n = 10, \qquad \zeta = 0.1$$

The frequency actually observed in the vibration is the damped natural frequency ω_d.

$$\omega_d = \omega_n\sqrt{1-\zeta^2} = 10\sqrt{1-0.01} = 9.95 \text{ rad/s}$$

In the present analysis, $\dot{x}(0)$ is given as zero. So referring to Eq. (7-13), solution $x(t)$ can be written

$$x(t) = x(0)e^{-\zeta\omega_n t}\left(\cos \omega_d t + \frac{\zeta}{\sqrt{1-\zeta^2}} \sin \omega_d t\right)$$

It follows that at $t = nT$, where $T = 2\pi/\omega_d$,

$$x(nT) = x(0)e^{-\zeta\omega_n nT}$$

Consequently, the amplitude four cycles later becomes

$$x(4T) = x(0)e^{-\zeta\omega_n 4T} = x(0)e^{-(0.1)(10)(4)(0.63)}$$

$$= 0.05e^{-2.52} = 0.05 \times 0.0804 = 0.00402 \text{ m}$$

Estimate of response time. The mass of the mechanical system shown in Fig. 7-6 is displaced $x(0)$ and released without initial velocity. The response is given by Eq. (7-14), rewritten thus:

$$x(t) = \frac{x(0)}{\sqrt{1-\zeta^2}}e^{-\zeta\omega_n t} \cos\left(\omega_d t - \tan^{-1}\frac{\zeta}{\sqrt{1-\zeta^2}}\right)$$

A typical response curve is shown in Fig. 7-9. Note that such a response curve is tangent to envelope exponentials $\pm[x(0)/\sqrt{1 - \zeta^2}]e^{-\zeta\omega_n t}$. The time constant T of these exponential curves is $1/(\zeta\omega_n)$.

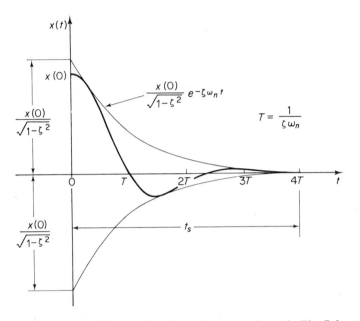

Fig. 7-9. Typical response curve of the system shown in Fig. 7-6 and its envelope exponentials.

The fact that the response curve $x(t)$ is tangent to the exponential curves enables us to estimate the response time of the second-order system such as shown in Fig. 7-6 in terms of the settling time t_s defined by

$$t_s = 4T = \frac{4}{\zeta\omega_n}$$

The settling time t_s can be considered as an approximate response time of the system, since for $t > t_s$ the response curve remains within 2% of the final value or 2% of the total change.

Comments. The preceding analysis, as well as the results derived, can be applied to any analogous systems having mathematical models in the form given by Eq. (7-10).

Step response of second-order systems. Let us consider next the step response of an electrical system and a mechanical system.

The electrical system shown in Fig. 7-10 is a typical second-order one. Assume that capacitor C has an initial charge q_0 and that switch S is closed at

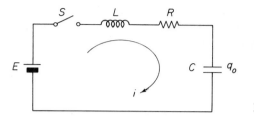

Fig. 7-10. Electrical system.

$t = 0$. Closing switch S and applying a constant voltage E to the circuit correspond to applying a step input to the system.

A mathematical model for the circuit is

$$L \frac{di}{dt} + Ri + \frac{1}{C} \int i \, dt = E$$

or in terms of the charge q, where $i = dq/dt$,

$$L\ddot{q} + R\dot{q} + \frac{1}{C}q = E$$

where the initial conditions are $q(0) = q_0$ and $\dot{q}(0) = i(0) = 0$. This last equation can be rewritten thus:

$$\ddot{q} + \frac{R}{L}\dot{q} + \frac{1}{LC}q = \frac{E}{L} \tag{7-15}$$

By defining

$$\omega_n = \sqrt{\frac{1}{LC}} = \text{undamped natural frequency, rad/s}$$

$$\zeta = \frac{R\sqrt{C}}{2\sqrt{L}} = \text{damping ratio}$$

Eq. (7-15) can be written

$$\ddot{q} + 2\zeta\omega_n\dot{q} + \omega_n^2 q = \frac{E}{L} \tag{7-16}$$

Next, define

$$q - \frac{1}{\omega_n^2}\frac{E}{L} = q - CE = x \tag{7-17}$$

Equation (7-16) can then be written in terms of the new variable x

$$\ddot{x} + 2\zeta\omega_n\dot{x} + \omega_n^2 x = 0 \tag{7-18}$$

with initial conditions $x(0) = q_0 - CE$ and $\dot{x}(0) = 0$. Since Eq. (7-18) is exactly the same as Eq. (7-10), the results obtained in the analysis of free vibrations of the spring-mass-damper system apply to this case. For instance, for the underdamped case ($0 < \zeta < 1$), the solution of Eq. (7-18) is given by Eq. (7-12), rewritten thus:

$$x(t) = e^{-\zeta\omega_n t}\left\{ x(0)\cos\omega_d t + \left[\frac{\zeta}{\sqrt{1 - \zeta^2}}x(0) + \frac{1}{\omega_d}\dot{x}(0) \right]\sin\omega_d t \right\}$$

By substituting $\omega_n = \sqrt{1/(LC)}$, $\zeta = R\sqrt{C}/(2\sqrt{L})$, $\omega_d = \omega_n\sqrt{1 - \zeta^2}$, $x(0) = q_0 - CE$, and $\dot{x}(0) = 0$ into this last equation and noting that $q(t) = x(t) + CE$ [Eq. (7-17)], we have the following solution.

$$q(t) = x(t) + CE$$

$$= CE\left\{1 + e^{-(R/2L)t}\left[\left(\frac{q_0}{CE} - 1\right)\cos\sqrt{\frac{1}{LC} - \frac{R^2}{4L^2}}\,t\right.\right.$$

$$\left.\left. + \frac{(q_0/CE) - 1}{\sqrt{[4L/(R^2C)] - 1}}\sin\sqrt{\frac{1}{LC} - \frac{R^2}{4L^2}}\,t\right]\right\}$$

For the critically damped and overdamped cases, the response $q(t)$ can be derived similarly.

It is important to note that the step response is basically the same as the response to the initial condition only. The difference between these two responses lies in the constant term in the solution.

It should also be noted that, as a matter of course, the same solution can be found by taking the Laplace transform of both sides of Eq. (7-16) and solving for $Q(s)$, where $Q(s) = \mathcal{L}[q(t)]$, and taking its inverse Laplace transform rather than by changing the variable from $q(t)$ to $x(t)$. When numerical values of ζ and ω_n are given and the initial conditions are zeros, this latter approach may be simpler than the one just presented. See Example 7-2.

Example 7-2. Let us consider the step response of a mechanical system (Fig. 7-11) in which a rigid, massless bar AA' is suspended from the ceiling through a spring and damper. Assume that at $t = 0$ a man weighing 193 lb$_f$ jumps up and grabs bar AA'. Neglecting the mass of the spring-damper device, what is the subsequent motion $x(t)$ of the bar AA'? Assume that viscous friction coefficient b is 2 lb$_f$-s/in. and that spring constant k is 20 lb$_f$/in.

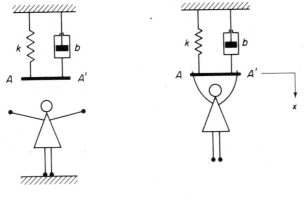

$t < 0$ $t > 0$

Fig. 7-11. Mechanical system.

The input to the system is a constant force mg, where m is the mass of the man. It acts as a step input to the system. The initial conditions are $x(0) = 0$ and $\dot{x}(0) = 0$. At $t = 0+$ the man already has a hold on the bar and starts to move up and down. The mathematical model or equation of motion is

$$m\ddot{x} + b\dot{x} + kx = mg$$

where

$m = 193\ \text{lb} = 6\ \text{slugs}$

$b = 2\ \text{lb}_f\text{-s/in.} = 24\ \text{lb}_f\text{-s/ft}$

$k = 20\ \text{lb}_f/\text{in.} = 240\ \text{lb}_f/\text{ft}$

By substituting the numerical values in the equation of motion, we find

$$6\ddot{x} + 24\dot{x} + 240x = 193$$

or

$$\ddot{x} + 4\dot{x} + 40x = 32.2$$

Then if we take the Laplace transform of this last equation and substitute the initial conditions $x(0) = 0$ and $\dot{x}(0) = 0$, the result is

$$s^2 X(s) + 4sX(s) + 40X(s) = \frac{32.2}{s}$$

Solving for $X(s)$,

$$X(s) = \frac{32.2}{(s^2 + 4s + 40)s}$$

$$= \frac{32.2}{40}\left(\frac{1}{s} - \frac{s+4}{s^2 + 4s + 40}\right)$$

$$= 0.805\left[\frac{1}{s} - \frac{s+2}{(s+2)^2 + 6^2} - \frac{1}{3}\frac{6}{(s+2)^2 + 6^2}\right]$$

The inverse Laplace transform of this last equation gives

$$x(t) = 0.805(1 - e^{-2t}\cos 6t - \tfrac{1}{3}e^{-2t}\sin 6t)\ \text{ft}$$

This solution depicts the up-and-down motion of the bar AA' and the man.

If we change the numerical values of m, b, k given in BES units to the SI system of units, we have

$m = 6\ \text{slugs} = 6 \times 14.594 = 87.56\ \text{kg}$

$b = 2\ \text{lb}_f\text{-s/in.} = 2 \times 4.4482 \times 12/0.3048 = 350.3\ \text{N-s/m}$

$k = 20\ \text{lb}_f/\text{in.} = 20 \times 4.4482 \times 12/0.3048 = 3503\ \text{N/m}$

Then the equation of motion for the system becomes

$$87.56\ddot{x} + 350.3\dot{x} + 3503x = 87.56 \times 9.81$$

which can be simplified to

$$\ddot{x} + 4\dot{x} + 40x = 9.81$$

The solution of this differential equation is

$$x(t) = 0.245(1 - e^{-2t}\cos 6t - \tfrac{1}{3}e^{-2t}\sin 6t)\ \text{m}$$

Impulse response. Let us turn next to the impulse response of mechanical systems. Such a response can be observed when a mechanical system is subjected to a very large force for a very short time—for instance, when the mass of a spring-mass-damper system is hit by a hammer or a bullet. Mathematically, such an impulse input can be expressed by an impulse function.

The unit impulse function defined in Section 6-4 is a mathematical function and, in actuality, does not exist. However, as shown in Fig. 7-12(a), if the actual input lasts for a short time duration (Δt second) but is of a large amplitude (h), so that the area ($h \Delta t$) in a time plot is not negligible, it can be approximated by an impulse function. The impulse input is usually denoted by a vertical arrow, as shown in Fig. 7-12(b), to indicate that it has a very short duration and a very large height.

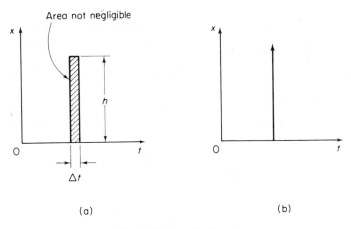

Fig. 7-12. Impulse inputs.

It should be noted that, in handling impulse functions, only the magnitude (or area) of the impulse function is important; its actual shape is immaterial. In other words, an input of amplitude $2h$ and duration $\Delta t/2$ can be considered the same impulse as that of amplitude h and duration Δt as long as Δt approaches zero and $h \Delta t$ is finite.

Before discussing the impulse response of mechanical systems, a review of the law of conservation of momentum, normally presented in college physics courses, is worthwhile.

Law of conservation of momentum. The momentum of a mass m moving at a velocity v is mv. According to Newton's second law,

$$F = ma = m\frac{dv}{dt} = \frac{d}{dt}(mv)$$

Hence

$$F\,dt = d(mv) \qquad (7\text{-}19)$$

Integrating both sides of Eq. (7-19), we have

$$\int_{t_1}^{t_2} F\,dt = \int_{v_1}^{v_2} d(mv) = mv_2 - mv_1$$

where $v_1 = v(t_1)$ and $v_2 = v(t_2)$. This last equation states that a change in momentum equals the integral of force between $t = t_1$ and $t = t_2$.

Momentum is a vector quantity. It has magnitude, direction, and sense. The direction of the change in momentum is in the direction of force.

In the absense of any external force Eq. (7-19) becomes

$$d(mv) = 0$$

or

$$mv = \text{constant}$$

Thus the total momentum of a system remains unchanged by any action that may take place within the system, provided that no external force is acting on the system. This is called the *law of conservation of momentum*.

Angular momentum of a rotating system is $J\omega$, where J is the moment of inertia of a body and ω is its angular velocity. In the absence of external torque the angular momentum $J\omega$ of a body remains unchanged. This is the *law of conservation of angular momentum*. So in the absence of external torque, if the moment of inertia of a body changes because of a change in the configuration of the body, as shown in Fig. 7-13, the angular velocity changes so as to keep $J\omega = \text{constant}$.

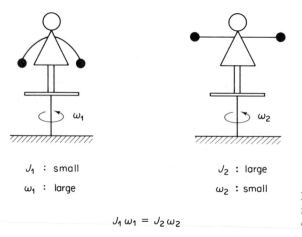

J_1 : small J_2 : large

ω_1 : large ω_2 : small

$$J_1\,\omega_1 = J_2\,\omega_2$$

Fig. 7-13. Figures illustrating the law of conservation of angular momentum.

Example 7-3. A bullet is fired horizontally into a wood block resting on a horizontal, frictionless surface. If mass m_1 of the bullet is 0.02 kg and velocity is 600 m/s, what is the velocity of the wood block after the bullet is imbedded in it? Assume that the wood block has a mass m_2 of 50 kg.

If we consider the bullet and wood block as constituting a system, no external force is acting on the system. Consequently, its total momentum remains unchanged.

$$\text{Momentum before impact} = m_1v_1 + m_2v_2$$

where v_1, the velocity of the bullet, is equal to 600 m/s and v_2, the velocity of the wood block before the impact, is equal to zero.

$$\text{Momentum after impact} = (m_1 + m_2)v$$

where v is the velocity of the wood block after the bullet is imbedded. (Velocities v_1 and v are in the same direction.)

The law of conservation of momentum states that

$$m_1v_1 + m_2v_2 = (m_1 + m_2)v$$

By substituting the given numerical values into this last equation, we have

$$0.02 \times 600 + 50 \times 0 = (0.02 + 50)v$$

or

$$v = 0.24 \text{ m/s}$$

Thus the wood block after the bullet is imbedded will move at the velocity of 0.24 m/s in the same direction as the original bullet velocity v_1.

Impulse response of mechanical system. Suppose that, for the mechanical system shown in Fig. 7-14, a bullet of mass m is shot into mass M (where $M \gg m$). It is assumed that when the bullet hits mass M, it will be imbedded there. Determine the response (displacement x) of mass M after being hit by the bullet.

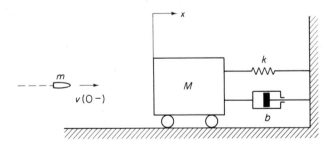

Fig. 7-14. Mechanical system subjected to an impulse input.

The input to the system in this case can be considered an impulse, the magnitude of which is equal to the rate of change of bullet momentum.

Assume that the bullet is shot at $t = 0-$ and that the initial velocity of the bullet is $v(0-)$. At the instant that the bullet hits mass M, the bullet velocity becomes the same as that of mass M, since the bullet is assumed to be imbedded in mass M.

Because we assumed $M \gg m$, the velocity $v(t)$ after the bullet hits mass M will be small compared to $v(0-)$. As a result, there is a sudden change in the velocity of the bullet, as Fig. 7-15(a) shows. Since the change in the velocity of the bullet occurs instantaneously, $\dot{v}$ has the form of an impulse as shown in Fig. 7-15(b). Note that $\dot{v}$ is negative. For $t > 0$, mass M and bullet m move as a combined mass $M + m$.

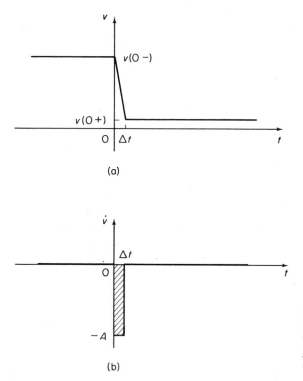

(a)

(b)

Fig. 7-15. (a) Change in velocity of the bullet when it hits the mass; (b) change in acceleration of the bullet when it hits the mass.

The equation of motion for the system is

$$M\ddot{x} + b\dot{x} + kx = F(t) \qquad (7\text{-}20)$$

where $F(t)$, an impulse force, is equal to $-m\dot{v}$. [Note that $-m\dot{v}$ is positive. The impulse force $F(t)$ is in the positive direction of x.] Referring to Fig. 7-15(b), the impulse force $F(t)$ can be written

$$F(t) = A \, \Delta t \, \delta(t)$$

where $A \, \Delta t$ is the magnitude of the impulse input. Thus

$$F(t) = A \, \Delta t \, \delta(t) = -m\dot{v}$$

from which

$$\int_{0-}^{0+} A \, \Delta t \, \delta(t) \, dt = -m \int_{0-}^{0+} \dot{v} \, dt$$

or

$$A \, \Delta t = mv(0-) - mv(0+) \qquad (7\text{-}21)$$

The momentum of the bullet is changed from $mv(0-)$ to $mv(0+)$. Since

$$v(0+) = \dot{x}(0+) = \text{initial velocity of mass } M$$

Eq. (7-21) can be written

$$A \, \Delta t = mv(0-) - m\dot{x}(0+)$$

Then Eq. (7-20) becomes

$$M\ddot{x} + b\dot{x} + kx = F(t) = [mv(0-) - m\dot{x}(0+)] \, \delta(t)$$

By taking the $\mathcal{L}_-$ transform of both sides of this last equation, we see that

$$M[s^2 X(s) - sx(0-) - \dot{x}(0-)] + b[sX(s) - x(0-)] + kX(s)$$
$$= mv(0-) - m\dot{x}(0+)$$

Also, noting that $x(0-) = 0$ and $\dot{x}(0-) = 0$, we have

$$(Ms^2 + bs + k)X(s) = mv(0-) - m\dot{x}(0+)$$

or

$$X(s) = \frac{mv(0-) - m\dot{x}(0+)}{Ms^2 + bs + k} \qquad (7\text{-}22)$$

In order to determine the value of $\dot{x}(0+)$, we can apply the initial value theorem.

$$\dot{x}(0+) = \lim_{t \to 0} \dot{x}(t) = \lim_{s \to \infty} s[sX(s)]$$

$$= \lim_{s \to \infty} \frac{s^2[mv(0-) - m\dot{x}(0+)]}{Ms^2 + bs + k}$$

$$= \frac{mv(0-) - m\dot{x}(0+)}{M}$$

from which

$$M\dot{x}(0+) = mv(0-) - m\dot{x}(0+)$$

or

$$\dot{x}(0+) = \frac{m}{M+m} v(0-)$$

It follows that

$$mv(0-) - m\dot{x}(0+) = \frac{Mm}{M+m} v(0-)$$

And so Eq. (7-22) becomes

$$X(s) = \frac{mv(0-)}{M+m} \frac{M}{Ms^2 + bs + k} \qquad (7\text{-}23)$$

The inverse Laplace transform of Eq. (7-23) gives the impulse response $x(t)$ or

$$x(t) = \mathcal{L}^{-1}[X(s)] = \frac{mv(0-)}{M+m} \mathcal{L}^{-1}\left[\frac{1}{s^2 + (b/M)s + (k/M)}\right]$$

The response $x(t)$ will exhibit damped vibrations if the system is under-damped. Otherwise it will reach a maximum displacement and then gradually creep back to the equilibrium position ($x = 0$) without vibration.

As an illustration, let us assume the following numerical values for M, m, b, k, and $v(0-)$ and determine the response $x(t)$.

$$M = 50 \text{ kg}$$

$$m = 0.01 \text{ kg}$$

$$b = 100 \text{ N-s/m}$$

$$k = 2500 \text{ N/m}$$

$$v(0-) = 800 \text{ m/s}$$

Substituting the given numerical values into Eq. (7-23) yields

$$X(s) = \frac{0.01 \times 800}{50 + 0.01} \frac{50}{50s^2 + 100s + 2500}$$

$$= \frac{8}{50.01} \frac{1}{s^2 + 2s + 50}$$

$$= \frac{8}{50.01 \times 7} \frac{7}{(s + 1)^2 + 7^2}$$

By taking the inverse Laplace transform of this last equation, we get

$$x(t) = 0.0229e^{-t} \sin 7t \text{ m}$$

The response $x(t)$ is thus a damped sinusoid as shown in Fig. 7-16.

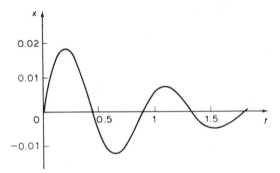

Fig. 7-16. Impulse response curve of the system shown in Fig. 7-14 with $M = 50$ kg, $m = 0.01$ kg, $b = 100$ N-s/m, $k = 2500$ N/m, and $v(0-) = 800$ m/s.

Comments. In Example 7-3 we regarded the wood block and bullet as constituting a system. If, however, we take the wood block alone as a system and the bullet impact as the external force, then we can proceed exactly as in the preceding analysis and derive Eq. (7-23). To check the result, substitute $m = m_1 = 0.02$ kg, $M = m_2 = 50$ kg, $b = 0$, $k = 0$, and $v(0-) = v_1 = 600$ m/s into Eq. (7-23).

$$X(s) = \frac{m_1 v_1}{m_1 + m_2} \frac{1}{s^2}$$

or

$$sX(s) = V(s) = \frac{m_1 v_1}{m_1 + m_2} \frac{1}{s}$$

The inverse Laplace transform of this last equation gives

$$v = \frac{m_1 v_1}{m_1 + m_2} = \frac{0.02 \times 600}{0.02 + 50} = 0.24 \text{ m/s}$$

which is the same as the result obtained in Example 7-3.

7-4 TRANSFER FUNCTIONS

In systems theory functions called "transfer functions" are frequently used to characterize the input-output relationships of linear, time-invariant, differential equation systems. We begin by defining a transfer function and follow with a derivation of the transfer functions of physical systems. Then analogous systems based on these functions are discussed.

Transfer functions. The *transfer function* of a linear, time-invariant, differential equation system is defined as the ratio of the Laplace transform of the output (response function) to the Laplace transform of the input (driving function) under the assumption that all initial conditions are zero.

Consider the linear system defined by the differential equation

$$\overset{(n)}{a_0 x} + \overset{(n-1)}{a_1 x} + \cdots + a_{n-1}\dot{x} + a_n x$$
$$= \overset{(m)}{b_0 p} + \overset{(m-1)}{b_1 p} + \cdots + b_{m-1}\dot{p} + b_m p \qquad (n \geq m)$$

where x is the output of the system and p is the input. The transfer function of this system is obtained by taking the Laplace transform of both sides of this last equation, under the assumption that all initial conditions are zero, or

$$\text{Transfer function} = G(s) = \frac{\mathcal{L}[\text{output}]}{\mathcal{L}[\text{input}]}\bigg|_{\text{zero initial conditions}}$$
$$= \frac{X(s)}{P(s)} = \frac{b_0 s^m + b_1 s^{m-1} + \cdots + b_{m-1}s + b_m}{a_0 s^n + a_1 s^{n-1} + \cdots + a_{n-1}s + a_n}$$

By using the concept of transfer function, it is possible to represent system dynamics by algebraic equations in s. If the highest power of s in the denominator of the transfer function is equal to n, the system is called an *nth-order system.*

Comments on transfer function. The applicability of the concept of the transfer function is limited to linear, time-invariant, differential equation systems. The transfer function approach, however, is extensively used in the analysis and design of such systems. In what follows, we shall list important comments concerning the transfer function. (Note that in the list, a system referred to is one described by a linear, time-invariant, differential equation.)

1. The transfer function of a system is a mathematical model in that it is an operational method of expressing the differential equation that relates the output variable to the input variable.

2. The transfer function is a property of a system itself, independent of the magnitude and nature of the input or driving function.

3. The transfer function includes the units necessary to relate the input to the output; however, it does not provide any information concerning the physical structure of the system. (The transfer functions of many physically different systems can be identical.)

4. If the transfer function of a system is known, the output or response can be studied for various forms of inputs with a view toward understanding the nature of the system.

5. If the transfer function of a system is unknown, it may be established experimentally by introducing known inputs and studying the output of the system. Once established, a transfer function gives a full description of the dynamic characteristics of the system, as distinct from its physical description.

Mechanical system. Consider the spring-mass-damper system shown in Fig. 7-17. Let us obtain the transfer function of this system by assuming that force $p(t)$ is the input and that displacement $x(t)$ of the mass is the output. Here we measure displacement x from the equilibrium position.

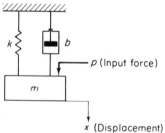

Fig. 7-17. Spring-mass-damper system.

To derive the transfer function, we proceed according to the following steps.

1. Write the differential equation of the system.
2. Take the Laplace transform of the differential equation, assuming all initial conditions to be zero.
3. Take the ratio of the output $X(s)$ to the input $P(s)$. This ratio is the transfer function.

Applying Newton's second law to the present system, we obtain

$$m\ddot{x} + b\dot{x} + kx = p$$

Taking the Laplace transform of both sides of this equation gives

$$m[s^2 X(s) - sx(0) - \dot{x}(0)] + b[sX(s) - x(0)] + kX(s) = P(s)$$

By equating all initial conditions to zero, this last equation simplifies to

$$(ms^2 + bs + k)X(s) = P(s)$$

Taking the ratio $X(s)$ to $P(s)$, we find that the transfer function of the system is

$$\text{Transfer function} = \frac{X(s)}{P(s)} = \frac{1}{ms^2 + bs + k}$$

Electrical circuit. Figure 7-18 shows an electrical circuit in which e_i is the input voltage and e_o the output voltage. The circuit consists of an inductance L (henry), a resistance R (ohm), and a capacitance C (farad). Applying Kirchhoff's voltage law to the system results in the following equations.

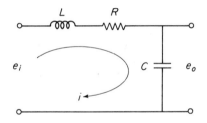

Fig. 7-18. Electrical circuit.

$$L\frac{di}{dt} + Ri + \frac{1}{C}\int i\,dt = e_i$$

$$\frac{1}{C}\int i\,dt = e_o$$

By taking the Laplace transforms of these two equations and assuming zero initial conditions, we have

$$LsI(s) + RI(s) + \frac{1}{C}\frac{1}{s}I(s) = E_i(s)$$

$$\frac{1}{C}\frac{1}{s}I(s) = E_o(s)$$

Therefore the transfer function of this system is

$$\text{Transfer function} = \frac{E_o(s)}{E_i(s)} = \frac{1}{LCs^2 + RCs + 1} \qquad (7\text{-}24)$$

Complex impedances. In deriving transfer functions for electrical circuits, it is often convenient to write the Laplace-transformed equations directly instead of writing the differential equations first. We can do so by using the concept of complex impedances.

The complex impedance $Z(s)$ of the two-terminal circuit in Fig. 7-19 is

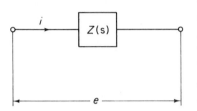

the ratio of $E(s)$, the Laplace transform of the voltage across the terminals, to $I(s)$, the Laplace transform of the current through the element, under the assumption that the initial conditions are zero, so that $Z(s) = E(s)/I(s)$. If the two-terminal element is a resistance R, capacitance C, or inductance L, then the respective complex impedance is given by R, $1/Cs$, or Ls. The general ralationship

Fig. 7-19. Electrical system with complex impedance $Z(s)$.

$$E(s) = Z(s)I(s)$$

corresponds to Ohm's law for purely resistive circuits. Note also that impedances can be combined in series and parallel just as resistances are.

Consider next the circuit shown in Fig. 7-20(a). The complex impedance $Z(s)$ is found from

$$E(s) = E_L(s) + E_R(s) + E_C(s) = \left(Ls + R + \frac{1}{Cs}\right)I(s)$$

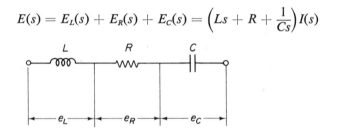

(a)

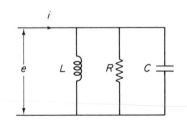

(b)

Fig. 7-20. Electrical circuits.

as

$$Z(s) = \frac{E(s)}{I(s)} = Ls + R + \frac{1}{Cs}$$

For the circuit shown in Fig. 7-20(b),

$$I(s) = \frac{E(s)}{Ls} + \frac{E(s)}{R} + \frac{E(s)}{1/Cs} = E(s)\left(\frac{1}{Ls} + \frac{1}{R} + Cs\right)$$

Consequently,

$$Z(s) = \frac{E(s)}{I(s)} = \frac{1}{\frac{1}{Ls} + \frac{1}{R} + Cs}$$

Deriving transfer functions of electrical circuits by use of complex impedances. The transfer function of an electrical circuit can be obtained as a ratio of complex impedances. For the circuit shown in Fig. 7-21, assume

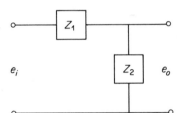

Fig. 7-21. Electrical circuit.

that voltages e_i and e_o are the input and output of the circuit, respectively. Then the transfer function of this circuit can be obtained as

$$\frac{E_o(s)}{E_i(s)} = \frac{Z_2(s)I(s)}{Z_1(s)I(s) + Z_2(s)I(s)} = \frac{Z_2(s)}{Z_1(s) + Z_2(s)}$$

As an example, consider the circuit shown in Fig. 7-18, where

$$Z_1(s) = Ls + R, \qquad Z_2(s) = \frac{1}{Cs}$$

The transfer function of this circuit can be found as

$$\frac{E_o(s)}{E_i(s)} = \frac{Z_2(s)}{Z_1(s) + Z_2(s)}$$

$$= \frac{1/Cs}{Ls + R + (1/Cs)} = \frac{1}{LCs^2 + RCs + 1}$$

which is, of course, identical to Eq. (7-24).

Transfer functions of nonloading elements in series. The transfer function of a system consisting of two nonloading cascaded elements can be found by eliminating the intermediate input and output. Consider, for instance, the system shown in Fig. 7-22(a). The transfer function of each element is

$$G_1(s) = \frac{X_2(s)}{X_1(s)}, \qquad G_2(s) = \frac{X_3(s)}{X_2(s)}$$

If the input impedance of the second element is infinite, the output of the first element is not affected by connecting it to the second element. Thus the transfer function of the whole system is

$$G(s) = \frac{X_3(s)}{X_1(s)} = \frac{X_2(s)}{X_1(s)} \frac{X_3(s)}{X_2(s)} = G_1(s)G_2(s)$$

The transfer function of the entire system is therefore the product of the transfer functions of the individual elements. This situation is shown in Fig. 7-22(b).

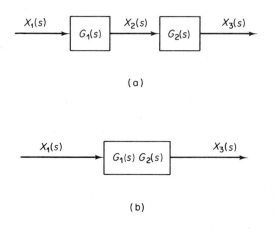

(a)

(b)

Fig. 7-22. (a) System consisting of two non-loading cascaded elements; (b) simplified diagram.

Let us look at another example, the system shown in Fig. 7-23. The insertion of an isolating amplifier between the circuits in order to obtain nonloading characteristics is frequently used in combining electronic circuits. Since both solid-state amplifiers and vacuum-tube amplifiers have very high input impedances, an isolating amplifier inserted between the two circuits justifies the nonloading assumption.

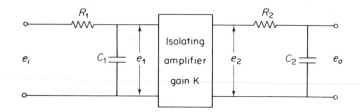

Fig. 7-23. Electrical system.

The two simple RC circuits isolated by an amplifier shown in Fig. 7-23 have negligible loading effects, and the transfer function of the entire circuit equals the product of the individual transfer functions. So in this case,

$$\frac{E_o(s)}{E_i(s)} = \frac{E_1(s)}{E_i(s)}\frac{E_2(s)}{E_1(s)}\frac{E_o(s)}{E_2(s)}$$

$$= \left(\frac{1}{R_1C_1s + 1}\right)(K)\left(\frac{1}{R_2C_2s + 1}\right)$$

$$= \frac{K}{(R_1C_1s + 1)(R_2C_2s + 1)}$$

Transfer functions of loading elements in series. Many systems, like the one illustrated in Fig. 7-24, have components that load each other. Let us

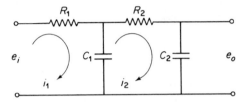

Fig. 7-24. Electrical system.

assume again that in this figure e_i is the input and e_o is the output. Here the second stage of the circuit (R_2C_2 portion) produces a loading effect on the first stage (R_1C_1 portion). The equations for this system are

$$R_1i_1 + \frac{1}{C_1}\int (i_1 - i_2)\, dt = e_i$$

$$\frac{1}{C_1}\int (i_2 - i_1)\, dt + R_2i_2 = -\frac{1}{C_2}\int i_2\, dt = -e_o$$

If we take the Laplace transforms of these two equations, assuming zero initial conditions, the results are

$$R_1I_1(s) + \frac{1}{C_1s}[I_1(s) - I_2(s)] = E_i(s)$$

$$\frac{1}{C_1s}[I_2(s) - I_1(s)] + R_2I_2(s) = -\frac{1}{C_2s}I_2(s) = -E_o(s)$$

Then eliminating $I_1(s)$ from these last two equations gives

$$E_i(s) = \frac{1}{C_2s}(R_1C_1R_2C_2s^2 + R_1C_1s + R_2C_2s + R_1C_2s + 1)I_2(s)$$

$$E_o(s) = \frac{1}{C_2s}I_2(s)$$

And so the transfer function between $E_o(s)$ and $E_i(s)$ is

$$\frac{E_o(s)}{E_i(s)} = \frac{1}{R_1C_1R_2C_2s^2 + (R_1C_1 + R_2C_2 + R_1C_2)s + 1} \qquad (7\text{-}25)$$

The term R_1C_2s in the denominator of the transfer function represents the interaction of two simple RC circuits. [Since $(R_1C_1 + R_2C_2 + R_1C_2)^2 > 4R_1C_1R_2C_2$, the two roots of the denominator of Eq. (7-25) are real.]

The present analysis shows that if two RC circuits are connected in cascade, so that the output from the first circuit is the input to the second, the overall transfer function is not the product of $1/(R_1C_1s + 1)$ and $1/(R_2C_2s + 1)$. This situation occurs because when we derive the transfer function for an isolated circuit, we implicitly assume that the output is unloaded. In other words, the load impedance is assumed to be infinite, which means that no power is being withdrawn at the output. Yet when the second circuit is connected to the output of the first, a certain amount of power is withdrawn, and thus the assumption of no loading is violated. The degree of the loading effect determines the amount of modification of the transfer function. Always remember that any loading effect must be accounted for when the transfer function is derived.

Analogous systems. In previous chapters we occasionally discussed analogous systems. Here we shall summarize what was discussed earlier.

Analogy, of course, is not limited to mechanical-electrical analogy, hydraulic-electric analogy, and similiar situations but includes any physical and nonphysical systems. Systems having an identical transfer function (or identical mathematical model) are analogous systems. (The transfer function is one of the simplest and most concise forms of mathematical models available today.)

Analogous systems will exhibit the same output in response to the same input. For any given physical system, the mathematical response can be given a physical interpretation.

The concept of analogy is useful in applying well-known results in one field to another. It proves particularly useful when a given physical system (mechanical, hydraulic, pneumatic, etc.) is complex, so that first analyzing an analogous electrical circuit is advantageous. Such an analogous electrical circuit can be built physically or can be simulated on an analog computer. (For electronic analog computers, refer to Sec. 7-7.)

For many engineers, electrical circuits or simulated systems on analog computers may be easier to analyze than hydraulic or pneumatic circuits. Consequently, the engineer should be able to obtain an analogous electrical circuit for a given physical system. In general, once the transfer function of a given physical system is found, it is not difficult to obtain an analogous electrical circuit or to simulate it on an analog computer.

Example 7-4. Let us derive the transfer functions of the systems shown in Fig. 7-25(a) and (b) and show that these systems are analogous.

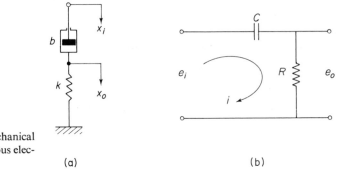

Fig. 7-25. (a) Mechanical system; (b) analogous electrical system.   (a)   (b)

For the mechanical system in Fig. 7-25(a), the equation of motion is

$$b(\dot{x}_i - \dot{x}_o) = kx_o$$

or

$$b\dot{x}_i = kx_o + b\dot{x}_o$$

By taking the Laplace transform of this last equation, assuming the zero initial conditions, we obtain

$$bsX_i(s) = (k + bs)X_o(s)$$

Hence the transfer function between $X_o(s)$ and $X_i(s)$ is

$$\frac{X_o(s)}{X_i(s)} = \frac{bs}{bs + k} = \frac{(b/k)s}{(b/k)s + 1}$$

For the electrical system in Fig. 7-25(b), we have

$$\frac{E_o(s)}{E_i(s)} = \frac{R}{(1/Cs) + R} = \frac{RCs}{RCs + 1}$$

Comparing the transfer functions obtained, we see that the two systems are analogous. Note that both b/k and RC have the dimension of time and are time constants of the respective systems. (For analogous quantities between the mechanical and electrical systems, see Sec. 3-5.)

7-5   FREQUENCY RESPONSE AND SINUSOIDAL TRANSFER FUNCTIONS

When a sinusoidal input is applied to a linear system, it will tend to vibrate at its own natural frequency as well as follow the frequency of the input. In the presence of damping that portion of motion not sustained by the sinusoidal input will gradually die out. As a result, the response at steady state is sinusoidal at the same frequency as the input. The steady-state output differs from the input only in the amplitude and phase angle. Thus the output-input amplitude ratio and the phase angle between the output and input sinusoid are the only two parameters needed to predict the output of a linear system when the input is a sinusoid. In general, the amplitude ratio and the phase angle depend on the input frequency.

Frequency response. The term *frequency response* refers to the steady-state response of a system to a sinusoidal input. For all frequencies from zero to infinity, the frequency response characteristics of a system can be completely described by the output-input amplitude ratio and the phase angle between the output and input sinusoid. In this method of systems analysis we vary the frequency of the input signal over a wide range and study the resulting response.

There are three principal reasons for emphasizing frequency response in systems analysis.

1. Many natural phenomena are sinusoidal in nature—for instance, simple harmonic motions are frequently generated in mechanical and electrical systems.
2. Any periodic signal can be represented by a series of sinusoidal components.
3. Sinusoidal signals are important in communications as well as in the generation and transmission of electric power.

Forced vibration without damping. Figure 7-26 illustrates a spring-mass system in which the mass is subjected to a sinusoidal input force $P \sin \omega t$. Let us find the response of the system when it is initially at rest.

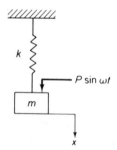

Fig. 7-26. Spring-mass system.

If we measure displacement x from the equilibrium position, the equation of motion for the system becomes

$$m\ddot{x} + kx = P \sin \omega t$$

or

$$\ddot{x} + \frac{k}{m}x = \frac{P}{m} \sin \omega t \qquad (7\text{-}26)$$

We shall first solve this problem by the conventional method. The solution of this equation consists of the vibration at its own natural frequency (complementary solution) and that at the forcing frequency (particular solution). Thus the solution $x(t)$ can be written

$$x(t) = \text{(complementary solution)} + \text{(particular solution)}$$

$$= \left(A \sin \sqrt{\frac{k}{m}}\, t + B \cos \sqrt{\frac{k}{m}}\, t \right) + (C \sin \omega t)$$

where A, B, and C are yet undetermined constants.

Assume that the input $P \sin \omega t$ is applied at $t = 0$. Since the system is initially at rest, we have the initial conditions $x(0) = 0$ and $\dot{x}(0) = 0$. Then

$$x(0) = B = 0$$

Consequently, $x(t)$ can be simplified as

$$x(t) = A \sin \sqrt{\frac{k}{m}}\, t + C \sin \omega t \qquad (7\text{-}27)$$

Noting that

$$\dot{x}(t) = A\sqrt{\frac{k}{m}} \cos \sqrt{\frac{k}{m}}\, t + C\omega \cos \omega t$$

we have

$$\dot{x}(0) = A\sqrt{\frac{k}{m}} + C\omega = 0$$

And so

$$C = -A\sqrt{\frac{k}{m}} \frac{1}{\omega}$$

The second derivative of $x(t)$ becomes

$$\ddot{x}(t) = -A\frac{k}{m} \sin \sqrt{\frac{k}{m}}t - C\omega^2 \sin \omega t \qquad (7\text{-}28)$$

Substitution of Eqs. (7-27) and (7-28) into Eq. (7-26) gives

$$\ddot{x} + \frac{k}{m}x = -C\omega^2 \sin \omega t + \frac{k}{m} C \sin \omega t = \frac{P}{m} \sin \omega t$$

from which

$$C\left(\frac{k}{m} - \omega^2\right) = \frac{P}{m}$$

or

$$C = \frac{P}{k - m\omega^2}$$

So it follows that

$$A = -\frac{C\omega}{\sqrt{k/m}} = -\frac{P\omega\sqrt{m/k}}{k - m\omega^2}$$

The solution now becomes

$$x(t) = -\frac{P\omega\sqrt{m/k}}{k - m\omega^2}\sin\sqrt{\frac{k}{m}}\,t + \frac{P}{k - m\omega^2}\sin\omega t \tag{7-29}$$

This is the complete solution (general solution). The first term is the complementary solution (which does not decay in this system), and the second term is the particular solution.

The same complete solution can be obtained by using the Laplace transform method. In fact, for more complicated systems (such as systems involving damping terms or systems having two or more degrees of freedom) the Laplace transform approach is much simpler than the conventional approach presented above. Let us demonstrate it for the present system. [Note that if we need only a steady-state solution (particular solution), the use of the sinusoidal transfer function simplifies the solution. The sinusoidal transfer function is discussed in detail in this section.]

By taking the Laplace transform of Eq. (7-26) and using the initial conditions $x(0) = 0$ and $\dot{x}(0) = 0$, we find

$$\left(s^2 + \frac{k}{m}\right)X(s) = \frac{P}{m}\frac{\omega}{s^2 + \omega^2}$$

Solving for $X(s)$,

$$X(s) = \frac{P}{m}\frac{\omega}{s^2 + \omega^2}\frac{1}{s^2 + (k/m)}$$

$$= \frac{-P\omega\sqrt{m/k}}{k - m\omega^2}\frac{\sqrt{k/m}}{s^2 + (k/m)} + \frac{P}{k - m\omega^2}\frac{\omega}{s^2 + \omega^2}$$

The inverse Laplace transform of this last equation gives

$$x(t) = -\frac{P\omega\sqrt{m/k}}{k - m\omega^2}\sin\sqrt{\frac{k}{m}}\,t + \frac{P}{k - m\omega^2}\sin\omega t$$

which is exactly the same as Eq. (7-29).

Let us examine the response of the system, Eq. (7-29). As the forcing frequency ω approaches zero, the amplitude of the vibration at its natural frequency $\sqrt{k/m}$ approaches zero and the amplitude of the vibration at the forcing frequency ω approaches P/k. This value P/k is the deflection of the mass that would result if the force P were applied steadily (at zero frequency). That is, P/k is the static deflection. As the frequency ω increases, the denominator of the solution, $k - m\omega^2$, becomes smaller and the amplitudes become larger. As the frequency ω is further increased and becomes equal to the natural frequency of the system, or $\omega = \omega_n = \sqrt{k/m}$, resonance occurs. At resonance the denominator of the solution, $k - m\omega^2$, becomes zero, and the amplitude of vibration will increase without bound. (When the sinusoidal input is applied at the natural frequency and in phase with the motion—that is, in the same direction as the velocity—the input force is actually doing work on the system and is adding to it energy that will appear as an increase in amplitudes.) As ω continues to increase past resonance, the denominator $k - m\omega^2$ becomes negative and assumes increasingly larger values, approaching negative infinity. Therefore the amplitudes of vibration (at the natural frequency and at the forcing frequency) approach zero from the negative side, starting at negative infinity when $\omega = \omega_n+$. In other words, if ω is below resonance, that part of the vibration at the forcing frequency (particular solution) is in phase with the forcing sinusoid. If ω is above resonance, this vibration becomes 180° out of phase.

Sinusoidal transfer function. The *sinusoidal transfer function* is defined as the transfer function $G(s)$ in which s is replaced by $j\omega$. When only the steady-state solution (particular solution) is wanted, the sinusoidal transfer function $G(j\omega)$ can simplify the solution. In the following discussion we shall consider the behaviour of stable linear systems under steady-state conditions —that is, after initial transients have died out. And we shall see that sinusoidal inputs will produce sinusoidal outputs in steady state with the amplitude and phase angle at each frequency ω determined by the magnitude and angle of $G(j\omega)$, respectively.

Deriving steady-state output to sinusoidal input. Let us see how the frequency response characteristics of a stable system can be derived directly from the sinusoidal transfer function. For the linear system $G(s)$ in Fig. 7-27,

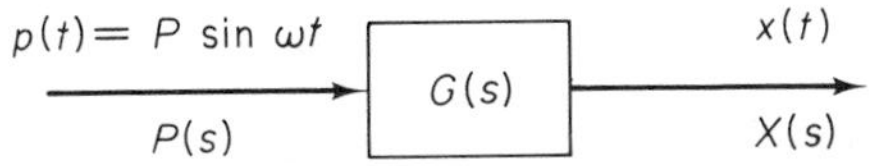

Fig. 7-27. Linear system.

the input and output are denoted by $p(t)$ and $x(t)$, respectively. The input $p(t)$ is sinusoidal and is given by

$$p(t) = P \sin \omega t$$

We shall show that the output $x(t)$ at steady state is given by

$$x(t) = |G(j\omega)| P \sin (\omega t + \phi)$$

where $|G(j\omega)|$ and ϕ are the magnitude and angle of $G(j\omega)$, respectively.

Suppose that the transfer function $G(s)$ can be written as a ratio of two polynomials in s—that is,

$$G(s) = \frac{K(s + z_1)(s + z_2)\cdots(s + z_m)}{(s + s_1)(s + s_2)\cdots(s + s_n)}$$

The Laplace transformed output $X(s)$ is

$$X(s) = G(s)P(s) \tag{7-30}$$

where $P(s)$ is the Laplace transform of the input $p(t)$.

Let us limit our discussion only to stable systems. For such systems, the real parts of the $-s_i$ are negative. The steady-state response of a stable linear system to a sinusoidal input does not depend on the initial conditions, and so they can be ignored.

If $G(s)$ has only distinct poles, then the partial-fraction expansion of Eq. (7-30) yields

$$X(s) = G(s)\frac{P\omega}{s^2 + \omega^2}$$

$$= \frac{a}{s + j\omega} + \frac{\bar{a}}{s - j\omega} + \frac{b_1}{s + s_1} + \frac{b_2}{s + s_2} + \cdots + \frac{b_n}{s + s_n} \quad (7\text{-}31)$$

where a and b_i (where $i = 1, 2, \ldots, n$) are constants and $\bar{a}$ is the complex conjugate of a. The inverse Laplace transform of Eq. (7-31) gives

$$x(t) = ae^{-j\omega t} + \bar{a}e^{j\omega t} + b_1 e^{-s_1 t} + b_2 e^{-s_2 t} + \cdots + b_n e^{-s_n t}$$

For a stable system, as t approaches infinity, the terms $e^{-s_1 t}, e^{-s_2 t}, \ldots, e^{-s_n t}$ approach zero, since $-s_1, -s_2, \ldots, -s_n$ have negative real parts. Thus all terms on the right-hand side of this last equation, except the first two, drop out at steady state.

If $G(s)$ involves k multiple poles s_j, then $x(t)$ will involve such terms as $t^h e^{-s_j t}$ (where $h = 0, 1, \ldots, k - 1$). Since the real part of the $-s_j$ is negative for a stable system, the terms $t^h e^{-s_j t}$ approach zero as t approaches infinity.

Regardless of whether the system involves multiple poles, the steady-state response thus becomes

$$x(t) = ae^{-j\omega t} + \bar{a}e^{j\omega t} \quad (7\text{-}32)$$

where the constants a and $\bar{a}$ can be evaluated from Eq. (7-31).

$$a = G(s)\frac{P\omega}{s^2 + \omega^2}(s + j\omega)\Big|_{s=-j\omega} = -\frac{P}{2j}G(-j\omega)$$

$$\bar{a} = G(s)\frac{P\omega}{s^2 + \omega^2}(s - j\omega)\Big|_{s=j\omega} = \frac{P}{2j}G(j\omega)$$

(Note that $\bar{a}$ is the complex conjugate of a.) Referring to Fig. 7-28, we can write

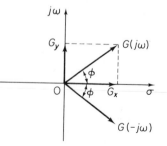

Fig. 7-28. Complex function and its complex conjugate.

$$G(j\omega) = G_x + jG_y$$
$$= |G(j\omega)|\cos\phi + j|G(j\omega)|\sin\phi$$
$$= |G(j\omega)|(\cos\phi + j\sin\phi)$$
$$= |G(j\omega)|e^{j\phi}$$

(Note that $\underline{/G(j\omega)} = \underline{/e^{j\phi}} = \phi$.) Similarly,

$$G(-j\omega) = |G(-j\omega)|e^{-j\phi} = |G(j\omega)|e^{-j\phi}$$

It follows that

$$a = -\frac{P}{2j}|G(j\omega)|e^{-j\phi}$$

$$\bar{a} = \frac{P}{2j}|G(j\omega)|e^{j\phi}$$

Then Eq. (7-32) can be written

$$x(t) = |G(j\omega)|P\frac{e^{j(\omega t+\phi)} - e^{-j(\omega t+\phi)}}{2j}$$

$$= |G(j\omega)|P\sin(\omega t+\phi)$$

$$= X\sin(\omega t+\phi) \qquad (7\text{-}33)$$

where $X = |G(j\omega)|P$ and $\phi = \underline{/G(j\omega)}$. We see that a stable linear system subjected to a sinusoidal input will, at steady state, have a sinusoidal output of the same frequency as the input. But the amplitude and phase angle of the output will, in general, differ from the input's. In fact, the output's amplitude is given by the product of the amplitude of the input and $|G(j\omega)|$, whereas the phase angle differs from that of the input by the amount $\phi = \underline{/G(j\omega)}$.

On the basis of the preceding analysis we are able to derive the following important result. For sinusoidal inputs,

$$|G(j\omega)| = \left|\frac{X(j\omega)}{P(j\omega)}\right| = \text{amplitude ratio of the output}$$
$$\text{sinusoid to the input sinusoid}$$

$$\underline{/G(j\omega)} = \underline{/\frac{X(j\omega)}{P(j\omega)}} = \tan^{-1}\left[\frac{\text{imaginary part of } G(j\omega)}{\text{real part of } G(j\omega)}\right]$$

$$= \text{phase shift of the output sinusoid}$$
$$\text{with respect to the input sinusoid}$$

Thus the steady-state response characteristics of a linear system to a sinusoidal input can be found directly from $G(j\omega)$, the ratio of $X(j\omega)$ to $P(j\omega)$.

Note that the sinusoidal transfer function $G(j\omega)$ is a complex quantity that can be represented by the magnitude and phase angle with frequency ω as a parameter. In order to characterize a linear system completely by the frequency response curves, we must specify both the amplitude ratio and the phase angle as a function of the frequency ω.

Comments. It should be pointed out that Eq. (7-33) is valid only if $G(s) = X(s)/P(s)$ is a stable system—that is, if all poles of $G(s)$ lie in the left half-plane. If a pole is at the origin and/or poles of $G(s)$ lie on the $j\omega$ axis (any poles on the $j\omega$ axis must occur as a pair of complex conjugates), the output $x(t)$ may be obtained by taking the inverse Laplace transform of the equation

$$X(s) = G(s)P(s) = G(s)\frac{P\omega}{s^2 + \omega^2}$$

or

$$x(t) = \mathcal{L}^{-1}[X(s)] = \mathcal{L}^{-1}\left[G(s)\frac{P\omega}{s^2 + \omega^2}\right]$$

Note that if a pole or poles of $G(s)$ lie in the right half-plane, the system is unstable and the response grows indefinitely. There is no steady state for such an unstable system.

Example 7-5. Consider the transfer function system

$$\frac{X(s)}{P(s)} = G(s) = \frac{1}{Ts + 1}$$

For the sinusoidal input $p(t) = P \sin \omega t$, what is the steady-state output $x(t)$?
    Substituting $j\omega$ for s in $G(s)$ yields

$$G(j\omega) = \frac{1}{Tj\omega + 1}$$

The output-input amplitude ratio is

$$|G(j\omega)| = \frac{1}{\sqrt{T^2\omega^2 + 1}}$$

whereas the phase angle ϕ is

$$\phi = \underline{/G(j\omega)} = -\tan^{-1} T\omega$$

So for the input $p(t) = P \sin \omega t$, the steady-state output $x(t)$ can be found as

$$x(t) = \frac{P}{\sqrt{T^2\omega^2 + 1}} \sin (\omega t - \tan^{-1} T\omega)$$

From this equation we see that, for small ω, the amplitude of the output $x(t)$ is almost equal to the amplitude of the input. For large ω, the amplitude of the output is small and almost inversely proportional to ω. The phase angle is 0° at $\omega = 0$ and approaches $-90°$ as ω increases indefinitely.

Example 7-6. Suppose that a sinusoidal force $p(t) = P \sin \omega t$ is applied to the mechanical system shown in Fig. 7-29. Assuming that displacement x is measured from the equilibrium position, find the steady-state output.
    The equation of motion for the system is

$$m\ddot{x} + b\dot{x} + kx = p(t)$$

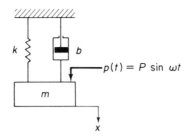

Fig. 7-29. Mechanical system.

The Laplace transform of this equation, assuming zero initial conditions, is

$$(ms^2 + bs + k)X(s) = P(s)$$

where $X(s) = \mathcal{L}[x(t)]$ and $P(s) = \mathcal{L}[p(t)]$. (Note that the initial conditions do not affect the steady-state output and so can be assumed zero.) The transfer function between displacement $X(s)$ and input force $P(s)$ is, therefore, obtained as

$$\frac{X(s)}{P(s)} = G(s) = \frac{1}{ms^2 + bs + k}$$

Since the input is a sinusoidal function $p(t) = P \sin \omega t$, we can use the sinusoidal transfer function to obtain the steady-state solution. The sinusoidal transfer function is

$$\frac{X(j\omega)}{P(j\omega)} = G(j\omega) = \frac{1}{-m\omega^2 + bj\omega + k} = \frac{1}{(k - m\omega^2) + jb\omega}$$

Referring to Eq. (7-33), the output $x(t)$ can be written

$$x(t) = |G(j\omega)| P \sin (\omega t + \phi)$$

where

$$\phi = \underline{/G(j\omega)} = \left/ \frac{1}{(k - m\omega^2) + jb\omega} \right. = -\tan^{-1} \frac{b\omega}{k - m\omega^2}$$

Thus

$$x(t) = \frac{P}{\sqrt{(k - m\omega^2)^2 + b^2\omega^2}} \sin \left(\omega t - \tan^{-1} \frac{b\omega}{k - m\omega^2} \right)$$

Since $k/m = \omega_n^2$ and $b/k = 2\zeta/\omega_n$, this equation can be written

$$x(t) = \frac{x_{st}}{\sqrt{[1 - (\omega^2/\omega_n^2)]^2 + (2\zeta\omega/\omega_n)^2}} \sin \left[\omega t - \tan^{-1} \frac{2\zeta\omega/\omega_n}{1 - (\omega^2/\omega_n^2)} \right] \quad (7\text{-}34)$$

where $x_{st} = P/k$ is the static deflection.

By writing the amplitude of $x(t)$ as X, we find that the amplitude ratio X/x_{st} is

$$\frac{X}{x_{st}} = \frac{1}{\sqrt{[1 - (\omega^2/\omega_n^2)]^2 + (2\zeta\omega/\omega_n)^2}}$$

Figure 7-30 shows the effects of the input frequency ω and the damping ratio ζ on the amplitude and phase angle of the steady-state output.

From the figure we see that as the damping ratio is increased, the amplitude

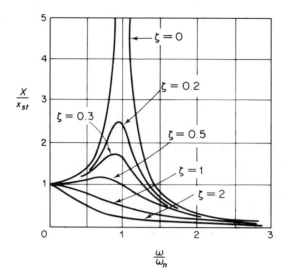

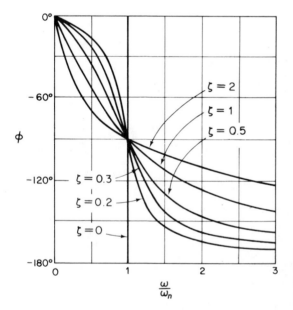

Fig. 7-30. Normalized amplitude versus normalized frequency curves and phase angle versus normalized frequency curves.

ratio decreases. The maximum amplitude ratio occurs at a frequency that is less than the undamped natural frequency ω_n. Notice that the frequency ω_m, at which the amplitude ratio is a maximum, occurs at

$$\omega_m = \sqrt{\frac{k}{m} - 2\left(\frac{b}{2m}\right)^2} = \omega_n\sqrt{1 - 2\zeta^2}$$

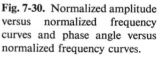

(This frequency is somewhat smaller than the damped natural frequency $\omega_d = \omega_n\sqrt{1 - \zeta^2}$.) The value of ω_m can be obtained as the frequency that minimizes

$$\left(1 - \frac{\omega^2}{\omega_n^2}\right)^2 + \left(2\zeta\frac{\omega}{\omega_n}\right)^2$$

That is, by differentiating this expression with respect to ω, substituting $\omega = \omega_m$, and setting the result equal to zero, we have

$$2\left(1 - \frac{\omega_m^2}{\omega_n^2}\right)\left(-\frac{2\omega_m}{\omega_n^2}\right) + 4\zeta^2\frac{2\omega_m}{\omega_n^2} = 0$$

Then solving for ω_m^2, we get

$$\frac{\omega_m^2}{\omega_n^2} = 1 - 2\zeta^2$$

which yields

$$\omega_m = \omega_n\sqrt{1 - 2\zeta^2}$$

Example 7-7. Referring to the system shown in Fig. 7-29, if the numerical values of m, b, k, P, and ω are given as $m = 10$ kg, $b = 30$ N-s/m, $k = 1000$ N/m, $P = 10$ N, and $\omega = 2$ rad/s, what is the steady-state output $x(t)$?

The system equation is

$$10\ddot{x} + 30\dot{x} + 1000x = 10 \sin 2t$$

The undamped natural frequency ω_n is 10 rad/s, the damping ratio ζ is 0.15, and the static deflection x_{st} is 0.01 m. By substituting the numerical values into Eq. (7-34), the steady-state output turns out to be

$$x(t) = \frac{0.01}{\sqrt{[1 - (2^2/10^2)]^2 + (2 \times 0.15 \times 2/10)^2}} \sin\left[2t - \tan^{-1}\frac{2 \times 0.15 \times 2/10}{1 - (2^2/10^2)}\right]$$

$$= 0.0104 \sin (2t - \tan^{-1} 0.0625)$$

$$= 0.0104 \sin (2t - 0.0625)$$

The steady-state output has the amplitude of 0.0104 m and lags the input (forcing function) by 0.0625 rad or 3.58°.

Example 7-8. Assume, in the circuit of Fig. 7-31, that voltage e_i is applied to the input terminals and voltage e_o appears at the output terminals. In addition, assume that the input is sinusoidal and is given by

$$e_i(t) = E_i \sin \omega t$$

What is the steady-state current $i(t)$?

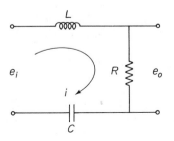

Fig. 7-31. Electrical circuit.

Applying Kirchhoff's voltage law to the circuit yields

$$L\frac{di}{dt} + Ri + \frac{1}{C}\int i\, dt = e_i$$

Then the Laplace transform of this last equation, assuming zero initial conditions, is

$$\left(Ls + R + \frac{1}{Cs}\right)I(s) = E_i(s)$$

Hence the transfer function between $I(s)$ and $E_i(s)$ becomes

$$\frac{I(s)}{E_i(s)} = \frac{1}{Ls + R + (1/Cs)} = \frac{Cs}{LCs^2 + RCs + 1}$$

The sinusoidal transfer function is

$$\frac{I(j\omega)}{E_i(j\omega)} = G(j\omega) = \frac{Cj\omega}{-LC\omega^2 + RCj\omega + 1}$$

Therefore the steady-state current $i(t)$ is given by

$$i(t) = |G(j\omega)|\,E_i \sin\left[\omega t + \underline{/G(j\omega)}\right]$$

$$= \frac{CE_i\omega}{\sqrt{(1 - LC\omega^2)^2 + (RC\omega)^2}} \sin\left(\omega t + 90° - \tan^{-1}\frac{RC\omega}{1 - LC\omega^2}\right)$$

$$= \frac{CE_i\omega}{\sqrt{(1 - LC\omega^2)^2 + (RC\omega)^2}} \cos\left(\omega t - \tan^{-1}\frac{RC\omega}{1 - LC\omega^2}\right)$$

*7-6   VIBRATION ISOLATION

Vibration is, in general, undesirable because it may cause parts to break down, generate noise, transmit forces to the foundation, and so on. In order to reduce the amount of force transmitted to the foundation as a result of a machine's vibration (force isolation) as much as possible, machines are usually mounted on vibration isolators that consist of springs and dampers. Similarly, in order to reduce the amount of motion transmitted to a delicate instrument by the motion of the foundation (motion isolation), instruments are mounted on isolators. In this section centripetal force, centrifugal force, and force due to a rotating unbalance are described. Afterward vibrations caused by the exciting force resulting from unbalance, vibration isolators, transmissibility, and, finally, dynamic vibration absorbers are analyzed.

Centripetal force and centrifugal force. Suppose that point mass m is moving in a circular path with a constant speed as in Fig. 7-32(a). The

---

*Starred sections deal with more challenging topics than the rest of the book. Depending on course objectives, these sections (although important) may be omitted from classroom discussions without losing the continuity of the main subject matter.

magnitudes of the velocities of the mass m at point A and point B are the same, but the directions are different. Referring to Fig. 7-32(b), the direction $\overrightarrow{PQ}$ becomes perpendicular to the direction $\overrightarrow{AP}$ (the direction of the velocity vector at point A) if points A and B are close to each other. This means that the point mass must possess an acceleration toward the center of rotation, point O. To produce this acceleration, a force of mass times acceleration is required. If the acceleration is toward the center, the reaction force is outward and its magnitude is equal to this centrally directed force. The force acting toward the center is called the *centripetal force* and the opposing inertia reaction force the *centrifugal force*.

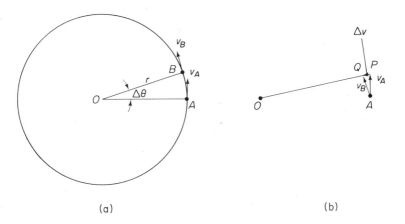

(a)   (b)

Fig. 7-32. (a) Point mass moving in a circular path; (b) velocity vector diagram.

The acceleration a acting toward the center of rotation is derived as follows. Noting that triangles OAB and APQ are similar, we have

$$\frac{|\Delta v|}{|v_A|} = \frac{r\,\Delta\theta}{r}$$

where $|\Delta v|$ and $|v_A|$ represent magnitudes of velocity Δv and velocity v_A, respectively. Noting that $|v_A| = \omega r$ and $\omega = \lim_{\Delta t \to 0}(\Delta\theta/\Delta t)$, we see that

$$a = \lim_{\Delta t \to 0}\frac{|\Delta v|}{\Delta t} = \lim_{\Delta t \to 0}\frac{|v_A|\,r\Delta\theta}{r\,\Delta t} = \omega^2 r$$

This acceleration acts toward the center of rotation, and the centripetal force is $m\omega^2 r$. The centrifugal force is the reaction force and is outward. Its magnitude is also $m\omega^2 r$.

<hr>

Example 7-9. A boy swings, in a circular arc, a stone of mass 0.01 kg attached to
the end of 1 m rope. Suppose that the speed of the stone at circular motion is 2 m/s.
What is the tension T in the rope?

$$T = m\omega^2 r = m\frac{v^2}{r} = 0.01\frac{2^2}{1} = 0.04 \text{ N}$$

<hr>

Vibration due to rotating unbalance. Force inputs that excite vibratory
motion often arise from rotating unbalance. Such a rotating unbalance exists
if the mass center of a rotating rigid body and the center of rotation do not
coincide. Figure 7-33 shows an unbalanced machine resting on shock
mounts. Assume that the rotor is rotating at a constant speed ω rad/s and that
the unbalanced mass m is located at a distance r from the center of rotation.
The unbalanced mass will produce a centrifugal force of magnitude $m\omega^2 r$.

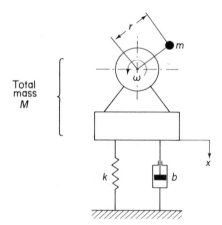

Total
mass
M

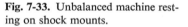

Fig. 7-33. Unbalanced machine resting on shock mounts.

In the present analysis, we limit the motion to the vertical direction only,
even though the rotating unbalance produces the horizontal component of
force. The vertical component of this force, $m\omega^2 r \sin \omega t$, acts on the bearings
and is thus transmitted to the foundation, thereby possibly causing the
machine to vibrate excessively. [Note that, for convenience, we arbitrarily
choose the time origin ($t = 0$) so that the unbalance force applied to the
system is $m\omega^2 r \sin \omega t$.]

Let us assume that the total mass of the system is M, which includes the
unbalanced mass m. Here we consider only vertical motion and measure
vertical displacement x from the equilibrium position in the absence of the
forcing function. Then the equation of motion for the system becomes

$$M\ddot{x} + b\dot{x} + kx = p(t) \tag{7-35}$$

where $p(t)$ is the force applied to the system and is given by

$$p(t) = m\omega^2 r \sin \omega t$$

By taking the Laplace transform of both sides of Eq. (7-35), assuming zero initial conditions, we have

$$(Ms^2 + bs + k)X(s) = P(s)$$

or

$$\frac{X(s)}{P(s)} = \frac{1}{Ms^2 + bs + k}$$

The sinusoidal transfer function is

$$\frac{X(j\omega)}{P(j\omega)} = G(j\omega) = \frac{1}{-M\omega^2 + bj\omega + k}$$

For the sinusoidal forcing function $p(t)$, the steady-state output is obtained from Eq. (7-33) as

$$x(t) = X \sin(\omega t + \phi)$$

$$= |G(j\omega)| m\omega^2 r \sin\left(\omega t - \tan^{-1} \frac{b\omega}{k - M\omega^2}\right)$$

$$= \frac{m\omega^2 r}{\sqrt{(k - M\omega^2)^2 + b^2\omega^2}} \sin\left(\omega t - \tan^{-1} \frac{b\omega}{k - M\omega^2}\right)$$

In this last equation, if we divide the numerator and denominator of the amplitude and those of the phase angle through by k and substitute $k/M = \omega_n^2$ and $b/M = 2\zeta\omega_n$ into the result, the steady-state output becomes

$$x(t) = \frac{m\omega^2 r/k}{\sqrt{[1 - (\omega^2/\omega_n^2)]^2 + (2\zeta\omega/\omega_n)^2}} \sin\left[\omega t - \tan^{-1}\frac{2\zeta\omega/\omega_n}{1 - (\omega^2/\omega_n^2)}\right]$$

From this expression we see that the amplitude of the steady-state output becomes large when the damping ratio ζ is small and the forcing frequency ω is close to the natural frequency ω_n.

Vibration isolators. Vibration isolation is a process by which vibratory effects are minimized or eliminated. The function of a vibration isolator is to reduce the magnitude of force transmitted from a machine to its foundation or to reduce the magnitude of motion transmitted from a vibratory foundation to a machine.

The concept is illustrated in Fig. 7-34(a) and (b). Here the system consists of a rigid body representing a machine connected to a foundation by an isolator that consists of a spring and a damper. Figure 7-34(a) illustrates the case in which the source of vibration is a vibrating force originating within the machine (force excitation). The isolator reduces the force transmitted to the foundation. In Fig. 7-34(b) the source of vibration is a vibrating motion of the foundation (motion excitation). The isolator reduces the vibration amplitude of the machine.

The isolator essentially consists of resilient load-supporting means (such as a spring) and energy-dissipating means (such as a damper). A

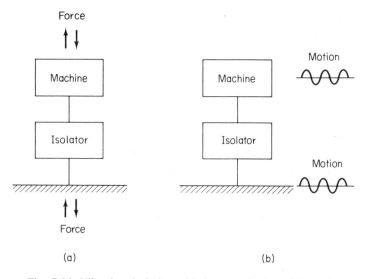

Fig. 7-34. Vibration isolation; (a) force excitation; (b) motion excitation.

typical vibration isolator appears in Fig. 7-35. (In a simple vibration isolator, a single element like synthetic rubber can perform the functions of both the load-supporting means and the energy-dissipating means.) In the analysis given here the machine and the foundation are assumed rigid and the isolator is assumed massless.

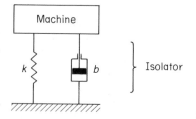

Fig. 7-35. Vibration isolator.

Transmissibility. Transmissibility is a measure of the reduction of transmitted force or motion afforded by an isolator. If the source of vibration is a vibrating force due to the unbalance of the machine (force excitation), transmissibility is the ratio of the force amplitude transmitted to the foundation to the amplitude of the exciting force. If the source of vibration is a vibratory motion of the foundation (motion excitation), transmissibility is the ratio of the vibration amplitude of the machine to the vibration amplitude of the foundation.

Transmissibility for force excitation. For the system shown in Fig. 7-33, the source of vibration is a vibrating force resulting from the unbalance of the machine. The transmissibility in this case is the force amplitude ratio and is given by

$$\text{Transmissibility} = \text{TR} = \frac{F_t}{F_0} = \frac{\text{amplitude of the transmitted force}}{\text{amplitude of the exciting force}}$$

Let us find the transmissibility of this system in terms of damping ratio ζ and frequency ratio $\beta = \omega/\omega_n$.

The exciting force (in the vertical direction) is caused by the unbalanced mass of the machine and is

$$p(t) = m\omega^2 r \sin \omega t = F_0 \sin \omega t$$

The equation of motion for the system is given by Eq. (7-35), rewritten thus:

$$M\ddot{x} + b\dot{x} + kx = p(t) \tag{7-36}$$

The force transmitted to the foundation is the sum of the damper and spring forces or

$$f(t) = b\dot{x} + kx = F_t \sin(\omega t + \phi) \tag{7-37}$$

Taking the Laplace transforms of Eqs. (7-36) and (7-37), assuming the zero initial conditions, gives

$$(Ms^2 + bs + k)X(s) = P(s)$$

$$(bs + k)X(s) = F(s)$$

where $X(s) = \mathcal{L}[x(t)]$, $P(s) = \mathcal{L}[p(t)]$, and $F(s) = \mathcal{L}[f(t)]$. Hence

$$\frac{X(s)}{P(s)} = \frac{1}{Ms^2 + bs + k}$$

$$\frac{F(s)}{X(s)} = bs + k$$

Elimination of $X(s)$ from the last two equations yields

$$\frac{F(s)}{P(s)} = \frac{F(s)}{X(s)}\frac{X(s)}{P(s)} = \frac{bs + k}{Ms^2 + bs + k}$$

The sinusoidal transfer function $F(j\omega)/P(j\omega)$ is

$$\frac{F(j\omega)}{P(j\omega)} = \frac{bj\omega + k}{-M\omega^2 + bj\omega + k} = \frac{(b/M)j\omega + (k/M)}{-\omega^2 + (b/M)j\omega + (k/M)}$$

By substituting $k/M = \omega_n^2$ and $b/M = 2\zeta\omega_n$ into this last equation and simplifying, we have

$$\frac{F(j\omega)}{P(j\omega)} = \frac{1 + j(2\zeta\omega/\omega_n)}{1 - (\omega^2/\omega_n^2) + j(2\zeta\omega/\omega_n)}$$

from which

$$\left|\frac{F(j\omega)}{P(j\omega)}\right| = \frac{\sqrt{1 + (2\zeta\omega/\omega_n)^2}}{\sqrt{[1 - (\omega^2/\omega_n^2)]^2 + (2\zeta\omega/\omega_n)^2}} = \frac{\sqrt{1 + (2\zeta\beta)^2}}{\sqrt{(1 - \beta^2)^2 + (2\zeta\beta)^2}}$$

where $\beta = \omega/\omega_n$.

Noting that the amplitude of the exciting force is $F_0 = |P(j\omega)|$ and that the amplitude of the transmitted force is $F_t = |F(j\omega)|$, we obtain the transmissibility as follows:

$$\text{TR} = \frac{F_t}{F_0} = \frac{|F(j\omega)|}{|P(j\omega)|} = \frac{\sqrt{1 + (2\zeta\beta)^2}}{\sqrt{(1 - \beta^2)^2 + (2\zeta\beta)^2}} \qquad (7\text{-}38)$$

From Eq. (7-38) we find that the transmissibility depends on both β and ζ. It is important to point out, however, that when $\beta = \sqrt{2}$, the transmissibility is equal to unity regardless of the value of the damping ratio ζ.

Figure 7-36 shows the transmissibility versus β (where $\beta = \omega/\omega_n$) curves. We see that all curves pass through a critical point, a point where $\text{TR} = 1$, $\beta = \sqrt{2}$. For $\beta < \sqrt{2}$, as the damping ratio ζ increases, the transmissibility at resonance decreases. For $\beta > \sqrt{2}$, as the damping ratio ζ increases, the transmissibility increases. Therefore for $\beta < \sqrt{2}$ or $\omega < \sqrt{2}\,\omega_n$ (the forcing frequency ω is smaller than $\sqrt{2}$ times the undamped natural frequency ω_n), increasing damping improves the vibration isolation. For $\beta > \sqrt{2}$ or $\omega > \sqrt{2}\,\omega_n$, increasing damping adversely affects the vibration isolation.

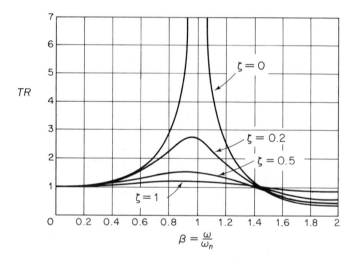

Fig. 7-36. Transmissibility TR versus β ($= \omega/\omega_n$) curves.

Note that since $|P(j\omega)| = F_0 = m\omega^2 r$, the amplitude of the force transmitted to the foundation is

$$F_t = |F(j\omega)| = \frac{m\omega^2 r\sqrt{1 + (2\zeta\beta)^2}}{\sqrt{(1 - \beta^2)^2 + (2\zeta\beta)^2}} \qquad (7\text{-}39)$$

Example 7-10. In the system shown in Fig. 7-33, if the numerical values of M, b, k, m, r, and ω are given as $M = 15$ kg, $b = 450$ N-s/m, $k = 6000$ N/m, $m = 0.005$ kg, $r = 0.2$ m, and $\omega = 16$ rad/s, what is the force transmitted to the foundation?

The equation of motion for the system is

$$15\ddot{x} + 450\dot{x} + 6000x = (0.005)(16)^2(0.2)\sin 16t$$

Consequently,

$$\omega_n = 20 \text{ rad/s}, \qquad \zeta = 0.75$$

and we find $\beta = \omega/\omega_n = 16/20 = 0.8$. Referring to Eq. (7-39), we have

$$
\begin{aligned}
F_t &= \frac{m\omega^2 r\sqrt{1 + (2\zeta\beta)^2}}{\sqrt{(1 - \beta^2)^2 + (2\zeta\beta)^2}} \\
&= \frac{(0.005)(16)^2(0.2)\sqrt{1 + (2 \times 0.75 \times 0.8)^2}}{\sqrt{(1 - 0.8^2)^2 + (2 \times 0.75 \times 0.8)^2}} = 0.319 \text{ N}
\end{aligned}
$$

The force transmitted to the foundation is sinusoidal and has the amplitude of 0.319 N.

Automobile suspension system. Figure 7-37(a) shows an automobile system. As the car moves along the road, the vertical displacements at the tires act as the motion excitation to the automobile suspension system. Figure 7-37(b) is a schematic diagram of an automobile suspension system. The motion of this system consists of a translational motion of the center of mass and a rotational motion about the center of mass. A complete analysis of this suspension system would be quite involved. A very simplified version appears

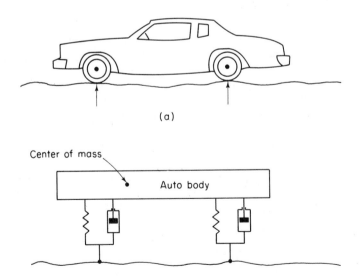

Fig. 7-37. (a) Automobile system; (b) schematic diagram of an automobile suspension system.

in Fig. 7-38. In the following pages we shall analyze this simple model when the motion input is sinusoidal and thus derive the transmissibility for the motion excitation system.

Transmissibility for motion excitation. In the system shown in Fig. 7-39 the motion of the body is in the vertical direction only. The motion y at point P is the input to the system; the vertical motion x of the body is the output. Displacement x is measured from the equilibrium position in the absence of input y. We assume that motion y is sinusoidal, or $y = Y \sin \omega t$. (Figure 7-39 can be considered as a simplified representation of a vehicle of mass m moving on a rough road with a spring and damper suspension between the mass and the wheel.)

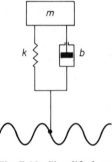

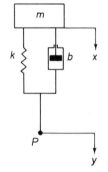

Fig. 7-38. Simplified version of the automobile suspension system.

Fig. 7-39. Mechanical system.

The equation of motion for the system is

$$m\ddot{x} + b(\dot{x} - \dot{y}) + k(x - y) = 0$$

or

$$m\ddot{x} + b\dot{x} + kx = b\dot{y} + ky$$

Then the Laplace transform of this last equation, assuming zero initial conditions, gives

$$(ms^2 + bs + k)X(s) = (bs + k)Y(s)$$

Hence

$$\frac{X(s)}{Y(s)} = \frac{bs + k}{ms^2 + bs + k}$$

The sinusoidal transfer function is

$$\frac{X(j\omega)}{Y(j\omega)} = \frac{bj\omega + k}{-m\omega^2 + bj\omega + k}$$

The steady-state output $x(t)$ has the amplitude $|X(j\omega)|$. The input amplitude is $|Y(j\omega)|$. The transmissibility in this case is the displacement amplitude ratio and is given by

$$\text{Transmissibility} = \text{TR} = \frac{\text{amplitude of the output displacement}}{\text{amplitude of the input displacement}}$$

Thus

$$\text{TR} = \frac{|X(j\omega)|}{|Y(j\omega)|} = \frac{\sqrt{b^2\omega^2 + k^2}}{\sqrt{(k - m\omega^2)^2 + b^2\omega^2}}$$

Noting that $k/m = \omega_n^2$ and $b/m = 2\zeta\omega_n$, the transmissibility is given, in terms of damping ratio ζ and undamped natural frequency ω_n, by

$$\text{TR} = \frac{\sqrt{1 + (2\zeta\beta)^2}}{\sqrt{(1 - \beta^2)^2 + (2\zeta\beta)^2}} \tag{7-40}$$

where $\beta = \omega/\omega_n$. This equation is identical to Eq. (7-38).

Example 7-11. A rigid body is mounted on an isolator in order to reduce the vibratory effect. Assume that the mass of the rigid body is 500 kg, the damping coefficient of the isolator is very small ($\zeta = 0.01$), and the effective spring constant of the isolator is 12 500 N/m. Find the percent of motion transmitted to the body if the frequency of the motion excitation of the base of the isolator is 20 rad/s.

The undamped natural frequency ω_n of the system is

$$\omega_n = \sqrt{\frac{12\ 500}{500}} = 5 \text{ rad/s}$$

And so

$$\beta = \frac{\omega}{\omega_n} = \frac{20}{5} = 4$$

By substituting $\zeta = 0.01$ and $\beta = 4$ into Eq. (7-40), we have

$$\text{TR} = \frac{\sqrt{1 + (2\zeta\beta)^2}}{\sqrt{(1 - \beta^2)^2 + (2\zeta\beta)^2}} = \frac{\sqrt{1 + (2 \times 0.01 \times 4)^2}}{\sqrt{(1 - 4^2)^2 + (2 \times 0.01 \times 4)^2}} = 0.0669$$

The effect of the isolator is that it reduces the vibratory motion of the rigid body to 6.69% of the vibratory motion of the base of the isolator.

Seismograph. Figure 7-40 is a schematic diagram of a *seismograph*, a device used to measure ground displacement during earthquakes. The displacement of mass m relative to inertial space is denoted by x and the displacement of the case relative to inertial space by y. Displacement x is measured from the equilibrium position when $y = 0$. Displacement y is the input to the system. This displacement, in the case of earthquakes, is approximately sinusoidal, or $y(t) = Y \sin \omega t$. In the seismograph we measure the relative displacement between x and y.

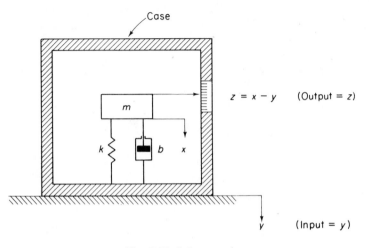

Fig. 7-40. Seismograph.

The equation of motion for the seismograph is

$$m\ddot{x} + b(\dot{x} - \dot{y}) + k(x - y) = 0 \tag{7-41}$$

Let us define the displacement of the mass m relative to the case as z—that is,

$$z = x - y$$

In terms of the relative displacement z, Eq. (7-41) becomes

$$m(\ddot{y} + \ddot{z}) + b\dot{z} + kz = 0$$

or

$$m\ddot{z} + b\dot{z} + kz = -m\ddot{y}$$

By taking the Laplace transform of this last equation and assuming zero initial conditions, we find that

$$(ms^2 + bs + k)Z(s) = -ms^2 Y(s)$$

Note that the input to the system is displacement y and that the output is the relative displacement z. The transfer function between $Z(s)$ and $Y(s)$ is

$$\frac{Z(s)}{Y(s)} = \frac{-ms^2}{ms^2 + bs + k}$$

The sinusoidal transfer function is

$$\frac{Z(j\omega)}{Y(j\omega)} = \frac{m\omega^2}{-m\omega^2 + bj\omega + k}$$

Substitution of $k/m = \omega_n^2$ and $b/m = 2\zeta\omega_n$ into this last equation gives

$$\frac{Z(j\omega)}{Y(j\omega)} = \frac{\omega^2}{-\omega^2 + 2\zeta\omega_n j\omega + \omega_n^2} = \frac{\beta^2}{1 - \beta^2 + j2\zeta\beta} \tag{7-42}$$

where $\beta = \omega/\omega_n$.

In the seismograph we want to determine accurately the input displacement $y(t)$ by measuring the relative displacement $z(t)$. By examining Eq. (7-42), we can do so easily if $\beta \gg 1$. If $\beta \gg 1$, Eq. (7-42) reduces to

$$\frac{Z(j\omega)}{Y(j\omega)} \doteq -\frac{\beta^2}{\beta^2} = -1$$

The seismograph measures and records the displacement of its case y accurately if $\beta \gg 1$ or $\omega \gg \omega_n$. In fact, for $\omega \gg \omega_n$, mass m tends to remain fixed in space, and the motion of the case can be seen as a relative motion between the mass and the case.

To accomplish the condition $\omega \gg \omega_n$, we choose the undamped natural frequency ω_n as low as possible (choose a relatively large mass and a spring as soft as the elastic and static deflection limits allow). Then the seismograph will measure and record the displacements of all frequencies well above the undamped natural frequency ω_n, which is very low.

Accelerometer. A schematic diagram of a translational *accelerometer* is given in Fig. 7-41. The system configuration is basically the same as the seismograph, but their essential difference lies in the choice of the undamped natural frequency ω_n. Let us denote the displacement of mass m relative to inertial space by x and that of the case relative to inertial space by y. Displacement x is measured from the equilibrium position when $y = 0$. The input to the translational accelerometer is the acceleration $\ddot{y}$. The output is the displacement of the mass m relative to the case, or $z = x - y$. (We measure and record the relative displacement z, not the absolute displacement x.)

The equation of motion for the system is

$$m\ddot{x} + b(\dot{x} - \dot{y}) + k(x - y) = 0$$

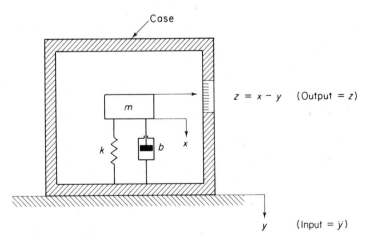

Fig. 7-41. Translational accelerometer.

In terms of the relative displacement z, this last equation becomes

$$m(\ddot{y} + \ddot{z}) + b\dot{z} + kz = 0$$

or

$$m\ddot{z} + b\dot{z} + kz = -m\ddot{y}$$

The Laplace transform of this last equation, assuming zero initial conditions, gives

$$(ms^2 + bs + k)Z(s) = -ms^2 Y(s)$$

The transfer function between output $Z(s)$ and input $s^2 Y(s)$ [the input is the accerelation $\ddot{y}$ and thus its Laplace transform is $s^2 Y(s)$] is

$$\frac{Z(s)}{s^2 Y(s)} = \frac{-m}{ms^2 + bs + k} = \frac{-1}{s^2 + 2\zeta\omega_n s + \omega_n^2} \tag{7-43}$$

From Eq. (7-43) we see that if the undamped natural frequency ω_n is sufficiently large compared with the frequencies of the input, then

$$\frac{Z(s)}{s^2 Y(s)} \doteq -\frac{1}{\omega_n^2}$$

Thus the displacement z is nearly proportional to $\ddot{y}$.

Dynamic vibration absorber. In many situations, rotating machines (such as turbines and compressors) cause vibrations and transmit large vibratory forces to the foundation. Vibratory forces may be caused by an unbalanced mass of the rotor. If the exciting frequency ω is equal to or nearly equal to the undamped natural frequency of the rotating machine on its mounts, then resonance occurs and large forces are transmitted to the foundation.

If the machine operates at nearly constant speed, a device called a *dynamic vibration absorber* can be attached to it to eliminate the large transmitted force. This device is usually in the form of a spring-mass system tuned to have a natural frequency equal to the operating frequency ω. When it is added to a one degree-of-freedom vibratory system, the entire system becomes a two degrees-of-freedom system with two natural frequencies. To reduce or nearly eliminate the transmitted force, one of the natural frequencies is set above the operating frequency, whereas the other is set below it.

Our discussion here focuses on a simple dynamic vibration absorber that will reduce the vertical force transmitted to the foundation. Note that only vertical motions are discussed.

Reducing vibrations by use of dynamic vibration absorber. A rotating machine, due to an unbalanced mass of the rotor, transmits a large vibratory force to the foundation. Let us assume that the machine is supported by a spring and a damper as shown in Fig. 7-42(a). The unbalanced rotor is represented by mass M, which includes the unbalanced mass, and is rotating at frequency ω. The excitation force is $p(t) = P \sin \omega t$, where $P = m\omega^2 r$. (Here m is the unbalanced mass and r is the distance of the unbalanced mass from the center of rotation.) Because of this force excitation, a sinusoidal force of amplitude

$$\frac{m\omega^2 r \sqrt{k^2 + b^2\omega^2}}{\sqrt{(k - M\omega^2)^2 + b^2\omega^2}}$$

is transmitted to the foundation. [To obtain this amplitude, substitute $\beta = \omega/\omega_n = \omega/\sqrt{k/M}$ and $\zeta = b/(2\sqrt{kM})$ into Eq. (7-39).]

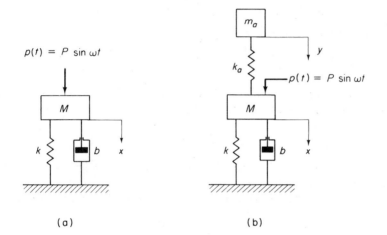

Fig. 7-42. (a) Machine supported by a spring and a damper; (b) machine with a dynamic vibration absorber.

If the viscous damping coefficient b is small and the natural frequency $\sqrt{k/M}$ of the system is equal to the excitation frequency, then resonance occurs and the machine is subjected to excessive vibration and the transmitted force becomes extremely large.

In the following analysis, we assume that b is very small and that the natural frequency $\sqrt{k/M}$ is very close to the excitation frequency ω. In such a case, in order to reduce the transmitted force, a dynamic vibration absorber consisting of a mass (m_a) and a spring (k_a) may be added to the machine as shown in Fig. 7-42(b).

The equations of motion for the system of Fig. 7-42(b) are

$$M\ddot{x} + b\dot{x} + kx + k_a(x - y) = p(t) = P \sin \omega t$$

$$m_a\ddot{y} + k_a(y - x) = 0$$

where x and y, the displacements of mass M and mass m_a, respectively, are measured from the respective equilibrium positions in the absence of excitation force $p(t)$. By taking the Laplace transforms of the last two equations, assuming zero initial conditions, we see that

$$(Ms^2 + bs + k + k_a)X(s) - k_aY(s) = P(s)$$

$$(m_as^2 + k_a)Y(s) - k_aX(s) = 0$$

Elimination of $Y(s)$ from these two equations yields

$$\left(Ms^2 + bs + k + k_a - \frac{k_a^2}{m_as^2 + k_a}\right)X(s) = P(s)$$

It follows that

$$\frac{X(s)}{P(s)} = \frac{m_as^2 + k_a}{(Ms^2 + bs + k + k_a)(m_as^2 + k_a) - k_a^2}$$

The sinusoidal transfer function is

$$\frac{X(j\omega)}{P(j\omega)} = \frac{-m_a\omega^2 + k_a}{(-M\omega^2 + bj\omega + k + k_a)(-m_a\omega^2 + k_a) - k_a^2}$$

If the viscous damping coefficient b is negligibly small, we may substitute $b = 0$ into this last equation. Then

$$\frac{X(j\omega)}{P(j\omega)} \doteq \frac{-m_a\omega^2 + k_a}{(-M\omega^2 + k + k_a)(-m_a\omega^2 + k_a) - k_a^2}$$

[Note that in the actual system free vibrations eventually die out due to damping (even though it may be negligibly small) and the forced vibration at steady state can be represented by this last equation.] The transmitted force $f(t)$ to the foundation is

$$f(t) = kx + b\dot{x} \doteq kx$$

Also, the amplitude of this transmitted force is $k\,|X(j\omega)|$, where $|X(j\omega)|$ is given [noting that $|P(j\omega)| = P = m\omega^2r$] as

$$|X(j\omega)| = \left|\frac{(k_a - m_a\omega^2)}{(k + k_a - M\omega^2)(k_a - m_a\omega^2) - k_a^2}\right||P(j\omega)|$$

$$= \left|\frac{m\omega^2r(k_a - m_a\omega^2)}{(k + k_a - M\omega^2)(k_a - m_a\omega^2) - k_a^2}\right| \tag{7-44}$$

In examining Eq. (7-44), notice that if m_a and k_a are chosen so that

$$k_a - m_a\omega^2 = 0$$

or $k_a/m_a = \omega^2$, then $|X(j\omega)| = 0$ and the force transmitted to the foundation is zero. So if the natural frequency $\sqrt{k_a/m_a}$ of the dynamic vibration absorber is made equal to the excitation frequency ω it is possible to eliminate the force transmitted to the foundation. In general, such a dynamic vibration absorber is used only when the natural frequency $\sqrt{k/M}$ of the original system is very close to the excitation frequency ω. (Without this device, the system would be in near resonance.)

Physically, the effect of the dynamic vibration absorber is to produce the spring force $k_a y$ such that it cancels the excitation force $p(t)$. To see this point, note first that if the viscous damping coefficient b is neglibibly small, then

$$\frac{Y(j\omega)}{P(j\omega)} = \frac{X(j\omega)}{P(j\omega)} \frac{Y(j\omega)}{X(j\omega)}$$

$$= \frac{k_a}{(-M\omega^2 + k + k_a)(-m_a\omega^2 + k_a) - k_a^2}$$

If m_a and k_a are chosen so that $k_a = m_a\omega^2$, we find

$$\frac{Y(j\omega)}{P(j\omega)} = \frac{k_a}{-k_a^2} = -\frac{1}{k_a}$$

Consequently,

$$y(t) = \left| -\frac{1}{k_a} \right| P \sin \left(\omega t + \Big/ -\frac{1}{k_a} \right)$$

$$= \frac{P}{k_a} \sin (\omega t - 180°)$$

$$= -\frac{P}{k_a} \sin \omega t$$

This means that the spring k_a gives a force $k_a y = -P \sin \omega t$ to mass M. The magnitude of this force is equal to the excitation force, and the phase angle lags $180°$ from the excitation force (mass m_a is vibrating in phase opposition to the excitation force) with the result that the spring force $k_a y$ and the excitation force $p(t)$ cancel each other and mass M stays stationary.

We have shown that the addition of a dynamic vibration absorber will reduce the vibration of the machine and the transmitted force to the foundation to zero when the machine is excited by the unbalanced mass (or other causes) at frequency ω. It can also be shown that there will now be two frequencies at which mass M will be in resonance. These two frequencies are the natural frequencies of this two-degrees-of-freedom system and can be found from the equation

$$(k + k_a - M\omega_i^2)(k_a - m_a\omega_i^2) - k_a^2 = 0 \qquad (i = 1, 2)$$

The two values of frequency, ω_1 and ω_2, that satisfy this last equation are the natural frequencies of the system with a dynamic vibration absorber. Figure 7-43(a) and (b) shows the amplitude $|X(j\omega)|$ versus frequency ω curves for the systems shown in Fig. 7-42(a) and (b), respectively, when b is negligibly small.

Note that addition of viscous damping in parallel with the absorber spring k_a relieves excessive vibrations at these two natural frequencies. That is, very large amplitudes at the two resonance frequencies may be reduced to smaller values.

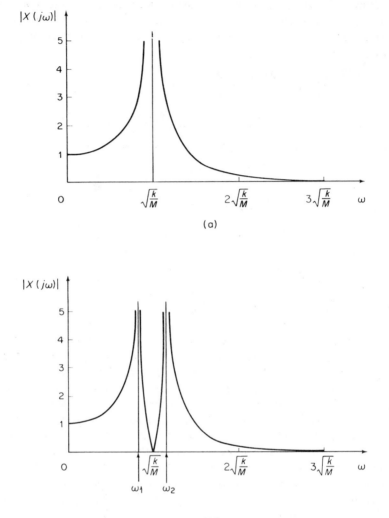

Fig. 7-43. (a) Amplitude versus frequency curve for the system shown in Fig. 7-42(a); (b) amplitude versus frequency curve for the system shown in Fig. 7-42(b).

7-7   ANALOG COMPUTERS

Practical dynamic systems may be described by differential equations of higher order. Solving such equations is generally a time-consuming process. The analog computer is quite useful as a differential equation solver in that it saves time, particularly when solutions for various different values of parameters are needed.

Another feature of the analog computer is that it can be used as a simulator. In fact, the simulation of physical systems is an important application of this type of computer. It can be used to simulate a component, components, or even an entire system. As a real-time simulator, the computer is set up to simulate a component or components of a system that has not yet been built. By using suitable transducers, the analog computer is connected to the rest of the real system that has been built. The composite system can then be tested as a unit and system performance evaluated, a procedure that is used extensively in industry. In particular, the analog computer has proven quite useful in determining the effects of parameter variations on system performance.

Here we present the operating principle of electronic analog computers and the techniques of setting up computer diagrams for solving differential equations and simulating physical systems. Only linear, time-invariant, differential equation systems are considered.

Operational amplifiers. Operational amplifiers, as used in analog computers, are capable of performing the mathematical functions of integration, summation, and sign inversion. An operational amplifier is a dc amplifier and has a very high gain, approximately 10^6 to 10^8. The current drawn at the input of an operational amplifier is negligibly small. The output voltage from the operational amplifier is usually limited to ± 100 V. (In small scale analog computers it is limited to ± 10 V.) Figure 7-44 is a schematic diagram of an operational amplifier. The output voltage e_o and input voltage e are related by

$$e_o = -Ke$$

where $K = 10^6$ to 10^8.

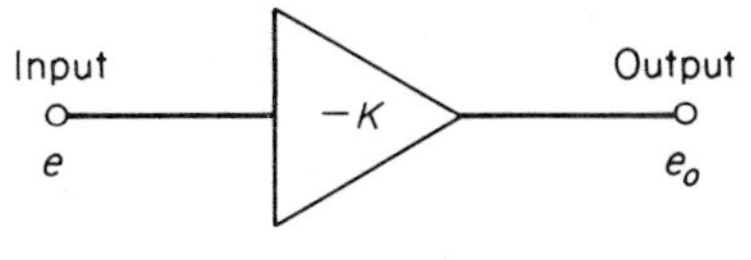

Input · · · Output
e $-K$ e_o

Fig. 7-44. Schematic diagram of an operational amplifier.

Sign inverters. Figure 7-45(a) is a schematic diagram of a sign inverter. An operational amplifier is in series with an input resistance R_i and is in parallel with a feedback resistance R_o. Because the internal impedance of the amplifier is very high, essentially there is negligible current i or

$$i \doteq 0$$

Therefore by Kirchhoff's current law

$$i_i = i_o$$

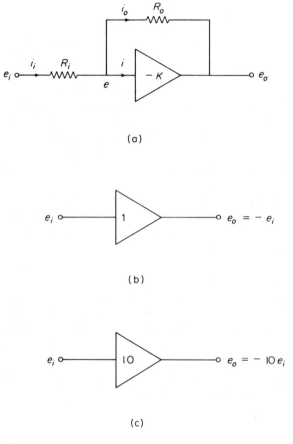

(a)

(b)

(c)

Fig. 7-45. (a) Schematic diagram of a sign inverter; (b) symbol for the sign inverter when $R_o/R_i = 1$; (c) symbol for the sign inverter when $R_o/R_i = 10$.

where

$$i_i = \frac{e_i - e}{R_i}, \qquad i_o = \frac{e - e_o}{R_o}$$

Consequently, we have

$$\frac{e_i - e}{R_i} = \frac{e - e_o}{R_o} \tag{7-45}$$

Noting that $e_o = -Ke$, Eq. (7-45) can be written

$$e_i = -\left(\frac{1}{K} + \frac{R_i}{KR_o} + \frac{R_i}{R_o}\right)e_o$$

Since K is a very large number (10^6 to 10^8) and R_i/R_o is on the order of 0.1 to 10, neglecting terms involving K on the right-hand side of this last equation, we find that

$$e_o = -\frac{R_o}{R_i}e_i \tag{7-46}$$

Note that Eq. (7-46) can also be obtained simply by substituting $e = 0$ into Eq. (7-45).

From Eq. (7-46) we see that the output voltage e_o is equal to the input voltage e_i times a constant $(-R_o/R_i)$, which is negative. The values of resistances R_i and R_o are normally 0.1 MΩ, 0.25 MΩ, and 1 MΩ. Thus several values of R_o/R_i are possible. In many analog computers, however, the values of R_o/R_i are fixed at 1, 4, or 10.

Figure 7-45(b) and (c) shows commonly used symbols for the sign inverter with $R_o/R_i = 1$ and $R_o/R_i = 10$, respectively.

Summers. The schematic diagram of a summer that adds n inputs is given in Fig. 7-46(a). In the summer, resistors are used as the input and feedback impedances of an operational amplifier. This circuit is the same as that of the sign inverter. In fact, any summer can be used as the sign inverter.

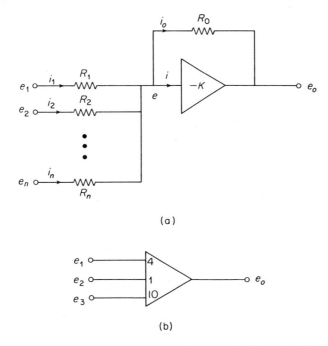

(a)

(b)

Fig. 7-46. (a) Schematic diagram of a summer; (b) symbol for the summer.

Noting that the current i is negligibly small ($i \doteq 0$), the equation for this circuit can be obtained as

$$i_1 + i_2 + \cdots + i_n = i_o$$

or

$$\frac{e_1 - e}{R_1} + \frac{e_2 - e}{R_2} + \cdots + \frac{e_n - e}{R_n} = \frac{e - e_o}{R_o}$$

By substituting $e = 0$ into this last equation, we have

$$e_o = -\left(\frac{R_o}{R_1}e_1 + \frac{R_o}{R_2}e_2 + \cdots + \frac{R_o}{R_n}e_n\right) \qquad (7\text{-}47)$$

Thus the circuit shown in Fig. 7-46(a) performs the weighted summation of n inputs. (Note that the summer changes the algebraic sign). If, for example, $R_o = 1$ MΩ, $R_1 = 0.25$ MΩ, $R_2 = 1$ MΩ, and $R_3 = 0.1$ MΩ, then Eq. (7-47) becomes

$$e_o = -(4e_1 + e_2 + 10e_3)$$

The commonly used symbol for the summer appears in Fig. 7-46(b).

Integrators. Figure 7-47(a) is a schematic diagram of the integrator. In this circuit a resistor is used as the input impedance and a capacitor as the feedback impedance.

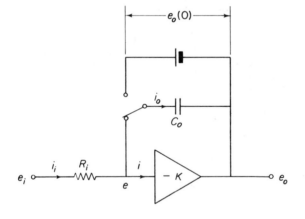

(a)

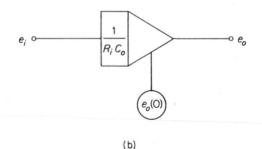

(b)

Fig. 7-47. (a) Schematic diagram of an integrator; (b) symbol for the integrator.

The equation for the circuit can be derived in the following way. Noting that current i is negligibly small, or $i \doteq 0$, we have

$$i_i = i_o$$

where

$$i_i = \frac{e_i - e}{R_i}, \qquad i_o = C_o \frac{d(e - e_o)}{dt}$$

Therefore

$$\frac{e_i - e}{R_i} = C_o \frac{d(e - e_o)}{dt}$$

Substituting $e = 0$ into this last equation yields

$$\frac{e_i}{R_i} = -C_o \frac{de_o}{dt}$$

or

$$\frac{de_o}{dt} = -\frac{1}{R_i C_o} e_i$$

Integrating both sides of this last equation from 0 to t, we find

$$e_o(t) - e_o(0) = -\frac{1}{R_i C_o} \int_0^t e_i(t)\, dt$$

or

$$e_o(t) = -\frac{1}{R_i C_o} \int_0^t e_i(t)\, dt + e_o(0) \qquad (7\text{-}48)$$

Equation (7-48) shows that the circuit of Fig. 7-47(a) is an integrator. The integrator must be initially biased by a dc voltage in order to give the necessary initial condition $e_o(0)$.

Figure 7-47(b) shows the commonly used symbol for the integrator. The initial condition $e_o(0)$ is indicated in the circle. Note that in many analog computers standard resistors of 0.1 MΩ, 0.25 MΩ, 1 MΩ, and a standard capacitor of 1 μF are used. In such a case, the values of $1/R_i C_o$ are equal only to 1, 4, or 10.

As in the summing operation, if two input signals are applied to the integrator as shown in Fig. 7-48(a), then the output $e_o(t)$ is given by the sum of two integrals and the initial condition $e_o(0)$ or

$$e_o(t) = -\frac{1}{R_1 C_o} \int_0^t e_1(t)\, dt - \frac{1}{R_2 C_o} \int_0^t e_2(t)\, dt + e_o(0) \qquad (7\text{-}49)$$

Equation (7-49) can be found by noting that

$$i_1 + i_2 = i_o$$

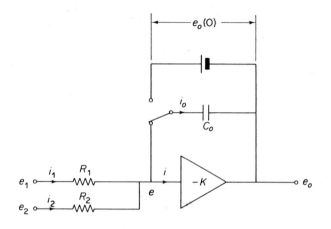

(a)

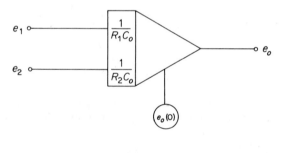

(b)

Fig. 7-48. (a) Schematic diagram of an integrator with two inputs;
(b) simplified diagram.

where

$$i_1 = \frac{e_1 - e}{R_1} \doteq \frac{e_1}{R_1}$$

$$i_2 = \frac{e_2 - e}{R_2} \doteq \frac{e_2}{R_2}$$

$$i_o = C_o \frac{d(e - e_o)}{dt} \doteq -C_o \frac{de_o}{dt}$$

A simplified diagram of Fig. 7-48(a) is shown in Fig. 7-48(b).

Multiplication by a fraction. The multiplication of e_i by a constant α, where $0 < \alpha < 1$, can be accomplished by using a potentiometer [see Fig. 7-49(a)]. The output e_o is

$$e_o = \frac{R_o}{R_i} e_i$$

Figure 7-49(b) illustrates the commonly used symbol for a potentiometer.

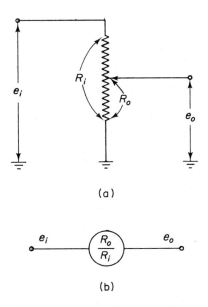

(a)

(b)

Fig. 7-49. (a) Potentiometer; (b) symbol for the potentiometer.

Solving differential equations. In solving differential equations by means of an analog computer, we always integrate derivatives rather than differentiate them. The reason for this fact is that spurious noise is always present in the analog computer system. Differentiation accentuates the noise effect, whereas integration smoothes it out and so analog computers use integration rather than differentiation as a basic operator.

Note that in order to solve linear, time-invariant differential equations like

$$\overset{(n)}{x} + a_1 \overset{(n-1)}{x} + \cdots + a_{n-1}\dot{x} + a_n x = p(t)$$

the components listed below are needed.

1. The integrator
2. The summer
3. The sign inverter
4. The potentiometer
5. The dc voltage source

Procedure for solving differential equations. As an illustration, consider the differential equation

$$\ddot{x} + 10\dot{x} + 16x = 0, \quad x(0) = 0, \quad \dot{x}(0) = 80 \qquad (7\text{-}50)$$

The first step in setting up a computer diagram is to assume that the highest-order derivative is available. Then, solve the differential equation for this highest-order derivative. In the present differential equation

$$\ddot{x} = -10\dot{x} - 16x$$

Noting that the variable $-\dot{x}$ can be obtained by integrating $\ddot{x}$ and also that x can be obtained by integrating $-\dot{x}$, we produce signals $-10\dot{x}$ and $-16x$ by use of two integrators and a sign inverter. The next step is to add these two signals, $-10\dot{x}$ and $-16x$, and equate the result to $\ddot{x}$, the highest-order derivative term that was originally assumed available. Finally, the initial conditions are set at the outputs of the integrators. (The initial conditions are indicated in the circles in the computer diagram.) Figure 7-50 shows a computer diagram for the system defined by Eq. (7-50).

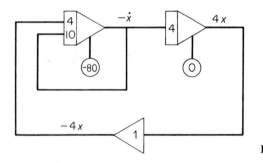

Fig. 7-50. Analog computer diagram.

It is important to remember that a sign change is associated with each operational amplifier. So if the number of operational amplifiers (integrators, summers, and sign inverters) in a loop is even, the output voltages will increase until they saturate. To eliminate any chance of unstable operation, the number of operational amplifiers in any loop should be odd in quantity. (In the computer diagram of Fig. 7-50 the inner loop has one operational amplifier and the outer loop three.) This requirement serves as a convenient check on whether any mistake was made in setting up the computer diagram.

Generating an exponential function. Let us demonstrate how to produce an exponential function $x(t) = 20e^{-0.5t}$. In order to set up the analog computer diagram, we first obtain the corresponding differential equation, the lowest-order differential equation whose solution is $x(t) = 20e^{-0.5t}$.

By differentiating $x(t)$ with respect to t, we have

$$\dot{x} = -10e^{-0.5t}$$

Hence the required differential equation is

$$\dot{x} + 0.5x = 0, \qquad x(0) = 20$$

Solving this differential equation for $\dot{x}$ yields

$$\dot{x} = -0.5x$$

Assuming that $-\dot{x}$ is available, x can be obtained by integrating $-\dot{x}$ once. Figure 7-51 shows an analog computer diagram for generating the given exponential function.

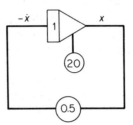

Fig. 7-51. Analog computer diagram.

Generating a sinusoidal function. Here we wish to produce a sinusoidal signal, such as 10 sin 3t. In order to set up the analog computer diagram, let us obtain the lowest-order differential equation whose solution is 10 sin 3t. Let

$$x(t) = 10 \sin 3t$$

Then

$$\ddot{x}(t) = -90 \sin 3t$$

So the required differential equation is

$$\ddot{x} + 9x = 0, \qquad x(0) = 0, \qquad \dot{x}(0) = 30$$

Solving this differential equation for the highest-order derivative, we have

$$\ddot{x} = -9x$$

Assuming that $\ddot{x}$ is available, x can be obtained by integrating $\ddot{x}$ twice. A computer diagram for this system is given in Fig. 7-52.

Note that the outputs of the first and second integrators oscillate between 30 and -30 V and between 10 and -10 V, respectively. The output

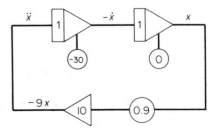

Fig. 7-52. Analog computer diagram.

of the sign inverter oscillates between 90 and -90 V. In order to have good accuracy, it is desirable to make the maximum swing of output voltage at any amplifier equal 80 to 90 V. This step can be accomplished by using proper magnitude scale factors. (The magnitude scale factors will be discussed in detail later in this section.)

Time scale factor. In solving a differential equation system, the actual solution time may be so fast that the recorder is unable to follow the response accurately. For physical phenomena that take place this rapidly, the speed at which they are simulated by the computer must be slowed down. On the other hand, in some cases, the actual solution may take an excessively long time. To avoid such inconveniences, the technique known as the *time-scaling technique* is needed.

Time scaling relates the independent variable of the physical system to the independent variable of the analog computer. The computer may be set up to run faster or slower than "real time" if convenient or necessary. Note that if actual parts of the system are to be used with the computer—that is, if the computer is used to simulate a component or components of the real system and is directly connected to real system hardware—time scaling must be one to one. In other words, the computer must work in real time.

Let the following equation relate the real time t in seconds to the computer time (or machine time) τ in seconds:

$$\tau = \lambda t$$

where λ is the time scale factor. If λ is chosen as 0.1, then 10 seconds of real time is equivalent to 1 computer second. This means that if the actual response takes 10 seconds of real time to complete, then the response is completed in 1 second on the computer. Conversely, if λ is chosen as 10, then 1 second of real time is equivalent to 10 seconds of computer time. Therefore, in order to speed up (slow down) the response on the computer, λ must be chosen to be less than (greater than) unity.

To illustrate, consider the differential equation

$$\frac{d^2x}{dt^2} + 10\frac{dx}{dt} + 100x = 0, \qquad x(0) = 10, \qquad \dot{x}(0) = 15 \qquad (7\text{-}51)$$

For this system, since the undamped natural frequency ω_n is equal to 10 rad/s and the damping ratio ζ is equal to 0.5, the settling time t_s is

$$t_s = \frac{4}{\zeta\omega_n} = \frac{4}{0.5 \times 10} = 0.8 \text{ s}$$

The response settles within the 2% of the final value in 0.8 second.

Suppose that we wish to slow down the response so that the settling time is 8 seconds. We can do so by choosing the time scale factor λ to be 10. Let us convert the independent variable t to τ. Since $\tau = \lambda t$, we obtain

$$\frac{dx}{dt} = \frac{dx}{d\tau}\frac{d\tau}{dt} = \lambda\frac{dx}{d\tau}$$

$$\frac{d^2x}{dt^2} = \lambda^2\frac{d^2x}{d\tau^2}$$

Equation (7-51) then becomes

$$\lambda^2\frac{d^2x}{d\tau^2} + 10\lambda\frac{dx}{d\tau} + 100\,x = 0$$

or

$$\frac{d^2x}{d\tau^2} + \frac{10}{\lambda}\frac{dx}{d\tau} + \frac{100}{\lambda^2}x = 0$$

To slow down the solution by a factor of 10, we substitute $\lambda = 10$ into this last equation. The computer equation is then

$$\frac{d^2x}{d\tau^2} + \frac{dx}{d\tau} + x = 0$$

The initial conditions are transformed to

$$x(0) = 10, \qquad \frac{dx}{d\tau}\bigg|_{\tau=0} = \frac{1}{\lambda}\frac{dx}{dt}\bigg|_{t=0} = \frac{1}{10}(15) = 1.5$$

Example 7-12. In the electrical system of Fig. 7-53 the capacitor is not charged initially. The switch S is closed at $t = 0$. Let us simulate this electrical system on an analog computer.

The equation for the circuit for $t > 0$ is

$$L\frac{di}{dt} + Ri + \frac{1}{C}\int i\,dt = E$$

By substituting $dq/dt = i$ into this last equation, we have

$$L\frac{d^2q}{dt^2} + R\frac{dq}{dt} + \frac{1}{C}q = E, \qquad q(0) = 0, \qquad \frac{dq}{dt}\bigg|_{t=0} = 0$$

Let us define $q/C = x$. Then this last equation becomes

$$LC\frac{d^2x}{dt^2} + RC\frac{dx}{dt} + x = E, \qquad x(0) = 0, \qquad \frac{dx}{dt}\bigg|_{t=0} = 0$$

$L = 5$ mH
$R = 5\,\Omega$
$C = 200\,\mu\text{F}$
$E = 24$ V

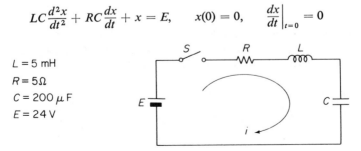

Fig. 7-53. Electrical system.

Substituting the given numerical values,

$$(5)(10^{-3})(200)(10^{-6})\frac{d^2x}{dt^2} + (5)(200)(10^{-6})\frac{dx}{dt} + x = 24$$

or

$$\frac{d^2x}{dt^2} + 10^3\frac{dx}{dt} + 10^6x = (24)(10^6)$$

The response of this system is very fast. (The undamped natural frequency ω_n is equal to 10^3 rad/s and the damping ratio ζ equal to 0.5.) Let us slow down the response on the analog computer by a factor of 10^3, or choose the time scale factor λ to be 10^3. Then by changing the independent variable from t to τ, where $\tau = \lambda t$, we see that

$$\frac{d^2x}{d\tau^2} + \frac{10^3}{\lambda}\frac{dx}{d\tau} + \frac{10^6}{\lambda^2}x = \frac{(24)(10^6)}{\lambda^2}$$

which, with the substitution of $\lambda = 10^3$, becomes

$$\frac{d^2x}{d\tau^2} + \frac{dx}{d\tau} + x = 24$$

A computer diagram for simulating this system is shown in Fig. 7-54. Note that $q = Cx = 2 \times 10^{-4}x$ and $i = dq/dt = \lambda(dq/d\tau) = \lambda C(dx/d\tau) = 0.2\,(dx/d\tau)$.

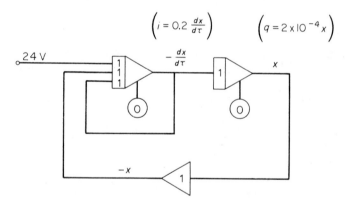

Fig. 7-54. Analog computer diagram for the system shown in Fig. 7-53.

Example 7-13. (Simulation of a dynamic vibration absorber.) For the mechanical system with a dynamic vibration absorber shown in Fig. 7-55, assume that all initial conditions are zero and that the input force $P \sin \omega t$ is given at $t = 0$. Let us simulate this system on an analog computer.

The equations for this system are

$$m\frac{d^2x_1}{dt^2} + b\frac{dx_1}{dt} + kx_1 + k_a(x_1 - x_2) = P \sin \omega t$$

$$m_a\frac{d^2x_2}{dt^2} + k_a(x_2 - x_1) = 0$$

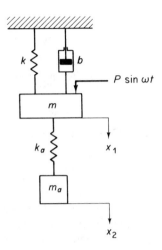

$m = 1\,\text{kg}$

$b = 5\,\text{N-s/m}$

$k = 500\,\text{N/m}$

$m_a = 0.5\,\text{kg}$

$k_a = 200\,\text{N/m}$

$P = 20\,\text{N}$

$\omega = 20\,\text{rad/s}$

$P \sin \omega t$

Fig. 7-55. Mechanical system with a dynamic vibration absorber.

By substituting the given numerical values into these equations, we have

$$\frac{d^2 x_1}{dt^2} + 5\frac{dx_1}{dt} + 500x_1 + 200(x_1 - x_2) = 20 \sin 20t$$

$$0.5\frac{d^2 x_2}{dt^2} + 200(x_2 - x_1) = 0$$

If we choose the time scale factor λ to be 10, the system equations become

$$\frac{d^2 x_1}{d\tau^2} + 0.5\frac{dx_1}{d\tau} + 5x_1 + 2(x_1 - x_2) = 0.2 \sin 2\tau$$

$$\frac{d^2 x_2}{d\tau^2} + 4(x_2 - x_1) = 0$$

Now define new variables y_1 and y_2 such that

$$y_1 = 100\,x_1, \qquad y_2 = 100\,x_2$$

The system equations then become

$$\frac{d^2 y_1}{d\tau^2} + 0.5\frac{dy_1}{d\tau} + 5y_1 + 2(y_1 - y_2) = 20 \sin 2\tau$$

$$\frac{d^2 y_2}{d\tau^2} + 4(y_2 - y_1) = 0$$

To simplify notations, let us write

$$\frac{dy_1}{d\tau} = \dot{y}_1, \qquad \frac{d^2 y_1}{d\tau^2} = \ddot{y}_1, \qquad \frac{dy_2}{d\tau} = \dot{y}_2, \qquad \frac{d^2 y_2}{d\tau^2} = \ddot{y}_2$$

Then, the system equations can be written

$$\ddot{y}_1 + 0.5\dot{y}_1 + 5y_1 + 2(y_1 - y_2) = 20 \sin 2\tau \qquad (7\text{-}52)$$

$$\ddot{y}_2 + 4(y_2 - y_1) = 0 \qquad (7\text{-}53)$$

The initial conditions are

$$y_1(0) = 0, \qquad \dot{y}_1(0) = 0, \qquad y_2(0) = 0, \qquad \dot{y}_2(0) = 0$$

We shall use variables y_1 and y_2 to simulate this mechanical system.

In simulating this system, we first produce the forcing function $20 \sin 2\tau$. Note that $p = 20 \sin 2\tau$ is the solution of

$$\ddot{p} + 4p = 0, \qquad p(0) = 0, \qquad \dot{p}(0) = 40$$

In the next step, we solve Eqs. (7-52) and (7-53) for the highest-order derivative terms, respectively.

$$\ddot{y}_1 = -0.5\dot{y}_1 - 7y_1 + 2y_2 + 20 \sin 2\tau$$

$$\ddot{y}_2 = 4y_1 - 4y_2$$

Then we assume that $\ddot{y}_1$ and $\ddot{y}_2$ are available and integrate these signals so as to obtain $-\dot{y}_1$ and $-\dot{y}_2$ and also integrate $-\dot{y}_1$ and $-\dot{y}_2$ in order to obtain y_1 and y_2. By feeding these lower-order terms into the appropriate components required by the system equations, we generate the highest-order derivative terms $\ddot{y}_1$ and $\ddot{y}_2$ and close the loop. Figure 7-56 is an analog computer diagram simulating the mechanical system with a dynamic vibration absorber considered.

Note that the output signals from the computer are given in voltages. So it is necessary to interpret the output voltages from amplifiers in terms of the original physical quantities. (A systematic method for correlating output voltages to physical quantities is available. For details, see magnitude scale factors presented below.) In the present example problem, if the instantaneous voltages of signals y_1 and y_2 are 5 V and 10 V, respectively, then displacements x_1 and x_2 are interpreted as 0.05 m and 0.1 m, respectively.

In the analog computer solution (Fig. 7-56), the amplitude of signal y_1 decreases to zero as steady state is reached. At steady state, signal y_1 is zero and signal $2y_2$ is $-20 \sin 2\tau$. Consequently, signal $2y_2$ cancels forcing function $p(\tau) = 20 \sin 2\tau$ at steady state and so Eq. (7-52) becomes

$$\ddot{y}_1 + 0.5\dot{y}_1 + 7y_1 = 0$$

Thus the system has no net forcing function at steady state and $y_1(\infty)$ becomes zero.

Magnitude scale factors. The magnitude of the output voltage of the amplifier depends a great deal on the accuracy of the circuit. When setting up the circuit, the voltage should be made as large as possible within the limits of the machine. The limits are usually ±100 V. (In certain small-scale analog computers the limits are ±10 V.)

Following the choice of a suitable time scale factor, attention must also be paid to scaling magnitudes. Since the computer manipulates voltages, it is necessary to transform the real system equations, which may involve, for instance, pressure, temperature, displacement, and similiar quantities, to analogous voltage equations. That is, for a pressure system, we must decide

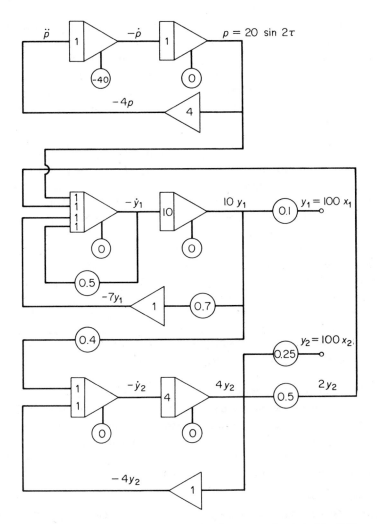

Fig. 7-56. Analog computer diagram for the system shown in Fig. 7-55.

on how many newtons per square meter in a real system should be represented by one volt in the computer. The magnitude scale factors relate the output voltages of amplifiers to the corresponding physical quantities.

In choosing magnitude scale factors, the following requirements should be kept in mind. The output voltage of any amplifier should not exceed the amplifier limits (usually ± 100 V) if saturation is to be avoided. Voltage saturation will cause errors in the solution. And in order to eliminate noise effect, the maximum voltage at any amplifier should not be too small. To ensure proper accuracy, preferably the maximum swing of output voltage at any amplifier should be about ± 80 to ± 90 V. In this respect, choosing proper magnitude scale factors is most important. Note that in most analog computers the order of magnitude of the error is normally a few percent. (Notice that some of the errors are roughly constant. For such errors large outputs result in smaller percentage errors.) This magnitude of error may be adequate, since simplifying assumptions in engineering analysis often involve even greater inaccuracy.

Procedure for determining magnitude scale factors. Because changing the time scale will alter the time derivatives of the dependent variables, the time scale factor should be decided on before the magnitude scale factors are determined. If the speed of the solutions of the real system is within a reasonable range of the computer, time scaling will not be necessary. The problem can be run in real time.

The first step in determining magnitude scale factors is to estimate the maximum magnitudes of variables that will occur in the physical system. In practice, variable ranges are usually unknown before the solution is obtained. Therefore a certain amount of trial and error is necessary in establishing proper magnitude scale factors. Such estimates may come from a knowledge of the actual system, from rough calculations, from a pure guess, or from a combination of them. (In many cases, estimates are made by neglecting the damping in the system.) Except for familiar problems, considerable guesswork may be necessary.

Once the initial estimates of the maximum magnitudes of variables are found, magnitude scale factors can be determined. The values so determined may be tested to see if they are appropriate by running the problem with assumed magnitude scale factors and noting if voltages are too large or too small. If assumed magnitude scale factors are not appropriate, they must be varied until satisfactory results are derived.

Example 7-14. Consider the system shown in Fig. 7-57. Assume that displacement x is measured from the equilibrium position. The initial conditions are given as

$$x(0) = 0 \text{ m}, \qquad \left.\frac{dx}{dt}\right|_{t=0} = 3 \text{ m/s}$$

Let us simulate this mechanical system on an analog computer.

The system equation is

$$m\frac{d^2x}{dt^2} + b\frac{dx}{dt} + kx = 0$$

By substituting the given numerical values for m, b, and k, we have

$$0.2\frac{d^2x}{dt^2} + 1.2\frac{dx}{dt} + 180\,x = 0$$

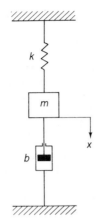

$m = 0.2$ kg

$b = 1.2$ N-s/m

$k = 180$ N/m

Fig. 7-57. Mechanical system.

or

$$\frac{d^2x}{dt^2} + 6\frac{dx}{dt} + 900\,x = 0$$

Since the settling time for the present system is

$$t_s = \frac{4}{\zeta\omega_n} = \frac{4}{0.1 \times 30} = 1.33 \text{ s}$$

let us slow down the response and make the new settling time to be 13.3 seconds. We can do so easily by choosing the time scale factor λ to be 10.

By changing the independent variable from t to τ, where $\tau = \lambda t = 10t$, we have

$$\frac{d^2x}{d\tau^2} + 0.6\frac{dx}{d\tau} + 9x = 0$$

where

$$x(0) = 0 \text{ m}, \qquad \frac{dx}{d\tau}\bigg|_{\tau=0} = 0.3 \text{ m/s}$$

For the sake of simplicity, let us write

$$\frac{dx}{d\tau} = \dot{x}, \qquad \frac{d^2x}{d\tau^2} = \ddot{x}$$

Then a properly time-scaled system equation becomes

$$\ddot{x} + 0.6\,\dot{x} + 9x = 0, \qquad x(0) = 0, \qquad \dot{x}(0) = 0.3 \qquad (7\text{-}54)$$

We shall use Eq. (7-54) as the starting equation for the determination of magnitude scale factors. Solving Eq. (7-54) for the highest-order derivative gives

$$\ddot{x} = -0.6\,\dot{x} - 9x \qquad (7\text{-}55)$$

Let us determine magnitude scale factors so that the maximum swing in each amplifier is 90 V. Define k_1 and k_2 as magnitude scale factors such that k_1 relates voltage to velocity (m/s) and k_2 relates voltage to displacement (m). Therefore k_1 has the dimension of volts per meter per second (V-s/m), and k_2 has the dimension of volts per meter (V/m). Let us rewrite Eq. (7-55) as

$$\ddot{x} = -\frac{0.6}{k_1}(k_1\dot{x}) - \frac{9}{k_2}(k_2 x)$$

In order to minimize the effect of noise and keep accuracy high, a minimum number of amplifiers should be used. (In any analog computer the number of amplifiers is limited. In solving a complex problem requiring many integrators and summers, a minimum number of amplifiers should be used for each equation in order to save components.) The present system is of second order, and so we need two integrators. Since the number of amplifiers in any loop must be odd, we need at least one sign inverter. Thus the minimum number of amplifiers needed is three. Figure 7-58 shows a computer diagram for the problem requiring a minimum number of amplifiers.

Referring to Fig. 7-58, the output voltage of the first integrator is $-k_1\dot{x}$. The output voltage of the second integrator is $k_2 x$. The output voltage of the sign inverter is $-k_2 x$. These output voltages must be limited to ± 90 V. (The absolute maximum voltage is ± 100 V, so ± 90 V is a conservative choice.) A second-order system such as that represented by Eq. (7-54) has its most violent motion when the damping term is removed. To obtain conservative or overly large estimates of maximum values, we can use the solution for

$$\ddot{x} + 9x = 0, \qquad x(0) = 0, \qquad \dot{x}(0) = 0.3$$

The solution of this simplified equation is

$$x(\tau) = 0.1 \sin 3\tau$$

Consequently,

$$\dot{x}(\tau) = 0.3 \cos 3\tau$$

From the present solution we can obtain conservative (overly large) estimates for the system defined by Eq. (7-54) such that

$$\text{Maximum value of } |x| = |x|_{\max} = 0.1$$

$$\text{Maximum value of } |\dot{x}| = |\dot{x}|_{\max} = 0.3$$

We shall choose k_1 and k_2 so that $|k_1\dot{x}| = |k_2 x| = 90$ V for the maximum values

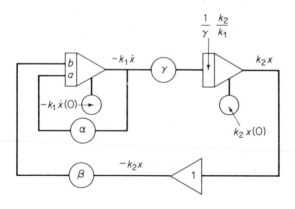

Fig. 7-58. Analog computer diagram for the determination of magnitude scale factors.

of $\dot{x}$ and x, respectively. Therefore the magnitude scale factors are determined as

$$k_1 = \frac{90}{|\dot{x}|_{max}} = \frac{90}{0.3} = 300 \text{ V-s/m}$$

$$k_2 = \frac{90}{|x|_{max}} = \frac{90}{0.1} = 900 \text{ V/m}$$

And so

$$\frac{k_2}{k_1} = 3$$

Note that from Fig. 7-58 we have

$$k_1\ddot{x} = a\alpha(-k_1\dot{x}) + b\beta(-k_2x)$$

or

$$\ddot{x} + a\alpha\dot{x} + b\beta\frac{k_2}{k_1}x = 0$$

Noting that $k_2/k_1 = 3$, this last equation becomes

$$\ddot{x} + a\alpha\dot{x} + 3b\beta x = 0 \tag{7-56}$$

By comparing Eqs. (7-54) and (7-56), we see that

$$a\alpha = 0.6, \qquad b\beta = 3$$

Let us choose $a = 1$, $\alpha = 0.6$, $b = 10$, and $\beta = 0.3$.

Next, we must determine the value of γ. The constant of the second integrator $(1/\gamma)(k_2/k_1)$ is generally set equal to 1 or 10. Since $k_2/k_1 = 3$, we choose

$$\frac{1}{\gamma}\frac{k_2}{k_1} = 10$$

This results in $\gamma = 0.3$. Then all the unknown constants in Fig. 7-58 are determined. A properly scaled computer diagram is shown in Fig. 7-59(a). The initial conditions are

$$-k_1\dot{x}(0) = -300 \times 0.3 = -90 \text{ V}$$

$$k_2x(0) = 900 \times 0 = 0 \text{ V}$$

The output of the second integrator is $900x(\tau)$.

It is important to note that the potentiometer denoted by γ in Fig. 7-58 can be eliminated, as Fig. 7-59(b) shows. (This situation is equivalent to setting the constant of the integrator equal to 3.)

Note that because we employed time scaling at the beginning of the solution of the problem, the time involved in the computer solution is the computer time τ (where $\tau = 10 t$ and t is the real time). Note also that in this analog computer solution the displacement and velocity are obtained in volts. The voltage values can be changed back into the corresponding physical quantities by referring to the definition of magnitude scale factors k_1 and k_2.

In this example, the displacement measured in volts can be transformed back into meters by use of the following relationship

$$1 \text{ V} \quad \text{corresponds to} \quad \frac{1}{900} \text{ m}$$

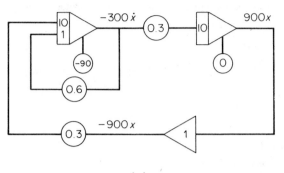

(a)

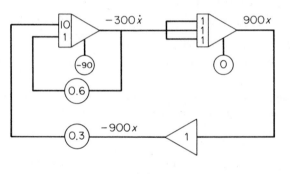

(b)

Fig. 7-59. Analog computer diagrams.

and the velocity measured in volts can be transformed back into meters per second by use of the following conversion

$$1 \text{ V} \quad \text{corresponds to} \quad \frac{1}{300} \frac{\text{meter}}{\text{computer second}}$$

Since in the present case,

$$10 \text{ computer second} = 1 \text{ real second}$$

we have

$$1 \text{ V} \quad \text{corresponds to} \quad \frac{1}{300} \frac{\text{meter}}{0.1 \text{ real second}} = \frac{1}{30} \frac{\text{meter}}{\text{real second}}$$

Summary of procedures for solving differential equations. Steps that we normally follow in solving differential equations can be summarized as follows.

1. Determine the time scale factor and magnitude scale factors as needed.

2. Solve the differential equation for the highest-order derivative. The right-hand side of the derived equation defines the inputs to the first integrator.

3. Integrate the highest-order derivative to obtain the lower-order derivatives and the variable itself.

4. Feed these lower-order derivative terms into appropriate components as called for by the system equations, thus generating the highest-order derivative and closing the loop.

5. Provide initial conditions as required.

Concluding comments. Analog computer simulation plays an important role in the analysis and design of complicated systems. The effects of changes in system parameters on the performance of the system can easily be determined. The advantage of analog simulation is that any convenient time scale can be used. Nevertheless, it does have the limitation that the analog computer can only solve specific equations with numerical initial conditions and that it gives the solution as a curve. The computer cannot give a general solution with arbitrary constants. Thus the computer solution differs in character from the analytical solution by exact methods.

In general, the precise mathematical representation of a complicated component is difficult. It is probable that some of the important features of the component may be overlooked in the simulation, a factor that may cause serious errors in the solution. In order to avoid such errors, the simulator may include actual system components. If such components are included, no important characteristics of the actual components are lost. The solution, however, must be obtained in real time.

Large-scale analog computers can be used for simulating nonlinear systems or solving nonlinear differential equations. Nonlinear operations such as the multiplication of two variables can be carried out easily with the electronic analog computer. Standard electronic circuits are available for simulating such commonly encountered nonlinearities as saturation, dead zone, and backlash. The input-output characteristic curves of these nonlinearities are shown in Fig. 7-60(a), (b), and (c). The use of the analog computer for nonlinear systems is not essentially different from that for linear systems described in this section.

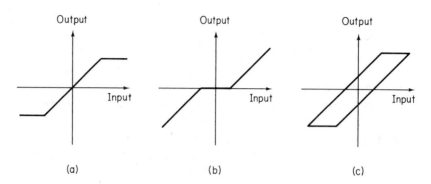

Fig. 7-60. Input-output characteristic curves; (a) saturation non-linearity; (b) dead zone nonlinearity; (c) backlash nonlinearity.

REFERENCES

7-1   Ashley, R. J., *Introduction to Analog Computation*, New York: John Wiley & Sons, Inc., 1963.

7-2   Cannon, R. H., *Dynamics of Physical Systems*, New York: McGraw-Hill Book Company, Inc., 1967.

7-3   Doebelin, E. O., *Dynamic Analysis and Feedback Control*, New York: McGraw-Hill Book Company, Inc., 1962.

7-4   Johnson, C. L., *Analog Computer Techniques*, 2nd ed., New York: McGraw-Hill Book Company, Inc., 1963.

7-5   Korn, G. A., and T. M. Korn, *Electronic Analog Computers*, 2nd ed., New York: McGraw-Hill Book Company, Inc., 1956.

7-6   Ogata, K., *Modern Control Engineering*, Englewood Cliffs, N.J.: Prentice-Hall, Inc., 1970.

7-7   Reswick, J. B., and C. K. Taft, *Introduction to Dynamic Systems*, Englewood Cliffs, N.J.: Prentice-Hall, Inc., 1967.

7-8   Shearer, J. L., A. T. Murphy, and H. H. Richardson, *Introduction to System Dynamics*, Reading, Mass.: Addison-Wesley Publishing Company, Inc., 1967.

EXAMPLE PROBLEMS AND SOLUTIONS

Problem A-7-1. Referring to the circuit shown in Fig. 7-61, assume that there is an initial charge q_0 on the capacitor just before switch S is closed at $t = 0$. Find current $i(t)$.

Solution. The equation for the circuit is

$$Ri + \frac{1}{C} \int i \, dt = E$$

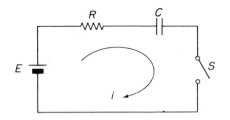

Fig. 7-61. Electrical circuit.

Taking the Laplace transform of this last equation,

$$RI(s) + \frac{1}{C}\frac{I(s) + \int i(t)\,dt\big|_{t=0}}{s} = \frac{E}{s}$$

Since

$$\int i(t)\,dt\bigg|_{t=0} = q(0) = q_0$$

we obtain

$$RI(s) + \frac{1}{C}\frac{I(s) + q_0}{s} = \frac{E}{s} \tag{7-57}$$

or

$$RCsI(s) + I(s) + q_0 = CE$$

Solving for $I(s)$, we have

$$I(s) = \frac{CE - q_0}{RCs + 1} = \frac{(E/R) - (q_0/RC)}{s + (1/RC)}$$

The inverse Laplace transform of this last equation gives the current $i(t)$.

$$i(t) = \left(\frac{E}{R} - \frac{q_0}{RC}\right)e^{-t/RC}$$

Note that the Laplace transform of the integral equation automatically includes the initial condition as seen above. If the equation for the circuit is written in the form

$$Ri + \frac{1}{C}\left(\int_0^t i\,dt + q_0\right) = E$$

then the Laplace transform of this equation gives

$$RI(s) + \frac{1}{C}\left[\frac{I(s)}{s} + \frac{q_0}{s}\right] = \frac{E}{s}$$

which is, of course, the same as Eq. (7-57) obtained above.

PROBLEM A-7-2. Suppose that a disk is rotated at a constant speed of 100 rad/s and we wish to stop it in 2 minutes. Assuming that the moment of inertia J of the disk is 6 kg-m^2, determine the necessary torque T to stop the rotation.

Solution. The necessary torque T must act so as to reduce the speed. Thus the equation of motion is

$$J\dot{\omega} = -T, \qquad \omega(0) = 100$$

By integrating this last equation with respect to t, we obtain

$$J\omega(t) = -Tt + k$$

The integration constant k is determined by use of the initial condition. Thus

$$J\omega(0) = k = 100J$$

And so

$$J\omega(t) = -Tt + 100J$$

At $t = 2$ min $= 120$ s, we want to stop, or $\omega(120)$ must equal zero. Therefore

$$J\omega(120) = 0 = -T \times 120 + 100 \times 6$$

Solving for T, we get

$$T = \frac{600}{120} = 5 \text{ N-m}$$

PROBLEM A-7-3. A mass m is attached to a string that is under tension T in the system of Fig. 7-62(a). We assume that tension T is to remain constant for small displacement x. Neglecting gravity, find the natural frequency of the vertical motion of mass m. What is displacement $x(t)$ when the mass is given initial conditions $x(0) = x_0$ and $\dot{x}(0) = 0$?

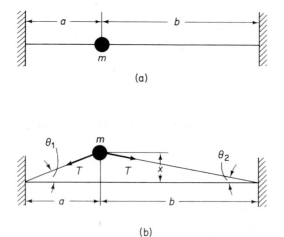

(a)

(b)

Fig. 7-62. (a) Mechanical vibratory system; (b) diagram showing tension forces.

Solution. Referring to Fig. 7-62(b), the vertical component of the force due to tension is

$$-T \sin \theta_1 - T \sin \theta_2$$

For small x, angles θ_1 and θ_2 are small and

$$\sin \theta_1 = \tan \theta_1 = \frac{x}{a}$$

$$\sin \theta_2 = \tan \theta_2 = \frac{x}{b}$$

The equation of motion for the system is

$$m\ddot{x} = -T \sin \theta_1 - T \sin \theta_2 = -T\frac{x}{a} - T\frac{x}{b}$$

or

$$m\ddot{x} + T\left(\frac{1}{a} + \frac{1}{b}\right)x = 0$$

Hence the natural frequency of the vertical motion of the mass is

$$\omega_n = \sqrt{\frac{T}{m}\left(\frac{1}{a} + \frac{1}{b}\right)}$$

The solution $x(t)$ is given by

$$x(t) = x_0 \cos \omega_n t$$

PROBLEM A-7-4. Derive the equation of motion for the pendulum system shown in Fig. 7-63 and obtain the natural frequency. Assume that when the pendulum is vertical, there is no spring force; also, assume that θ is small. Finally, determine $\theta(t)$ when the pendulum is given initial conditions $\theta(0) = \theta_0$ and $\dot{\theta}(0) = 0$.

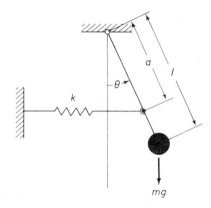

Fig. 7-63. Pendulum system. mg

Solution. Two torques are acting on this system—one due to the gravitational force and the other due to the spring force. By applying Newton's second law, the equation of motion for the system becomes

$$J\ddot{\theta} = -mgl \sin \theta - (ka \sin \theta)(a \cos \theta)$$

where $J = ml^2$. By rewriting this last equation,

$$ml^2\ddot{\theta} + mgl \sin \theta + ka^2 \sin \theta \cos \theta = 0$$

For small θ, we have $\sin \theta = \theta$ and $\cos \theta = 1$. So the equation of motion can be simplified to

$$ml^2\ddot{\theta} + (mgl + ka^2)\theta = 0$$

or

$$\ddot{\theta} + \left(\frac{g}{l} + \frac{ka^2}{ml^2}\right)\theta = 0$$

The natural frequency ω_n of the system is

$$\omega_n = \sqrt{\frac{g}{l} + \frac{ka^2}{ml^2}}$$

The solution $\theta(t)$ is given by

$$\theta(t) = \theta_0 \cos \omega_n t$$

PROBLEM A-7-5. Two masses m_1 and m_2 are connected by a spring of constant k in Fig. 7-64. Assuming no friction, derive the equation of motion. In addition, find $x_1(t)$ and $x_2(t)$ when external force F is constant. Assume that $x_1(0) = 0$, $\dot{x}_1(0) = 0$ and $x_2(0) = 0$, $\dot{x}_2(0) = 0$.

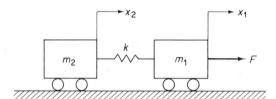

Fig. 7-64. Mechanical system.

Solution. The equations of motion are

$$m_1 \ddot{x}_1 = -k(x_1 - x_2) + F$$
$$m_2 \ddot{x}_2 = -k(x_2 - x_1)$$

Rewriting,

$$m_1 \ddot{x}_1 + k(x_1 - x_2) = F \tag{7-58}$$
$$m_2 \ddot{x}_2 + k(x_2 - x_1) = 0 \tag{7-59}$$

From Eqs. (7-58) and (7-59) we obtain

$$m_1 m_2 (\ddot{x}_1 - \ddot{x}_2) + (km_2 + km_1)(x_1 - x_2) = m_2 F$$

If we define $x_1 - x_2 = x$, then this last equation simplifies to

$$m_1 m_2 \ddot{x} + k(m_1 + m_2)x = m_2 F$$

It follows that

$$\ddot{x} + \frac{k(m_1 + m_2)}{m_1 m_2} x = \frac{F}{m_1} \tag{7-60}$$

Let us define

$$\omega_n^2 = \frac{k(m_1 + m_2)}{m_1 m_2}$$

Then Eq. (7-60) becomes

$$\ddot{x} + \omega_n^2 x = \frac{F}{m_1}$$

By taking the Laplace transform of this last equation, substituting the initial conditions $x(0) = 0$ and $\dot{x}(0) = 0$, and noting that F is a constant, we have

$$(s^2 + \omega_n^2)X(s) = \frac{F}{m_1 s}$$

or

$$X(s) = \frac{F}{m_1 s} \frac{1}{s^2 + \omega_n^2} = \frac{F}{m_1 \omega_n^2}\left(\frac{1}{s} - \frac{s}{s^2 + \omega_n^2}\right)$$

The inverse Laplace transform of $X(s)$ gives

$$x(t) = \frac{F}{m_1 \omega_n^2}(1 - \cos \omega_n t) \tag{7-61}$$

Now we shall determine $x_2(t)$. From Eqs. (7-59) and (7-61) we find

$$m_2 \ddot{x}_2 = kx = \frac{kF}{m_1 \omega_n^2}(1 - \cos \omega_n t)$$

Since $F = $ constant, we can easily integrate the right-hand side of this last equation. Noting that $x_2(0) = 0$ and $\dot{x}_2(0) = 0$, we get

$$m_2 \dot{x}_2 = \frac{kF}{m_1 \omega_n^2}\left(t - \frac{1}{\omega_n} \sin \omega_n t\right)$$

and

$$m_2 x_2 = \frac{kF}{m_1 \omega_n^2}\left(\frac{t^2}{2} + \frac{1}{\omega_n^2} \cos \omega_n t\right) - \frac{kF}{m_1 \omega_n^4}$$

Thus x_2 is obtained as

$$x_2(t) = \frac{F}{m_1 + m_2} \frac{t^2}{2} - \frac{Fm_1 m_2}{k(m_1 + m_2)^2}\left[1 - \cos \sqrt{\frac{k(m_1 + m_2)}{m_1 m_2}}\,t\right] \tag{7-62}$$

Then the solution $x_1(t)$ is obtained from

$$x_1(t) = x(t) + x_2(t)$$

By substituting Eqs. (7-61) and (7-62) into this last equation and simplifying,

$$x_1(t) = \frac{F}{m_1 + m_2} \frac{t^2}{2} + \frac{Fm_2^2}{k(m_1 + m_2)^2}\left[1 - \cos \sqrt{\frac{k(m_1 + m_2)}{m_1 m_2}}\,t\right]$$

PROBLEM A-7-6. In Fig. 7-65 the system is at rest initially. At $t = 0$ a unit-step displacement input is applied to point A. Assuming that the system remains linear throughout the response period and is underdamped, find the response $x(t)$ as well as the values of $x(0+)$, $\dot{x}(0+)$, and $x(\infty)$.

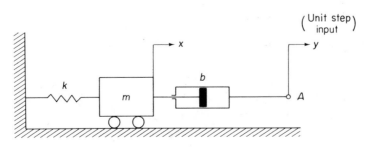

Fig. 7-65. Mechanical system.

Solution. The equation of motion for the system is

$$m\ddot{x} + b(\dot{x} - \dot{y}) + kx = 0$$

or

$$m\ddot{x} + b\dot{x} + kx = b\dot{y}$$

Noting that $x(0-) = 0$, $\dot{x}(0-) = 0$, and $y(0-) = 0$, the $\mathscr{L}_-$ transform of this last equation gives

$$(ms^2 + bs + k)X(s) = bs\,Y(s)$$

Thus

$$\frac{X(s)}{Y(s)} = \frac{bs}{ms^2 + bs + k}$$

Since the input y is a unit step, $Y(s) = 1/s$. Consequently,

$$X(s) = \frac{bs}{ms^2 + bs + k}\frac{1}{s} = \frac{b}{ms^2 + bs + k}$$

$$= \frac{2\zeta\omega_n}{s^2 + 2\zeta\omega_n s + \omega_n^2}$$

$$= \frac{2\zeta}{\sqrt{1 - \zeta^2}}\frac{\omega_n\sqrt{1 - \zeta^2}}{(s + \zeta\omega_n)^2 + (\omega_n\sqrt{1 - \zeta^2})^2}$$

where we used relationships $k/m = \omega_n^2$ and $b/m = 2\zeta\omega_n$. The inverse Laplace transform of $X(s)$ is

$$x(t) = \frac{2\zeta}{\sqrt{1 - \zeta^2}}e^{-\zeta\omega_n t}\sin\omega_n\sqrt{1 - \zeta^2}\,t$$

Although the values of $x(0+)$, $\dot{x}(0+)$, and $x(\infty)$ can be found easily from this last equation, the initial and final value theorems will be used here instead in order to demonstrate their application.

By applying the initial value theorem to this problem, the initial values $x(0+)$ and $\dot{x}(0+)$ can be found as

$$x(0+) = \lim_{s\to\infty} sX(s) = \lim_{s\to\infty}\frac{s2\zeta\omega_n}{s^2 + 2\zeta\omega_n s + \omega_n^2} = 0$$

$$\dot{x}(0+) = \lim_{s\to\infty} s^2 X(s) = \lim_{s\to\infty}\frac{s^2 2\zeta\omega_n}{s^2 + 2\zeta\omega_n s + \omega_n^2} = 2\zeta\omega_n$$

The final value $x(\infty)$ is obtained by use of the final value theorem

$$x(\infty) = \lim_{s\to 0} sX(s) = \lim_{s\to 0}\frac{s2\zeta\omega_n}{s^2 + 2\zeta\omega_n s + \omega_n^2} = 0$$

Thus mass m returns to the original position as time elapses.

Problem A-7-7. Consider the rotating system shown in Fig. 7-66 and assume that torque T applied to the rotor is of short duration but of a large amplitude so that it can be considered an impulse input. Let us assume that initially the angular velocity is zero, or $\omega(0-) = 0$. Given the numerical values

$$J = 10 \text{ kg-m}^2$$

$$b = 2 \text{ N-s/m}$$

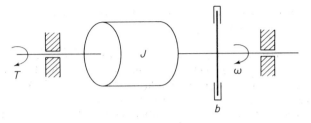

Fig. 7-66. Mechanical rotating system.

find the response $\omega(t)$. Assume that the amplitude of torque T is 300 N-m/s and that the duration of torque is 0.1 s—that is, the magnitude of torque T is $300 \times 0.1 = 30$ N-m.

Solution. The equation of motion for the system is

$$J\dot{\omega} + b\omega = T, \qquad \omega(0-) = 0$$

Let us consider the impulsive torque of magnitude 1 N-m to be $\delta(t)$. Then by substituting the given numerical values into this last equation, we obtain

$$10\dot{\omega} + 2\omega = 30\,\delta(t)$$

Taking the $\mathcal{L}_-$ transform of this last equation,

$$10[s\Omega(s) - \omega(0-)] + 2\Omega(s) = 30$$

or

$$\Omega(s) = \frac{30}{10s + 2} = \frac{3}{s + 0.2}$$

The inverse Laplace transform of $\Omega(s)$ is

$$\omega(t) = 3e^{-0.2t} \qquad\qquad (7\text{-}63)$$

Note that $\omega(0+) = 3$ rad/s. The angular velocity of the rotor is thus changed instantaneously from $\omega(0-) = 0$ to $\omega(0+) = 3$ rad/s.

If the system is only subjected to the initial condition $\omega(0) = 3$ rad/s and no external torque, or $T = 0$, then the equation of motion becomes

$$10\dot{\omega} + 2\omega = 0, \qquad \omega(0) = 3$$

By taking the Laplace transform of this last equation,

$$10[s\Omega(s) - \omega(0)] + 2\Omega(s) = 0$$

or

$$\Omega(s) = \frac{10\omega(0)}{10s + 2} = \frac{30}{10s + 2} = \frac{3}{s + 0.2}$$

The inverse Laplace transform of $\Omega(s)$ gives

$$\omega(t) = 3e^{-0.2t}$$

which is identical to Eq. (7-63).

From the preceding analysis we see that the response of a first-order system to an impulse input is identical to the motion from the initial condition at $t = 0+$.

That is, the effect of the impulse input to the first-order system is to generate the nonzero initial condition at $t = 0+$.

PROBLEM A-7-8. Referring to Fig. 7-67, a man drops a steel ball of mass m onto the center of mass M from a height d and catches it on the first bounce. Assuming that the system is initially at rest, what is the motion of mass M after it is hit by the steel ball? Assume that the impact is perfectly elastic. In addition, assume that the numerical values of M, m, b, k and d are given as $M = 1$ kg, $m = 0.1$ kg, $b = 4$ N-s/m, $k = 125$ N/m, and $d = 1$ m. The displacement x of mass M is measured from the equilibrium position before the ball hits it. The initial conditions are $x(0-) = 0$ and $\dot{x}(0-) = 0$.

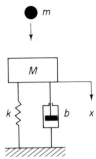

Fig. 7-67. Mechanical system subjected to an impulse input.

Solution. The equation of motion for the system is

$$M\ddot{x} + b\dot{x} + kx = p(t) \qquad (7\text{-}64)$$

Since the impact is assumed to be perfectly elastic, the momentum of the ball is changed from mv downward (at $t = 0-$) to mv upward (at $t = 0+$), or a total change of $2mv$, where v is the velocity of the ball just before it hits mass M. The impact by the steel ball is an impulse input to mass M. The magnitude or area of the impulse input $p(t)$ is

$$\int_{0-}^{0+} p(t)\, dt = 2mv$$

Thus

$$p(t) = 2mv\, \delta(t)$$

And so Eq. (7-64) can be written

$$M\ddot{x} + b\dot{x} + kx = 2mv\, \delta(t)$$

Noting that $x(0-) = 0$ and $\dot{x}(0-) = 0$, the $\mathcal{L}_-$ transform of this last equation gives

$$(Ms^2 + bs + k)X(s) = 2mv$$

Solving for $X(s)$, we obtain

$$X(s) = \frac{2mv}{Ms^2 + bs + k}$$

Since the velocity v of the ball after falling a distance d is

$$v = \sqrt{2gd}$$

it follows that

$$X(s) = \frac{2m\sqrt{2gd}}{Ms^2 + bs + k}$$

By substituting the given numerical values into this last equation, we have

$$X(s) = \frac{2 \times 0.1\sqrt{2 \times 9.81 \times 1}}{s^2 + 4s + 125} = \frac{0.886}{(s + 2)^2 + 11^2}$$

$$= 0.0805 \frac{11}{(s + 2)^2 + 11^2}$$

The inverse Laplace transform of $X(s)$ gives

$$x(t) = 0.0805 e^{-2t} \sin 11t \ \text{m}$$

Thus the response of mass M is a damped sinusoidal motion.

PROBLEM A-7-9. Find the transfer function $E_o(s)/E_i(s)$ of the electrical circuit shown in Fig. 7-68.

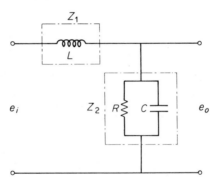

Fig. 7-68. Electrical circuit.

Solution. Complex impedances Z_1 and Z_2 are

$$Z_1 = Ls$$

$$\frac{1}{Z_2} = \frac{1}{R} + Cs = \frac{1 + RCs}{R}$$

Therefore

$$\frac{E_o(s)}{E_i(s)} = \frac{Z_2}{Z_1 + Z_2} = \frac{\dfrac{R}{1 + RCs}}{Ls + \dfrac{R}{1 + RCs}} = \frac{R}{LRCs^2 + Ls + R}$$

PROBLEM A-7-10. An external force $p(t)$ is applied to mass m_2 in the mechanical system shown in Fig. 7-69. Derive the transfer function $X(s)/P(s)$. In the diagram displacements x and y are measured from their respective equilibrium positions.

Solution. The equations of motion for the system are

$$m_2\ddot{x} + b_1(\dot{x} - \dot{y}) + k_1(x - y) + k_2x = p$$

$$m_1\ddot{y} + b_1(\dot{y} - \dot{x}) + k_1(y - x) = 0$$

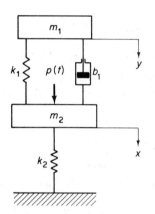

Fig. 7-69. Mechanical system.

Taking the Laplace transforms of these two equations and assuming zero initial conditions, we have

$$(m_2s^2 + b_1s + k_1 + k_2)X(s) = (b_1s + k_1)Y(s) + P(s)$$

$$(m_1s^2 + b_1s + k_1)Y(s) = (b_1s + k_1)X(s)$$

By eliminating $Y(s)$ from the last two equations, the transfer function $X(s)/P(s)$ is found to be

$$\frac{X(s)}{P(s)} = \frac{m_1s^2 + b_1s + k_1}{(m_2s^2 + k_2)m_1s^2 + (m_1s^2 + m_2s^2 + k_2)(b_1s + k_1)}$$

PROBLEM A-7-11. Derive the transfer functions $X_o(s)/X_i(s)$ and $E_o(s)/E_i(s)$ of the systems shown in Fig. 7-70(a) and (b), respectively, and show that the systems are analogous.

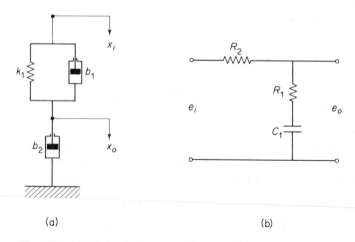

(a)   (b)

Fig. 7-70. (a) Mechanical system; (b) analogous electrical system.

Solution. The equation of motion for the mechanical system in Fig. 7-70(a) is

$$b_1(\dot{x}_i - \dot{x}_o) + k_1(x_i - x_o) = b_2 \dot{x}_o$$

So by taking the Laplace transform of this equation, assuming zero initial conditions, we have

$$(b_1 s + k_1)X_i(s) = (b_1 s + k_1 + b_2 s)X_o(s)$$

The transfer function $X_o(s)/X_i(s)$ is

$$\frac{X_o(s)}{X_i(s)} = \frac{b_1 s + k_1}{(b_1 + b_2)s + k_1} = \frac{\dfrac{b_1}{k_1}s + 1}{\dfrac{b_1 + b_2}{k_1}s + 1}$$

Next, consider the electrical system shown in Fig. 7-70(b). Using complex impedances, the transfer function $E_o(s)/E_i(s)$ is obtained as

$$\frac{E_o(s)}{E_i(s)} = \frac{R_1 + \dfrac{1}{C_1 s}}{R_2 + R_1 + \dfrac{1}{C_1 s}} = \frac{R_1 C_1 s + 1}{(R_1 + R_2)C_1 s + 1}$$

Comparing the transfer functions obtained, we see that they have the same form and hence are analogous systems.

PROBLEM A-7-12. Find the transfer function $X_o(s)/X_i(s)$ of the mechanical system shown in Fig. 7-71(a) and show that this system is analogous to the electrical system in Fig. 7-71(b).

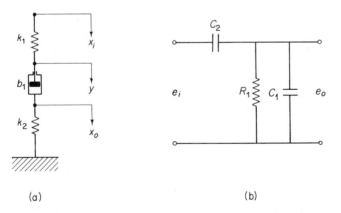

(a)   (b)

Fig. 7-71. (a) Mechanical system; (b) analogous electrical system.

Solution. The equations of motion for the mechanical system in Fig. 7-71(a) are

$$k_1(x_i - y) = b_1(\dot{y} - \dot{x}_o)$$
$$b_1(\dot{y} - \dot{x}_o) = k_2 x_o$$

(Note that equal force is transmitted through each component.) By taking the Laplace transforms of these two equations, assuming zero initial conditions, we have

$$k_1[X_i(s) - Y(s)] = b_1[s\,Y(s) - sX_o(s)]$$
$$b_1[s\,Y(s) - sX_o(s)] = k_2 X_o(s)$$

So by eliminating $Y(s)$ from the last two equations, we obtain $X_o(s)/X_i(s)$ as

$$\frac{X_o(s)}{X_i(s)} = \frac{\dfrac{b_1}{k_2}\,s}{b_1\left(\dfrac{1}{k_1} + \dfrac{1}{k_2}\right)s + 1}$$

Next, consider the electrical system shown in Fig. 7-71(b). Using complex impedances, the transfer function $E_o(s)/E_i(s)$ can be obtained as

$$\frac{E_o(s)}{E_i(s)} = \frac{\dfrac{1}{(1/R_1) + C_1 s}}{\dfrac{1}{C_2 s} + \dfrac{1}{(1/R_1) + C_1 s}} = \frac{R_1 C_2 s}{R_1(C_1 + C_2)s + 1}$$

Comparison of these two transfer functions shows that the two systems are analogous.

PROBLEM A-7-13. After deriving the transfer functions $X_o(s)/X_i(s)$ and $E_o(s)/E_i(s)$ of the systems shown in Fig. 7-72(a) and (b), show that they are analogous systems.

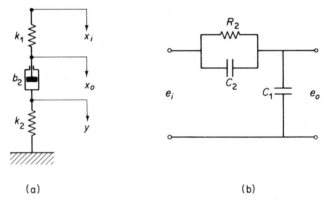

(a)   (b)

Fig. 7-72. (a) Mechanical system; (b) analogous electrical system.

Solution. The equations of motion for the system in Fig. 7-72(a) are

$$k_1(x_i - x_o) = b_2(\dot{x}_o - \dot{y})$$
$$b_2(\dot{x}_o - \dot{y}) = k_2 y$$

By taking the Laplace transforms of these two equations, assuming zero initial conditions, we obtain

$$k_1[X_i(s) - X_o(s)] = b_2[sX_o(s) - s\,Y(s)]$$
$$b_2[sX_o(s) - s\,Y(s)] = k_2\,Y(s)$$

By eliminating $Y(s)$ from the last two equations, the transfer function becomes

$$\frac{X_o(s)}{X_i(s)} = \frac{\dfrac{b_2}{k_2}s + 1}{b_2\left(\dfrac{1}{k_1} + \dfrac{1}{k_2}\right)s + 1}$$

The transfer function $E_o(s)/E_i(s)$ of the electrical system in Fig. 7-72(b) can be obtained as

$$\frac{E_o(s)}{E_i(s)} = \frac{\dfrac{1}{C_1 s}}{\dfrac{R_2}{R_2 C_2 s + 1} + \dfrac{1}{C_1 s}} = \frac{R_2 C_2 s + 1}{R_2(C_1 + C_2)s + 1}$$

Comparing the transfer functions of the mechanical and electrical systems, we see that they are analogous systems.

PROBLEM A-7-14. Find the transfer function $X_o(s)/X_i(s)$ of the mechanical system in Fig. 7-73(a) and show that it is analogous to the electrical system in Fig. 7-73(b).

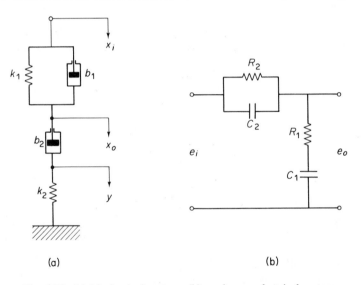

(a)   (b)

Fig. 7-73. (a) Mechanical system; (b) analogous electrical system.

Solution. The equations of motion for the mechanical system in Fig. 7-73(a) are

$$b_1(\dot{x}_i - \dot{x}_o) + k_1(x_i - x_o) = b_2(\dot{x}_o - \dot{y})$$

$$b_2(\dot{x}_o - \dot{y}) = k_2 y$$

By taking the Laplace transforms of these two equations, assuming zero initial conditions, we have

$$b_1[sX_i(s) - sX_o(s)] + k_1[X_i(s) - X_o(s)] = b_2[sX_o(s) - sY(s)]$$

$$b_2[sX_o(s) - sY(s)] = k_2 Y(s)$$

If we eliminate $Y(s)$ from the last two equations, the transfer function $X_o(s)/X_i(s)$ becomes

$$\frac{X_o(s)}{X_i(s)} = \frac{\left(\frac{b_1}{k_1}s + 1\right)\left(\frac{b_2}{k_2}s + 1\right)}{\left(\frac{b_1}{k_1}s + 1\right)\left(\frac{b_2}{k_2}s + 1\right) + \frac{b_2}{k_1}s}$$

For the electrical system in Fig. 7-73(b), the transfer function $E_o(s)/E_i(s)$ is found to be

$$\frac{E_o(s)}{E_i(s)} = \frac{R_1 + \dfrac{1}{C_1 s}}{\dfrac{1}{(1/R_2) + C_2 s} + R_1 + \dfrac{1}{C_1 s}}$$

$$= \frac{(R_1 C_1 s + 1)(R_2 C_2 s + 1)}{(R_1 C_1 s + 1)(R_2 C_2 s + 1) + R_2 C_1 s}$$

A comparison of the transfer functions shows that the systems in Fig. 7-73(a) and (b) are analogous.

PROBLEM A-7-15. Find the period of the conical pendulum in which a ball of mass m revolves about a fixed vertical axis at a constant speed as shown in Fig. 7-74.

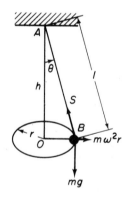

Fig. 7-74. Conical pendulum.

Solution. As the ball stands out at a constant angle, the vertical component of the tension S in the cord balances with the gravitational force mg, and the horizontal component of S balances with the centrifugal force $m\omega^2 r$. Hence, from geometry,

$$\frac{m\omega^2 r}{mg} = \frac{r}{h}$$

or

$$\omega^2 = \frac{g}{h}$$

Therefore period T is

$$T = \frac{2\pi}{\omega} = 2\pi\sqrt{\frac{h}{g}}$$

PROBLEM A-7-16. A boy riding a bicycle with a constant speed of 800 m/min around a horizontal circular path of radius $r = 50$ m leans inward at angle θ with the vertical as in Fig. 7-75. Determine the angle of inclination θ needed in order to maintain a steady-state circular motion.

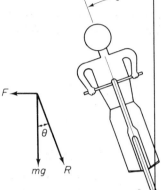

Fig. 7-75. Boy riding a bicycle around a circular path.

Solution. The centripetal force that is necessary for a circular motion is

$$m\omega^2 r = m\frac{v^2}{r}$$

The gravitational force mg can be resolved into two component forces, F and R as shown in Fig. 7-75. The horizontal force $F = mg \tan \theta$ must provide the necessary centripetal force mv^2/r. (Note that the horizontal force F can be supplied by friction if the surface is sufficiently rough. If it is not, the boy must reduce the speed if he is to avoid slipping.) Hence

$$mg \tan \theta = m\frac{v^2}{r}$$

or

$$\tan \theta = \frac{v^2}{gr}$$

By substituting the given numerical values into this last equation, we find

$$\tan \theta = \frac{(800/60)^2}{9.81 \times 50} = 0.3625$$

or

$$\theta = 19.9°$$

PROBLEM A-7-17. In rotating systems, if shafts rotate at critical speeds, large vibrations may develop as a result of resonance effects. Referring to Fig. 7-76(a), where the disk of mass m is mounted on an elastic shaft whose mass is negligible compared to that of the disk and is placed midway between bearings, assume that the disk is not perfectly symmetrical and that there is an eccentricity e from the center of the disk. The geometrical center of the disk, the center of mass of the disk, and the center of rotation are denoted by points O, G, and R, respectively. Distance between points R and O is r and that between points O and G is e. Assume that the equivalent spring constant of the elastic shaft is k, so that the restoring force due to the elastic shaft is kr. What is the critical speed of the system?

Solution. Referring to the system in Fig. 7-76(a), the centrifugal force acting on the shaft is $m\omega^2(e + r)$. This force balances with the restoring force of the elastic shaft, kr. So

$$m\omega^2(e + r) = kr \tag{7-65}$$

or

$$\omega^2(e + r) = \omega_n^2 r$$

where $\omega_n = \sqrt{k/m}$. Solving for r,

$$r = \frac{e}{(\omega_n^2/\omega^2) - 1}$$

The deflection r tends to increase rapidly as ω approaches ω_n. At $\omega = \omega_n$, resonance occurs. The deflection r increases until Eq. (7-65) no longer holds true. The critical

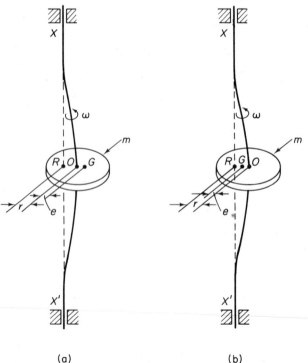

Fig. 7-76. Rotating system where (a) angular speed is lower than the critical speed; (b) angular speed is higher than the critical speed.

(a)   (b)

speed of the shaft is thus

$$\omega_{cr} = \omega_n = \sqrt{\frac{k}{m}}$$

At speeds higher than the critical, the center of gravity G will be situated as shown in Fig. 7-76(b), and the centrifugal force becomes

$$m\omega^2(r - e)$$

and this force balances with the restoring force of the elastic shaft kr. Hence

$$m\omega^2(r - e) = kr$$

Solving for r and noting that $k/m = \omega_n^2$, we have

$$r = \frac{e}{1 - (\omega_n^2/\omega^2)}$$

For $\omega > \omega_n$, the deflection r decreases and approaches e with increasing ω. For $\omega \gg \omega_n$, the center of gravity of the disk moves toward the line XX', and in this case the disk does not whirl, but the deflected shaft whirls about the center of gravity G.

PROBLEM A-7-18. Consider the spring-mass system shown in Fig. 7-77. The system is initially at rest, or $x(0) = 0$ and $\dot{x}(0) = 0$. At $t = 0$ a force $p(t) = P \sin \omega t$ is applied to the mass. Using the Laplace transform method, determine $x(t)$ for $t \geq 0$. When the numerical values of m, k, P, and ω are given as $m = 1$ kg, $k = 100$ N/m, $P = 50$ N, and $\omega = 5$ rad/s, find the solution $x(t)$.

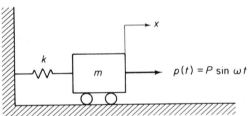

Fig. 7-77. Spring-mass system.

Solution. The equation of motion for the system is

$$m\ddot{x} + kx = P \sin \omega t$$

By defining $\omega_n = \sqrt{k/m}$, this last equation can be written

$$\ddot{x} + \omega_n^2 x = \frac{P}{m} \sin \omega t$$

The Laplace transform of this last equation, using initial conditions $x(0) = 0$ and $\dot{x}(0) = 0$, is

$$(s^2 + \omega_n^2)X(s) = \frac{P}{m} \frac{\omega}{s^2 + \omega^2}$$

Consequently,

$$X(s) = \frac{P\omega}{m} \frac{1}{(s^2 + \omega_n^2)(s^2 + \omega^2)}$$

$$= \frac{P\omega}{m} \left(\frac{1}{\omega^2 - \omega_n^2} \frac{1}{s^2 + \omega_n^2} - \frac{1}{\omega^2 - \omega_n^2} \frac{1}{s^2 + \omega^2} \right)$$

The inverse Laplace transform of this last equation is

$$x(t) = \frac{P}{m} \frac{1}{\omega^2 - \omega_n^2} \left(\frac{\omega}{\omega_n} \sin \omega_n t - \sin \omega t \right)$$

From the given numerical values we find $\omega_n = \sqrt{k/m} = \sqrt{100/1} = 10$ rad/s, $P/m = 50$ N/kg, and $\omega/\omega_n = 5/10 = 0.5$. Substituting these numerical values into the last equation, we have

$$x(t) = -\tfrac{1}{3} \sin 10t + \tfrac{2}{3} \sin 5t \text{ m}$$

PROBLEM A-7-19. Assuming that the mechanical system in Fig. 7-78 is at rest before excitation force $P \sin \omega t$ is given, derive the complete solution $x(t)$ and the steady-state solution $x_{ss}(t)$. The displacement x is measured from the equilibrium position.

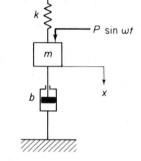

Fig. 7-78. Mechanical system.

Solution. The equation of motion for the system is

$$m\ddot{x} + b\dot{x} + kx = P \sin \omega t$$

Noting that $x(0) = 0$ and $\dot{x}(0) = 0$, the Laplace transform of this equation is

$$(ms^2 + bs + k)X(s) = P \frac{\omega}{s^2 + \omega^2}$$

or

$$X(s) = \frac{P\omega}{(s^2 + \omega^2)(ms^2 + bs + k)}$$

$$= \frac{P\omega}{m} \frac{1}{s^2 + \omega^2} \frac{1}{s^2 + 2\zeta\omega_n s + \omega_n^2}$$

where $\omega_n = \sqrt{k/m}$ and $\zeta = b/(2\sqrt{mk})$. $X(s)$ can be expanded as

$$X(s) = \frac{P\omega}{m} \left(\frac{as + c}{s^2 + \omega^2} + \frac{-as + d}{s^2 + 2\zeta\omega_n s + \omega_n^2} \right)$$

By simple calculations it can be found that

$$a = \frac{-2\zeta\omega_n}{(\omega_n^2 - \omega^2)^2 + 4\zeta^2\omega_n^2\omega^2}$$

$$c = \frac{(\omega_n^2 - \omega^2)}{(\omega_n^2 - \omega^2)^2 + 4\zeta^2\omega_n^2\omega^2}$$

$$d = \frac{4\zeta^2\omega_n^2 - (\omega_n^2 - \omega^2)}{(\omega_n^2 - \omega^2)^2 + 4\zeta^2\omega_n^2\omega^2}$$

Hence

$$X(s) = \frac{P\omega}{m} \frac{1}{(\omega_n^2 - \omega^2)^2 + 4\zeta^2\omega_n^2\omega^2}$$

$$\times \left[\frac{-2\zeta\omega_n s + (\omega_n^2 - \omega^2)}{s^2 + \omega^2} + \frac{2\zeta\omega_n(s + \zeta\omega_n) + 2\zeta^2\omega_n^2 - (\omega_n^2 - \omega^2)}{s^2 + 2\zeta\omega_n s + \omega_n^2} \right]$$

The inverse Laplace transform of $X(s)$ gives

$$x(t) = \frac{P\omega}{m[(\omega_n^2 - \omega^2)^2 + 4\zeta^2\omega_n^2\omega^2]} \left[-2\zeta\omega_n \cos \omega t + \frac{(\omega_n^2 - \omega^2)}{\omega} \sin \omega t \right.$$

$$\left. + 2\zeta\omega_n e^{-\zeta\omega_n t} \cos \omega_n \sqrt{1 - \zeta^2}\, t + \frac{2\zeta^2\omega_n^2 - (\omega_n^2 - \omega^2)}{\omega_n\sqrt{1 - \zeta^2}} e^{-\zeta\omega_n t} \sin \omega_n \sqrt{1 - \zeta^2}\, t \right]$$

At steady state ($t \to \infty$) the terms involving $e^{-\zeta\omega_n t}$ approach zero. Thus at steady state

$$x_{ss}(t) = \frac{P\omega}{m[(\omega_n^2 - \omega^2)^2 + 4\zeta^2\omega_n^2\omega^2]} \left(-2\zeta\omega_n \cos \omega t + \frac{\omega_n^2 - \omega^2}{\omega} \sin \omega t \right)$$

$$= \frac{P\omega}{(k - m\omega^2)^2 + b^2\omega^2} \left(-b \cos \omega t + \frac{k - m\omega^2}{\omega} \sin \omega t \right)$$

$$= \frac{P}{\sqrt{(k - m\omega^2)^2 + b^2\omega^2}} \sin \left(\omega t - \tan^{-1} \frac{b\omega}{k - m\omega^2} \right)$$

PROBLEM A-7-20. Consider the mechanical system shown in Fig. 7-79. If excitation force $p(t) = P \sin \omega t$, where $P = 1$ N and $\omega = 2$ rad/s, is applied, the steady-state amplitude of $x(t)$ is found to be 0.05 m. If the forcing frequency is changed to $\omega = 10$ rad/s, the steady-state amplitude of $x(t)$ is found to be 0.02 m. Determine the values of b and k.

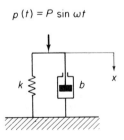

$p(t) = P \sin \omega t$

Fig. 7-79. Mechanical system.

Solution. The equation of motion for the system is

$$b\dot{x} + kx = p(t)$$

The transfer function is

$$\frac{X(s)}{P(s)} = \frac{1}{bs + k}$$

Hence the sinusoidal transfer function is

$$\frac{X(j\omega)}{P(j\omega)} = \frac{1}{bj\omega + k}$$

The amplitude ratio is

$$\left|\frac{X(j\omega)}{P(j\omega)}\right| = \frac{1}{\sqrt{b^2\omega^2 + k^2}}$$

And so

$$|X(j\omega)| = \frac{|P(j\omega)|}{\sqrt{b^2\omega^2 + k^2}}$$

From the problem statement, if $p(t) = P \sin \omega t = \sin 2t$, the amplitude of $x(t)$ is 0.05 m. Therefore

$$0.05 = \frac{1}{\sqrt{b^2 \times 2^2 + k^2}}$$

or

$$4b^2 + k^2 = 400 \tag{7-66}$$

If $p(t) = P \sin \omega t = \sin 10t$, then the amplitude of $x(t)$ is 0.02 m. Hence

$$0.02 = \frac{1}{\sqrt{b^2 \times 10^2 + k^2}}$$

or

$$100b^2 + k^2 = 2500 \tag{7-67}$$

From Eqs. (7-66) and (7-67) we obtain

$$96b^2 = 2100$$

or

$$b = 4.68 \text{ N-s/m}$$

Also,

$$k^2 = 312.5$$

or

$$k = 17.7 \text{ N/m}$$

PROBLEM A-7-21. Referring to the system shown in Fig. 7-80, assume that the input and output are displacement y and displacement x, respectively. Suppose that $y(t) = Y \sin \omega t$. What is the output $x(t)$ at steady state?

Solution. The equation of motion for the system is

$$m\ddot{x} + b(\dot{x} - \dot{y}) + kx = 0$$

or

$$m\ddot{x} + b\dot{x} + kx = b\dot{y}$$

Hence the transfer function between $X(s)$ and $Y(s)$ is

$$\frac{X(s)}{Y(s)} = \frac{bs}{ms^2 + bs + k}$$

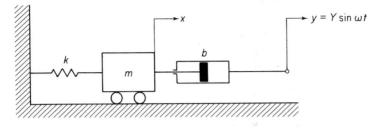

Fig. 7-80. Mechanical system.

Then the sinusoidal transfer function is

$$\frac{X(j\omega)}{Y(i\omega)} = \frac{bj\omega}{-m\omega^2 + bj\omega + k}$$

Thus

$$\left|\frac{X(j\omega)}{Y(j\omega)}\right| = \frac{b\omega}{\sqrt{(k - m\omega^2)^2 + b^2\omega^2}}$$

and

$$\phi = \left/\frac{X(j\omega)}{Y(j\omega)}\right. = \tan^{-1}\frac{b\omega}{0} - \tan^{-1}\frac{b\omega}{k - m\omega^2}$$

$$= 90° - \tan^{-1}\frac{b\omega}{k - m\omega^2}$$

Noting that $|Y(j\omega)| = Y$, the output is obtained as

$$x(t) = |X(j\omega)|\sin(\omega t + \phi)$$

$$= \frac{b\omega Y}{\sqrt{(k - m\omega^2)^2 + b^2\omega^2}}\sin\left(\omega t + 90° - \tan^{-1}\frac{b\omega}{k - m\omega^2}\right)$$

The angle $\tan^{-1}[b\omega/(k - m\omega^2)]$ varies from 0 to 180° as ω increases from zero to infinity. So for small ω the output leads the input by almost 90°, for large ω the output lags the input by almost 90°.

PROBLEM A-7-22. Find the steady-state displacements $x_1(t)$ and $x_2(t)$ of the system shown in Fig. 7-81. Assume that the viscous damping coefficients b_1 and b_2 are positive but negligibly small. (This means that in obtaining system equations, we may assume $b_1 \doteq 0$, $b_2 \doteq 0$. Since b_1 and b_2 are positive, however small, the system is a stable one and Eq. (7-33) can be used to find the steady-state solution.) The displacements x_1 and x_2 are measured from respective equilibrium positions in the absence of the excitation force.

Solution. The equations of motion for the system are

$$m_1\ddot{x}_1 + b_1\dot{x}_1 + k_1x_1 + b_2(\dot{x}_1 - \dot{x}_2) + k_2(x_1 - x_2) = p(t) = P\sin\omega t$$

$$m_2\ddot{x}_2 + b_2(\dot{x}_2 - \dot{x}_1) + k_2(x_2 - x_1) = 0$$

Since b_1 and b_2 are negligibly small, let us substitute $b_1 = 0$ and $b_2 = 0$ into the

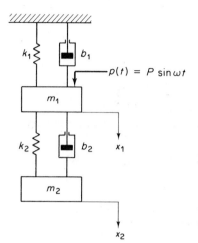

Fig. 7-81. Mechanical system.

equations of motion. Then

$$m_1 \ddot{x}_1 + k_1 x_1 + k_2 (x_1 - x_2) = p(t)$$

$$m_2 \ddot{x}_2 + k_2 (x_2 - x_1) = 0$$

By taking the Laplace transforms of these two equations, assuming zero initial conditions, we have

$$(m_1 s^2 + k_1 + k_2) X_1(s) - k_2 X_2(s) = P(s)$$

$$(m_2 s^2 + k_2) X_2(s) - k_2 X_1(s) = 0$$

from which

$$\frac{X_2(s)}{X_1(s)} = \frac{k_2}{m_2 s^2 + k_2}$$

and

$$\frac{X_1(s)}{P(s)} = \frac{m_2 s^2 + k_2}{(m_1 s^2 + k_1 + k_2)(m_2 s^2 + k_2) - k_2^2}$$

Since the system is basically a stable one, Eq. (7-33) can be applied. In applying Eq. (7-33) to the present problem the amplitudes $|X_1(j\omega)|$ and $|X_2(j\omega)|$ are obtained from the sinusoidal transfer functions as follows:

$$\frac{X_1(j\omega)}{P(j\omega)} = \frac{k_2 - m_2 \omega^2}{(k_1 + k_2 - m_1 \omega^2)(k_2 - m_2 \omega^2) - k_2^2}$$

$$\frac{X_2(j\omega)}{X_1(j\omega)} = \frac{k_2}{k_2 - m_2 \omega^2}$$

Thus the steady-state solution $x_1(t)$ is

$$x_1(t) = |X_1(j\omega)| \sin \left[\omega t + \left/ \frac{X_1(j\omega)}{P(j\omega)} \right. \right]$$

$$= \frac{(k_2 - m_2 \omega^2) P}{(k_1 + k_2 - m_1 \omega^2)(k_2 - m_2 \omega^2) - k_2^2} \sin \omega t$$

The steady-state solution $x_2(t)$ is

$$x_2(t) = |X_2(j\omega)| \sin\left[\omega t + \left|\frac{X_2(j\omega)}{P(j\omega)}\right.\right]$$

$$= \frac{k_2}{k_2 - m_2\omega^2} |X_1(j\omega)| \sin\left[\omega t + \left|\frac{X_2(j\omega)X_1(j\omega)}{X_1(j\omega)P(j\omega)}\right.\right]$$

$$= \frac{k_2 P}{(k_1 + k_2 - m_1\omega^2)(k_2 - m_2\omega^2) - k_2^2} \sin \omega t$$

Note that angles $\underline{/X_1(j\omega)/P(j\omega)}$ and $\underline{/X_2(j\omega)/P(j\omega)}$ are either 0° or 180°. The motions of masses m_1 and m_2 are either in phase or 180° out of phase with the excitation.

Note also that masses m_1 and m_2 move in the same direction if $\omega < \sqrt{k_2/m_2}$ and in the opposite direction if $\omega > \sqrt{k_2/m_2}$. If $\omega = \sqrt{k_2/m_2}$, mass m_1 stays still, but mass m_2 moves sinusoidally.

PROBLEM A-7-23. Figure 7-82 is a schematic diagram of an accelerometer. Assume that the case of the accelerometer is attached to an aircraft frame. The accelerometer indicates the acceleration of its case with respect to inertial space. The tilt angle θ measured from the horizontal line is assumed to be constant during the measurement period. Show that, for low-frequency inputs, the acceleration of the case relative to inertial space can be determined by the displacement of mass m with respect to its case. In the diagram x is the displacement of mass m relative to inertial space and is measured from the position where the spring is neither compressed nor stretched and y is the displacement of the case relative to inertial space.

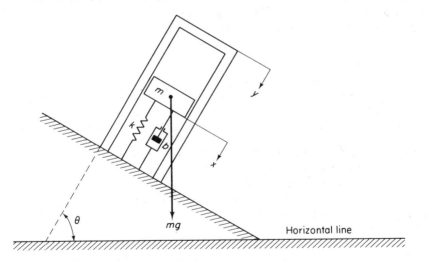

Fig. 7-82. Schematic diagram of an accelerometer system.

Solution. The equation of motion for the system is

$$m\ddot{x} + b(\dot{x} - \dot{y}) + k(x - y) = mg \sin \theta$$

In terms of a relative displacement $x - y$, this last equation becomes

$$m(\ddot{x} - \ddot{y}) + b(\dot{x} - \dot{y}) + k(x - y) = mg \sin \theta - m\ddot{y} \qquad (7\text{-}68)$$

Since θ is assumed to be constant during the measurement period, $mg \sin \theta$ is constant. Therefore it is possible to calibrate the displacement and define a new variable z such that

$$z = x - y - \frac{mg}{k} \sin \theta$$

Then Eq. (7-68) can be written

$$m\ddot{z} + b\dot{z} + kz = -m\ddot{y}$$

If acceleration $\ddot{y}$ (the acceleration of the case relative to inertial space) is taken to be the input to the system and variable z is taken to be the output, the transfer function of the system becomes

$$\frac{Z(s)}{s^2 Y(s)} = \frac{-m}{ms^2 + bs + k} = \frac{-1}{s^2 + (b/m)s + (k/m)}$$

The sinusoidal transfer function is

$$\frac{Z(j\omega)}{-\omega^2 Y(j\omega)} = \frac{-1}{-\omega^2 + (b/m)j\omega + (k/m)}$$

If the input frequency ω is very low compared to $\sqrt{k/m}$, then

$$\frac{Z(j\omega)}{-\omega^2 Y(j\omega)} \doteq -\frac{m}{k}$$

which means that $z = x - y - (mg/k) \sin \theta$ is nearly proportional to the slowly varying input acceleration $\ddot{y}$. Thus, for low-frequency inputs, the acceleration of the case relative to inertial space can be given by

$$\ddot{y} = -\frac{k}{m} z = \frac{k}{m}\left(y - x + \frac{mg}{k} \sin \theta\right)$$

And so, for low-frequency inputs, the acceleration $\ddot{y}$ of the case relative to inertial space can be determined by the displacement of mass m with respect to its case.

Note that such an accelerometer must have a sufficiently high undamped natural frequency $\sqrt{k/m}$ compared to the highest input frequency to be measured.

Problem A-7-24. A rotating machine with a mass of 100 kg and mounted on an isolator rotates at a constant speed of 10 Hz. An unbalanced mass m located at a distance r from the center of the rotor is exciting vibrations at a frequency ω that is very close to the natural frequency ω_n of the system with the result that the machine vibrates violently and a large vibratory force is transmitted to the foundation.

Design a dynamic vibration absorber to reduce the vibration. When the dynamic vibration absorber is added to the rotating machine as shown in Fig. 7-83, the entire system will become a two-degrees-of-freedom system. Determine mass m_a and spring constant k_a of the dynamic vibration absorber such that the lower natural frequency is 20% off the operating frequency. Determine also the higher natural frequency of the system. Assume that the values of b (viscous friction coeffi-

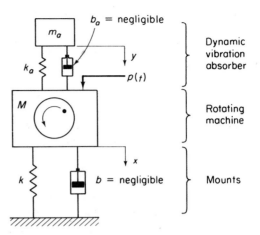

Fig. 7-83. Rotating machine with a dynamic vibration absorber.

cient of the isolator) and b_a (viscous friction coefficient of the dynamic vibration absorber) are positive but negligibly small. (Note that since the values of b and b_a are positive, however small, the system is a stable one. Therefore the steady-state displacements can be obtained by use of the sinusoidal transfer function.)

Solution. The equations of motion for the system are

$$M\ddot{x} + b\dot{x} + kx + b_a(\dot{x} - \dot{y}) + k_a(x - y) = p(t) = m\omega^2 r \sin \omega t$$

$$m_a\ddot{y} + b_a(\dot{y} - \dot{x}) + k_a(y - x) = 0$$

where x and y are displacements of mass M and mass m_a, respectively, and both x and y are measured from the respective equilibrium positions. Since $b \doteq 0$ and $b_a \doteq 0$, the last two equations may be simplified to

$$M\ddot{x} + kx + k_a(x - y) = p(t)$$

$$m_a\ddot{y} + k_a(y - x) = 0$$

When the viscous frictions are neglected, the system becomes the same as that shown in Fig. 7-42(b) with $b = 0$. Therefore from Eq. (7-44) we obtain the amplitude of $x(t)$ as

$$|X(j\omega)| = \left| \frac{m\omega^2 r(k_a - m_a\omega^2)}{(k + k_a - M\omega^2)(k_a - m_a\omega^2) - k_a^2} \right| \quad (7\text{-}69)$$

To make this amplitude equal to zero, we choose

$$k_a = m_a\omega^2$$

Since the operating speed is 10 Hz, we have

$$\omega = 10 \times 2\pi = 62.8 \text{ rad/s}$$

And so

$$\frac{k_a}{m_a} = 62.8^2 = 3944$$

The two natural frequencies ω_1 and ω_2 (where $\omega_1 < \omega_2$) of the entire system can be found from the characteristic equation. [The denominator of Eq. (7-69) is the characteristic polynomial.]

$$(k + k_a - M\omega_i^2)(k_a - m_a\omega_i^2) - k_a^2 = 0 \qquad (i = 1, 2)$$

or

$$\left(1 + \frac{k_a}{k} - \frac{M}{k}\omega_i^2\right)\left(1 - \frac{m_a}{k_a}\omega_i^2\right) - \frac{k_a}{k} = 0 \qquad (7\text{-}70)$$

Note that in the present system, since the natural frequency $\omega_n = \sqrt{k/M}$ is very close to the operating frequency $\omega = \sqrt{k_a/m_a}$, we can set

$$\sqrt{\frac{k}{M}} = \sqrt{\frac{k_a}{m_a}} = \omega = 62.8$$

From the problem statement the lower natural frequency must be 20% off the operating frequency ω. Since $\omega_1 < \omega$, this means that

$$\omega_1 = 0.8\omega = 0.8 \times 62.8$$

By substituting $\omega_i = \omega_1$ and $k/M = k_a/m_a = \omega^2$ into Eq. (7-70), we have

$$\left(1 + \frac{k_a}{k} - \frac{\omega_1^2}{\omega^2}\right)\left(1 - \frac{\omega_1^2}{\omega^2}\right) - \frac{k_a}{k} = 0$$

Substituting $\omega_1/\omega = 0.8$ into this last equation and simplifying yield

$$\left(1 + \frac{k_a}{k} - 0.8^2\right)(1 - 0.8^2) - \frac{k_a}{k} = 0$$

Solving for k_a/k,

$$\frac{k_a}{k} = 0.2025$$

It follows that

$$\frac{m_a}{M} = \frac{k_a}{k} = 0.2025$$

Since $M = 100$ kg, we have

$$m_a = 0.2050 \times 100 = 20.25 \text{ kg}$$

Since

$$\frac{k}{M} = \frac{k_a}{m_a} = 62.8^2$$

we obtain

$$k_a = (62.8)^2 m_a = (62.8)^2(20.25) = 79.9 \times 10^3 \text{ N/m}$$

Thus the mass and spring constant of the dynamic vibration absorber are $m_a = 20.25$ kg and $k_a = 79.9 \times 10^3$ N/m, respectively.

The two natural frequencies ω_1 and ω_2 can be determined by substituting $k_a/k = 0.2025$ into the equation

$$\left(1 + \frac{k_a}{k} - \frac{\omega_i^2}{\omega^2}\right)\left(1 - \frac{\omega_i^2}{\omega^2}\right) - \frac{k_a}{k} = 0 \qquad (i = 1, 2)$$

or

$$\left(1 + 0.2025 - \frac{\omega_i^2}{\omega^2}\right)\left(1 - \frac{\omega_i^2}{\omega^2}\right) - 0.2025 = 0$$

Solving for ω_i/ω,

$$\frac{\omega_i^2}{\omega^2} = 0.64 \quad \text{or} \quad 1.5625$$

Since $\omega_1 < \omega_2$,

$$\frac{\omega_1^2}{\omega^2} = 0.64, \qquad \frac{\omega_2^2}{\omega^2} = 1.5625$$

Therefore

$$\omega_1 = 0.8\omega = 0.8 \times 62.8 = 50.24 \text{ rad/s} = 8 \text{ Hz} = 480 \text{ cpm}$$

and

$$\omega_2 = 1.25\omega = 1.25 \times 62.8 = 78.5 \text{ rad/s} = 12.5 \text{ Hz} = 750 \text{ cpm}$$

PROBLEM A-7-25. Consider a differential equation in real time t:

$$\frac{d^2x}{dt^2} + 0.01\frac{dx}{dt} + 0.0001x = 0, \qquad x(0) = 10, \qquad \frac{dx}{dt}\bigg|_{t=0} = 0$$

Assuming that we solve it by using an analog computer, determine the time scale factor λ (where $\tau = \lambda t$) such that the settling time τ_s is 10 seconds.

Solution. By changing real time t to computer time τ, where $\tau = \lambda t$, the given differential equation can be written

$$\frac{d^2x}{d\tau^2} + \frac{0.01}{\lambda}\frac{dx}{d\tau} + \frac{0.0001}{\lambda^2}x = 0$$

From this last equation we obtain $\omega_n = 0.01/\lambda$ and $\zeta = 0.5$. The settling time τ_s is four times the time constant, or $4/\zeta\omega_n$. Hence

$$\tau_s = \frac{4}{\zeta\omega_n} = \frac{4\lambda}{0.005}$$

By setting $\tau_s = 10$, we obtain $\lambda = 0.0125$.

PROBLEM A-7-26. In the differential equation system

$$\ddot{x} + 0.4\dot{x} + 4x = 40 \cdot 1(t)$$

the right-hand side of the equation represents the forcing function, a step function of magnitude 40 occurring at $t = 0$. The initial conditions are $x(0) = 0$ and $\dot{x}(0) = 0$. Draw an analog computer diagram for obtaining the response $x(t)$. Make the maximum output voltage for each amplifier ± 80 V.

Solution. For $t > 0$, we obtain

$$\ddot{x} + 0.4\dot{x} + 4x = 40$$

Solving this last equation for the highest-order derivative gives

$$\ddot{x} = -0.4\dot{x} - 4x + 40 \tag{7-71}$$

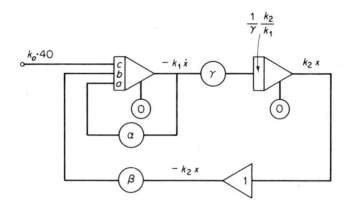

Fig. 7-84. Analog computer diagram for the determination of magnitude scale factors.

Let us define k_0, k_1, and k_2 as magnitude scale factors and rewrite Eq. (7-71) as

$$\ddot{x} = -\frac{0.4}{k_1}(k_1\dot{x}) - \frac{4}{k_2}(k_2x) + \frac{1}{k_0}(k_0 \cdot 40)$$

Referring to Fig. 7-84, the input voltage is $k_0 \cdot 40$, the output voltage of the first integrator is $-k_1\dot{x}$, and the output voltage of the second integrator is k_2x. (The output voltage of the sign inverter is $-k_2x$). From the problem statement the maximum output voltages must be ± 80 V.

Conservative estimates of the maximum values of x and $\dot{x}$ can be obtained by neglecting the damping term in the system equation. The simplified equation is

$$\ddot{x} + 4x = 40, \qquad x(0) = 0, \qquad \dot{x}(0) = 0$$

The solution for this simplified equation is

$$x(t) = 10 - 10\cos 2t$$

So

$$\dot{x}(t) = 20\sin 2t$$

The maximum values are

$$|x|_{\max} = 20$$
$$|\dot{x}|_{\max} = 20$$

We shall now choose k_0, k_1, and k_2 so that the maximum values of $k_0 \cdot 40, |-k_1\dot{x}|$, and $|k_2x|$ are 80 V. Thus we have

$$k_0 = \frac{80}{40} = 2$$

$$k_1 = \frac{80}{|\dot{x}|_{\max}} = \frac{80}{20} = 4$$

$$k_2 = \frac{80}{|x|_{\max}} = \frac{80}{20} = 4$$

It follows that

$$\frac{k_2}{k_1} = \frac{4}{4} = 1$$

Since $k_2/k_1 = 1$, we choose $\gamma = 1$. (The potentiometer γ can therefore be eliminated.) The constant of the second integrator $(1/\gamma)(k_2/k_1)$ becomes 1. Note that from Fig. 7-84 we obtain

$$k_1\ddot{x} = a\alpha(-k_1\dot{x}) + b\beta(-k_2x) + ck_040$$

or

$$\ddot{x} = -a\alpha\dot{x} - b\beta\frac{k_2}{k_1}x + c\frac{k_0}{k_1}40$$

Noting that $k_2/k_1 = 1$ and $k_0/k_1 = 0.5$, this last equation becomes

$$\ddot{x} = -a\alpha\dot{x} - b\beta x + 20c \qquad (7\text{-}72)$$

A comparison of Eqs. (7-71) and (7-72) shows that

$$a\alpha = 0.4, \qquad b\beta = 4, \qquad 20c = 40$$

Therefore we may choose $a = 1$, $\alpha = 0.4$, $b = 4$, $\beta = 1$, and $c = 2$. Then all the unknown constants in Fig. 7-84 are determined. Figure 7-85 is the computer diagram for obtaining the response $x(t)$. [The output of the second integrator gives $4x(t)$.]

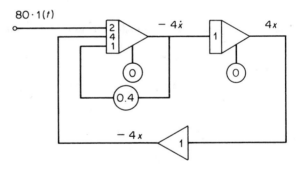

Fig. 7-85. Analog computer diagram.

PROBLEM A-7-27. Find the differential equation that is represented by the analog computer diagram shown in Fig. 7-86.

Solution. Referring to the diagram, we see that

$$-\dot{x} = -0.5u + y + 1.5x$$

$$-\dot{y} = u - x$$

Eliminating y and $\dot{y}$ from the two equations results in the differential equation for the system.

$$\ddot{x} + 1.5\dot{x} + x = u + 0.5\dot{u}$$

where $x(0) = 0$ and $\int (x - u)\, dt \Big|_{t=0} = 0$.

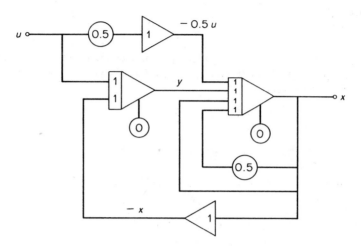

Fig. 7-86. Analog computer diagram.

PROBLEM A-7-28. Draw an analog computer diagram for the following transfer function system, which involves a numerator dynamics.

$$\frac{X(s)}{U(s)} = \frac{5s + 1}{s^2 + 3s + 2}$$

Solution. By rewriting the transfer function, we obtain

$$s^2 X(s) + 3sX(s) + 2X(s) = 5sU(s) + U(s)$$

Solving for $s^2 X(s)$,

$$s^2 X(s) = s[5U(s) - 3X(s)] + U(s) - 2X(s)$$

or

$$X(s) = \frac{1}{s}[5U(s) - 3X(s)] + \frac{1}{s^2}[U(s) - 2X(s)]$$

$$= \frac{1}{s}\left\{5U(s) - 3X(s) + \frac{1}{s}[U(s) - 2X(s)]\right\}$$

An analog computer diagram for this last equation is shown in Fig. 7-87.

PROBLEM A-7-29. Draw an analog computer diagram for the transfer function system

$$\frac{X(s)}{U(s)} = \frac{b_0 s^3 + b_1 s^2 + b_2 s + b_3}{s^3 + a_1 s^2 + a_2 s + a_3}$$

Solution. By rewriting the transfer function, we have

$$(s^3 + a_1 s^2 + a_2 s + a_3)X(s) = (b_0 s^3 + b_1 s^2 + b_2 s + b_3)U(s)$$

Solving for $s^3 X(s)$, we get

$$s^3 X(s) = (b_0 s^3 + b_1 s^2 + b_2 s + b_3)U(s) - (a_1 s^2 + a_2 s + a_3)X(s)$$

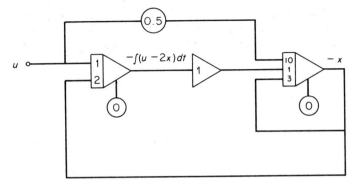

Fig. 7-87. Analog computer diagram.

Hence

$$X(s) = b_0 U(s) + \frac{1}{s}[b_1 U(s) - a_1 X(s)] + \frac{1}{s^2}[b_2 U(s) - a_2 X(s)]$$

$$+ \frac{1}{s^3}[b_3 U(s) - a_3 X(s)]$$

$$= b_0 U(s) + \frac{1}{s}\Big([b_1 U(s) - a_1 X(s)] + \frac{1}{s}\Big\{[b_2 U(s) - a_2 X(s)]$$

$$+ \frac{1}{s}[b_3 U(s) - a_3 X(s)]\Big\}\Big)$$

An analog computer diagram to represent this system is shown in Fig. 7-88.

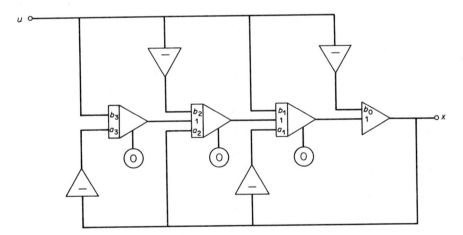

Fig. 7-88. Analog computer diagram.

PROBLEM A-7-30. Draw an analog computer diagram for simulating the mechanical system of Fig. 7-89. Assume that $x(0) = 0$ and $y(0) = 0$. The numerical values for b_1, b_2, k_1, and k_2 are given as $b_1 = 20$ N-s/m, $b_2 = 30$ N-s/m, $k_1 = 100$ N/m, and $k_2 = 60$ N/m.

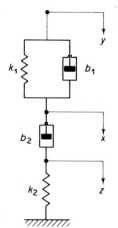

Fig. 7-89. Mechanical system.

Solution. The equations of motion for the system are

$$b_1(\dot{y} - \dot{x}) + k_1(y - x) = b_2(\dot{x} - \dot{z})$$

$$b_2(\dot{x} - \dot{z}) = k_2 z$$

By taking the Laplace transforms of these two equations, substituting zero initial conditions, and eliminating $Z(s)$, we find

$$\frac{X(s)}{Y(s)} = \frac{(b_1 s + k_1)(b_2 s + k_2)}{(b_1 s + k_1)(b_2 s + k_2) + b_2 k_2 s}$$

Substituting the given numerical values into this last equation gives

$$\frac{X(s)}{Y(s)} = \frac{(20s + 100)(30s + 60)}{(20s + 100)(30s + 60) + 30 \times 60s}$$

$$= \frac{s^2 + 7s + 10}{s^2 + 10s + 10}$$

Consequently,

$$(s^2 + 10s + 10)X(s) = (s^2 + 7s + 10)Y(s)$$

Solving for $s^2 X(s)$, we have

$$s^2 X(s) = (s^2 + 7s + 10)Y(s) - (10s + 10)X(s)$$

Thus

$$X(s) = Y(s) + \frac{1}{s}\left\{[7Y(s) - 10X(s)] + \frac{1}{s}[10Y(s) - 10X(s)]\right\}$$

An analog computer diagram for this equation is shown in Fig. 7-90.

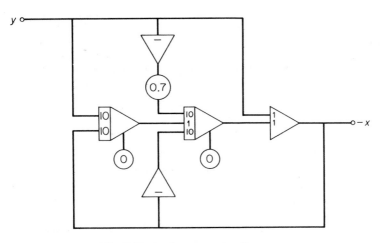

Fig. 7-90. Analog computer diagram.

PROBLEMS

PROBLEM B-7-1. In the system of Fig. 7-91 switch S is closed at $t = 0$. Find the voltage $e_o(t)$. Assume that the capacitor is initially uncharged.

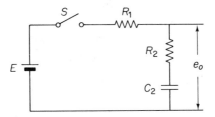

Fig. 7-91. Electrical system.

PROBLEM B-7-2. Referring to Fig. 7-92, the voltage source E is suddenly connected by means of switch S at time $t = 0$. Assume that capacitor C is initially uncharged and that inductance L carries no initial current. What is the current $i(t)$?

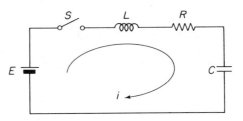

Fig. 7-92. Electrical system.

PROBLEM B-7-3. Mass m ($m = 1$ kg) is vibrating initially in the mechanical system shown in Fig. 7-93. At $t = 0$ we hit the mass with an impulsive force $p(t)$ whose strength is 10 N. Assuming that spring constant k is 100 N/m and that $x(0-) = 0.1$ m, $\dot{x}(0-) = 1$ m/s, find the displacement $x(t)$ as a function of time t. Displacement $x(t)$ is measured from the equilibrium position in the absence of excitation force.

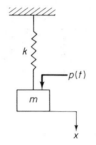

Fig. 7-93. Mechanical system.

PROBLEM B-7-4. A free vibration of the mechanical system in Fig. 7-94(a) indicates that the amplitude of vibration decreases to 25% of the value at $t = t_0$ after four consecutive cycles of motion, as Fig. 7-94(b) shows. Determine the viscous friction coefficient b of the system if $m = 1$ kg and $k = 500$ N/m.

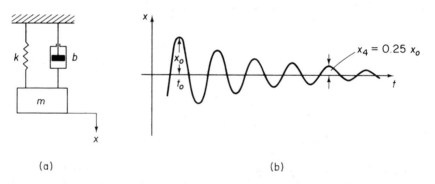

(a)   (b)

Fig. 7-94. (a) Mechanical system; (b) portion of a free vibration curve.

PROBLEM B-7-5. A mass of 20 kg is supported by a spring and damper as shown in Fig. 7-95(a). When a mass of 2 kg is added to the 20 kg mass, the system vibrates as shown in Fig. 7-95(b). Determine the spring constant k and the viscous friction coefficient b. [Note that $(0.02/0.08) \times 100 = 25\%$ maximum overshoot corresponds to $\zeta = 0.4$.]

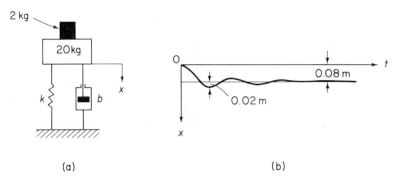

(a)   (b)

Fig. 7-95. (a) Mechanical system; (b) step response curve.

PROBLEM B-7-6. Consider the mechanical system shown in Fig. 7-96. The pendulum m_2 is supported by mass m_1, which vibrates because of an elastic connection. Derive the equations of motion for the system.

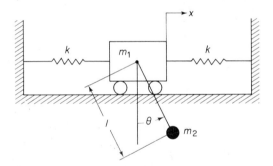

Fig. 7-96. Mechanical system.

PROBLEM B-7-7. The system shown in Fig. 7-97 is at rest initially. At $t = 0$ mass m is set into motion by an impulsive force whose strength is unity. Can it be stopped by another such impulsive force?

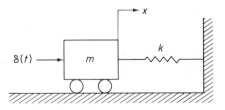

Fig. 7-97. Mechanical system.

PROBLEM B-7-8. Figure 7-98 shows a system that consists of a mass and a damper. The system is initially at rest. When it is set into motion by an impulsive force whose strength is unity, find the response $x(t)$. Determine the initial velocity of mass m.

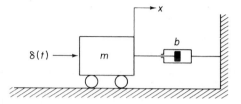

Fig. 7-98. Mechanical system.

PROBLEM B-7-9. Find the transfer functions $X_o(s)/X_i(s)$ and $E_o(s)/E_i(s)$ of the mechanical and electrical systems shown in Fig. 7-99(a) and (b), respectively.

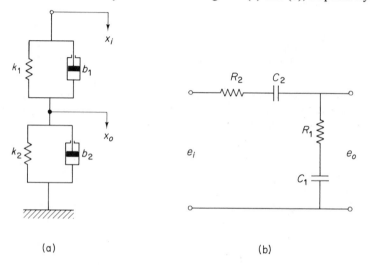

(a)   (b)

Fig. 7-99. (a) Mechanical system; (b) electrical system.

PROBLEM B-7-10. Derive the transfer functions $X_o(s)/X_i(s)$ and $E_o(s)/E_i(s)$ of the systems shown in Fig. 7-100(a) and (b) and show that they are analogous systems.

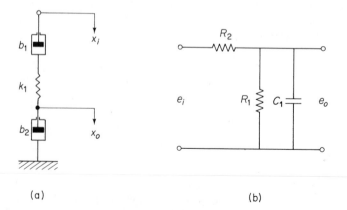

(a)   (b)

Fig. 7-100. (a) Mechanical system; (b) analogous electrical system.

PROBLEM B-7-11. After finding the transfer function $X_o(s)/X_i(s)$ of the mechanical system shown in Fig. 7-101, derive an analogous electrical system.

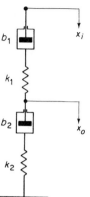

Fig. 7-101. Mechanical system.

PROBLEM B-7-12. Find the transfer function $E_o(s)/E_i(s)$ of the electrical system shown in Fig. 7-102. In addition, find an analogous mechanical system.

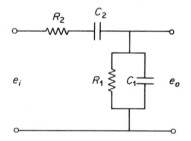

Fig. 7-102. Electrical system.

PROBLEM B-7-13. Obtain both the transfer function $X_o(s)/X_i(s)$ of the mechanical system shown in Fig. 7-103 and an analogous electrical system.

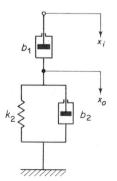

Fig. 7-103. Mechanical system.

PROBLEM B-7-14. In the thermal system shown in Fig. 7-104(a) it is assumed that the tank is insulated to eliminate heat loss to the surrounding air, that there is no heat storage in the insulation, and that the liquid in the tank is perfectly mixed so that it is at a uniform temperature. (Thus a single temperature can be used to denote the temperature of the liquid in the tank and of the outflowing liquid.) It is further assumed that the flow rate of liquid into and out of the tank is constant and that the inflow temperature is constant at $\bar{\Theta}_i$ K. For $t < 0$ the system is at steady state and the heater supplies heat at the rate $\bar{H}$ J/s. At $t = 0$ the heat input rate is changed from $\bar{H}$ to $\bar{H} + h$ J/s. This change causes the outflow liquid temperature to change from $\bar{\Theta}_0$ to $\bar{\Theta}_0 + \theta$ K. Suppose that the change in temperature, θ K, is the output and that the change in the heat input, h J/s, is the input to the system. Determine the transfer function $\Theta(s)/H(s)$, where $\Theta(s) = \mathcal{L}[\theta(t)]$ and $H(s) = \mathcal{L}[h(t)]$. Show that the thermal system is analogous to the electrical system shown in Fig. 7-104(b), where voltage e_o is the output and current i is the input.

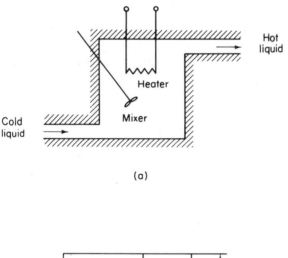

(a)

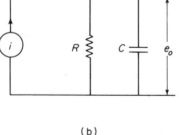

(b)

Fig. 7-104. (a) Thermal system; (b) analogous electrical system.

PROBLEM B-7-15. A stone of mass 0.1 kg is attached to the end of 1 m cord and, rotated at an angular speed of 1 Hz. Find the tension in the cord. If the maximum allowable tension in the cord is 40 N, what is the maximum angular speed (in Hz) that can be attained without breaking the cord?

PROBLEM B-7-16. In the speed regulator of Fig. 7-105, what is the frequency ω needed to maintain the configuration shown in the diagram?

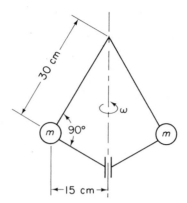

Fig. 7-105. Speed regulator system.

PROBLEM B-7-17. The spring-mass system shown in Fig. 7-106 is initially at rest. If mass m is excited by a sinusoidal force $p(t) = P \sin \omega t$, what is the response $x(t)$? Assume that $m = 1$ kg, $k = 100$ N/m, $P = 5$ N, and $\omega = 2$ rad/s.

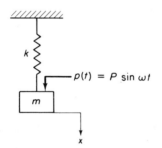

Fig. 7-106. Spring-mass system.

PROBLEM B-7-18. A rotating machine of mass $M = 100$ kg has an unbalanced mass $m = 0.2$ kg at a distance $r = 0.5$ m from the center of rotation. (The mass M includes mass m.) The operating speed is 10 Hz. Suppose that this machine is mounted on an isolator consisting of a spring and damper as shown in Fig. 7-107. If it is

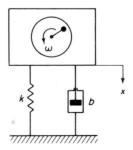

Fig. 7-107. Rotating machine mounted on a vibration isolator.

desired to have $\zeta = 0.2$, specify the spring constant k such that only 10% of the excitation force is transmitted to the foundation. Determine the amplitude of the transmitted force.

PROBLEM B-7-19. In Fig. 7-108 an instrument is attached to a base whose motion is to be measured. The relative motion between mass m and the base recorded by a rotating drum will indicate the motion of the base. Assume that x is the displacement of the mass, y is the displacement of the base, and $z = x - y$ is the motion of the pen relative to the base. If the motion of the base is $y = Y \sin \omega t$, what is the steady-state amplitude ratio of z to y? Show that if $\omega \gg \omega_n$, where $\omega_n = \sqrt{k/m}$, the device can be used for measuring the displacement of the base and that, if $\omega \ll \omega_n$, it can be used for measuring the acceleration of the base.

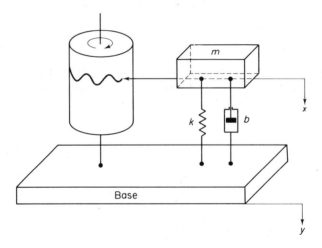

Fig. 7-108. Motion or acceleration measuring instrument.

PROBLEM B-7-20. Figure 7-109 shows a machine m mounted on an isolator in which spring k_1 is the load-carrying spring and viscous damper b_2 is in series with

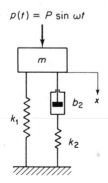

$p(t) = P \sin \omega t$

Fig. 7-109. Machine mounted on a vibration isolator.

spring k_2. When mass m is subjected to a sinusoidal excitation force $p(t) = P \sin \omega t$, determine the force transmissibility. Determine also the amplitude of the force transmitted to the foundation.

PROBLEM B-7-21. A machine m is mounted on an isolator in Fig. 7-110. If the foundation is vibrating according to $y = Y \sin \omega t$, where y is the displacement of the foundation, find the vibration amplitude of the machine. Determine the motion transmissibility.

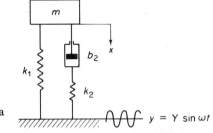

Fig. 7-110. Machine mounted on a vibration isolator.

PROBLEM B-7-22. Figure 7-111 shows a machine with a dynamic vibration absorber. The undamped natural frequency of the system in the absence of the dynamic vibration absorber is $\omega_n = \sqrt{k/m}$. Suppose that the operating frequency ω is close to ω_n. If the dynamic vibration absorber is tuned so that $\sqrt{k_a/m_a} = \omega$, what is the amplitude of mass m_a of the vibration absorber?

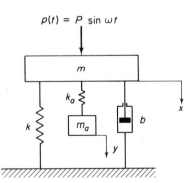

Fig. 7-111. Machine with a dynamic vibration absorber.

PROBLEM B-7-23. In solving the following differential equation by means of an analog computer

$$50\ddot{x} + 2\dot{x} + 0.02x = \sin t$$

it is desirable to employ time scaling in order to reduce the variation in the magnitudes of the coefficients and adjust response speed. Determine a suitable time scale factor λ so that the settling time is 50 seconds.

PROBLEM B-7-24. Obtain the transfer function $X(s)/U(s)$ of the system shown in Fig. 7-112.

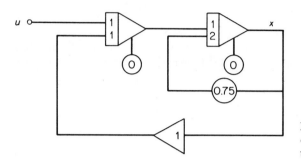

Fig. 7-112. Analog computer diagram for simulating a system.

PROBLEM B-7-25. Draw an analog computer diagram for generating a signal

$$x(t) = 80e^{-t} \cos t$$

Use the minimal number of operational amplifiers.

PROLBEM B-7-26. Draw an analog computer diagram to solve the equation

$$\ddot{x} + 2\dot{x} + 3x = 10 \cdot 1(t), \qquad x(0) = 0, \qquad \dot{x}(0) = 0$$

Determine magnitude scale factors so that the maximum output voltage of each amplifier is ± 90 V.

PROBLEM B-7-27. Find the transfer function $X(s)/U(s)$ of the system shown in Fig. 7-113.

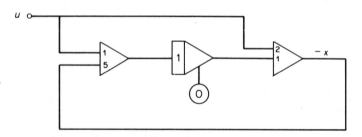

Fig. 7-113. Analog computer diagram for simulating a system.

PROBLEM B-7-28. Determine the transfer function $X(s)/U(s)$ from the analog computer diagram of Fig. 7-114.

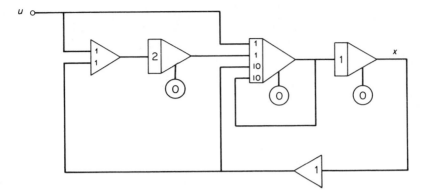

Fig. 7-114. Analog computer diagram for simulating a system.

PROBLEM B-7-29. Figure 7-115 is an analog computer diagram for simulating a certain system. Obtain the transfer function $X(s)/U(s)$ of the system.

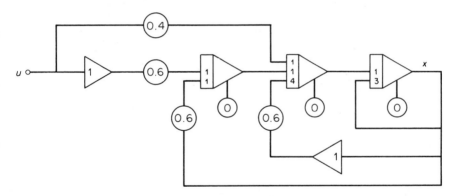

Fig. 7-115. Analog computer diagram for simulating a system.

PROBLEM B-7-30. Referring to the mechanical vibratory system of Fig. 7-116, draw an analog computer diagram for simulating this system. Assume that displacement

Input force

$p(t) = P \sin \omega t$

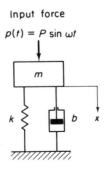

Fig. 7-116. Mechanical vibratory system.

x is measured from the equilibrium position in the absence of the sinusoidal excitation force. The initial conditions are $x(0) = 0$ and $\dot{x}(0) = 0$, and the input force $P \sin \omega t$ is given at $t = 0$. The numerical values of m, b, k, P, and ω are given as $m = 2$ kg, $b = 0.2$ N-s/m, $k = 200$ N/m, $P = 5$ N, and $\omega = 3$ rad/s.

8

CONTROL SYSTEMS
ANALYSIS

8-1 INTRODUCTION

Only introductory material on control systems is presented in this chapter. Our discussion is limited to time-domain analysis or transient response analysis. We begin by defining the terminologies that are necessary to describe control systems, such as plants, disturbances, feedback control, and feedback control systems, and then follow with a description of closed-loop and open-loop control systems. Finally, the advantages and disadvantages of closed-loop and open-loop control systems are compared.

Plants. A *plant* is a piece of equipment, perhaps a set of machine parts functioning together, the purpose of which is to perform a particular operation. In this book we shall call any physical object to be controlled a plant.

Disturbances. A *disturbance* is a signal that tends to affect the value of the output of a system adversely. If the disturbance is generated within the system, it is called *internal*; an *external* disturbance is generated outside the system and is an input.

Feedback control. *Feedback control* refers to an operation that, in the presence of disturbances, tends to reduce the difference between the output of a system and some reference input and that does so on the basis of this difference. Here only unpredictable disturbances are so specified, since predictable or known disturbances can always be compensated for within the system.

Feedback control systems. A system that maintains a prescribed relationship between the output and some reference input by comparing them and using the difference as a means of control is called a *feedback control system*. An example would be a room-temperature control system. By measuring the actual room temperature and comparing it with the reference temperature (desired temperature), the thermostat turns the heating or cooling equipment on or off in such a way as to ensure that the room temperature remains at a comfortable level regardless of outside conditions.

Feedback control systems, of course, are not limited to engineering but can be found in various nonengineering fields as well. The human body, for instance, is a highly advanced feedback control system. Both body temperature and blood pressure are kept constant by means of physiological feedback. In fact, feedback performs a vital function: It makes the human body relatively insensitive to external disturbances, thus enabling it to function properly in a changing environment.

As another example, consider the control of automobile speed by a human operator. The driver decides on an appropriate speed for the situation, which may be the posted speed limit on the road or highway involved. This speed acts as the reference speed. The driver observes the actual speed by looking at the speedometer. If he is traveling too slowly, he depresses the accelerator and the car speeds up. If the actual speed is too high, he releases the pressure on the accelerator and the car slows down. This is a feedback control system with a human operator. The human operator here can easily be replaced by a mechanical, electrical, or similiar device. Instead of the driver observing the speedometer, an electric generator can be used to produce a voltage that is proportional to the speed. This voltage can be compared with a reference voltage that corresponds to the desired speed. The difference in the voltages can then be used as the error signal to position the throttle to increase or decrease the speed as needed.

Closed-loop control systems. Feedback control systems are often referred to as *closed-loop control systems*. In practice, the terms feedback control and closed-loop control are used interchangeably. In a closed-loop control system the actuating error signal, which is the difference between the input signal and the feedback signal (which may be the output signal itself or a function of the output signal and its derivatives), is fed to the controller so as to reduce the error and bring the output of the system to a desired value. The term closed-loop control always implies the use of feedback control action in order to reduce system error.

Open-loop control systems. Those systems in which the output has no effect on the control action are called *open-loop control systems*. In other words, in an open-loop control system the output is neither measured nor fed back for comparison with the input. One practical example is a washing machine. Soaking, washing, and rinsing in the washer operate on a time basis. The machine does not measure the output signal—that is, the cleanliness of the clothes.

In any open-loop control system the output is not compared with the reference input. Thus to each reference input there corresponds a fixed operating condition; as a result, the accuracy of the system depends on calibration. In the presence of disturbances an open-loop system will not perform the desired task. Open-loop control can be used, in practice, only if the relationship between the input and output is known and if there are neither internal nor external disturbances. Clearly, such systems are not feedback control systems. Note that any control system that operates on a time basis is open-loop. For example, traffic control by means of signals operated on a time basis is another example of open-loop control.

Closed-loop versus open-loop control systems. An advantage of the closed-loop control system is the fact that the use of feedback makes the system response relatively insensitive to external disturbances and internal variations in system parameters. It is thus possible to use relatively inaccurate and inexpensive components to obtain the accurate control of a given plant, whereas doing so is impossible in the open-loop case.

From the point of view of stability, the open-loop control system is easier to build because system stability is not a major problem. On the other hand, stability *is* a major problem in the closed-loop control system, which may tend to overcorrect errors that can cause oscillations of constant or changing amplitude.

General requirements of control systems. Any control system must be stable. This is a primary requirement. In addition to absolute stability, a control system must have a reasonable relative stability; that is, the response must show reasonable damping. Moreover, the speed of response must be reasonably fast. A control system must also be capable of reducing errors to zero or to some small tolerable value. Any useful control system must satisfy these requirements.

Because the need for reasonable relative stability and for steady-state accuracy tend to be incompatible, it is necessary in designing control systems to make the most effective compromise between the two.

Outline of the chapter. As noted, this chapter presents introductory material on control systems analysis. In addition to the necessary definitions, several examples of control systems have been given in Sec. 8-1. In Sec. 8-2 we deal with block diagrams of control systems and their components. After first describing the control actions generally found in industrial automatic controllers, Sec. 8-3 then presents the standard techniques for obtaining various control actions through the use of pneumatic, hydraulic, and electronic components. Next, the transient response analysis of control systems is covered in Sec. 8-4. Here the response of first- and second-order systems to aperiodic inputs is presented, and the effects of various control actions on the transient response characteristics of control systems are discussed. Section 8-5 is concerned with transient response specifications. Methods to improve transient response characteristics are given in Sec. 8-6. The chapter ends with a simple design problem in Sec. 8-7.

8-2   BLOCK DIAGRAMS

A system may consist of a number of components. In order to show the functions performed by each component, a diagram called the block diagram is frequently used in the analysis and design of systems. This section explains what a block diagram is, presents a method for obtaining block diagrams for physical systems, and, finally, discusses techniques to simplify such diagrams.

Block diagrams. A *block diagram* of a system is a pictorial representation of the functions performed by each component and of the flow of signals. Such a diagram depicts the interrelationships that exist among the various components. Differing from a purely abstract mathematical representation, a block diagram has the advantage of indicating more realistically the signal flows of the actual system.

In a block diagram all system variables are linked to each other through functional blocks. The *functional block* or simply *block* is a symbol for the mathematical operation on the input signal to the block that produces the output. The transfer functions of the components are usually entered in the corresponding blocks, which are connected by arrows to indicate the direction of the flow of signals. Note that the signal can pass only in the direction of the arrows. Thus a block diagram of a control system explicitly shows a unilateral property.

Figure 8-1 shows an element of the block diagram. The arrowhead pointing toward the block indicates the input, and the arrowhead leading

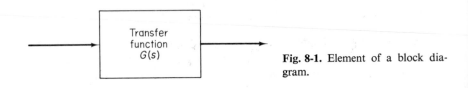

Fig. 8-1. Element of a block diagram.

away from the block represents the output. Such arrows are referred to as *signals*.

Note that the dimensions of the output signal from the block are the dimensions of the input signal multiplied by the dimensions of the transfer function in the block.

The advantages of the block diagram representation of a system lie in the fact that it is easy to form the overall block diagram for the entire system by merely connecting the blocks of the components according to the signal flow and that it is possible to evaluate the contribution of each component to the overall performance of the system.

In general, the functional operation of the system can be visualized more readily by examining the block diagram than by examining the physical system itself. A block diagram contains information concerning dynamic behavior, but it does not include any information on the physical construction of the system. Consequently, many dissimilar and unrelated systems can be represented by the same block diagram.

It should be noted that in a block diagram the main source of energy is not explicitly shown and that the block diagram of a given system is not unique. A number of different block diagrams can be drawn for a system, depending on the point of view of the analysis.

Summing point. Referring to Fig. 8-2, a circle with a cross is the symbol that indicates a summing operation. The plus or minus sign at each arrowhead indicates whether that signal is to be added or subtracted. It is important that the quantities being added or subtracted have the same dimensions and the same units.

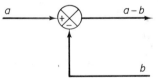

Fig. 8-2. Summing point.

Branch point. A *branch point* is a point from which the signal from a block goes concurrently to other blocks or summing points.

Block diagram of closed-loop system. Figure 8-3 is an example of a block diagram of a closed-loop system. The output $\Theta_o(s)$ is fed back to the summing point, where it is compared with the input $\Theta_i(s)$. The closed-loop nature of the system is clearly indicated by the figure. The output of the block $\Theta_o(s)$ in this case is obtained by multiplying the transfer function $G(s)$ by the input to the block, $E(s)$.

Any linear system can be represented by a block diagram consisting of blocks, summing points, and branch points. When the output is fed back to the summing point for comparison with the input, it is necessary to convert

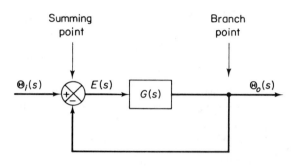

Fig. 8-3. Block diagram of a closed-loop system.

the form of the output signal to that of the input signal. This conversion is accomplished by the feedback element whose transfer function is $H(s)$, as shown in Fig. 8-4. Another important role of the feedback element is to modify the output before it is compared with the input. In the present example the feedback signal that is fed back to the summing point for comparison with the input is $B(s) = H(s)\Theta_o(s)$.

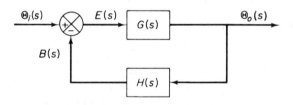

Fig. 8-4. Block diagram of a closed-loop system.

Open-loop transfer function and feedforward transfer function. The ratio of the feedback signal $B(s)$ to the actuating error signal $E(s)$ is called the *open-loop transfer function*. That is,

$$\text{Open-loop transfer function } = \frac{B(s)}{E(s)} = G(s)H(s)$$

The ratio of the output $\Theta_o(s)$ to the actuating error signal $E(s)$ is called the *feedforward transfer function*, so that

$$\text{Feedforward transfer function } = \frac{\Theta_o(s)}{E(s)} = G(s)$$

If the feedback transfer function is unity, then the open-loop transfer function and the feedforward transfer function are the same.

Closed-loop transfer function. For the system shown in Fig. 8-4, the output $\Theta_o(s)$ and input $\Theta_i(s)$ are related as follows:

$$\Theta_o(s) = G(s)E(s)$$
$$E(s) = \Theta_i(s) - B(s) = \Theta_i(s) - H(s)\Theta_o(s)$$

Eliminating $E(s)$ from these equations gives

$$\Theta_o(s) = G(s)[\Theta_i(s) - H(s)\Theta_o(s)]$$

or

$$\frac{\Theta_o(s)}{\Theta_i(s)} = \frac{G(s)}{1 + G(s)H(s)} \tag{8-1}$$

The transfer function relating $\Theta_o(s)$ to $\Theta_i(s)$ is called the *closed-loop transfer function*. This transfer function relates the closed-loop system dynamics to the dynamics of the feedforward elements and feedback elements. Since the Laplace transform of the output $\Theta_o(s)$ is given by Eq. (8-1) as

$$\Theta_o(s) = \frac{G(s)}{1 + G(s)H(s)}\Theta_i(s)$$

the behavior of a given closed-loop system clearly depends on both the closed-loop transfer function and the nature of the input.

Procedures for drawing a block diagram. In order to draw a block diagram for a system, first write the equations that describe the dynamic behavior of each component. Then take the Laplace transforms of these equations, assuming zero initial conditions, and represent each Laplace-transformed equation individually in block form. Finally, assemble the elements into a complete block diagram.

As an example, consider the RC circuit shown in Fig. 8-5(a). The equations for this circuit are

$$i = \frac{e_i - e_o}{R} \tag{8-2}$$

$$e_o = \frac{1}{C}\int i\, dt \tag{8-3}$$

The Laplace transforms of Eqs. (8-2) and (8-3), with zero initial condition, become

$$I(s) = \frac{1}{R}[E_i(s) - E_o(s)] \tag{8-4}$$

$$E_o(s) = \frac{1}{Cs}I(s) \tag{8-5}$$

Equation (8-4) represents a summing operation, and the corresponding diagram is shown in Fig. 8-5(b). Equation (8-5) represents the block as shown in Fig. 8-5(c). Assembling these two elements, we obtain the overall block diagram for the system as shown in Fig. 8-5(d).

Block diagram reduction. It should be noted that blocks can be connected in series only if the output of one block is not affected by the next following block. If there are any loading effects between the components, these components must be combined into a single block.

Any number of cascaded blocks representing nonloading components

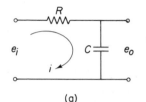

(a)

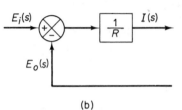

(b)

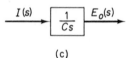

(c)

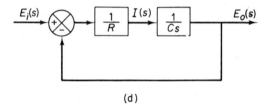

(d)

Fig. 8-5. (a) *RC* circuit; (b) block diagram corresponding to Eq. (8-4); (c) block diagram corresponding to Eq. (8-5); (d) block diagram for the *RC* circuit.

can be replaced by a single block, the transfer function of which is simply the product of the individual transfer functions.

A complicated block diagram involving many feedback loops can be simplified by a step-by-step rearrangement, using rules of block diagram algebra. Some of these important rules are given in Table 8-1. They are obtained by writing the same equation in a different way. Simplification of the block diagram by rearrangements and substitutions considerably reduces the labor needed for subsequent mathematical analysis. It should be noted, however, that as the block diagram is simplified, the transfer functions in new blocks become more complex because new poles and new zeros are generated.

In simplifying a block diagram, remember the following.

1. The product of the transfer functions in the feedforward direction must remain the same.

2. The product of the transfer functions around the loop must remain
the same.

A general rule for simplifying a block diagram is to move branch points
and summing points, interchange summing points, and then reduce internal
feedback loops.

Example 8-1. Consider the system shown in Fig. 8-6(a). Simplify this diagram by
using the rules given in Table 8-1.

By moving the summing point of the negative feedback loop containing H_2
outside the positive feedback loop containing H_1, we obtain Fig. 8-6(b). Eliminating
the positive feedback loop, we have Fig. 8-6(c). Then elimination of the loop containing H_2/G_1 gives Fig. 8-6(d). Finally, eliminating the feedback loop results in
Fig. 8-6(e).

Notice that the numerator of the closed-loop transfer function $\Theta_o(s)/\Theta_i(s)$ is
the product of the transfer functions of the feedforward path. The denominator of
$\Theta_o(s)/\Theta_i(s)$ is equal to

$$1 - \sum \text{(product of the transfer functions around each loop)}$$

$$= 1 - (G_1H_1 - G_2H_2 - G_1G_2)$$

$$= 1 - G_1H_1 + G_2H_2 + G_1G_2$$

(The positive feedback loop yields a negative term in the denominator.)

Example 8-2. Draw a block diagram for the circuit shown in Fig. 8-7. Then simplify the block diagram and obtain the transfer function between $E_o(s)$ and $E_i(s)$.

Let us define the voltage across capacitance C_1 as e_1. Then the equations for
the circuit are

$$e_i - e_1 = R_1 i_1$$

$$e_1 - e_o = R_2 i_2$$

$$e_1 = \frac{1}{C_1} \int (i_1 - i_2)\, dt$$

$$e_o = \frac{1}{C_2} \int i_2\, dt$$

Rewriting these four equations in Laplace-transformed forms,

$$I_1(s) = \frac{1}{R_1}[E_i(s) - E_1(s)] \tag{8-6}$$

$$I_2(s) = \frac{1}{R_2}[E_1(s) - E_o(s)] \tag{8-7}$$

$$E_1(s) = \frac{1}{C_1 s}[I_1(s) - I_2(s)] \tag{8-8}$$

$$E_o(s) = \frac{1}{C_2 s}I_2(s) \tag{8-9}$$

Table 8-1. RULES OF BLOCK DIAGRAM ALGEBRA

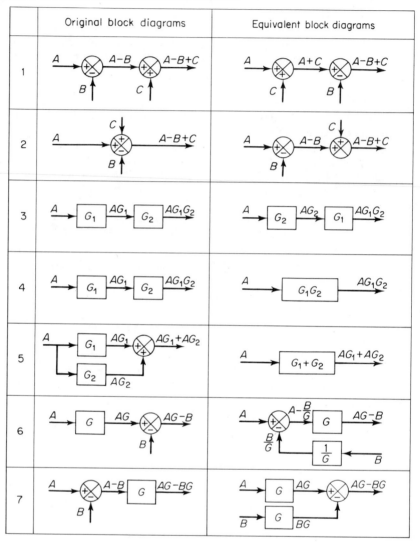

Table 8-1.   (CONTINUED)

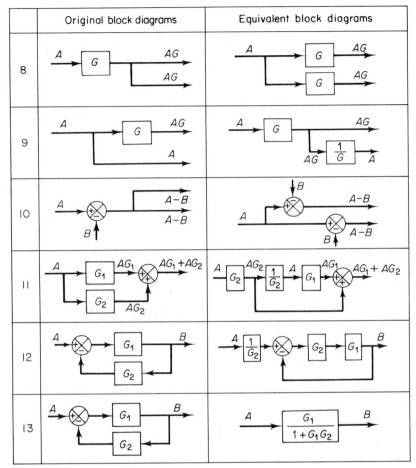

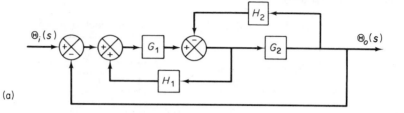

(a)

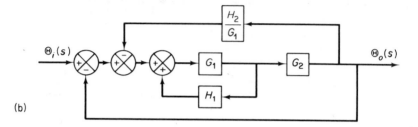

(b)

(c)

(d)

(e)

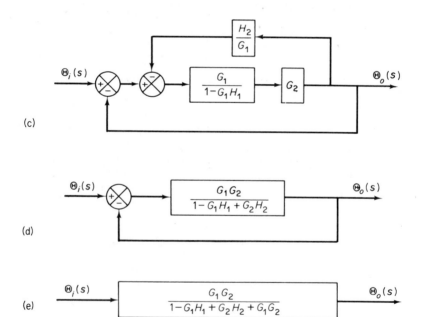

Fig. 8-6. Block diagram of a system and simplified diagrams.

From Eqs. (8-6) through (8-9) we obtain elements of the block diagram as shown in Fig. 8-8(a). By connecting signals properly, we can construct a block diagram as in Fig. 8-8(b). The interaction of the two simple *RC* circuits can be seen clearly in

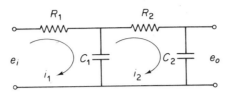

Fig. 8-7. Electrical circuit.

the diagram. Using rules of block diagram algebra given in Table 8-1, this diagram can be simplified into the one shown in Fig. 8-8(c). Further simplification results in Fig. 8-8(d) and (e). The transfer function between $E_o(s)$ and $E_i(s)$ is thus

$$\frac{E_o(s)}{E_i(s)} = \frac{1}{R_1 C_1 R_2 C_2 s^2 + (R_1 C_1 + R_2 C_2 + R_1 C_2)s + 1}$$

8-3   INDUSTRIAL AUTOMATIC CONTROLLERS

An automatic controller compares the actual value of the plant output with the desired value, determines the deviation, and produces a control signal that will reduce the deviation to zero or a small value. The way in which the automatic controller produces the control signal is called the *control action.*

Here we describe the fundamental control actions commonly used in industrial automatic controllers, followed by the basic principles of pneumatic controllers, hydraulic controllers, and electronic controllers.

Control actions. The control actions normally found in industrial automatic controllers consist of: two-position or on-off, proportional, integral, derivative, and combinations of proportional, integral, or derivative. A good understanding of the basic properties of various control actions is necessary in order for the engineer to select the one best suited to his particular application.

Classifications of industrial automatic controllers. Industrial automatic controllers can be classified according to their control action as

1. Two-position or on-off controllers
2. Proportional controllers
3. Integral controllers
4. Proportional-plus-integral controllers
5. Proportional-plus-derivative controllers
6. Proportional-plus-derivative-plus-integral controllers

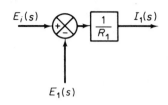

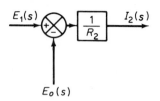

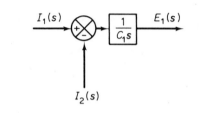

(a)

$E_i(s)$ → $\dfrac{1}{R_1}$ → $I_1(s)$ with $E_1(s)$

$E_1(s)$ → $\dfrac{1}{R_2}$ → $I_2(s)$ with $E_o(s)$

$I_1(s)$ → $\dfrac{1}{C_1 s}$ → $E_1(s)$ with $I_2(s)$

$I_2(s)$ → $\dfrac{1}{C_2 s}$ → $E_o(s)$

(b)

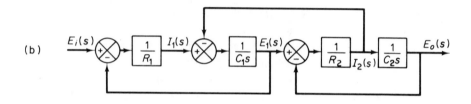

(c)

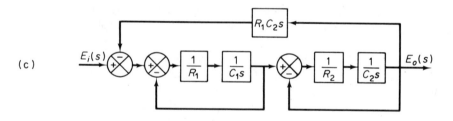

(d)

$E_i(s)$ → $\dfrac{1}{R_1 C_1 s + 1}$ → $\dfrac{1}{R_2 C_2 s + 1}$ → $E_o(s)$, feedback $R_1 C_2 s$

(e)

$$E_i(s) \rightarrow \boxed{\dfrac{1}{R_1 C_1 R_2 C_2 s^2 + (R_1 C_1 + R_2 C_2 + R_1 C_2)s + 1}} \rightarrow E_o(s)$$

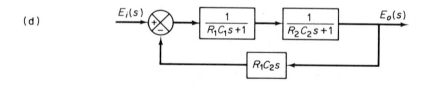

Fig. 8-8. (a) Elements of a block diagram; (b) block diagram as a result of combining elements; (c), (d), (e) simplified block diagrams.

490

Automatic controller, actuator, and measuring element. Figure 8-9 is a block diagram of an industrial control system, which consists of an automatic controller, an actuator, a plant, and a measuring element. The controller detects the actuating error signal, which is usually at a very low power level, and amplifies it to a sufficiently high level. (Thus the automatic controller comprises an error detector and amplifier.) Quite often a suitable feedback circuit, together with an amplifier, is used to alter the actuating error signal to produce a better control signal.

The actuator is an element that produces the input to the plant according to the control signal so that the feedback signal will correspond to the reference input signal.

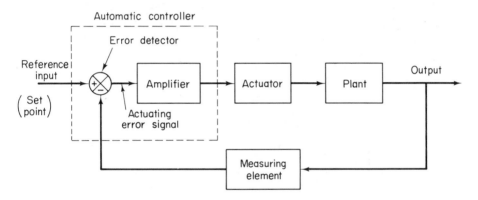

Fig. 8-9. Block diagram of an industrial control system, which consists of an automatic controller, an actuator, a plant, and a measuring element.

The measuring element is a device that converts the output variable into another suitable variable, such as a displacement, pressure, or voltage, which can be used to compare the output to the reference input signal. This element is in the feedback path of the closed-loop system. The set point of the controller must be converted to a reference input of the same units as the feedback signal from the measuring element.

Two-position or on-off control action. In a two-position control system the actuating element has only two fixed positions, which are, in many cases, simply on and off. Two-position or on-off control is simple and inexpensive and, for this reason, is extensively used in both industrial and household control systems.

To explain the concept let the output signal from the controller be $m(t)$ and the actuating error signal be $e(t)$. In two-position control the signal $m(t)$ remains at either a maximum or minimum value, depending on whether the actuating error signal is positive or negative, so that

$$m(t) = M_1 \quad \text{for} \quad e(t) > 0$$
$$= M_2 \quad \text{for} \quad e(t) < 0$$

where M_1 and M_2 are constants. The minimum value M_2 is generally either zero or $-M_1$. As a rule, two-position controllers are electrical devices, and an electric solenoid-operated valve is widely used in such controllers. Pneumatic proportional controllers with very high gains act as two-position controllers and are sometimes called pneumatic two-position controllers.

Figure 8-10(a) and (b) shows the block diagrams for this type of controller. The range through which the actuating error signal must move before the switching occurs is called the *differential gap* [see Fig. 8-10(b)]. Such a gap causes the controller output $m(t)$ to maintain its present value until the actuating error signal has moved slightly beyond the zero value. In some cases, the differential gap is a result of unintentional friction and lost motion; however, quite often it is intentionally provided in order to prevent too frequent operation of the on-off mechanism.

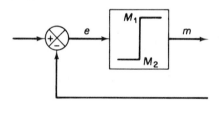

(a)

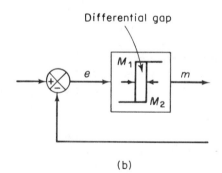

(b)

Fig. 8-10. (a) Block diagram of a two-position controller; (b) block diagram of a two-position controller with a differential gap.

Let us look at the liquid-level control system of Fig. 8-11. With two-position control the input valve is either open or closed, and so the liquid inflow rate is either a positive constant or zero. As shown in Fig. 8-12, the output signal continuously moves between the two limits required, thereby causing the actuating element to move from one fixed position to the other.

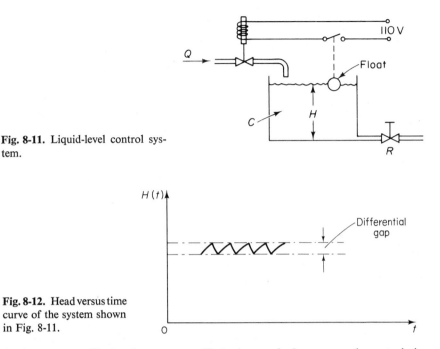

Fig. 8-11. Liquid-level control system.

Fig. 8-12. Head versus time curve of the system shown in Fig. 8-11.

Such output oscillation between two limits is a typical response characteristic of a system under two-position control.

From Fig. 8-12 we see that the amplitude of the output oscillation can be reduced by decreasing the differential gap. This step, however, increases the number of on-off switchings per unit time and reduces the useful life of the component. The magnitude of the differential gap must be determined from such factors as the accuracy required and the life of the component.

Proportional, integral, and derivative control actions. In addition to two-position or on-off control action, proportional, integral, and derivative control actions are basic control actions found in industrial automatic controllers. For each control action the relationship between the output of the controller $M(s)$ and the actuating error signal $E(s)$ is related by a transfer function of specific form. In what follows we illustrate transfer functions $M(s)/E(s)$ for proportional control action, proportional-plus-integral control action, proportional-plus-derivative control action, and proportional-plus-derivative-plus-integral control action.

Referring to the controller shown in Fig. 8-13, for proportional control action, $M(s)$ and $E(s)$ are related by

$$\frac{M(s)}{E(s)} = G_c(s) = K_p$$

where K_p is termed the *proportional gain.*

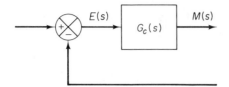

Fig. 8-13. Block diagram of a controller.

For integral control action, the relationship between $M(s)$ and $E(s)$ is

$$\frac{M(s)}{E(s)} = G_c(s) = \frac{K_i}{s}$$

where K_i is a constant.

For proportional-plus-integral control action, $M(s)$ and $E(s)$ are related by

$$\frac{M(s)}{E(s)} = G_c(s) = K_p\left(1 + \frac{1}{T_i s}\right)$$

where K_p is the proportional gain and T_i is a constant called the *integral time*.

For proportional-plus-derivative control action, $M(s)$ and $E(s)$ are related by

$$\frac{M(s)}{E(s)} = G_c(s) = K_p(1 + T_d s)$$

where K_p is the proportional gain and T_d is a constant called the *derivative time*.

Similarly, for proportional-plus-derivative-plus-integral control action, $M(s)$ and $E(s)$ are related by

$$\frac{M(s)}{E(s)} = G_c(s) = K_p\left(1 + T_d s + \frac{1}{T_i s}\right)$$

Pneumatic controllers. The past decades have seen a great development in low-pressure pneumatic controllers for industrial control systems, and today they are used extensively in industrial processes. Reasons for their broad appeal include an explosion-proof character, simplicity, and ease of maintenance.

Pneumatic nozzle-flapper amplifiers. A schematic diagram of a pneumatic nozzle-flapper amplifier appears in Fig. 8-14(a). In this system pressurized air is fed through the orifice, and the air is ejected from the nozzle toward the flapper. Generally the supply pressure P_s for such a controller is 1.38×10^5 N/m² gage (1.4 kg$_f$/cm² gage or 20 psig). The diameter of the orifice is on the order of 0.25 mm (or 0.01 in.) and that of the nozzle is on the order of 0.4 mm (or 0.016 in.). To ensure proper functioning of the amplifier, the nozzle diameter must be larger than the orifice diameter.

In operating this system, the flapper is positioned against the nozzle opening. The nozzle back pressure P_b is controlled by the nozzle-flapper distance. As the flapper approaches the nozzle, the opposition to the flow of

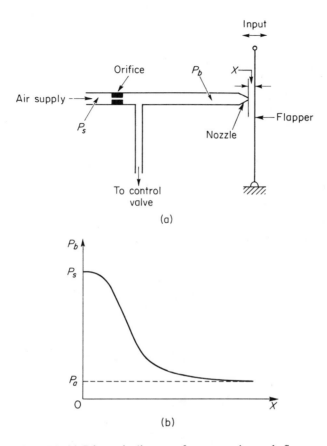

Fig. 8-14. (a) Schematic diagram of a pneumatic nozzle-flapper amplifier; (b) nozzle back pressure versus nozzle-flapper distance curve.

air through the nozzle increases with the result that nozzle back pressure P_b increases. If the nozzle is completely closed by the flapper, nozzle back pressure P_b becomes equal to supply pressure P_s. If the flapper is moved away from the nozzle, so that the nozzle-flapper distance is wide (on the order of 0.25 mm or 0.01 in.), then there is almost no restriction to flow, and nozzle back pressure P_b takes on a minimum value that depends on the nozzle-flapper device. (The lowest possible pressure is ambient pressure P_a.) A typical curve relating nozzle back pressure P_b to nozzle-flapper distance X is shown in Fig. 8-14(b). The steep and almost linear part of the curve is utilized in the actual operation of the nozzle-flapper amplifier.

The nozzle-flapper amplifier converts displacement into a pressure signal. Since industrial control systems require large output power to operate large pneumatic actuating valves, the power amplification of the nozzle-flapper amplifier is usually insufficient. Consequently, a pneumatic relay often serves as a power amplifier in connection with the nozzle-flapper amplifier.

Pneumatic relays. In practice, in a pneumatic controller, a nozzle-flapper amplifier acts as the first-stage amplifier and a pneumatic relay as the second-stage amplifier. The pneumatic relay is capable of handling a large quantity of airflow.

A schematic diagram of a pneumatic relay is shown in Fig. 8-15(a). As nozzle back pressure P_b increases, the ball valve is forced toward the lower seat, thereby decreasing control pressure P_c. Such a relay is called a *reverse acting relay*. When the ball valve is at the top of the seat, the atmospheric opening is closed and control pressure P_c becomes equal to supply pressure P_s. When the ball valve is at the bottom of its seat, it shuts off the air supply and control pressure P_c drops to the ambient pressure. The control pressure P_c can thus be made to vary from zero gage pressure to full supply pressure (from 0 N/m^2 gage to 1.38×10^5 N/m^2 gage, or from 0 psig to 20 psig). The

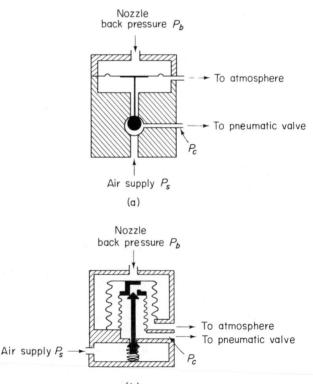

Fig. 8-15. (a) Schematic diagram of a bleed-type pneumatic relay; (b) schematic diagram of a nonbleed-type pneumatic relay.

total movement of the ball valve between the upper and lower seat is small (on the order of 0.25 mm or 0.01 in.). In all positions of the ball valve except at the top seat, air continues to bleed into the atmosphere, even after an equilibrium condition is attained between the nozzle back pressure and the control pressure. Thus the relay shown in Fig. 8-15(a) is called a *bleed-type relay.*

There is another type of relay, the *nonbleed* type. Here the air bleed stops when an equilibrium condition is reached, and so there is no loss of pressurized air at steady-state operation. A schematic diagram of a nonbleed-type relay is shown in Fig. 8-15(b).

Pneumatic proportional controllers. In the schematic diagram of a pneumatic proportional controller shown in Fig. 8-16, the nozzle-flapper amplifier constitutes the first-stage amplifier, and the nozzle back pressure is controlled by the nozzle-flapper distance. The relay-type amplifier constitutes the second-stage amplifier. The nozzle back pressure determines the position of the ball valve for the second-stage amplifier.

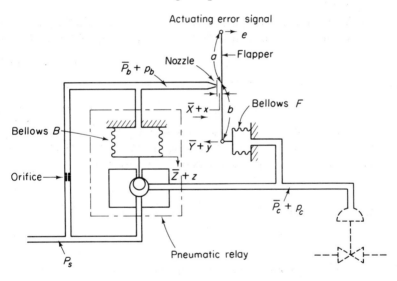

Fig. 8-16. Pneumatic proportional controller.

This controller operates as follows. The input signal to the two-stage pneumatic amplifier is the actuating error signal. Increasing the actuating error signal moves the flapper to the right. This step will, in turn, decrease the nozzle back pressure and bellows B will contract, which results in an upward movement of the ball valve. Consequently, there is more air flow to the pneumatic valve and control pressure increases. This increase will cause bellows F to expand and move the flapper to the left, thus closing the nozzle. Because of this feedback, the nozzle-flapper displacement is very small, but the change in the control pressure can be large. If the actuating error decreases, the nozzle back pressure increases and the ball valve moves down, thereby resulting in a decrease in control flow and an increase in bleeding to the atmosphere. This situation will cause the control pressure to decrease.

It should be noted that proper operation of the controller requires that the feedback bellows move the flapper less than that movement caused by the error signal alone. (If these two movements were equal, no control action would result.)

Equations for this controller can be derived as follows. When the actuating error is zero, or $e = 0$, an equilibrium state exists with the nozzle-flapper distance equal to $\bar{X}$, the displacement of bellows F equal to $\bar{Y}$, the displacement of bellows B equal to $\bar{Z}$, the nozzle back pressure equal to $\bar{P}_b$, and the control pressure equal to $\bar{P}_c$. When an actuating error exists, the nozzle-flapper distance, the displacements of bellows F and B, the nozzle back pressure, and the control pressure deviate from their respective equilibrium values. Let these deviations be x, y, z, p_b, and p_c, respectively. (The positive direction for each displacement variable is indicated by an arrowhead in the diagram.)

Assuming that the relationship between the variation in the nozzle back pressure and the variation in the nozzle-flapper distance is linear, we have

$$p_b = -K_1 x \tag{8-10}$$

where K_1 is a constant. For bellows B,

$$p_b = K_2 z \tag{8-11}$$

where K_2 is a constant. The position of the ball valve, which depends on the displacement of bellows B, determines the control pressure. If the ball valve is such that the relationship between p_c and z is linear, then

$$p_c = -K_3 z \tag{8-12}$$

where K_3 is a constant. From Eqs. (8-10), (8-11), and (8-12), we obtain

$$p_c = -\frac{K_3}{K_2} p_b = Kx \tag{8-13}$$

where $K = K_1 K_3 / K_2$ is a constant. For the flapper movement, we have

$$x = \frac{b}{a+b} e - \frac{a}{a+b} y \tag{8-14}$$

Bellows F acts like a spring, and the following equation holds.

$$A p_c = k_s y \tag{8-15}$$

Here A is the effective area of bellows F, and k_s is the equivalent spring constant—that is, the stiffness due to the action of the corrugated side of the bellows.

Assuming that all variations in the variables are within a linear range, we can obtain a block diagram for this system from Eqs. (8-13), (8-14), and

(8-15) as shown in Fig. 8-17(a). From Fig. 8-17(a) it can clearly be seen that the pneumatic controller shown in Fig. 8-16 is itself a feedback system. The transfer function between p_c and e is given by

$$\frac{P_c(s)}{E(s)} = \frac{\dfrac{b}{a+b}K}{1 + K\dfrac{a}{a+b}\dfrac{A}{k_s}} = K_p$$

A simplified block diagram is given in Fig. 8-17(b). Since p_c and e are proportional, the pneumatic controller in Fig. 8-16 is a proportional controller. In commercial proportional controllers, adjusting mechanisms are provided to vary gain K_p.

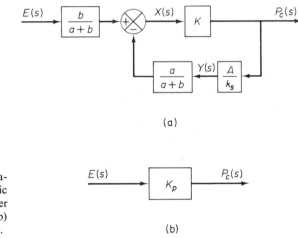

(a)

Fig. 8-17. (a) Block diagram for the pneumatic proportional controller shown in Fig. 8-16; (b) simplified block diagram.

(b)

As noted earlier, the actuating error signal moved the flapper in one direction, and the feedback bellows moved the flapper in the opposite direction but to a smaller degree. The effect of the feedback bellows is thus to reduce the sensitivity of the controller. The principle of feedback is commonly used to obtain wide proportional-band controllers.

Pneumatic controllers that do not have feedback mechanisms (which means that one end of the flapper is fixed) have high sensitivity and are called *pneumatic two-position controllers* or *pneumatic on-off controllers*. In such a controller, only a small motion between the nozzle and the flapper is required to give a complete change from the maximum to the minimum control pressure.

Obtaining derivative and integral control actions. The basic principle in generating a desired control action is to insert the inverse of the desired transfer function in the feedback path. For the system shown in Fig. 8-18, the

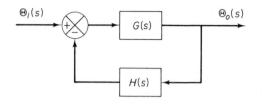

$\Theta_i(s)$ $\Theta_o(s)$

Fig. 8-18. Block diagram of a closed-loop system.

closed-loop transfer function is

$$\frac{\Theta_o(s)}{\Theta_i(s)} = \frac{G(s)}{1 + G(s)H(s)}$$

If $|G(s)H(s)| \gg 1$, then this closed-loop transfer function can be modified to

$$\frac{\Theta_o(s)}{\Theta_i(s)} = \frac{1}{H(s)}$$

So if proportional-plus-derivative control action is desired, we insert an element having the transfer function $1/(Ts + 1)$ in the feedback path; and if proportional-plus-integral control action is wanted, we insert an element having the transfer function $Ts/(Ts + 1)$ in the feedback path.

Pneumatic proportional-plus-derivative controllers. In the pneumatic bellows system shown in Fig. 8-19 the resistance of the restriction (valve) is denoted by R and the capacitance of the bellows by C. Referring to Sec. 5-5, the equation for this system can be given by Eq. (5-40), rewritten thus:

$$RC\frac{dp_o}{dt} + p_o = p_i$$

The transfer function of this bellows system is therefore

$$\frac{P_o(s)}{P_i(s)} = \frac{1}{RCs + 1}$$

Hence if the bellows system shown in Fig. 8-19 is inserted in the feedback path of the pneumatic proportional controller, the controller will become a proportional-plus-derivative one.

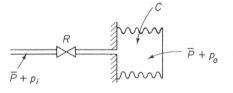

R

$\bar{P} + p_o$

$\bar{P} + p_i$

Fig. 8-19. Pneumatic bellows system.

Consider the pneumatic controller shown in Fig. 8-20(a). Assuming small changes in the actuating error, nozzle-flapper distance, and control pressure once more, a block diagram for this controller can be drawn as in Fig. 8-20(b). In the block diagram K is a constant, A is the area of the bellows,

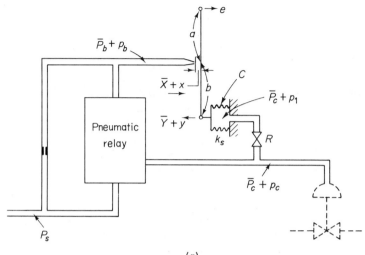

(a)

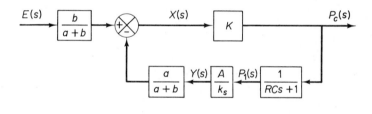

(b)

Fig. 8-20. (a) Pneumatic proportional-plus-derivative controller; (b) block diagram.

and k_s is the equivalent spring constant of the bellows. The transfer function between p_c and e can be found from the block diagram as follows:

$$\frac{P_c(s)}{E(s)} = \frac{\dfrac{b}{a+b}K}{1 + \dfrac{Ka}{a+b}\dfrac{A}{k_s}\dfrac{1}{RCs+1}}$$

Note that in such a controller the loop gain $|KaA/[(a+b)k_s(RCs+1)]|$ is usually made very much greater than unity. The transfer function $P_c(s)/E(s)$

can therefore be simplified to

$$\frac{P_c(s)}{E(s)} = \frac{bk_s}{aA}(RCs + 1) = K_p(T_d s + 1)$$

where

$$K_p = \frac{bk_s}{aA}, \qquad T_d = RC$$

Thus delayed negative feedback, or the transfer function $1/(RCs + 1)$, in the feedback path modifies the proportional controller to a proportional-plus-derivative controller.

Note that if the feedback valve is fully opened so that resistance R is negligible, or $R \doteq 0$, the control action becomes proportional. If the feedback valve is fully closed, so that $R = \infty$, the control action becomes two-position or on-off.

Pneumatic proportional-plus-integral controllers. Referring to the pneumatic controller shown in Fig. 8-21(a), the bellows denoted by I is connected to the control pressure source without any restriction. The bellows denoted by II is connected to the control pressure through a restriction (valve).

A block diagram for this controller under the assumption of small variations in the variables appears in Fig. 8-21(b). A simplification of this block diagram yields Fig. 8-21(c). The transfer function of this controller is

$$\frac{P_c(s)}{E(s)} = \frac{\dfrac{b}{a + b}K}{1 + \dfrac{Ka}{a + b}\dfrac{A}{k_s}\left(1 - \dfrac{1}{RCs + 1}\right)}$$

where K is a constant, A the area of the bellows, and k_s the equivalent spring constant of the bellows. If $|KaARCs/[(a + b)k_s(RCs + 1)]| \gg 1$, which is usually the case, the transfer function can be simplified to

$$\frac{P_c(s)}{E(s)} = \frac{bk_s}{aA}\frac{(RCs + 1)}{RCs} = K_p\left(1 + \frac{1}{T_i s}\right)$$

where

$$K_p = \frac{bk_s}{aA}, \qquad T_i = RC$$

Thus, the controller shown in Fig. 8-21(a) is a proportional-plus-integral controller.

Pneumatic proportional-plus-derivative-plus-integral controllers. A combination of the pneumatic controllers in Figs. 8-20(a) and 8-21(a) yields a proportional-plus-derivative-plus-integral controller. Figure 8-22(a) is a schematic diagram of this type. Here resistances R_i and R_d are chosen such that $R_i \gg R_d$. Figure 8-22(b) shows a block diagram of the controller under the assumption of small variations in the variables.

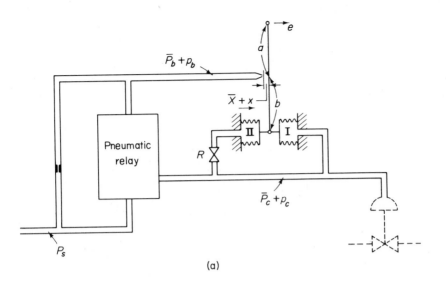

(a)

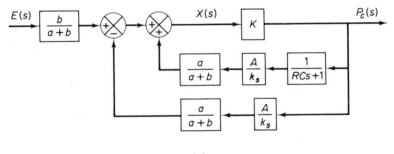

(b)

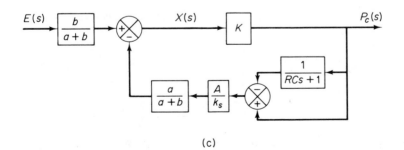

(c)

Fig. 8-21. (a) Pneumatic proportional-plus-integral controller; (b) block diagram; (c) simplified block diagram.

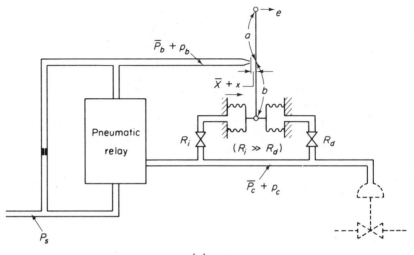

(a)

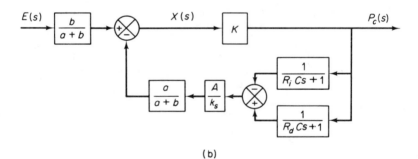

(b)

Fig. 8-22. (a) Pneumatic proportional-plus-derivative-plus-integral controller; (b) block diagram.

The transfer function of this controller is

$$\frac{P_c(s)}{E(s)} = \frac{\dfrac{b}{a+b}K}{1 + \dfrac{Ka}{a+b}\dfrac{A}{k_s}\dfrac{(R_iC - R_dC)s}{(R_dCs+1)(R_iCs+1)}} \qquad (8\text{-}16)$$

Note that generally $|KaA(R_iC - R_dC)s/[(a+b)k_s(R_dCs+1)(R_iCs+1)]|$ $\gg 1$. By defining

$$T_i = R_iC, \qquad T_d = R_dC$$

and noting that $T_i \gg T_d$, Eq. (8-16) simplifies to

$$\frac{P_c(s)}{E(s)} \doteq \frac{bk_s}{aA} \frac{(T_d s + 1)(T_i s + 1)}{(T_i - T_d)s}$$

$$\doteq \frac{bk_s}{aA} \frac{T_d T_i s^2 + T_i s + 1}{T_i s}$$

$$= K_p \left(1 + T_d s + \frac{1}{T_i s} \right) \tag{8-17}$$

where

$$K_p = \frac{bk_s}{aA}$$

Equation (8-17) indicates that the controller shown in Fig. 8-22(a) is a proportional-plus-derivative-plus-integral controller.

Hydraulic controllers. Like pneumatic controllers, hydraulic controllers are also used extensively in industry. High-pressure hydraulic systems enable very large forces to be derived. Moreover, these systems permit a rapid and accurate positioning of loads. And frequently a combination of electronic and hydraulic systems is found because of the advantages resulting from a mixture of both electronic control and hydraulic power.

Hydraulic integral controllers. The hydraulic servomotor shown in Fig. 8-23 is essentially a pilot-valve-controlled hydraulic power amplifier and actuator. The pilot valve is a balanced valve in the sense that the pressure forces acting on it are all balanced. A very large power output can be controlled by a pilot valve, which can be positioned with very little power.

In the present analysis, we assume that hydraulic fluid is incompressible and that the inertia force of the power piston and load is negligible compared to the hydraulic force at the power piston. We also assume that the pilot

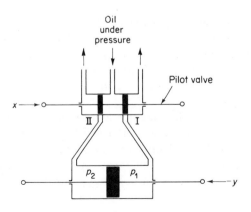

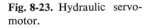
Fig. 8-23. Hydraulic servo-motor.

valve is a zero-lapped valve and the oil flow rate is proportional to the pilot valve displacement.

Operation of this hydraulic servomotor is as follows. If input x moves the pilot valve to the right, port I is uncovered, and so high-pressure oil enters the right-hand side of the power piston. Since port II is connected to the drain port, the oil in the left-hand side of the power piston is returned to the drain. The oil flowing into the power cylinder is at high pressure; the oil flowing out from the power cylinder into the drain is at low pressure. The resulting difference in pressure on both sides of the power piston will cause it to move to the left.

Note that the rate of flow of oil q (kg/s) times dt (s) is equal to the power piston displacement dy (m) times the piston area A (m²) times the density of oil ρ (kg/m³). Therefore

$$A\rho \, dy = q \, dt \qquad (8\text{-}18)$$

Because of the assumption that the oil flow rate q is proportional to the pilot valve displacement x, we have

$$q = K_1 x \qquad (8\text{-}19)$$

where K_1 is a proportionality constant. From Eqs. (8-18) and (8-19) we obtain

$$A\rho \frac{dy}{dt} = K_1 x$$

The Laplace transform of this last equation, assuming a zero initial condition, gives

$$A\rho s \, Y(s) = K_1 X(s)$$

or

$$\frac{Y(s)}{X(s)} = \frac{K_1}{A\rho s} = \frac{K}{s}$$

where $K = K_1/(A\rho)$. Thus the hydraulic servomotor shown in Fig. 8-23 acts as an integral controller.

Hydraulic proportional controllers. The hydraulic servomotor of Fig. 8-23 can be modified to a proportional controller by means of a feedback link. Consider the hydraulic controller shown in Fig. 8-24(a). The left-hand side of the pilot valve is joined to the left-hand side of the power piston by a link ABC. This link is a floating link rather than one moving about a fixed pivot.

The controller here operates in the following way. If input x moves the pilot valve to the right, port I will be uncovered and high-pressure oil will flow through port I into the right-hand side of the power piston and force this piston to the left. The power piston, in moving to the left, will carry the feedback link ABC with it, thereby moving the pilot valve to the left. This

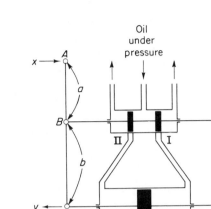

(a)

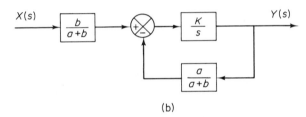

Fig. 8-24. (a) Hydraulic proportional controller; (b) block diagram.

(b)

action continues until the pilot valve again covers ports I and II. A block diagram of the system can be drawn as in Fig. 8-24(b). The transfer function between y and x is given by

$$\frac{Y(s)}{X(s)} = \frac{\dfrac{b}{a+b}\dfrac{K}{s}}{1 + \dfrac{K}{s}\dfrac{a}{a+b}} = \frac{bK}{s(a+b)+Ka} \tag{8-20}$$

Noting that under the normal operating conditions we have $|Ka/[s(a+b)]| \gg 1$, and so Eq. (8-20) can be simplified to

$$\frac{Y(s)}{X(s)} = \frac{b}{a} = K_p$$

The transfer function between y and x becomes a constant. Thus, the hydraulic controller shown in Fig. 8-24(a) acts as a proportional controller, the gain of which is K_p. This gain can be adjusted by effectively changing the lever ratio b/a. (The adjusting mechanism is not shown in the diagram.)

Electronic proportional controllers. An electronic proportional controller is an amplifier that receives a small voltage signal and produces a voltage output at a higher power level. A schematic diagram of such a controller is shown in Fig. 8-25. For this controller,

$$e_o = K\left(e_i - e_o\frac{R_2}{R_1}\right)$$

where $R_2 > 0$ and $KR_2/R_1 \gg 1$. Consequently,

$$\frac{E_o(s)}{E_i(s)} = \frac{R_1}{R_2} = K_p$$

where K_p is the gain of the amplifier or proportional controller. The gain K_p can be adjusted by changing the ratio of resistances (the value of R_1/R_2) in the feedback circuit.

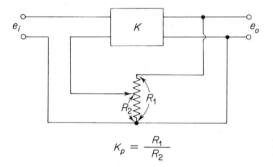

$$K_p = \frac{R_1}{R_2}$$

Fig. 8-25. Electronic proportional controller.

Obtaining derivative and integral control actions in electronic controllers. In the following discussion the principle involved in obtaining derivative and integral control actions in electronic controllers is described. Essentially, we insert an appropriate circuit in the feedback path in order to generate the proportional-plus-derivative control action, proportional-plus-integral control action, or proportional-plus-derivative-plus-integral control action. For the controller shown in Fig. 8-26,

$$\frac{E_f(s)}{E_o(s)} = \frac{1}{R_d C_d s + 1}$$

$$[E_i(s) - E_f(s)]K = E_o(s)$$

So for $|K/(R_d C_d s + 1)| \gg 1$, which is usually the case,

$$\frac{E_o(s)}{E_i(s)} = \frac{K(R_d C_d s + 1)}{R_d C_d s + 1 + K} \doteq R_d C_d s + 1 = T_d s + 1$$

where $T_d = R_d C_d$. Thus the controller shown in Fig. 8-26 is a proportional-plus-derivative controller.

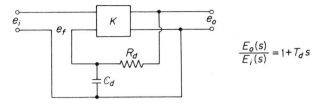

Fig. 8-26. Electronic proportional-plus-derivative controller.

Similarly, for the controller shown in Fig. 8-27,

$$\frac{E_f(s)}{E_o(s)} = \frac{R_i C_i s}{R_i C_i s + 1}$$

$$[E_i(s) - E_f(s)]K = E_o(s)$$

And so for $|KR_i C_i s/(R_i C_i s + 1)| \gg 1$, which is usually the case,

$$\frac{E_o(s)}{E_i(s)} = \frac{K(R_i C_i s + 1)}{KR_i C_i s + R_i C_i s + 1} \doteq \frac{R_i C_i s + 1}{R_i C_i s} = 1 + \frac{1}{T_i s}$$

where $T_i = R_i C_i$. Consequently, the controller shown in Fig. 8-27 is a proportional-plus-integral controller.

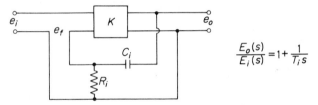

Fig. 8-27. Electronic proportional-plus-integral controller.

8-4   TRANSIENT RESPONSE ANALYSIS

In this section we are concerned with the transient response analysis of control systems and with the effects of integral and derivative control actions on the transient response performance. We begin with an analysis of the proportional control of a first-order system, followed by description of the effects of integral and derivative control actions on the transient performance. Then we present the proportional control of a system with an inertia load and illustrate the fact that adding derivative control action markedly improves the transient performance.

Proportional control of first-order system. Suppose that the controller in the liquid-level control system of Fig. 8-28 is a proportional one. Suppose also that the reference input for the system is $\bar{X}$. At $t = 0$ a change in the reference input is made from $\bar{X}$ to $\bar{X} + x$. Assume that all the variables shown in the diagram—x, q_i, h, and q_o—are measured from their respective steady-state values $\bar{X}$, $\bar{Q}$, $\bar{H}$, and $\bar{Q}$. We also assume that the magnitudes of the variables x, q_i, h, and q_o are sufficiently small, which means that the system can be approximated by a linear mathematical model.

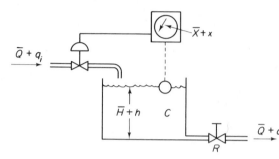

Fig. 8-28. Liquid-level control system.

Referring to Sec. 4-5, the equation for the liquid-level system can be derived as

$$RC\frac{dh}{dt} + h = Rq_i \qquad (8\text{-}21)$$

[Refer to Eq. (4-17).] So the transfer function between $H(s)$ and $Q_i(s)$ is found from Eq. (8-21) as

$$\frac{H(s)}{Q_i(s)} = \frac{R}{RCs + 1}$$

Here we assume that the gain K_v of the control valve is constant near the steady-state operating condition. Then since the controller is a proportional one, the change in inflow rate q_i is proportional to the actuating error e or

$$q_i = K_p K_v e \qquad (8\text{-}22)$$

where K_p is the gain of the proportional controller. In terms of Laplace-transformed quantities, Eq. (8-22) becomes

$$Q_i(s) = K_p K_v E(s)$$

A block diagram for this system appears in Fig. 8-29(a). To simplify our analysis, we assume that x and h are the same kind of signal with the same units and so they can be compared directly. (Otherwise we must insert a feedback transfer function K_b in the feedback path.) A simplified block diagram is given in Fig. 8-29(b), where $K = K_p K_v$.

In the following material we shall investigate the response $h(t)$ to a change in the reference input. We shall assume a unit-step change in $x(t)$. The closed-loop transfer function between $H(s)$ and $X(s)$ is given by

$$\frac{H(s)}{X(s)} = \frac{KR}{RCs + 1 + KR} \qquad (8\text{-}23)$$

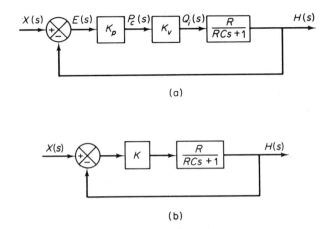

(a)

(b)

Fig. 8-29. (a) Block diagram of the liquid-level control system shown in Fig. 8-28; (b) simplified block diagram.

Since the Laplace transform of the unit-step function is $1/s$, substituting $X(s) = 1/s$ into Eq. (8-23) gives

$$H(s) = \frac{KR}{RCs + 1 + KR} \frac{1}{s}$$

Then the expansion of $H(s)$ into partial fractions results in

$$H(s) = \frac{KR}{1 + KR} \frac{1}{s} - \frac{KR}{1 + KR} \frac{1}{s + [(1 + KR)/RC]} \qquad (8\text{-}24)$$

Next, by taking the inverse Laplace transforms of both sides of Eq. (8-24), we obtain the time solution $h(t)$.

$$h(t) = \frac{KR}{1 + KR}(1 - e^{-t/T_1}) \qquad \text{for } t \geq 0 \qquad (8\text{-}25)$$

where

$$T_1 = \frac{RC}{1 + KR}$$

Notice that the time constant T_1 of the closed-loop system is different from the time constant RC of the liquid-level system alone.

The response curve $h(t)$ is plotted in Fig. 8-30. From Eq. (8-25) we see that as t approaches infinity, the value of $h(t)$ approaches $KR/(1 + KR)$ or

$$h(\infty) = \frac{KR}{1 + KR}$$

Since $x(\infty) = 1$, there is a steady-state error of magnitude $1/(1 + KR)$. Such an error is called *offset*. The value of the offset becomes smaller as the gain K becomes larger.

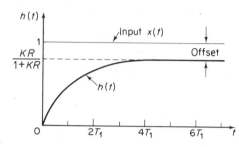

Fig. 8-30. Unit step response curve for the system shown in Fig. 8-29(b).

Eliminating offset by use of integral control. In the proportional control of a plant whose transfer function does not possess an integrator $1/s$ (thus the feedforward transfer function does not involve an integrator or integrators), there is a steady-state error, or offset, in the response to a step input. Such an offset can be eliminated if integral control action is included in the controller.

Under integral control action the control signal (the output signal from the controller) at any instant is the area under the actuating-error-signal curve up to that instant. The control signal $m(t)$ can have a nonzero value when the actuating error signal $e(t)$ is zero, as Fig. 8-31(a) shows. This situation is impossible in the case of the proportional controller, since a nonzero control signal requires a nonzero actuating error signal. (A nonzero actuating error signal at steady state means that there is an offset.) Figure 8-31(b) shows the curve $e(t)$ versus t and the corresponding curve $m(t)$ versus t when the controller is of the proportional type.

Note that integral control action improves steady-state accuracy by removing offset or steady-state error. Yet it may lead to an oscillatory response of slowly decreasing amplitude or even increasing amplitude, both of which are usually undesirable.

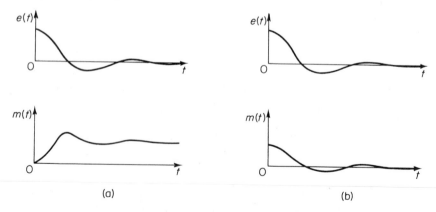

(a)   (b)

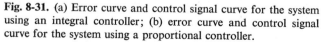

Fig. 8-31. (a) Error curve and control signal curve for the system using an integral controller; (b) error curve and control signal curve for the system using a proportional controller.

Integral control of a liquid-level system. Figure 8-32(a) shows a liquid-level control system. Here we assume that the controller is an integral one. We also assume that the variables x, q_i, h, and q_o, which are measured from their respective steady-state values $\bar{X}$, $\bar{Q}$, $\bar{H}$, and $\bar{Q}$, are small quantities; therefore the system can be considered linear. Under these assumptions the block diagram of the system can be obtained as shown in Fig. 8-32(b). From this diagram the closed-loop transfer function between $H(s)$ and $X(s)$ is

$$\frac{H(s)}{X(s)} = \frac{KR}{RCs^2 + s + KR}$$

It follows that

$$\frac{E(s)}{X(s)} = \frac{X(s) - H(s)}{X(s)} = \frac{RCs^2 + s}{RCs^2 + s + KR}$$

Since the system is a stable one, the steady-state error e_{ss} for the unit-step response is found by applying the final value theorem.

$$e_{ss} = \lim_{s \to 0} sE(s)$$
$$= \lim_{s \to 0} \frac{s(RCs^2 + s)}{RCs^2 + s + KR} \frac{1}{s}$$
$$= 0$$

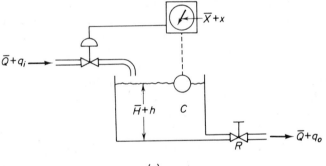

(a)

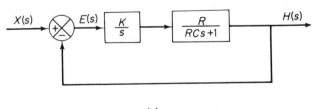

(b)

Fig. 8-32. (a) Liquid-level control system; (b) block diagram.

Integral control of the liquid-level system thus eliminates the steady-state error in the response to the step input, thereby improving steady-state accuracy. This is an important improvement over proportional control alone, which gives offset.

It should be noted that proportional-plus-integral control action gives just as good a steady-state accuracy as integral control action alone. In fact, the use of proportional-plus-integral control action will enable the transient response to decay faster.

Derivative control action. Derivative control action, when added to a proportional controller, provides a means of obtaining a controller with high sensitivity. An advantage of using derivative action is that it responds to the rate of change of the actuating error and can produce a significant correction before the magnitude of the actuating error becomes too large. Derivative control thus anticipates the actuating error, initiates an early corrective action, and tends to increase the stability of the system.

Although derivative control does not affect the steady-state error directly, it adds damping to the system, and therefore permits the use of a larger value of the system gain, a factor that will result in improving steady-state accuracy.

Notice that because derivative control operates on the rate of change of the actuating error and not on the actuating error itself, this mode is never used alone. It is always used in combination with proportional or proportional-plus-integral control action.

Proportional control of a system with inertia load. Before considering the effect of derivative control action on system performance, let us discuss the proportional control of an inertia load.

In the position control system of Fig. 8-33(a) the box with the transfer function K_p represents a proportional controller. Its output is a torque signal T, which is applied to an inertia element J. The output of the system is the angular displacement θ_o of the inertia element. For the inertia element, we have

$$J\ddot{\theta}_o = T$$

The Laplace transform of this last equation, assuming zero initial conditions, becomes

$$Js^2\Theta_o(s) = T(s)$$

Hence

$$\frac{\Theta_o(s)}{T(s)} = \frac{1}{Js^2}$$

The diagram of Fig. 8-33(a) can be redrawn as in Fig. 8-33(b). From this

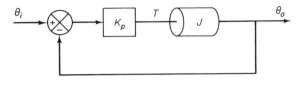

(a)

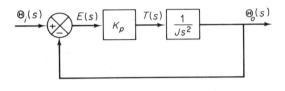

(b)

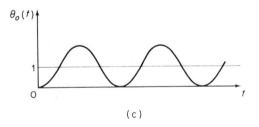

Fig. 8-33. (a) Position control system; (b) block diagram; (c) unit step response
curve.

(c)

diagram the closed-loop transfer function can be obtained as

$$\frac{\Theta_o(s)}{\Theta_i(s)} = \frac{K_p}{Js^2 + K_p}$$

Since the roots of the characteristic equation

$$Js^2 + K_p = 0$$

are imaginary, the response to a unit-step input continues to oscillate indefinitely as shown in Fig. 8-33(c).

Control systems exhibiting such sustained oscillations are not acceptable. We shall see that the addition of derivative control will stabilize the system.

Proportional-plus-derivative control of a system with inertia load. Let us modify the proportional controller to a proportional-plus-derivative controller whose transfer function is $K_p(1 + T_d s)$. The torque developed by the controller is proportional to $K_p(e + T_d \dot{e})$, where e is the actuating error signal. Derivative control action is essentially anticipatory; it measures the instantaneous error velocity and predicts the large overshoot ahead of time in order to produce an appropriate counteraction before too large an overshoot occurs.

For the system shown in Fig. 8-34(a), the closed-loop transfer function is given by

$$\frac{\Theta_o(s)}{\Theta_i(s)} = \frac{K_p(1 + T_d s)}{Js^2 + K_p T_d s + K_p}$$

The characteristic equation

$$Js^2 + K_p T_d s + K_p = 0$$

now has two roots with negative real parts, for positive values of J, K_p, and T_d. Thus derivative control action introduces a damping effect. A typical response curve $\theta_o(t)$ to a unit-step input is given in Fig. 8-34(b). Clearly, the response curve shows a marked improvement over the original response curve, shown in Fig. 8-33(c).

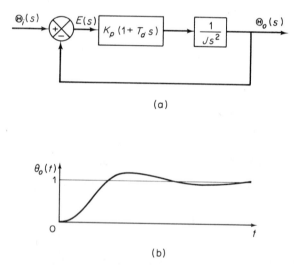

(a)

(b)

Fig. 8-34. (a) Block diagram of a position control system using a proportional-plus-derivative controller; (b) unit step response curve.

8-5 TRANSIENT RESPONSE SPECIFICATIONS

Because systems with energy storages cannot respond instantaneously, they will exhibit transient response when subjected to inputs or disturbances. Consequently, the transient response characteristics constitute one of the most important factors in system design.

In many practical cases, the desired performance characteristics of control systems can be given in terms of transient response specifications.

Frequently, such performance characteristics are specified in terms of the transient response to a unit-step input, since it is easy to generate and is sufficiently drastic. (If the response of a linear system to a step input is known, it is mathematically possible to compute the response to any input.)

The transient response of a system to a unit-step input depends on initial conditions. For convenience in comparing the transient responses of various systems, it is common practice to use a standard initial condition: The system is at rest initially with output and all time derivatives thereof zero. Then the response characteristics can be easily compared.

Transient response specifications. The transient response of a practical control system often exhibits damped oscillations before reaching steady state. In specifying the transient response characteristics of a control system to a unit-step input, it is common to name the following.

1. Delay time, t_d
2. Rise time, t_r
3. Peak time, t_p
4. Maximum overshoot, M_p
5. Settling time, t_s

These specifications are defined in what follows and are graphically shown in Fig. 8-35.

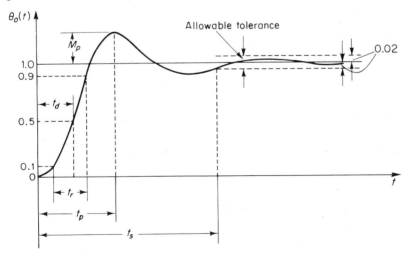

Fig. 8-35. Diagram showing transient response specifications.

Delay time. The delay time t_d is the time needed for the response to reach half the final value the very first time.

Rise time. The rise time t_r is the time required for the response to rise from 10 to 90%, or 5 to 95%, or 0 to 100% of its final value. For underdamped second-order systems, the 0 to 100% rise time is normally used. For over-damped systems, the 10 to 90% rise time is common.

Peak time. The peak time t_p is the time required for the response to reach the first peak of the overshoot.

Maximum (percent) overshoot. The maximum overshoot M_p is the maximum peak value of the response curve [$\theta_o(t)$ vs. t curve] measured from unity. If the final steady-state value of the response differs from unity, then it is common practice to use the maximum percent overshoot. It is defined by

$$\text{Maximum percent overshoot} = \frac{\theta_o(t_p) - \theta_o(\infty)}{\theta_o(\infty)} \times 100\%$$

The amount of the maximum (percent) overshoot directly indicates the relative stability of the system.

Settling time. The settling time t_s is the time required for the response curve to reach and stay within 2% of the final value. (Note that in some cases 5%, instead of 2%, is used as the percentage of the final value.) The settling time is related to the largest time constant of the system.

Comments. Note that if we specify the values of t_d, t_r, t_p, t_s, and M_p, the shape of the response curve is virtually determined. This fact can be seen clearly from Fig. 8-36.

In addition, note that not all these specifications necessarily apply to any given case. For instance, for an overdamped system, the terms peak time and maximum overshoot do not apply.

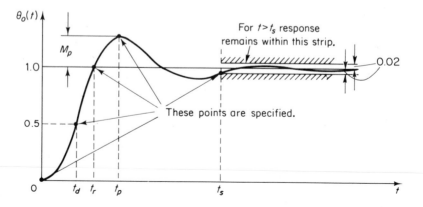

Fig. 8-36. Specifications of transient response curve.

Position control system. The position control system (servomechanism) shown in Fig. 8-37(a) consists of a proportional controller and load elements (inertia and viscous friction elements). Suppose that we wish to control the output position θ_o in accordance with the input position θ_i.

The equation for the load elements is

$$J\ddot{\theta}_o + b\dot{\theta}_o = T$$

where T is the torque produced by the proportional controller whose gain constant is K. By taking Laplace transforms of both sides of this last equation, assuming the zero initial conditions, we find

$$Js^2\Theta_o(s) + bs\Theta_o(s) = T(s)$$

So the transfer function between $\Theta_o(s)$ and $T(s)$ is

$$\frac{\Theta_o(s)}{T(s)} = \frac{1}{s(Js + b)}$$

By using this transfer function, Fig. 8-37(a) can be redrawn as in Fig. 8-37(b). The closed-loop transfer function is then obtained as

$$\frac{\Theta_o(s)}{\Theta_i(s)} = \frac{K}{Js^2 + bs + K} = \frac{K/J}{s^2 + (b/J)s + (K/J)}$$

or

$$\frac{\Theta_o(s)}{\Theta_i(s)} = \frac{\omega_n^2}{s^2 + 2\zeta\omega_n s + \omega_n^2} \tag{8-26}$$

where

$$\omega_n = \sqrt{\frac{K}{J}} = \text{undamped natural frequency}$$

$$\zeta = \frac{b}{2\sqrt{KJ}} = \text{damping ratio}$$

In terms of ζ and ω_n, the block diagram of Fig. 8-37(b) can be redrawn as in Fig. 8-37(c).

Next, let us consider the unit-step response of this system when $0 < \zeta < 1$. For a unit-step input, we have $\Theta_i(s) = 1/s$. Then

$$\Theta_o(s) = \frac{\omega_n^2}{s^2 + 2\zeta\omega_n s + \omega_n^2} \frac{1}{s}$$

$$= \frac{1}{s} - \frac{s + 2\zeta\omega_n}{s^2 + 2\zeta\omega_n s + \omega_n^2}$$

$$= \frac{1}{s} - \frac{s + \zeta\omega_n}{(s + \zeta\omega_n)^2 + \omega_d^2} - \frac{\zeta\omega_n}{(s + \zeta\omega_n)^2 + \omega_d^2} \tag{8-27}$$

where $\omega_d = \omega_n\sqrt{1 - \zeta^2}$. The inverse Laplace transform of Eq. (8-27) gives

$$\theta_o(t) = 1 - e^{-\zeta\omega_n t}\cos\omega_d t - \frac{\zeta}{\sqrt{1 - \zeta^2}}e^{-\zeta\omega_n t}\sin\omega_d t$$

$$= 1 - e^{-\zeta\omega_n t}\left(\cos\omega_d t + \frac{\zeta}{\sqrt{1 - \zeta^2}}\sin\omega_d t\right) \tag{8-28}$$

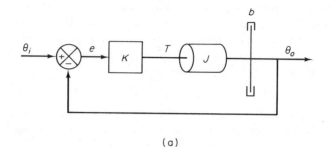

(a)

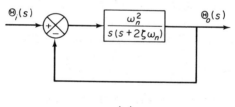

(b)

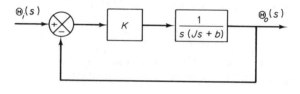

(c)

Fig. 8-37. (a) Position control system; (b) block diagram; (c) block diagram of a second-order system in standard form.

or

$$\theta_o(t) = 1 - \frac{e^{-\zeta\omega_n t}}{\sqrt{1-\zeta^2}} \sin\left(\omega_d t + \tan^{-1}\frac{\sqrt{1-\zeta^2}}{\zeta}\right) \qquad (8\text{-}29)$$

A family of curves $\theta_o(t)$ with various values of ζ is shown in Fig. 8-38, where the abscissa is the dimensionless variable $\omega_n t$. The curves are functions only of ζ.

A few comments on transient response specifications. Except in certain applications where oscillations cannot be tolerated, it is preferable that the transient response be sufficiently fast as well as reasonably damped. So in order to get a desirable transient response for a second-order system, the

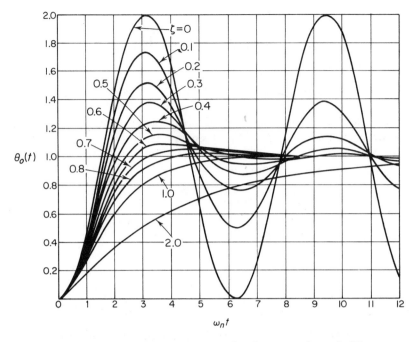

Fig. 8-38. Unit step response curves for the system shown in Fig. 8-37(c).

damping ratio ζ may be chosen between 0.4 and 0.8. Small values of ζ ($\zeta < 0.4$) yield excessive overshoot in the transient response, and a system with a large value of ζ ($\zeta > 0.8$) responds sluggishly.

Later on we shall see that the maximum overshoot and the rise time conflict with each other. In other words, both the maximum overshoot and the rise time cannot be made smaller simultaneously. If one is made smaller, the other necessarily becomes larger.

Second-order systems and transient response specifications. In the following pages we shall obtain the rise time, peak time, maximum overshoot, and settling time of the second-order system given by Eq. (8-26). These values will be derived in terms of ζ and ω_n. The system is assumed to be underdamped.

Rise time t_r. Referring to Eq. (8-28), we find rise time t_r by letting $\theta_o(t_r) = 1$ or

$$\theta_o(t_r) = 1 = 1 - e^{-\zeta\omega_n t_r}\left(\cos \omega_d t_r + \frac{\zeta}{\sqrt{1-\zeta^2}} \sin \omega_d t_r\right) \qquad (8\text{-}30)$$

Since $e^{-\zeta\omega_n t_r} \neq 0$, we can obtain from Eq. (8-30) the following equation.

$$\cos \omega_d t_r + \frac{\zeta}{\sqrt{1-\zeta^2}} \sin \omega_d t_r = 0$$

or

$$\tan \omega_d t_r = -\frac{\sqrt{1 - \zeta^2}}{\zeta}$$

Thus rise time t_r is

$$t_r = \frac{1}{\omega_d} \tan^{-1} \left(-\frac{\sqrt{1 - \zeta^2}}{\zeta} \right) = \frac{\pi - \beta}{\omega_d} \tag{8-31}$$

where β is defined in Fig. 8-39. Note that the value of $\tan^{-1}(-\sqrt{1 - \zeta^2}/\zeta)$ lies between $\frac{1}{2}\pi$ and π. If $\zeta = 0+$, then $\tan^{-1}(-\sqrt{1 - \zeta^2}/\zeta) = \frac{1}{2}\pi+$; if $\zeta = 1-$, then $\tan^{-1}(-\sqrt{1 - \zeta^2}/\zeta) = \pi-$. Clearly, in order to obtain a small value of t_r, we must have a large ω_d.

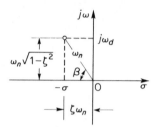

Fig. 8-39. Definition of angle β.

Peak time t_p. Referring to Eq. (8-28), we can obtain the peak time by differentiating $\theta_o(t)$ with respect to time t and letting this derivative equal zero or

$$\frac{d\theta_o}{dt} = \frac{\omega_n}{\sqrt{1 - \zeta^2}} e^{-\zeta \omega_n t} \sin \omega_d t = 0$$

It follows that

$$\sin \omega_d t = 0$$

or

$$\omega_d t = 0, \pi, 2\pi, 3\pi, \ldots$$

Since the peak time corresponds to the first peak overshoot, we have $\omega_d t_p = \pi$. Then

$$t_p = \frac{\pi}{\omega_d} \tag{8-32}$$

The peak time t_p corresponds to one half cycle of the frequency of damped oscillation.

Maximum overshoot M_p. The maximum overshoot occurs at the peak time or at $t = t_p = \pi/\omega_d$. Thus from Eq. (8-28) M_p is obtained as

$$M_p = \theta_o(t_p) - 1$$
$$= -e^{-\zeta \omega_n (\pi/\omega_d)} \left(\cos \pi + \frac{\zeta}{\sqrt{1 - \zeta^2}} \sin \pi \right)$$
$$= e^{-\zeta \pi / \sqrt{1 - \zeta^2}} \tag{8-33}$$

The maximum percent overshoot is $e^{-\zeta \pi / \sqrt{1 - \zeta^2}} \times 100\%$.

Settling time t_s. For an underdamped second-order system, the transient response for a unit-step input is given by Eq. (8-29). Notice that the response curve $\theta_o(t)$ always remains within a pair of the envelope curves as shown in Fig. 8-40. [The curves $1 \pm (e^{-\zeta\omega_n t}/\sqrt{1-\zeta^2})$ are the envelope curves of the transient response to a unit-step input.] The time constant of these envelope curves is $1/\zeta\omega_n$. The settling time t_s is four times this time constant or

$$t_s = \frac{4}{\zeta\omega_n} \tag{8-34}$$

Note that the settling time is inversely proportional to the undamped natural frequency of the system. Since the value of ζ is usually determined from the requirement of permissible maximum overshoot, the settling time is determined primarily by the undamped natural frequency ω_n. In other words, the duration of the transient period can be varied, without changing the maximum overshoot, by adjusting the undamped natural frequency ω_n.

From the preceding analysis it is clear that ω_d must be large if we are to have a rapid response. In order to limit maximum overshoot M_p and make the settling time small, the damping ratio ζ should not be too small. The relationship between maximum percent overshoot $M_p\%$ and damping ratio ζ is presented in Fig. 8-41. Note that if the damping ratio is between 0.4 and

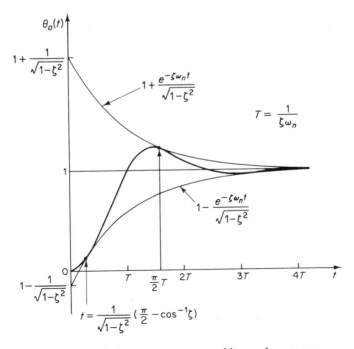

Fig. 8-40. Unit step response curve and its envelope curves.

0.8, then the maximum percent overshoot for a step response is between 25 and 2.5%.

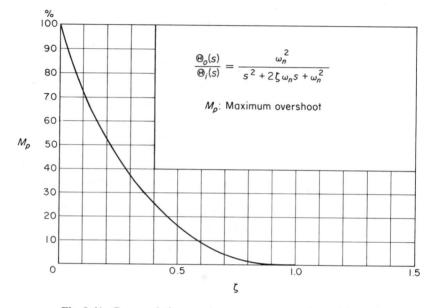

Fig. 8-41. Curve relating maximum percent overshoot M_p and damping ratio ζ.

Example 8-3. Determine the rise time, peak time, maximum overshoot, and settling time when the system in Fig. 8-42 is subjected to a unit-step input.

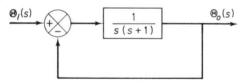

Fig. 8-42. Control system.

Notice that $\omega_n = 1$ rad/s and $\zeta = 0.5$ for this system. So

$$\omega_d = \omega_n\sqrt{1 - \zeta^2} = \sqrt{1 - 0.5^2} = 0.866$$

Rise time t_r: Referring to Eq. (8-31) the rise time t_r is

$$t_r = \frac{\pi - \beta}{\omega_d}$$

where $\beta = \sin^{-1} 0.866 = 1.05$ rad. Therefore

$$t_r = \frac{3.14 - 1.05}{0.866} = 2.41 \text{ s}$$

Peak time t_p: The peak time t_p is given by Eq. (8-32).

$$t_p = \frac{\pi}{\omega_d} = \frac{3.14}{0.866} = 3.63 \text{ s}$$

Maximum overshoot M_p: Referring to Eq. (8-33) the maximum overshoot M_p is

$$M_p = e^{-\zeta\pi/\sqrt{1-\zeta^2}} = e^{-0.5 \times 3.14/0.866} = e^{-1.81} = 0.163$$

Settling time t_s: The settling time t_s is defined by Eq. (8-34) and is

$$t_s = \frac{4}{0.5 \times 1} = 8 \text{ s}$$

8-6 IMPROVING TRANSIENT RESPONSE CHARACTERISTICS

In Sec. 8-5 we considered the step response of position control systems. It was shown that a small damping ratio will make the maximum overshoot in the step response large and the settling time large as well. Such features are generally undesirable. This section begins with a method for improving the damping characteristics of second-order systems through velocity feedback (tachometer feedback). Then the response of second-order systems to ramp inputs is considered. Here we present a method to improve steady-state behavior for ramp response by means of proportional-plus-derivative control action. Finally, another method to improve such steady-state behavior through a proportional-plus-derivative type prefilter is described.

Tachometers. A dc tachometer is a generator that produces a voltage proportional to its rotating speed. It is used as a transducer, converting the velocity of the rotating shaft into a proportional dc voltage. If the input to the tachometer is the shaft position θ and the output is the voltage e, then the transfer function of the dc tachometer is

$$\frac{E(s)}{\Theta(s)} = K_h s$$

where $E(s) = \mathcal{L}[e]$, $\Theta(s) = \mathcal{L}[\theta]$, and K_h is a constant. Inclusion of such a tachometer in the feedback path of the position control system will improve the damping characteristics of the system. The use of a tachometer to obtain a velocity feedback signal is referred to as *velocity feedback* or *tachometer feedback*.

Position control systems with velocity feedback. Systems with a small damping ratio exhibit a large maximum overshoot and a long sustained oscillation in the step response. To increase the effective damping of the system and thus improve transient response characteristics, velocity feedback is frequently employed.

Consider the position control system with velocity feedback shown in Fig. 8-43(a). Let us assume that the viscous friction coefficient b is small so that the damping ratio in the absence of the tachometer is quite small. In the present system the velocity signal, together with the positional signal, is fed back to the input to produce the actuating error signal. (Note that in obtaining the velocity signal, it is preferable to use a tachometer rather than physically differentiate the output position signal because differentiation always accentuates noise signals.)

The block diagram in Fig. 8-43(a) can be simplified as in Fig. 8-43(b), giving

$$\frac{\Theta_o(s)}{\Theta_i(s)} = \frac{K}{Js^2 + (b + KK_h)s + K}$$

The damping ratio ζ for this system is

$$\zeta = \frac{b + KK_h}{2\sqrt{KJ}} \tag{8-35}$$

and the undamped natural frequency ω_n is

$$\omega_n = \sqrt{\frac{K}{J}}$$

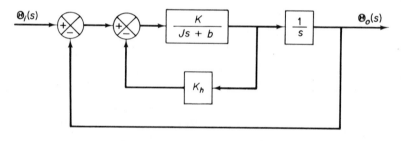

(a)

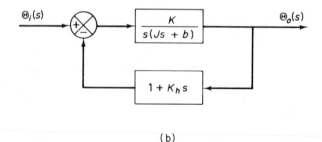

(b)

Fig. 8-43. (a) Block diagram of a position control system with velocity feedback; (b) simplified block diagram.

Notice that the undamped natural frequency ω_n is not affected by velocity feedback. Given the values of J and b, the value of K is determined from the requirement on the undamped natural frequency ω_n. The velocity feedback (tachometer feedback) constant K_h is then adjusted so that ζ is between 0.4 and 0.8.

Remember that velocity feedback (tachometer feedback) has the effect of increasing the damping ratio without affecting the undamped natural frequency of the second-order system.

Example 8-4. Assume, for the system shown in Fig. 8-43(b), that the numerical values of J and b are

$$J = 1 \text{ kg-m}^2$$

$$b = 1 \text{ N-m-s}$$

We wish to determine the values of gain K and velocity feedback constant K_h so that the maximum overshoot is 0.2 and the peak time is 1 s.

The maximum overshoot M_p is given by Eq. (8-33) as

$$M_p = e^{-\zeta\pi/\sqrt{1-\zeta^2}}$$

This value must be 0.2. Hence

$$e^{-\zeta\pi/\sqrt{1-\zeta^2}} = 0.2$$

which yields

$$\zeta = 0.456$$

The peak time t_p is specified as 1 s. Therefore from Eq. (8-32)

$$t_p = \frac{\pi}{\omega_d} = 1$$

or

$$\omega_d = 3.14$$

Since ζ is 0.456, ω_n is

$$\omega_n = \frac{\omega_d}{\sqrt{1-\zeta^2}} = 3.53$$

The undamped natural frequency ω_n is equal to $\sqrt{K/J} = \sqrt{K/1} = \sqrt{K}$, and so

$$K = \omega_n^2 = 12.5 \text{ N-m}$$

Then K_h is obtained from Eq. (8-35) as

$$K_h = \frac{2\sqrt{K}\,\zeta - 1}{K} = 0.178 \text{ s}$$

Steady-state error in ramp responses. Position control systems may be subjected to changing inputs that can be approximated by a series of piecewise ramp inputs. In such ramp response, steady-state error in following inputs must be small.

Consider the system shown in Fig. 8-44. The transient response of this system when subjected to a ramp input can be found by a straightforward method. In the present analysis, we shall examine the steady-state error when the system is subjected to such an input.

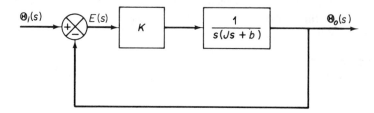

Fig. 8-44. Block diagram of a position control system with a proportional controller.

From the block diagram we have

$$\frac{E(s)}{\Theta_i(s)} = \frac{\Theta_i(s) - \Theta_o(s)}{\Theta_i(s)} = 1 - \frac{\Theta_o(s)}{\Theta_i(s)} = \frac{Js^2 + bs}{Js^2 + bs + K}$$

The steady-state error for the unit-ramp response can be obtained as follows. For a unit-ramp input $\theta_i(t) = t$, we have $\Theta_i(s) = 1/s^2$. The steady-state error e_{ss} is then obtained as

$$e_{ss} = \lim_{s \to 0} sE(s) = \lim_{s \to 0} s \frac{Js^2 + bs}{Js^2 + bs + K} \frac{1}{s^2}$$

$$= \lim_{s \to 0} \frac{s^2(Js + b)}{s^2(Js^2 + bs + K)} = \frac{b}{K}$$

In order to ensure small steady-state error in following a ramp input, the value of K must be large and the value of b small. However, a large value of K and a small value of b will make the damping ratio ζ small and will, in general, result in undesirable transient response characteristics. Consequently, some means of improving steady-state behavior in following ramp inputs without adversely affecting transient response behavior is necessary. Two such means are discussed below.

Proportional-plus-derivative control of second-order systems. A compromise between acceptable transient response behavior and acceptable steady-state behavior can be achieved through proportional-plus-derivative control action.

For the system in Fig. 8-45, the closed-loop transfer function is

$$\frac{\Theta_o(s)}{\Theta_i(s)} = \frac{K_p + K_d s}{Js^2 + (b + K_d)s + K_p} \tag{8-36}$$

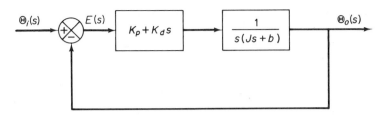

Fig. 8-45. Block diagram of a position control system with a proportional-plus-derivative controller.

Therefore

$$\frac{E(s)}{\Theta_i(s)} = \frac{\Theta_i(s) - \Theta_o(s)}{\Theta_i(s)} = \frac{Js^2 + bs}{Js^2 + (b + K_d)s + K_p}$$

For a unit-ramp input, $\Theta_i(s) = 1/s^2$. So it follows that

$$E(s) = \frac{Js^2 + bs}{Js^2 + (b + K_d)s + K_p}\frac{1}{s^2}$$

The steady-state error for a unit-ramp response is

$$e_{ss} = \lim_{s \to 0} sE(s) = \lim_{s \to 0} s \frac{Js^2 + bs}{Js^2 + (b + K_d)s + K_p}\frac{1}{s^2} = \frac{b}{K_p}$$

The characteristic equation is

$$Js^2 + (b + K_d)s + K_p = 0$$

The effective damping of this system is thus $b + K_d$ rather than b. Since the damping ratio ζ of this system is

$$\zeta = \frac{b + K_d}{2\sqrt{K_p J}}$$

it is possible to have both a small steady-state error e_{ss} for a ramp response and a reasonable damping ratio by making b small, K_p large, and choosing K_d large enough so that ζ is between 0.4 and 0.8.

Let us examine the unit-step response of this system. Define

$$\omega_n = \sqrt{\frac{K_p}{J}}, \qquad z = \frac{K_p}{K_d}$$

In terms of ω_n, ζ, and z, Eq. (8-36) can be written

$$\frac{\Theta_o(s)}{\Theta_i(s)} = \left(1 + \frac{s}{z}\right)\frac{\omega_n^2}{s^2 + 2\zeta\omega_n s + \omega_n^2}$$

Notice that if a zero $s = -z$ is located near the closed-loop poles, the transient response behavior differs considerably from that of a second-order system without a zero. Typical step response curves of this system with $\zeta = 0.5$ and various values of $z/\zeta\omega_n$ are shown in Fig. 8-46. From these curves

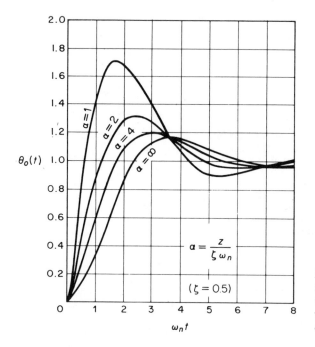

Fig. 8-46. Unit step response curves for the system shown in Fig. 8-45 with the damping ratio ζ equal to 0.5.

we see that proportional-plus-derivative control action will make the rise time smaller and the maximum overshoot larger.

Second-order systems with proportional-plus-derivative type prefilter. The steady-state error for following a ramp input can be eliminated if the input is introduced to the system through a proportional-plus-derivative type prefilter, as shown in Fig. 8-47, and if the value of k is properly set.

The transfer function $\Theta_o(s)/\Theta_i(s)$ for this system is

$$\frac{\Theta_o(s)}{\Theta_i(s)} = \frac{(1 + ks)K}{Js^2 + bs + K}$$

Therefore the difference between $\Theta_i(s)$ and $\Theta_o(s)$ is

$$E(s) = \Theta_i(s) - \Theta_o(s) = \left[1 - \frac{\Theta_o(s)}{\Theta_i(s)}\right]\Theta_i(s)$$

$$= \frac{Js^2 + (b - Kk)s}{Js^2 + bs + K}\Theta_i(s)$$

The steady-state error for ramp response is

$$e_{ss} = \lim_{s \to 0} sE(s) = \lim_{s \to 0} s\frac{Js^2 + (b - Kk)s}{Js^2 + bs + K}\frac{1}{s^2} = \frac{b - Kk}{K}$$

So if k is chosen as

$$k = \frac{b}{K}$$

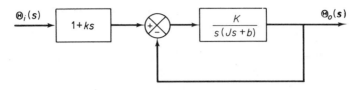

Fig. 8-47. Block diagram of a position control system with a proportional-plus-derivative type prefilter.

the steady-state error for following a ramp input can be made equal to zero.

Given the values of J and b, the value of K is normally determined from the requirement on $\omega_n = \sqrt{K/J}$. Once the value of K is determined, b/K is a constant and the value of $k = b/K$ becomes a constant. Use of such a prefilter eliminates steady-state error for ramp response.

It should be noted that the transient response of this system to a unit-step input will exhibit a smaller rise time and a larger maximum overshoot than the corresponding system without the prefilter.

It is worth while pointing out that the block diagram of the system with a proportional-plus-derivative controller shown in Fig. 8-45 can be redrawn as in Fig. 8-48. From this diagram it can be seen that the proportional-plus-derivative controller, in fact, is a combination of the prefilter and velocity feedback in which the values of both k and K_h are chosen to be K_d/K_p.

If the prefilter and velocity feedback are provided separately, the values of k and K_h can be chosen independently of each other. A proper choice of these values may enable the engineer to compromise between the acceptable steady-state error in following ramp input and acceptable transient response behavior to step input.

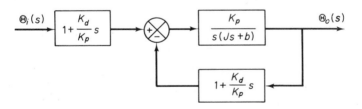

Fig. 8-48. Modified block diagram of the system shown in Fig. 8-45.

8-7   A DESIGN PROBLEM

In concluding this chapter, let us turn to the problem of designing a control system.

An inverted pendulum mounted on a motor-driven cart is shown in

Fig. 8-49. This is a model of the attitude control of a space booster on takeoff. The objective of the attitude control problem is to keep the space booster in a vertical position. The actual space booster (or the inverted pendulum in this problem) is unstable in that it may fall over any time in any direction unless a suitable control force is applied. Here we consider only a two-dimensional problem so that the pendulum in Fig. 8-49 moves only in the plane of the page.

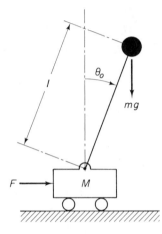

Fig. 8-49. Inverted pendulum mounted on a motor-driven cart.

Let us define the angle of the rod from the vertical line as θ_o. The positive direction of θ_o is shown in the diagram. The positive direction of the force F to the cart needed to control the pendulum position also appears in the diagram. Note that if angle θ_o is positive, then in order to reduce angle θ_o and keep the inverted pendulum vertical, we must apply a force F in the positive direction.

Assuming the numerical values

$$M = 1000 \text{ kg}$$

$$m = 200 \text{ kg}$$

$$l = 10 \text{ m}$$

design a suitable controller such that the system has a damping ratio ζ of 0.7 and an undamped natural frequency ω_n of 0.5 rad/s. Assume also that the weight of the rod is negligible, that external forces such as gusts of wind are neglibible, and that there is neither friction at the pivot nor slip at the wheels of the cart.

Since the inverted pendulum is an unstable plant, derivative control action must be included in the controller. (Note that derivative control action responds to the rate of change of the actuating error and can initiate an early corrective action in order to increase the stability of the system.) Since the derivative control action cannot be used alone, let us use a proportional-plus-derivative controller in the present problem.

Figure 8-50 shows a possible block diagram for this control system. Note that the fact that the reference input θ_i is zero means that we want to keep the inverted pendulum vertical. The proportional-plus-derivative controller has the transfer function $K_p(1 + K_d s)$. This controller produces a force $- F$ where

$$F = K_p(\theta_o + K_d \dot{\theta}_o) \qquad (8\text{-}37)$$

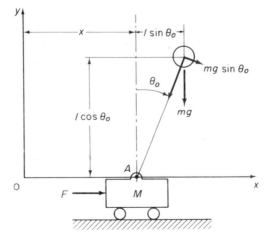

Fig. 8-50. Block diagram for the control system in which the inverted pendulum system shown in Fig. 8-49 is the plant.

Let us define the x-y coordinate system as shown in Fig. 8-51 and derive the equations of motion for this system. The horizontal position of the cart is x. The horizontal and vertical positions of the pendulum mass are $x + l \sin \theta_o$ and $l \cos \theta_o$, respectively.

Applying Newton's second law to the x direction of motion yields

$$M\frac{d^2 x}{dt^2} + m\frac{d^2}{dt^2}(x + l \sin \theta_o) = F \qquad (8\text{-}38)$$

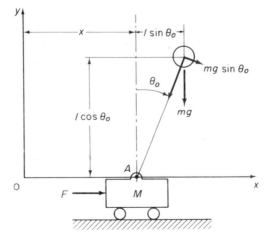

Fig. 8-51. Inverted pendulum system with the x-y coordinate system.

Noting that

$$\frac{d}{dt} \sin \theta_o = (\cos \theta_o)\dot{\theta}_o$$

$$\frac{d^2}{dt^2} \sin \theta_o = (-\sin \theta_o)\dot{\theta}_o^2 + (\cos \theta_o)\ddot{\theta}_o$$

$$\frac{d}{dt} \cos \theta_o = (-\sin \theta_o)\dot{\theta}_o$$

$$\frac{d^2}{dt^2} \cos \theta_o = (-\cos \theta_o)\dot{\theta}_o^2 - (\sin \theta_o)\ddot{\theta}_o$$

where a dot denotes differentiation with respect to time, Eq. (8-38) can be rewritten thus:

$$(M + m)\ddot{x} - ml(\sin \theta_o)\dot{\theta}_o^2 + ml(\cos \theta_o)\ddot{\theta}_o = F \qquad (8\text{-}39)$$

Next, consider the rotational motion of the inverted pendulum about point A in Fig. 8-51. Applying Newton's second law for the rotational motion, we have

$$\left[m\frac{d^2}{dt^2}(x + l \sin \theta_o) \right] l \cos \theta_o - \left[m\frac{d^2}{dt^2}(l \cos \theta_o) \right] l \sin \theta_o = mgl \sin \theta_o$$

This last equation can be simplified as follows:

$$m[\ddot{x} - l(\sin \theta_o)\dot{\theta}_o^2 + l(\cos \theta_o)\ddot{\theta}_o]l \cos \theta_o$$
$$- m[-l(\cos \theta_o)\dot{\theta}_o^2 - l(\sin \theta_o)\ddot{\theta}_o]l \sin \theta_o = mgl \sin \theta_o$$

and further simplified to

$$m\ddot{x} \cos \theta_o + ml\ddot{\theta}_o = mg \sin \theta_o \qquad (8\text{-}40)$$

Equations (8-39) and (8-40) are nonlinear differential equations. Since in this problem we must keep the inverted pendulum vertical, we may assume that $\theta_o(t)$ and $\dot{\theta}_o(t)$ are small. Under this assumption Eqs. (8-39) and (8-40) can be linearized. By substituting $\sin \theta_o \doteq \theta_o$ and $\cos \theta_o \doteq 1$ into Eqs. (8-39) and (8-40) and neglecting the term that involves $\theta_o\dot{\theta}_o^2$, we can derive the linearized equations of motion for the system.

$$(M + m)\ddot{x} + ml\ddot{\theta}_o = F \qquad (8\text{-}41)$$
$$m\ddot{x} + ml\ddot{\theta}_o = mg\theta_o \qquad (8\text{-}42)$$

These linearized equations are valid as long as θ_o and $\dot{\theta}_o$ are small. Subtracting Eq. (8-42) from Eq. (8-41) yields

$$M\ddot{x} = F - mg\theta_o$$

or

$$\ddot{x} = -\frac{mg}{M}\theta_o + \frac{F}{M} \qquad (8\text{-}43)$$

By substituting Eq. (8-43) into Eq. (8-42) and rearranging, we find

$$Ml\ddot{\theta}_o - (M + m)g\theta_o = -F \tag{8-44}$$

Referring to Eq. (8-44), the transfer function between $\Theta_o(s)$ and $-F(s)$ is

$$\frac{\Theta_o(s)}{-F(s)} = \frac{1}{Mls^2 - (M + m)g}$$

By using this transfer function, the block diagram of Fig. 8-50 can be modified to yield Fig. 8-52.

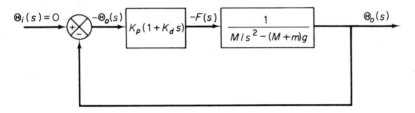

Fig. 8-52. Block diagram of the inverted pendulum control system.

Substitution of Eq. (8-37) into Eq. (8-44) gives

$$Ml\ddot{\theta}_o - (M + m)g\theta_o = -K_p(\theta_o + K_d\dot{\theta}_o)$$

Simplifying, we get

$$\ddot{\theta}_o + \frac{K_pK_d}{Ml}\dot{\theta}_o + \left[\frac{K_p}{Ml} - \left(1 + \frac{m}{M}\right)\frac{g}{l}\right]\theta_o = 0$$

from which

$$\omega_n^2 = \frac{K_p}{Ml} - \left(1 + \frac{m}{M}\right)\frac{g}{l}$$

$$2\zeta\omega_n = \frac{K_pK_d}{Ml}$$

Hence

$$K_p = \omega_n^2 Ml + (M + m)g$$

$$K_d = \frac{2\zeta\omega_n Ml}{K_p}$$

To determine the numerical values of K_p and K_d, we substitute $M = 1000$ kg, $m = 200$ kg, $\omega_n = 0.5$ rad/s, $\zeta = 0.7$, and $l = 10$ m into the equations for K_p and K_d, obtaining

$$K_p = 0.5^2 \times 1000 \times 10 + (1000 + 200) \times 9.81 = 14.27 \times 10^3 \text{ N/rad}$$

$$K_d = \frac{2 \times 0.7 \times 0.5 \times 1000 \times 10}{14.27 \times 10^3} = 0.491 \text{ s}$$

The controller designed here is a proportional-plus-derivative one and has the transfer function $G_c(s)$, where

$$G_c(s) = K_p(1 + K_d s) = 14.27 \times 10^3(1 + 0.491s)$$

and Fig. 8-52 is the block diagram of the designed system. Note that if K_p and K_d assume the values thus designed, any small tilt can be recovered without having the pendulum fall over. If an unknown disturbance gives a small step change in the angle θ_o—for instance, a step change of 0.1 rad—then the designed controller will produce the corrective force required to bring the inverted pendulum to a vertical position, and the resulting response $\theta_o(t)$ will exhibit a curve like the one shown in Fig. 8-53. Noting that the settling time for this response is $t_s = 4/(\zeta\omega_n) = 4/(0.7 \times 0.5) = 11.4$ seconds, any small step disturbance can be corrected in about 11 seconds. This completes our design problem.

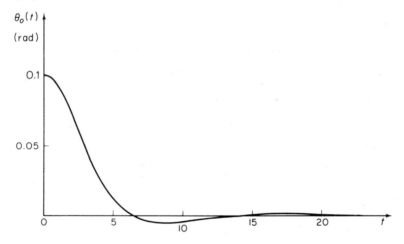

Fig. 8-53. Step response curve of the inverted pendulum control system.

REFERENCES

8-1 HEALEY, M., *Principles of Automatic Control*, Princeton, N.J.: D. Van Nostrand Company, Inc., 1967.

8-2 HIGDON, D. T., AND R. H. CANNON, JR., "On the Control of Unstable Multiple-Output Mechanical Systems," *ASME Paper No. 63-WA-148*, 1963.

8-3 LAJOY, M. H., *Industrial Automatic Controls*, Englewood Cliffs, N.J.: Prentice-Hall, Inc., 1954.

8-4 MERRITT, H. E., *Hydraulic Control Systems*, New York: John Wiley & Sons, Inc., 1967.

8-5 OGATA, K., *Modern Control Engineering*, Englewood Cliffs, N.J.: Prentice-Hall, Inc., 1970.

EXAMPLE PROBLEMS AND SOLUTIONS

PROBLEM A-8-1. Simplify the block diagram shown in Fig. 8-54.

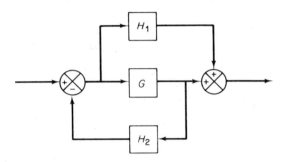

Fig. 8-54. Block diagram of a
system.

Solution. First, move the branch point of the path involving H_1 outside the loop involving H_2 as shown in Fig. 8-55(a). Then eliminating two loops results in Fig. 8-55(b). Combining two blocks into one gives Fig. 8-55(c).

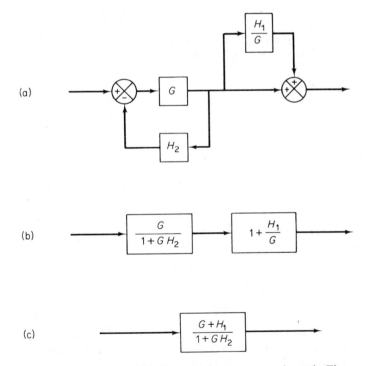

Fig. 8-55. Simplified block diagrams for the system shown in Fig. 8-54.

PROBLEM A-8-2. For the block diagram shown in Fig. 8-56, derive the transfer function relating $\Theta_o(s)$ and $\Theta_i(s)$.

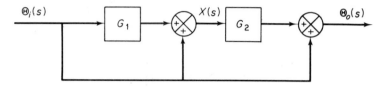

Fig. 8-56. Block diagram of a system.

Solution. Signal $X(s)$ is the sum of two signals $G_1\Theta_i(s)$ and $\Theta_i(s)$. Therefore

$$X(s) = G_1\Theta_i(s) + \Theta_i(s)$$

The output signal $\Theta_o(s)$ is the sum of $G_2 X(s)$ and $\Theta_i(s)$. Hence

$$\Theta_o(s) = G_2 X(s) + \Theta_i(s) = G_2[G_1\Theta_i(s) + \Theta_i(s)] + \Theta_i(s)$$

And so we have

$$\frac{\Theta_o(s)}{\Theta_i(s)} = G_1 G_2 + G_2 + 1$$

PROBLEM A-8-3. In the pneumatic pressure system of Fig. 8-57(a), assume that, for $t < 0$, the system is at steady state and that the pressure of the entire system is $\bar{P}$. Also, assume that the two bellows are identical. At $t = 0$ the input pressure is changed from $\bar{P}$ to $\bar{P} + p_i$. Then the pressures in bellows 1 and 2 will change from $\bar{P}$ to $\bar{P} + p_1$ and from $\bar{P}$ to $\bar{P} + p_2$, respectively. The capacity (volume) of each bellows is 5×10^{-4} m³, and the operating pressure difference Δp (difference between p_i and p_1 or difference between p_i and p_2) is between -0.5×10^5 N/m² and 0.5×10^5 N/m². The corresponding mass flow rates (kg/s) through the valves are shown in Fig. 8-57(b). Assume that the bellows expand or contract linearly with the air pressures applied to them, that the equivalent spring constant of the bellows system is $k = 1 \times 10^5$ N/m, and that each bellows has area $A = 15 \times 10^{-4}$ m².

Defining the displacement of the midpoint of the rod that connects two bellows as x, find the transfer function $X(s)/P_i(s)$. Assume that the expansion process is isothermal and that the temperature of the entire system stays at 30°C.

Solution. Referring to Section 8-3, transfer function $P_1(s)/P_i(s)$ can be obtained as

$$\frac{P_1(s)}{P_i(s)} = \frac{1}{R_1 Cs + 1} \tag{8-45}$$

Similarly, transfer function $P_2(s)/P_i(s)$ is

$$\frac{P_2(s)}{P_i(s)} = \frac{1}{R_2 Cs + 1} \tag{8-46}$$

The force acting on bellows 1 in the x direction is $A(\bar{P} + p_1)$, and the force acting on bellows 2 in the negative x direction is $A(\bar{P} + p_2)$. The resultant force balances with kx, the equivalent spring force of the corrugated sides of the bellows.

$$A(p_1 - p_2) = kx$$

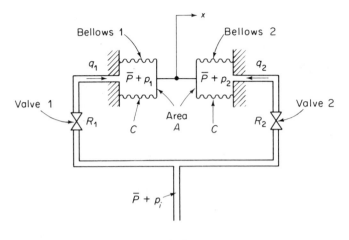

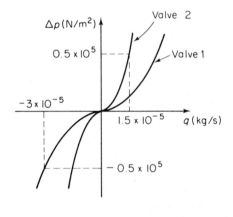

Fig. 8-57. (a) Pneumatic pressure system; (b) pressure difference versus mass flow rate curves.

or

$$A[P_1(s) - P_2(s)] = kX(s) \tag{8-47}$$

Referring to Eqs. (8-45) and (8-46), we see that

$$P_1(s) - P_2(s) = \left(\frac{1}{R_1 Cs + 1} - \frac{1}{R_2 Cs + 1}\right)P_i(s)$$

$$= \frac{R_2 Cs - R_1 Cs}{(R_1 Cs + 1)(R_2 Cs + 1)}P_i(s)$$

By substituting this last equation into Eq. (8-47) and rewriting, the transfer function $X(s)/P_i(s)$ is obtained as

$$\frac{X(s)}{P_i(s)} = \frac{A}{k}\frac{(R_2 C - R_1 C)s}{(R_1 Cs + 1)(R_2 Cs + 1)} \tag{8-48}$$

The numerical values of average resistances R_1 and R_2 are

$$R_1 = \frac{d\,\Delta p}{dq_1} = \frac{0.5 \times 10^5}{3 \times 10^{-5}} = 0.167 \times 10^{10}\,\frac{N/m^2}{kg/s}$$

$$R_2 = \frac{d\,\Delta p}{dq_2} = \frac{0.5 \times 10^5}{1.5 \times 10^{-5}} = 0.333 \times 10^{10}\,\frac{N/m^2}{kg/s}$$

The numerical value of capacitance C of each bellows is

$$C = \frac{V}{nR_{air}T} = \frac{5 \times 10^{-4}}{1 \times 287 \times (273 + 30)} = 5.75 \times 10^{-9}\,\frac{kg}{N/m^2}$$

Consequently,

$$R_1 C = 0.167 \times 10^{10} \times 5.75 \times 10^{-9} = 9.60\text{ s}$$

$$R_2 C = 0.333 \times 10^{10} \times 5.75 \times 10^{-9} = 19.2\text{ s}$$

By substituting the numerical values for A, k, $R_1 C$, and $R_2 C$ into Eq. (8-48), we have

$$\frac{X(s)}{P_i(s)} = \frac{1.44 \times 10^{-7}s}{(9.6s + 1)(19.2s + 1)}$$

Problem A-8-4. Figure 8-58 is a schematic diagram of a pneumatic diaphragm valve. At steady state the control pressure from a controller is $\bar{P}_c$, the pressure in the valve is also $\bar{P}_c$, and the valve stem displacement is $\bar{X}$. Assume that at $t = 0$ the control pressure is changed from $\bar{P}_c$ to $\bar{P}_c + p_c$. Then the valve pressure will be

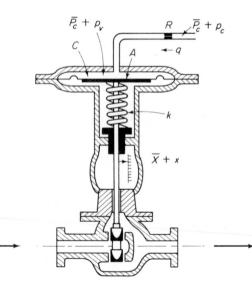

Fig. 8-58. Pneumatic diaphragm valve.

changed from $\bar{P}_c$ to $\bar{P}_c + p_v$. The change in valve pressure p_v will cause the valve stem displacement to change from $\bar{X}$ to $\bar{X} + x$. Find the transfer function between the change in valve stem displacement x and the change in control pressure p_c.

Solution. Let us define the air flow rate to the diaphragm valve through resistance R as q. Then

$$q = \frac{p_c - p_v}{R}$$

For the air chamber in the diaphragm valve, we have

$$C \, dp_v = q \, dt$$

Consequently,

$$C \frac{dp_v}{dt} = q = \frac{p_c - p_v}{R}$$

from which

$$RC \frac{dp_v}{dt} + p_v = p_c$$

Noting that

$$A p_v = kx$$

we have

$$\frac{k}{A} \left(RC \frac{dx}{dt} + x \right) = p_c$$

The transfer function between x and p_c is

$$\frac{X(s)}{P_c(s)} = \frac{A/k}{RCs + 1}$$

PROBLEM A-8-5. Suppose that, for the jet pipe hydraulic controller shown in Fig. 8-59, the power piston is connected to a light load so that the inertia force of the load element is negligible compared to the hydraulic force developed by the power piston. What type of control action does this controller produce?

Solution. Define the displacement of the jet nozzle from the neutral position as x and the displacement of the power piston as y. If the jet nozzle is moved to the right by a small displacement x, the oil flows to the right-hand side of the power piston, and the oil in the left-hand side of the power piston is returned to the drain. The oil flowing into the power cylinder is at high pressure; the oil flowing out from the power cylinder into the drain is at low pressure. The resulting pressure difference causes the power piston to move to the left.

For a small jet nozzle displacement x, the flow rate q to the power cylinder is proportional to x—that is,

$$q = K_1 x$$

For the power cylinder,

$$A\rho \, dy = q \, dt$$

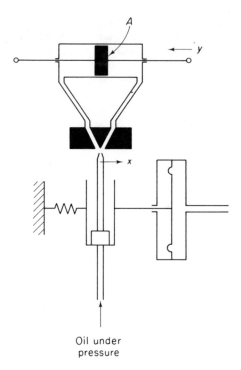

Oil under
pressure

Fig. 8-59. Jet pipe hydraulic controller.

where A is the power piston area and ρ is the density of oil. Hence

$$\frac{dy}{dt} = \frac{q}{A\rho} = \frac{K_1}{A\rho} x = Kx$$

where $K = K_1/(A\rho)$ = constant. The transfer function $Y(s)/X(s)$ is thus

$$\frac{Y(s)}{X(s)} = \frac{K}{s}$$

The controller produces the integral control action.

PROBLEM A-8-6. In Fig. 8-60 we have a schematic diagram of a hydraulic controller. Draw a block diagram for the controller and determine the transfer function $Y(s)/E(s)$.

Solution. Consider first the damper b and spring k. The equation for this part of the controller is

$$b(\dot{y} - \dot{z}) = kz$$

or

$$b\dot{y} = b\dot{z} + kz$$

So the transfer function $Z(s)/Y(s)$ becomes

$$\frac{Z(s)}{Y(s)} = \frac{bs}{bs + k}$$

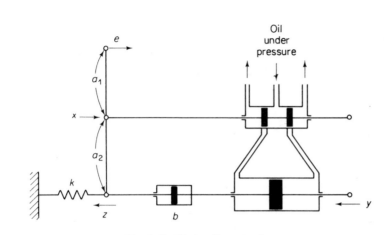

Fig. 8-60. Hydraulic controller.

This transfer function is in the feedback path of the controller. The block diagram for the controller is shown in Fig. 8-61. The transfer function $Y(s)/E(s)$ can be obtained from the block diagram as

$$\frac{Y(s)}{E(s)} = \frac{\dfrac{a_2}{a_1 + a_2} \dfrac{K}{s}}{1 + \dfrac{Ka_1}{a_1 + a_2} \dfrac{b}{bs + k}}$$

In such a controller the gain $|Ka_1b/[(a_1 + a_2)(bs + k)]|$ is usually very large compared with unity, and so the transfer function $Y(s)/E(s)$ can be simplified to

$$\frac{Y(s)}{E(s)} = \frac{a_2}{a_1}\left(1 + \frac{k}{bs}\right) = K_p\left(1 + \frac{1}{T_i s}\right)$$

where $K_p = a_2/a_1$ and $T_i = b/k$. Thus the controller shown in Fig. 8-60 is a proportional-plus-integral controller.

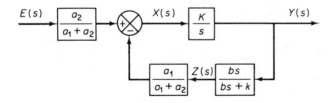

Fig. 8-61. Block diagram for the hydraulic controller shown in Fig. 8-60.

PROBLEM A-8-7. Draw a block diagram for the hydraulic controller shown in Fig. 8-62. What type of control action does this controller produce?

Solution. Assume that, for $t < 0$, the pressures in bellows 1 and 2 are the same at some pressure $\bar{P}$. At $t = 0$ input e is given in the positive direction shown in the

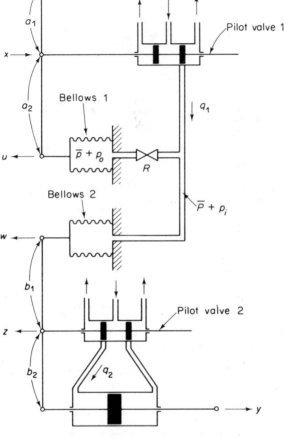
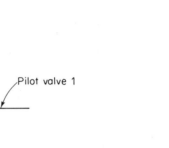

Fig. 8-62. Hydraulic controller.

diagram. Then pilot valve 1 is moved to the right by x. Displacement x will cause the changes in the pressures in bellows 1 and 2 from $\bar{P}$ to $\bar{P} + p_o$ and $\bar{P} + p_i$, respectively. Assume that the pressure change p_i is proportional to the pilot valve displacement x. Assume also that displacements u and w of bellows 1 and 2 are proportional to pressure changes p_o and p_i, respectively, and that all changes in the variables are relatively small.

For bellows 1, we have

$$A_1 p_o = k_1 u$$

where A_1 is the area of the bellows and k_1 is the spring constant of the bellows. Then

$$\frac{U(s)}{P_o(s)} = \frac{A_1}{k_1}$$

The pressures p_o and p_i are related by

$$\frac{P_o(s)}{P_i(s)} = \frac{1}{RCs + 1}$$

where R is the resistance of the valve and C is the capacitance of bellows 1. For bellows 2,

$$A_2 p_i = k_2 w$$

where A_2 is the area of the bellows and k_2 is the spring constant of the bellows. Hence

$$\frac{W(s)}{P_i(s)} = \frac{A_2}{k_2}$$

Then the block diagram for the controller can be drawn as shown in Fig. 8-63(a).

Note that the block diagrams for the first- and second-stage amplifiers can be simplified, assuming $|K_1 a_1 A_1/[(a_1 + a_2)k_1(RCs + 1)]| \gg 1$ and $|K_2 b_1/[s(b_1 + b_2)]| \gg 1$, as shown in Fig. 8-63(b). This diagram can be further simplified as shown in Fig. 8-63(c). The transfer function between $Y(s)$ and $E(s)$ is thus

$$\frac{Y(s)}{E(s)} = \frac{a_2 b_2 k_1 A_2}{a_1 b_1 A_1 k_2}(RCs + 1) = K_p(1 + T_d s)$$

where

$$K_p = \frac{a_2 b_2 k_1 A_2}{a_1 b_1 A_1 k_2}, \qquad T_d = RC$$

Therefore the controller shown in Fig. 8-62 is a proportional-plus-derivative controller.

PROBLEM A-8-8. In the system of Fig. 8-64 $x(t)$ is the input displacement and $\theta(t)$ is the output angular displacement. Assume that the masses involved are negligibly small and that all motions are restricted to be small; therefore the system can be considered a linear one. The initial conditions for x and θ are zeros, or $x(0-) = 0$ and $\theta(0-) = 0$. Show that this system is an differentiating element. Then obtain the response $\theta(t)$ when $x(t)$ is a unit-step input.

Solution. The equation for the system is

$$b(\dot{x} - l\dot{\theta}) = kl\theta$$

or

$$l\dot{\theta} + \frac{k}{b} l\theta = \dot{x}$$

The Laplace transform of this last equation, using zero initial conditions, gives

$$\left(ls + \frac{k}{b}l\right)\Theta(s) = sX(s)$$

And so

$$\frac{\Theta(s)}{X(s)} = \frac{1}{l}\frac{s}{s + (k/b)}$$

Thus the system is a differentiating system.

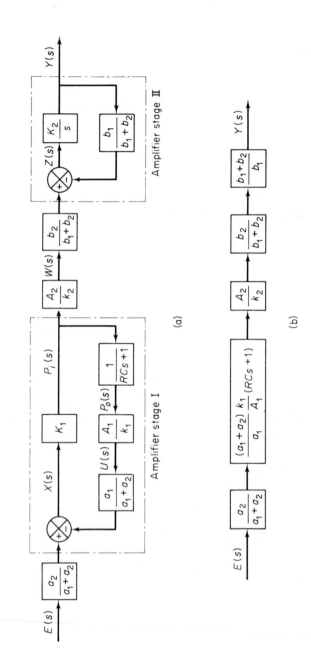

Fig. 8-63. (a) Block diagram for the hydraulic controller shown in Fig. 8-62; (b), (c) simplified block diagrams.

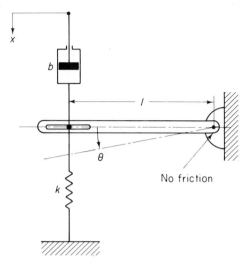

Fig. 8-64. Mechanical system.

For the unit-step input $X(s) = 1/s$, the output $\Theta(s)$ becomes

$$\Theta(s) = \frac{1}{l}\,\frac{1}{s + (k/b)}$$

The inverse Laplace transform of $\Theta(s)$ gives

$$\theta(t) = \frac{1}{l}e^{-(k/b)t}$$

Note that if the value of k/b is large, the response $\theta(t)$ approaches a pulse signal as shown in Fig. 8-65.

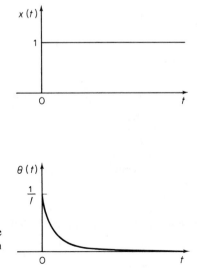

Fig. 8-65. Unit step input and the response of the mechanical system shown in Fig. 8-64.

PROBLEM A-8-9. Consider the system shown in Fig. 8-66(a). The massless bar AA' is displaced 0.05 m by a constant force of 100 N. Suppose that the system is at rest before the force is abruptly released. The time response curve, when the force is abruptly released at $t = 0$, is shown in Fig. 8-66(b). Determine the numerical values of b and k.

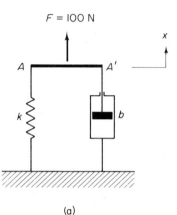

(a)

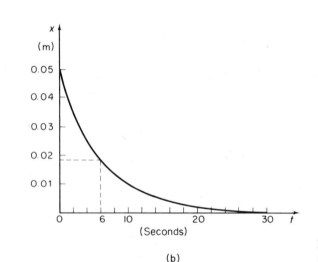

(b)

Fig. 8-66. (a) Mechanical system; (b) response curve.

Solution. Since the system is at rest before the force is abruptly released, the equation of motion is

$$kx = F \qquad (t \le 0)$$

Note that the effect of force F is to give the initial condition $x(0)$ or

$$x(0) = \frac{F}{k}$$

Since $x(0) = 0.05$ m, we have

$$k = \frac{F}{x(0)} = \frac{100}{0.05} = 2000 \text{ N/m}$$

At $t = 0$ force F is abruptly released, and so, for $t > 0$, the equation of motion becomes

$$b\dot{x} + kx = 0 \qquad (t > 0)$$

The solution of this last equation is

$$x(t) = x(0)e^{-(k/b)t} = 0.05e^{-(2000/b)t}$$

Since the solution is an exponential function, at $t = $ time constant $= b/2000$ the response x becomes

$$x\left(\frac{b}{2000}\right) = 0.05 \times 0.368 = 0.0184 \text{ m}$$

From Fig. 8-66(b) $x = 0.0184$ m occurs at $t = 6$ s. Hence

$$\frac{b}{2000} = 6$$

from which

$$b = 12\,000 \text{ N-s/m}$$

PROBLEM A-8-10. For the system shown in Fig. 8-67, discuss the effects that varying the values of K and b have on the steady-state error in a unit-ramp response. Sketch typical unit-ramp response curves for a small value, medium value, and large value of K.

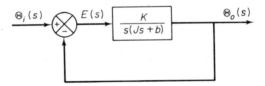

Fig. 8-67. Closed-loop system.

Solution. The closed-loop transfer function is

$$\frac{\Theta_o(s)}{\Theta_i(s)} = \frac{K}{Js^2 + bs + K}$$

Therefore

$$\frac{E(s)}{\Theta_i(s)} = \frac{\Theta_i(s) - \Theta_o(s)}{\Theta_i(s)} = \frac{Js^2 + bs}{Js^2 + bs + K}$$

For a unit-ramp input, $\Theta_i(s) = 1/s^2$. Thus

$$E(s) = \frac{Js^2 + bs}{Js^2 + bs + K}\frac{1}{s^2}$$

The steady-state error is

$$e_{ss} = \lim_{s \to 0} sE(s) = \frac{b}{K}$$

We see that we can reduce the steady-state error e_{ss} by increasing gain K or decreasing viscous-friction coefficient b. However, increasing the gain or decreasing the viscous-friction coefficient causes the damping ratio to decrease with the result that the transient response of the system becomes more oscillatory. Doubling K decreases e_{ss} to half its original value, whereas ζ is decreased to 0.707 of its original value, since ζ is inversely proportional to the square root of K. On the other hand, decreasing b to half its original value decreases both e_{ss} and ζ to half their original values. So it is advisable to increase the value of K rather than decrease the value of b.

After the transient response has died out and a steady state has been reached, the output velocity becomes the same as the input velocity. But there is a steady-state positional error between the input and the output. Examples of the unit-ramp response of the system for three different values of K are illustrated in Fig. 8-68.

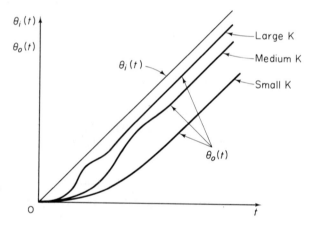

Fig. 8-68. Unit ramp response curves for the system shown in Fig. 8-67.

PROBLEM A-8-11. What is the unit-step response of the system shown in Fig. 8-69?

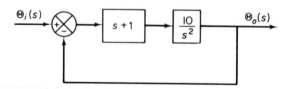

Fig. 8-69. Closed-loop system.

Solution. The closed-loop transfer function is

$$\frac{\Theta_o(s)}{\Theta_i(s)} = \frac{10s + 10}{s^2 + 10s + 10}$$

So for the unit-step input $[\Theta_i(s) = 1/s]$, we have

$$\Theta_o(s) = \frac{10s + 10}{s^2 + 10s + 10} \frac{1}{s}$$

$$= \frac{10s + 10}{(s + 5 + \sqrt{15})(s + 5 - \sqrt{15})s}$$

$$= \frac{-4 - \sqrt{15}}{3 + \sqrt{15}} \frac{1}{s + 5 + \sqrt{15}} + \frac{-4 + \sqrt{15}}{3 - \sqrt{15}} \frac{1}{s + 5 - \sqrt{15}} + \frac{1}{s}$$

The inverse Laplace transform of $\Theta_o(s)$ gives

$$\theta_o(t) = -\frac{4 + \sqrt{15}}{3 + \sqrt{15}} e^{-(5+\sqrt{15})t} + \frac{4 - \sqrt{15}}{-3 + \sqrt{15}} e^{-(5-\sqrt{15})t} + 1$$

$$= -1.15e^{-8.87t} + 0.145e^{-1.13t} + 1$$

PROBLEM A-8-12. Referring to the system in Fig. 8-70, determine the values of K and k such that the system has a damping ratio ζ of 0.7 and an undamped natural frequency ω_n of 4 rad/s.

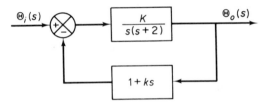

Fig. 8-70. Closed-loop system.

Solution. The closed-loop transfer function is

$$\frac{\Theta_o(s)}{\Theta_i(s)} = \frac{K}{s^2 + (2 + Kk)s + K}$$

Noting that

$$\omega_n = \sqrt{K}, \qquad 2\zeta\omega_n = 2 + Kk$$

we obtain

$$K = \omega_n^2 = 4^2 = 16$$

and

$$2 + Kk = 2\zeta\omega_n = 2 \times 0.7 \times 4 = 5.6$$

Thus

$$Kk = 3.6$$

or

$$k = \frac{3.6}{16} = 0.225$$

PROBLEM A-8-13. Determine the values of K and k of the closed-loop system shown in Fig. 8-71 so that the maximum overshoot in unit-step response is 25% and the peak time is 2 s. Assume that $J = 1$ kg-m^2.

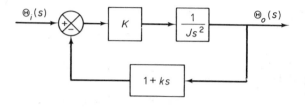

Fig. 8-71. Closed-loop system.

Solution. The closed-loop transfer function is

$$\frac{\Theta_o(s)}{\Theta_i(s)} = \frac{K}{Js^2 + Kks + K}$$

By substituting $J = 1$ kg-m² into this last equation, we have

$$\frac{\Theta_o(s)}{\Theta_i(s)} = \frac{K}{s^2 + Kks + K}$$

Note that

$$\omega_n = \sqrt{K}, \qquad 2\zeta\omega_n = Kk$$

The maximum overshoot M_p is

$$M_p = e^{-\zeta\pi/\sqrt{1-\zeta^2}}$$

which is specified as 25%. Hence

$$e^{-\zeta\pi/\sqrt{1-\zeta^2}} = 0.25$$

from which

$$\frac{\zeta\pi}{\sqrt{1-\zeta^2}} = 1.386$$

or

$$\zeta = 0.404$$

The peak time t_p is specified as 2 s. And so

$$t_p = \frac{\pi}{\omega_d} = 2$$

or

$$\omega_d = 1.57$$

Then the undamped natural frequency ω_n is

$$\omega_n = \frac{\omega_d}{\sqrt{1-\zeta^2}} = \frac{1.57}{\sqrt{1-0.404^2}} = 1.72$$

Therefore we obtain

$$K = \omega_n^2 = 1.72^2 = 2.95 \text{ N-m}$$

$$k = \frac{2\zeta\omega_n}{K} = \frac{2 \times 0.404 \times 1.72}{2.95} = 0.471 \text{ s}$$

PROBLEM A-8-14. When the system shown in Fig. 8-72(a) is subjected to a unit-step input, the system output responds as shown in Fig. 8-72(b). Determine the values of K and T from the response curve.

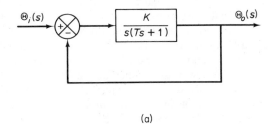

(a)

Fig. 8-72. (a) Closed-loop system;
(b) unit step response curve.   (b)

Solution. The maximum overshoot of 25.4% corresponds to $\zeta = 0.4$. From the response curve we have

$$t_p = 3$$

Consequently,

$$t_p = \frac{\pi}{\omega_d} = \frac{\pi}{\omega_n\sqrt{1 - \zeta^2}} = \frac{\pi}{\omega_n\sqrt{1 - 0.4^2}} = 3$$

It follows that

$$\omega_n = 1.14$$

From the block diagram we have

$$\frac{\Theta_o(s)}{\Theta_i(s)} = \frac{K}{Ts^2 + s + K}$$

from which

$$\omega_n = \sqrt{\frac{K}{T}}, \qquad 2\zeta\omega_n = \frac{1}{T}$$

Therefore the values of T and K are determined as

$$T = \frac{1}{2\zeta\omega_n} = \frac{1}{2 \times 0.4 \times 1.14} = 1.09$$

$$K = \omega_n^2 T = 1.14^2 \times 1.09 = 1.42$$

PROBLEM A-8-15. Figure 8-73 shows a spring-controlled governor that consists of two flyballs, a spring-loaded sleeve, and connecting linkages. Assume that the arms are vertical when the shaft is rotating at a reference speed $\bar{\Omega}$, the mass of each flyball is m, the mass of the sleeve is M, the mass of all other parts is negligible, the spring constant is k, and the coefficient of the viscous friction at the sleeve is b.

Find the transfer function relating x, a small change in the vertical displacement of the sleeve, and ω, a small change in the angular speed. In addition, find the condition on the spring constant k for stable operation of the governor. (Refer to Fig. 8-86 for a schematic diagram of a speed control system.)

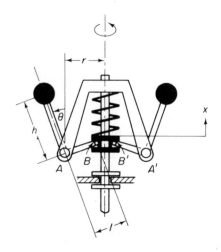

Fig. 8-73. Spring-controlled governor.

Solution. Assume that when the shaft is rotating at a reference speed $\bar{\Omega}$, the spring exerts a downward force $\frac{1}{2}F$ at point B and a similar force $\frac{1}{2}F$ at point B', where $F = kx_o$ and x_o is the displacement of the sleeve from a reference position. (Assume that the effect of the gravitational force mg is taken care of in choosing the reference position for the sleeve displacement.)

At the steady-state operation at $\bar{\Omega}$ the torques acting about point A consist of

$$\text{Torque due to the spring force} = \frac{1}{2}Fl$$

$$\text{Torque due to the centrifugal force} = m\bar{\Omega}^2 rh$$

So the torque balance equation is

$$\frac{1}{2}Fl - m\bar{\Omega}^2 rh = 0$$

or

$$\frac{1}{2}F = \frac{m\bar{\Omega}^2 rh}{l} \tag{8-49}$$

Equation (8-49) gives the spring force acting at point B. A similar force acts at point B'.

Assume that at $t = 0$ the shaft speed is increased from $\bar{\Omega}$ to $\bar{\Omega} + \omega$. This step will cause the sleeve to move upward by a small displacement x. Then the spring force acting downward at point B becomes $\frac{1}{2}F + \frac{1}{2}kx$ and that at point B' also becomes $\frac{1}{2}F + \frac{1}{2}kx$. One-half the inertia force of the sleeve and one-half the viscous-friction force act at point B, and the other halves act at point B'. When the sleeve moves upward by x, the radius at which the flyballs rotate changes from r to $r + h \sin \theta$. The torques acting about point A (or about point A') are

Torque due to the spring force $= (\frac{1}{2}F + \frac{1}{2}kx)l \cos \theta$

Torque due to the inertia force of sleeve $= \frac{1}{2}M\ddot{x}l \cos \theta$

Torque due to the viscous-friction force at the sleeve $= \frac{1}{2}b\dot{x}l \cos \theta$

Torque due to the centrigufal force $= m(\bar{\Omega} + \omega)^2(r + h \sin \theta)h \cos \theta$

Torque due to the inertia force of flyball $= mh \cos \theta \dfrac{d^2}{dt^2} (h \sin \theta)$

For a small change in the angular speed, angle θ is also small. By substituting $\sin \theta \doteq \theta = x/l$, $\cos \theta \doteq 1$, $\dot{\theta}^2 \doteq 0$, $\omega^2 \doteq 0$, and $x\omega \doteq 0$ into the preceding torque expressions, we have

Torque due to the spring force $= \frac{1}{2}(F + kx)l$

Torque due to the inertia force of sleeve $= \frac{1}{2}M\ddot{x}l$

Torque due to the viscous-friction force at the sleeve $= \frac{1}{2}b\dot{x}l$

Torque due to the centrifugal force $= m\left(\bar{\Omega}^2 r + 2\bar{\Omega}\omega r + \bar{\Omega}^2 \dfrac{h}{l}x\right)h$

Torque due to the inertia force of flyball $= mh^2\ddot{\theta} = \dfrac{mh^2}{l}\ddot{x}$

Then the torque balance equation becomes

$$\frac{mh^2}{l}\ddot{x} + \frac{1}{2}M\ddot{x}l + \frac{1}{2}b\dot{x}l + \frac{1}{2}(F + kx)l - m\left(\bar{\Omega}^2 r + 2\bar{\Omega}\omega r + \bar{\Omega}^2\frac{h}{l}x\right)h = 0$$

By substituting Eq. (8-49) into this last equation and simplifying, we find

$$\frac{mh^2}{l}\ddot{x} + \frac{1}{2}Ml\ddot{x} + \frac{1}{2}bl\dot{x} + \frac{1}{2}klx - m\bar{\Omega}^2\frac{h^2}{l}x = 2m\bar{\Omega}\omega rh$$

or

$$\ddot{x} + \frac{\frac{1}{2}bl^2}{mh^2 + \frac{1}{2}Ml^2}\dot{x} + \frac{\frac{1}{2}kl^2 - m\bar{\Omega}^2 h^2}{mh^2 + \frac{1}{2}Ml^2}x = \frac{2m\bar{\Omega}rhl}{mh^2 + \frac{1}{2}Ml^2}\omega \qquad (8\text{-}50)$$

By defining

$$\omega_n^2 = \frac{\frac{1}{2}kl^2 - m\bar{\Omega}^2 h^2}{mh^2 + \frac{1}{2}Ml^2}$$

$$2\zeta\omega_n = \frac{\frac{1}{2}bl^2}{mh^2 + \frac{1}{2}Ml^2}$$

$$K = \frac{2m\bar{\Omega}rhl}{mh^2 + \frac{1}{2}Ml^2}$$

Eq. (8-50) can be written

$$\ddot{x} + 2\zeta\omega_n\dot{x} + \omega_n^2 x = K\omega$$

Therefore the transfer function between $X(s)$ and $\Omega(s)$, where $X(s) = \mathcal{L}[x]$ and $\Omega(s) = \mathcal{L}[\omega]$, becomes

$$\frac{X(s)}{\Omega(s)} = \frac{K}{s^2 + 2\zeta\omega_n s + \omega_n^2}$$

Note that for stable operation of the governor, $\omega_n^2 > 0$, or

$$\frac{kl^2}{2mh^2} > \bar{\Omega}^2$$

Consequently, the spring constant k must satisfy the inequality

$$k > \frac{2mh^2}{l^2}\bar{\Omega}^2$$

If this condition is satisfied, a small disturbance will cause the sleeve to exhibit a damped oscillation about the operating point.

PROBLEMS

PROBLEM B-8-1. Simplify the block diagram shown in Fig. 8-74 and obtain the transfer function $\Theta_o(s)/\Theta_i(s)$.

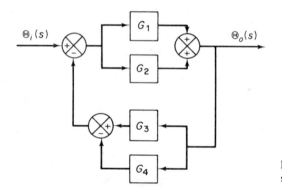

Fig. 8-74. Block diagram of a system.

PROBLEM B-8-2. Simplify the block diagram shown in Fig. 8-75 and obtain the transfer function $\Theta_o(s)/\Theta_i(s)$.

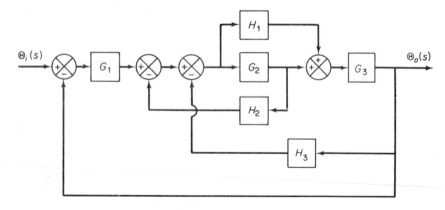

Fig. 8-75. Block diagram of a system.

PROBLEM B-8-3. Figure 8-76 shows an electric-pneumatic transducer. Show that the change in the output pressure is proportional to the change in the input current.

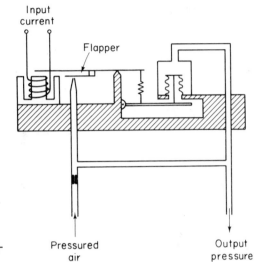

Fig. 8-76. Electric-pneumatic transducer.

PROBLEM B-8-4. Figure 8-77 shows a pneumatic controller. What type of control action does this controller produce?

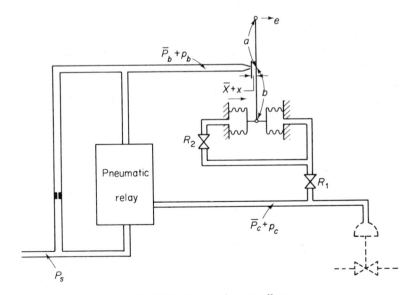

Fig. 8-77. Pneumatic controller.

Problem B-8-5. Figure 8-78 is a schematic diagram of an aircraft elevator control system. The input to the system is the deflection angle θ of the control lever, and the output is the elevator angle ϕ. Assume that angles θ and ϕ are relatively small. Show that for each angle θ of the control lever there is a corresponding (steady-state) elevator angle ϕ.

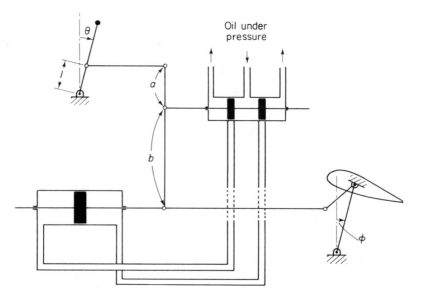

Fig. 8-78. Aircraft elevator control system.

Problem B-8-6. Consider industrial automatic controllers whose control actions are proportional, integral, proportional-plus-integral, proportional-plus-derivative, and proportional-plus-derivative-plus-integral. The transfer functions of these controllers can be given, respectively, by

$$\frac{M(s)}{E(s)} = K_p$$

$$\frac{M(s)}{E(s)} = \frac{K_i}{s}$$

$$\frac{M(s)}{E(s)} = K_p\left(1 + \frac{1}{T_i s}\right)$$

$$\frac{M(s)}{E(s)} = K_p(1 + T_d s)$$

$$\frac{M(s)}{E(s)} = K_p\left(1 + T_d s + \frac{1}{T_i s}\right)$$

where $M(s)$ is the Laplace transform of $m(t)$, the controller output, and $E(s)$ the Laplace transform of $e(t)$, the actuating error signal. Sketch $m(t)$ versus t curves for each of the five types of controllers when the actuating error signal is

1. $e(t) =$ unit step function

2. $e(t) =$ unit ramp function

In sketching curves, assume that the numerical values of K_p, K_i, T_i, and T_d are given as

$$K_p = \text{proportional gain} = 4$$
$$K_i = \text{integral gain} = 2$$
$$T_i = \text{integral time} = 2 \text{ s}$$
$$T_d = \text{derivative time} = 0.8 \text{ s}$$

PROBLEM B-8-7. Consider the hydraulic servo system shown in Fig. 8-79. Assuming that signal $e(t)$ is the input and power piston displacement $y(t)$ the output, find the transfer function $Y(s)/E(s)$.

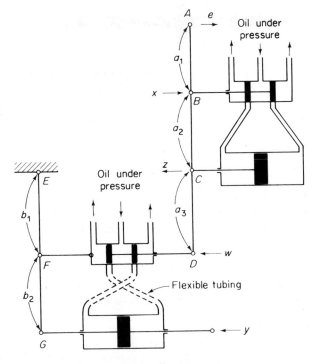

Fig. 8-79. Hydraulic servo system.

PROBLEM B-8-8. Many machines, such as lathes, milling machines, and grinders, are provided with tracers to reproduce the contour of templates. Figure 8-80 is a schematic diagram of a hydraulic tracing system in which the tool duplicates the shape of the template on the workpiece. Explain the operation of the system.

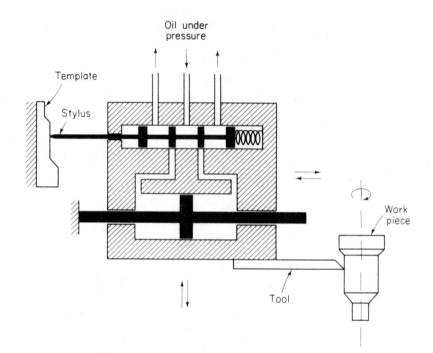

Fig. 8-80. Hydraulic tracing system.

PROBLEM B-8-9. Consider a glass-wall mercury thermometer. If the thermal capacitance of the glass of the thermometer is negligible, then it may be considered a first-order system and its transfer function may be given by

$$\frac{\Theta(s)}{\Theta_b(s)} = \frac{1}{Ts + 1}$$

where $\Theta(s)$ is the Laplace transform of the thermometer temperature θ and $\Theta_b(s)$ is the Laplace transform of the bath temperature θ_b, both temperatures measured from the ambient temperature.

Assume that a glass-wall mercury thermometer is used to measure the temperature of a bath and that the thermal capacitance of the glass is negligible. Assume also that the time constant of the thermometer is not known. So it is experimentally determined by lowering it into a pail of water held at 10°C. Figure 8-81 shows the temperature response observed in the test. Find the time constant. If this thermometer is placed in a bath, the temperature of which is increasing linearly at a rate of 10°C/min, how much steady-state error does the thermometer show?

If the thermal capacitance of the glass of a mercury thermometer is not negligible, it may be considered a second-order system and the transfer function may be modified to

$$\frac{1}{(T_1s + 1)(T_2s + 1)}$$

where T_1 and T_2 are time constants. Sketch a typical temperature response curve (θ versus t curve) when such a thermometer with two time constants is placed in a bath held at a constant temperature θ_b, where both thermometer temperature θ and bath temperature θ_b are measured from the ambient temperature.

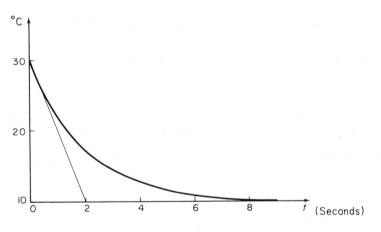

Fig. 8-81. Response curve for a thermometer system.

PROLBEM B-8-10. Referring to the system shown in Fig. 8-82, find the response $\theta_o(t)$ when input $\theta_i(t)$ is a unit ramp. Also, find the steady-state error for a ramp response. Assume that the system is underdamped.

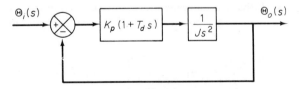

Fig. 8-82. Block diagram of a system.

PROBLEM B-8-11. In Fig. 8-83 a spring-mass-damper system is attached to a cart. For $t < 0$, the cart is standing still and the entire system is at rest. At $t = 0$ the cart is moved at a constant speed, or $\dot{x}_i = r = $ constant. What is the motion $x_o(t)$ relative to the ground? Assume that the spring-mass-damper system is underdamped.

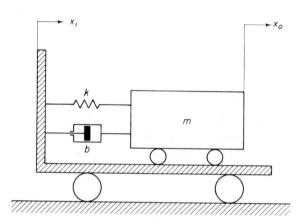

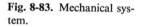

Fig. 8-83. Mechanical system.

PROBLEM B-8-12. Consider a system defined by

$$\frac{\Theta_o(s)}{\Theta_i(s)} = \frac{\omega_n^2}{s^2 + 2\zeta\omega_n s + \omega_n^2}$$

Determine the values of ζ and ω_n so that the system responds to a step input with approximately 5% overshoot and with a settling time of 2 seconds.

PROBLEM B-8-13. Figure 8-84 shows a position control system with velocity feedback. What is the response $\theta_o(t)$ to the unit-step input?

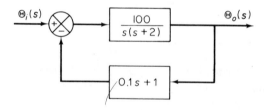

Fig. 8-84. Block diagram of a position control system with velocity feedback.

PROBLEM B-8-14. Consider the system shown in Fig. 8-85. Determine the value of k such that the damping ratio ζ is 0.5. Then obtain the rise time t_r, peak time t_p, maximum overshoot M_p, and settling time t_s in the unit-step response.

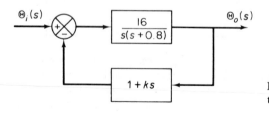

Fig. 8-85. Block diagram of a system.

PROBLEM **B-8-15.** Explain the operation of the speed-control system shown in Fig. 8-86.

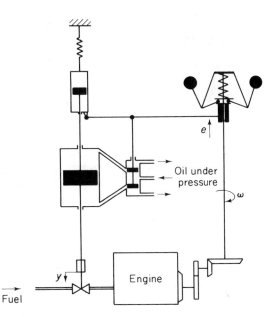

Fig. 8-86. Speed-control system.

SYSTEMS OF UNITS

In the following pages we first review the customary systems of units (the cgs system of units, mks system of units, etc.) and then present the International System of units (SI).

Units. A physical quantity can be measured only by a comparison with a like quantity. A distinct amount of a physical quantity is called a *unit*. (To be useful, the unit should be a convenient practical size.) Any physical quantity of the same kind can be compared with it, and its value can be stated in terms of a ratio number and the unit used.

Basic units and derived units. The general unit of a physical quantity is defined as its dimension. A unit system can be developed by choosing, for each basic dimension of the system, a specific unit (for example, the meter for length, the kilogram for mass, and the second for time). Such a unit is called a *basic unit*. The corresponding physical quantity is called a *basic quantity*. All units that are not basic are called *derived units*.

Systematic units. *Systematic units* are systematically derived units within a unit system. They can be obtained by replacing the general units (dimensions) by the basic units of the system.

If we define the dimensions of length, mass, and time as $[L]$, $[M]$, and $[T]$, respectively, then physical quantities may be expressed as $[L]^x[M]^y[T]^z$. For instance, the dimension of acceleration is $[L][T]^{-2}$ and that of force is $[L][M][T]^{-2}$. In the mks system of units, the systematic unit of acceleration is therefore 1 m/s² and that of force is 1 kg-m/s².

Absolute systems of units and gravitational systems of units. Systems of units in which the mass is taken as a basic unit are called *absolute systems of units,* whereas those in which the force rather than the mass is taken as a basic unit are called *gravitational systems of units.*

The cgs system of units. The *cgs system of units* is an absolute system of units and is based on the centimeter, gram mass, and second as basic units. This system has been widely used in science. Disadvantages include the fact that the derived units for force and energy are too small for practical purposes and that the system does not combine with the practical electrical units to form a comprehensive unit system.

The mks system of units. The *mks system of units* is an absolute system of units and is based on the meter, kilogram mass, and second as basic units. In this system the derived units for force and energy are a convenient size in an engineering sense, and all the practical electrical units fit in as the natural units to form a comprehensive unit system.

The metric engineering system of units. The *metric engineering system of units* is a gravitational system of units and is based on the meter, kilogram force, and second as basic units. (Since the standard of force is defined as the weight of the prototype standard mass of the kilogram, the basic unit of force is variable, but this factor is not a serious disadvantage.)

The British engineering system of units. The *British engineering system of units* is a gravitational system of units and is based on the foot, pound force, and second as basic units. This system is the one that has been used in the United States up to the present. The derived unit of mass is lb_f-s²/ft and is called a slug (1 slug = 1 lb_f-s²/ft).

The International System of units (SI). The *International System of units* (abbreviated SI) is the internationally agreed system of units for expressing the values of physical quantities. (See Table A-1.) In this system four basic units are added to the customary three basic units (meter, kilogram, second) of the mks absolute system of units. The four added basic units are the ampere as the unit of electric current, the kelvin as the unit of thermo-dynamic temperature, the candela as the unit of luminous intensity, and the

Table A-1. INTERNATIONAL SYSTEM OF UNITS (SI)

	Quantity	Unit	Symbol	Dimension
Basic units	Length	meter	m	
	Mass	kilogram	kg	
	Time	second	s	
	Electric current	ampere	A	
	Temperature	kelvin	K	
	Luminous intensity	candela	cd	
	Amount of substance	mole	mol	
Auxiliary units	Plane angle	radian	rad	
	Solid angle	steradian	sr	
Derived units	Acceleration	meter per second squared	m/s^2	
	Activity (of radioactive source)	1 per second	s^{-1}	
	Angular acceleration	radian per second squared	rad/s^2	
	Angular velocity	radian per second	rad/s	
	Area	square meter	m^2	
	Density	kilogram per cubic meter	kg/m^3	
	Dynamic viscosity	newton second per square meter	$N\text{-}s/m^2$	$m^{-1}\ kg\ s^{-1}$
	Electric capacitance	farad	F	$m^{-2}\ kg^{-1}\ s^4\ A^2$
	Electric charge	coulomb	C	$A\ s$
	Electric field strength	volt per meter	V/m	$m\ kg\ s^{-3}\ A^{-1}$
	Electric resistance	ohm	Ω	$m^2\ kg\ s^{-3}\ A^{-2}$
	Entropy	joule per kelvin	J/K	$m^2\ kg\ s^{-2}\ K^{-1}$

Table A-1. (CONTINUED)

	Quantity	Unit	Symbol	Dimension
	Force	newton	N	$m\,kg\,s^{-2}$
	Frequency	hertz	Hz	s^{-1}
	Illumination	lux	lx	$m^{-2}\,cd\,sr$
	Inductance	henry	H	$m^2\,kg\,s^{-2}\,A^{-2}$
	Kinematic viscosity	square meter per second	m^2/s	
	Luminance	candela per square meter	cd/m^2	
	Luminous flux	lumen	lm	$cd\,sr$
	Magnetomotive force	ampere turn	A	A
	Magnetic field strength	ampere per meter	A/m	
	Magnetic flux	weber	Wb	$m^2\,kg\,s^{-2}\,A^{-1}$
Derived units (cont.)	Magnetic flux density	tesla	T	$kg\,s^{-2}\,A^{-1}$
	Power	watt	W	$m^2\,kg\,s^{-3}$
	Pressure	pascal (newton per square meter)	Pa (N/m^2)	$m^{-1}\,kg\,s^{-2}$
	Radiant intensity	watt per steradian	W/sr	$m^2\,kg\,s^{-3}\,sr^{-1}$
	Specific heat	joule per kilogram kelvin	J/kg-K	$m^2\,s^{-2}\,K^{-1}$
	Thermal conductivity	watt per meter kelvin	W/m-K	$m\,kg\,s^{-3}\,K^{-1}$
	Velocity	meter per second	m/s	
	Volume	cubic meter	m^3	
	Voltage	volt	V	$m^2\,kg\,s^{-3}\,A^{-1}$
	Wave number	1 per meter	m^{-1}	
	Work, energy, quantity of heat	joule	J	$m^2\,kg\,s^{-2}$

mole as the unit of amount of substance. Thus in SI units the meter, kilogram, second, ampere, kelvin, candela, and mole constitute the seven basic units. There are two auxiliary units in the SI units—the radian, which is the unit of a plane angle, and the steradian, which is the unit of a solid angle. Table A-1 lists 7 basic units, 2 auxiliary units, and some of the derived units of the International System of units (SI). [Multiples and sub-multiples of the units are indicated by a series of sixteen prefixes for various powers of ten. (See page 10.)]

In SI units the seven basic units are defined in the following way.

Meter: The meter is the length equal to 1 650 763.73 wavelengths of radiation in vacuum corresponding to the unperturbed transition between levels 2 P_{10} and 5 d_5 of the atom of krypton 86, the orange-red line.

Kilogram: The kilogram is the mass of a particular cylinder (of diameter 39 mm and height 39 mm) of platinum-iridium alloy, called the International Prototype Kilogram, which is preserved in a vault at Sévres, France, by the International Bureau of Weights and Measures.

Second: The second is the duration of 9 192 631 770 periods of the radiation corresponding to the transition between the two hyperfine levels of the fundamental state of the atom of cesium 133.

Ampere: The ampere is a constant current that, if maintained in two straight, parallel conductors of infinite length, of negligible circular cross-sections, and placed 1 meter apart in a vacuum, will produce between these conductors a force equal to 2×10^{-7} newton per meter of length.

Kelvin: The kelvin is the fraction 1/273.16 of the thermodynamic temperature of the triple point of water. (Note that the triple point of water is 0.01°C.)

Candela: The candela is the luminous intensity, in the direction of the normal, of a blackbody surface 1/600 000 square meter in area, at the temperature of solidification of platinum under a pressure of 101 325 newtons per square meter.

Mole: The mole is the amount of substance of a system which contains as many elementary entities as there are atoms in 0.012 kilogram of carbon 12.

The two auxiliary units of SI (radian and steradian) are defined as follows:

Radian: The radian is a unit of plane angular measurement equal to the angle at the center of a circle subtended by an arc equal in length to the radius. (The dimension of the radian is zero, since it is a ratio of the quantities of the same dimension.)

Steradian: The steradian is a unit of measure of solid angles that is expressed as the solid angle subtended at the center of the sphere by a portion of the surface whose area is equal to the square of the radius of the sphere. (The dimension of the steradian is also zero, since it is a ratio of the quantities of the same dimension.)

B

CONVERSION TABLES

Conversion tables for mass, length, area, volume, pressure, energy, power, and temperature are presented in Tables B-1 through B-9.

Table B-1. CONVERSION TABLE FOR MASS

g	kg	lb	oz	grain	slug
1	10^{-3}	2.205×10^{-3}	3.527×10^{-2}	15.432	6.852×10^{-5}
10^3	1	2.205	35.27	15.432×10^3	6.852×10^{-2}
453.6	0.4536	1	16	7000	3.108×10^{-2}
28.35	2.835×10^{-2}	0.0625	1	437.5	1.943×10^{-3}
6.480×10^{-2}	6.480×10^{-5}	1.429×10^{-4}	2.286×10^{-3}	1	4.440×10^{-6}
1.459×10^4	14.59	32.17	514.78	2.252×10^5	1

Table B-2. CONVERSION TABLES FOR LENGTH

cm	m	in.	ft	yd
1	0.01	0.3937	0.03281	0.01094
100	1	39.37	3.281	1.0936
2.54	0.0254	1	0.08333	0.02778
30.48	0.3048	12	1	0.3333
91.44	0.9144	36	3	1

km	mile	nautical mile	ft
1	0.6214	0.5400	3280.84
1.6093	1	0.8690	5280
1.852	1.151	1	6076

Table B-3. CONVERSION TABLE FOR LENGTH (FROM in. TO mm)

in.	mm	in.	mm	in.	mm	in.	mm
1/32	0.794	9/32	7.144	17/32	13.494	25/32	19.844
1/16	1.587	5/16	7.937	9/16	14.287	13/16	20.638
3/32	2.381	11/32	8.731	19/32	15.081	27/32	21.431
1/8	3.175	3/8	9.525	5/8	15.875	7/8	22.225
5/32	3.969	13/32	10.319	21/32	16.669	29/32	23.019
3/16	4.762	7/16	11.112	11/16	17.462	15/16	23.812
7/32	5.556	15/32	11.906	23/32	18.256	31/32	24.606
1/4	6.350	1/2	12.700	3/4	19.050	1	25.400

Table B-4. CONVERSION TABLES FOR AREA

cm²	m²	in.²	ft²	yd²
1	10^{-4}	0.155	1.0764×10^{-3}	1.196×10^{-4}
10^4	1	1550	10.764	1.196
6.452	6.452×10^{-4}	1	6.944×10^{-3}	7.716×10^{-4}
929.0	0.09290	144	1	0.1111
8361	0.8361	1296	9	1

km²	mile²
1	0.3861
2.590	1

Table B-5. CONVERSION TABLES FOR VOLUME

mm³	cm³	in.³
1	10^{-3}	6.102×10^{-5}
10^3	1	6.102×10^{-2}
1.639×10^4	16.39	1

m³	ft³	yd³
1	35.315	1.308
2.832×10^{-2}	1	3.704×10^{-2}
0.7646	27	1

U.S. gallon	liter	barrel
1	3.785	2.381×10^{-2}
0.2642	1	0.6290×10^{-2}
42	159	1

Table B-6. CONVERSION TABLE FOR PRESSURE

Pa or N/m²	bar (10^5 N/m²)	kg_f/cm^2	$lb_f/in.^2$	atm (standard atmospheric pressure)	mm Hg	in. Hg	m H_2O
1	1×10^{-5}	1.0197×10^{-5}	1.450×10^{-4}	9.869×10^{-6}	7.501×10^{-3}	2.953×10^{-4}	1.0197×10^{-4}
1×10^5	1	1.0197	14.50	0.9869	750.1	29.53	10.197
9.807×10^4	0.9807	1	14.22	0.9678	735.6	28.96	10.000
6.895×10^3	0.06895	0.07031	1	0.06805	51.71	2.036	0.7031
1.0133×10^5	1.0133	1.0332	14.70	1	760	29.92	10.33
1.3332×10^2	1.3332×10^{-3}	1.3595×10^{-3}	19.34×10^{-3}	1.3158×10^{-3}	1	3.937×10^{-2}	1.360×10^{-2}
3.386×10^3	0.03386	0.03453	0.4912	0.03342	25.4	1	0.3453
9.807×10^3	0.09807	0.10000	1.422	0.09678	73.55	2.896	1

Table B-7. Conversion table for energy

J	kg$_f$-m	ft-lb$_f$	kWh	kcal	Btu
1	0.10197	0.7376	2.778×10^{-7}	2.389×10^{-4}	9.480×10^{-4}
9.807	1	7.233	2.724×10^{-6}	2.343×10^{-3}	9.297×10^{-3}
1.356	0.1383	1	3.766×10^{-7}	3.239×10^{-4}	1.285×10^{-3}
3.600×10^6	3.671×10^5	2.655×10^6	1	860	3413
4186	426.9	3087	1.163×10^{-3}	1	3.968
1055	107.6	778	2.930×10^{-4}	0.2520	1

Table B-8. Conversion table for power

kW	kg$_f$-m/s	ft-lb$_f$/s	British horse power hp	kcal/s	Btu/s
1	101.97	737.6	1.341	0.2389	0.9480
9.807×10^{-3}	1	7.233	1.315×10^{-2}	2.343×10^{-3}	9.297×10^{-3}
1.356×10^{-3}	0.1383	1	1.818×10^{-3}	3.239×10^{-4}	1.285×10^{-3}
0.7457	76.04	550	1	0.1782	0.7069
4.186	426.9	3087	5.613	1	3.968
1.055	107.6	778.0	1.414	0.2520	1

Table B-9. CONVERSION TABLE FOR TEMPERATURE

°C	°F	°C	°F	°C	°F
−50	−58	16	60.8	44	111.2
−40	−40	18	64.4	46	114.8
−30	−22	20	68.0	48	118.4
−20	−4	22	71.6	50	122.0
−10	14	24	75.2	55	131.0
−5	23	26	78.8	60	140.0
0	32	28	82.4	65	149.0
2	35.6	30	86.0	70	158.0
4	39.2	32	89.6	75	167.0
6	42.8	34	93.2	80	176.0
8	46.4	36	96.8	85	185.0
10	50.0	38	100.4	90	194.0
12	53.6	40	104.0	95	203.0
14	57.2	42	107.6	100	212.0

To convert from Fahrenheit to Celsius, subtract 32 and multiply by 5/9.

$$t_C = \frac{5}{9}(t_F - 32)$$

To convert from Celsius to Fahrenheit, multiply by 9/5 and add 32.

$$t_F = \frac{9}{5}t_C + 32$$

Absolute zero temperature occurs at −273.15° on the Celsius scale and at −459.67° on the Fahrenheit scale. Absolute temperatures on the two scales are $t_C + 273.15$ and $t_F + 459.67$. Note that in most calculations the constants used are 273 and 460. Note also that

$$t_C \text{ degrees Celsius} = (t_C + 273.15) \text{ kelvin}$$

C

LAGRANGE'S EQUATIONS OF MOTION

Mathematical models for physical systems (mechanical, electrical, etc.) may be derived from energy considerations without applying Newton's or Kirchhoff's laws to them. In Sec. 2-5 we presented a simple method for deriving equations of motion of mechanical systems from a knowledge of the system's potential and kinetic energies. In this appendix we present a more versatile energy method, due to Lagrange, for deriving mathematical models.

In deriving the equations of motion of a complicated mechanical system, it is advisable to do so by using two different methods (one based on Newton's second law and the other on energy considerations) to ensure that they are correct. In this regard, Lagrange's method is a convenient recourse for deriving system equations.

In deriving Lagrange's equations of motion, we need to define generalized coordinates and the Lagrangian and to state Hamilton's principle.

Generalized coordinates. The generalized coordinates of a system are a set of independent coordinates that is necessary to describe the motion of the system completely. The number of generalized coordinates needed to describe the system motion is equal to the number of degrees of freedom.

If a system requires n generalized coordinates $q_1, q_2, \ldots, q_n$, we can consider such n generalized coordinates as coordinates of an n-dimensional coordinate system in an n-dimensional space. Then at any given instant of time the system is characterized by a point in this n-dimensional space. As time elapses, the system point in the n-dimensional space moves and describes a curve in the space. (The curve represents the motion of the system point.)

Lagrangian. The Lagrangian L of a system is defined by

$$L = T - U \qquad \text{(C-1)}$$

where T is the kinetic energy and U is the potential energy of the system. The Lagrangian L in a general form is a function of $q_i, \dot{q}_i$ ($i = 1, 2, \ldots, n$), and time t or

$$L = L(q_i, \dot{q}_i\, t)$$

Hamilton's principle. Hamilton's principle states that the motion of the system point in the n-dimensional space from $t = t_1$ to $t = t_2$ is such that the integral

$$I = \int_{t_1}^{t_2} L(q_i, \dot{q}_i, t)\, dt \qquad (i = 1, 2, \ldots, n) \qquad \text{(C-2)}$$

is an extremum (minimum or maximum) for the path of motion.

Lagrange's equations of motion for conservative systems. If no energy is dissipated in a system, it is called a *conservative system.* A conservative mechanical system is one in which energy appears only as kinetic energy and potential energy.

Consider the case where the Lagrangian L is a function of one generalized coordinate q, its time derivative $\dot{q}$, and time t, or $L = L(q, \dot{q}, t)$. Then Eq. (C-2) becomes

$$I = \int_{t_1}^{t_2} L(q, \dot{q}, t)\, dt \qquad \text{(C-3)}$$

Let q be the function for which I is an extremum. Let us assume that δq is an arbitrary function that is continuous in $t_1 \leq t \leq t_2$, has a continuous derivative $\delta \dot{q}$ in $t_1 \leq t \leq t_2$, and vanishes at $t = t_1$ and $t = t_2$, or $\delta q(t_1) = \delta q(t_2) = 0$. Let us give small variations in q and $\dot{q}$ and evaluate the difference ΔI between two Lagrange integrals—that is,

$$\Delta I = \int_{t_1}^{t_2} L(q + \delta q, \dot{q} + \delta \dot{q}, t)\, dt - \int_{t_1}^{t_2} L(q, \dot{q}, t)\, dt$$

$$= \int_{t_1}^{t_2} [L(q + \delta q, \dot{q} + \delta \dot{q}, t) - L(q, \dot{q}, t)]\, dt$$

By expanding the integrand of the right-hand side of this last equation into a

Taylor series about point $(q, \dot{q})$, we obtain

$$\Delta I = \int_{t_1}^{t_2} \left(\frac{\partial L}{\partial q} \delta q + \frac{\partial L}{\partial \dot{q}} \delta \dot{q} \right) dt$$

$$+ \frac{1}{2} \int_{t_1}^{t_2} \left[\frac{\partial^2 L}{\partial q^2} (\delta q)^2 + 2 \frac{\partial^2 L}{\partial q \, \partial \dot{q}} \delta q \, \delta \dot{q} + \frac{\partial^2 L}{\partial \dot{q}^2} (\delta \dot{q})^2 \right] dt + \cdots$$

The first term on the right-hand side here is called the *first variation of I.*

It is well known in the theory of variations that the necessary condition for I to be an extremum is that the first variation of I or δI be zero.

$$\delta I = \int_{t_1}^{t_2} \left(\frac{\partial L}{\partial q} \delta q + \frac{\partial L}{\partial \dot{q}} \delta \dot{q} \right) dt = \int_{t_1}^{t_2} \frac{\partial L}{\partial q} \delta q \, dt + \int_{t_1}^{t_2} \frac{\partial L}{\partial \dot{q}} \frac{d \, \delta q}{dt} \, dt = 0$$

Since this last equation can be written

$$\int_{t_1}^{t_2} \frac{\partial L}{\partial q} \delta q \, dt + \frac{\partial L}{\partial \dot{q}} \delta q \bigg|_{t_1}^{t_2} - \int_{t_1}^{t_2} \frac{d}{dt} \left(\frac{\partial L}{\partial \dot{q}} \right) \delta q \, dt = 0$$

we have

$$\frac{\partial L}{\partial \dot{q}} \delta q(t_2) - \frac{\partial L}{\partial \dot{q}} \delta q(t_1) + \int_{t_1}^{t_2} \left(\frac{\partial L}{\partial q} - \frac{d}{dt} \frac{\partial L}{\partial \dot{q}} \right) \delta q \, dt = 0$$

Noting that $\delta q(t_1) = \delta q(t_2) = 0$, the first two terms in this last equation are equal to zero, and so we obtain

$$\int_{t_1}^{t_2} \left(\frac{\partial L}{\partial q} - \frac{d}{dt} \frac{\partial L}{\partial \dot{q}} \right) \delta q \, dt = 0$$

This last equation must hold true for any δq satisfying the condition that it is continuous and vanishes at $t = t_1$ and $t = t_2$. Then according to the theory of variations, the integrand must be identically zero or

$$\frac{d}{dt} \left(\frac{\partial L}{\partial \dot{q}} \right) - \frac{\partial L}{\partial q} = 0 \tag{C-4}$$

Equation (C-4) is known as the *Euler-Lagrange equation.* As will be seen later, Eq. (C-4) reduces to the equation of motion of the system that we would obtain were we to use Newton's second law (or Kirchhoff's laws, etc.). Therefore it is also known as *Lagrange's equation of motion* for the conservative system.

If the Lagrangian L is a function of n generalized coordinates, n generalized velocities, and time t, then

$$L = L(q_1, q_2, \ldots, q_n; \dot{q}_1, \dot{q}_2, \ldots, \dot{q}_n; t)$$

and the corresponding Euler-Lagrange equations are

$$\frac{d}{dt} \left(\frac{\partial L}{\partial \dot{q}_i} \right) - \frac{\partial L}{\partial q_i} = 0 \qquad (i = 1, 2, \ldots, n) \tag{C-5}$$

The n equations given by Eq. (C-5) are Lagrange's equations of motion for the conservative system.

Example C-1. **Simple pendulum.** Consider the simple pendulum shown in Fig. C-1. This is a one-degree-of-freedom system, and the angle θ is the only generalized coordinate.

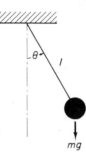

Fig. C-1. Simple pendulum.

The kinetic energy of the system is

$$T = \tfrac{1}{2}m(l\dot{\theta})^2$$

Assuming the position of mass when $\theta = 0$ as the datum line, the potential energy U can be written

$$U = mgl(1 - \cos\theta)$$

The Lagrangian L is

$$L = T - U = \tfrac{1}{2}m(l\dot{\theta})^2 - mgl(1 - \cos\theta)$$

Therefore Lagrange's equation

$$\frac{d}{dt}\left(\frac{\partial L}{\partial \dot{\theta}}\right) - \frac{\partial L}{\partial \theta} = 0$$

becomes

$$\frac{d}{dt}(ml^2\dot{\theta}) + mgl\sin\theta = 0$$

or

$$\ddot{\theta} + \frac{g}{l}\sin\theta = 0$$

which is the equation of motion for the system.

Example C-2. **Spring-loaded pendulum.** Consider the spring-loaded pendulum shown in Fig. C-2 and assume that the spring force is zero when the pendulum is vertical, or $\theta = 0$. This is also a one-degree-of-freedom system, and the angle θ is the only generalized coordinate.

The kinetic energy of the system is

$$T = \tfrac{1}{2}m(l\dot{\theta})^2$$

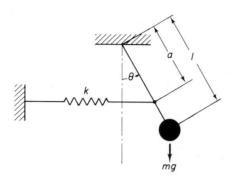

Fig. C-2. Spring-loaded pendulum.

and the potential energy of the system is

$$U = mgl(1 - \cos \theta) + \tfrac{1}{2}k(a \sin \theta)^2$$

The Lagrangian L is

$$L = T - U = \tfrac{1}{2}m(l\dot{\theta})^2 - mgl(1 - \cos \theta) - \tfrac{1}{2}ka^2 \sin^2 \theta$$

Hence Lagrange's equation

$$\frac{d}{dt}\left(\frac{\partial L}{\partial \dot{\theta}}\right) - \frac{\partial L}{\partial \theta} = 0$$

becomes

$$\frac{d}{dt}(ml^2\dot{\theta}) + mgl \sin \theta + ka^2 \sin \theta \cos \theta = 0$$

or

$$\ddot{\theta} + \frac{g}{l} \sin \theta + \frac{ka^2}{ml^2} \sin \theta \cos \theta = 0$$

This is the equation of motion for the system. For small values of θ, this last equation can be simplified to

$$\ddot{\theta} + \left(\frac{g}{l} + \frac{ka^2}{ml^2}\right)\theta = 0$$

(Refer to Problem A-7-4, which gives the derivation of this last equation by using Newton's second law.)

Example C-3. **Double pendulum.** Consider the double pendulum shown in Fig. C-3. This is a two-degrees-of-freedom system. The angles θ_1 and θ_2 are the generalized coordinates for the system.

The kinetic energy of the system is

$$T = \tfrac{1}{2}m_1 v_1^2 + \tfrac{1}{2}m_2 v_2^2$$

where v_1 and v_2 are the absolute velocities of masses m_1 and m_2, respectively. Note that

$$v_1 = l_1\dot{\theta}_1$$

The absolute velocity v_2 is not obvious from the diagram and so will be derived from

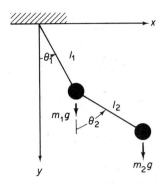

Fig. C-3. Double pendulum.

what follows. Since it is easier to obtain the absolute velocity v_2 in the rectangular x-y coordinate system, we shall first write the x and y coordinates of mass m_2 and then differentiate them to obtain $\dot{x}$ and $\dot{y}$.

$$x = l_1 \sin \theta_1 + l_2 \sin \theta_2$$
$$y = l_1 \cos \theta_1 + l_2 \cos \theta_2$$

The velocities $\dot{x}$ and $\dot{y}$ are

$$\dot{x} = l_1 \cos \theta_1 \dot{\theta}_1 + l_2 \cos \theta_2 \dot{\theta}_2$$
$$\dot{y} = -l_1 \sin \theta_1 \dot{\theta}_1 - l_2 \sin \theta_2 \dot{\theta}_2$$

Noting that

$$v_2^2 = \dot{x}^2 + \dot{y}^2$$

we obtain

$$v_2^2 = (l_1 \cos \theta_1 \dot{\theta}_1 + l_2 \cos \theta_2 \dot{\theta}_2)^2 + (-l_1 \sin \theta_1 \dot{\theta}_1 - l_2 \sin \theta_2 \dot{\theta}_2)^2$$
$$= l_1^2 \dot{\theta}_1^2 + l_2^2 \dot{\theta}_2^2 + 2 l_1 l_2 \dot{\theta}_1 \dot{\theta}_2 \cos (\theta_2 - \theta_1)$$

Hence the kinetic energy T is

$$T = \tfrac{1}{2} m_1 (l_1 \dot{\theta}_1)^2 + \tfrac{1}{2} m_2 [l_1^2 \dot{\theta}_1^2 + l_2^2 \dot{\theta}_2^2 + 2 l_1 l_2 \dot{\theta}_1 \dot{\theta}_2 \cos (\theta_1 - \theta_2)]$$

The potential energy U of the system is

$$U = m_1 g l_1 (1 - \cos \theta_1) + m_2 g [l_1 (1 - \cos \theta_1) + l_2 (1 - \cos \theta_2)]$$

where the potential energy of the system when $\theta_1 = 0$ and $\theta_2 = 0$ is taken as zero. The Lagrangian L of the system is

$$L = T - U = \tfrac{1}{2} m_1 (l_1 \dot{\theta}_1)^2 + \tfrac{1}{2} m_2 [l_1^2 \dot{\theta}_1^2 + l_2^2 \dot{\theta}_2^2 + 2 l_1 l_2 \dot{\theta}_1 \dot{\theta}_2 \cos (\theta_2 - \theta_1)]$$
$$- m_1 g l_1 (1 - \cos \theta_1) - m_2 g [l_1 (1 - \cos \theta_1) + l_2 (1 - \cos \theta_2)]$$

Lagrange's equations for this system are

$$\frac{d}{dt}\left(\frac{\partial L}{\partial \dot{\theta}_1}\right) - \frac{\partial L}{\partial \theta_1} = 0$$

$$\frac{d}{dt}\left(\frac{\partial L}{\partial \dot{\theta}_2}\right) - \frac{\partial L}{\partial \theta_2} = 0$$

Noting that

$$\frac{\partial L}{\partial \dot{\theta}_1} = (m_1 + m_2)l_1^2\dot{\theta}_1 + m_2 l_1 l_2 \dot{\theta}_2 \cos(\theta_2 - \theta_1)$$

$$\frac{\partial L}{\partial \theta_1} = m_2 l_1 l_2 \dot{\theta}_1 \dot{\theta}_2 \sin(\theta_2 - \theta_1) - (m_1 + m_2)g l_1 \sin \theta_1$$

$$\frac{\partial L}{\partial \dot{\theta}_2} = m_2 l_2^2 \dot{\theta}_2 + m_2 l_1 l_2 \dot{\theta}_1 \cos(\theta_2 - \theta_1)$$

$$\frac{\partial L}{\partial \theta_2} = -m_2 l_1 l_2 \dot{\theta}_1 \dot{\theta}_2 \sin(\theta_2 - \theta_1) - m_2 g l_2 \sin \theta_2$$

Lagrange's equations become

$$(m_1 + m_2)l_1\ddot{\theta}_1 + m_2 l_2[\ddot{\theta}_2 \cos(\theta_2 - \theta_1) - \dot{\theta}_2^2 \sin(\theta_2 - \theta_1)]$$
$$+ (m_1 + m_2)g \sin \theta_1 = 0$$

$$l_2\ddot{\theta}_2 + l_1[\ddot{\theta}_1 \cos(\theta_2 - {}_1\theta) + \dot{\theta}_1^2 \sin(\theta_2 - \theta_1)] + g \sin \theta_2 = 0$$

or

$$\ddot{\theta}_1 + \left(\frac{m_2}{m_1 + m_2}\right)\left(\frac{l_2}{l_1}\right)[\ddot{\theta}_2 \cos(\theta_2 - \theta_1) - \dot{\theta}_2^2 \sin(\theta_2 - \theta_1)] + \frac{g}{l_1} \sin \theta_1 = 0$$

$$\ddot{\theta}_2 + \left(\frac{l_1}{l_2}\right)[\ddot{\theta}_1 \cos(\theta_2 - \theta_1) + \dot{\theta}_1^2 \sin(\theta_2 - \theta_1)] + \frac{g}{l_2} \sin \theta_2 = 0$$

The last two equations are the equations of motion for the double pendulum system.

Example C-4. Moving pendulum. Consider the moving pendulum shown in Fig. C-4. This is a two-degrees-of-freedom system. The generalized coordinates are x and θ.

The kinetic energy of the system is

$$T = \tfrac{1}{2}Mv_1^2 + \tfrac{1}{2}mv_2^2$$

where $v_1 = \dot{x}$ and v_2 is the absolute velocity of mass m. Similar to the case of the double pendulum system, the square of velocity v_2 of mass m can be obtained as

$$v_2^2 = (\dot{x} + l\cos\theta\dot{\theta})^2 + (l\sin\theta\dot{\theta})^2$$
$$= \dot{x}^2 + l^2\dot{\theta}^2 + 2\dot{x}l\cos\theta\dot{\theta}$$

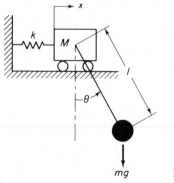

Fig. C-4. Moving pendulum.

So the kinetic energy is

$$T = \tfrac{1}{2}M\dot{x}^2 + \tfrac{1}{2}m(\dot{x}^2 + l^2\dot{\theta}^2 + 2\dot{x}l\cos\theta\dot{\theta})$$

The potential energy of the system is

$$U = mgl(1 - \cos\theta) + \tfrac{1}{2}kx^2$$

where the potential energy when $x = 0$ and $\theta = 0$ is taken as zero.
The Lagrangian L is

$$L = T - U$$
$$= \tfrac{1}{2}M\dot{x}^2 + \tfrac{1}{2}m(\dot{x}^2 + l^2\dot{\theta}^2 + 2\dot{x}l\cos\theta\dot{\theta}) - mgl(1 - \cos\theta) - \tfrac{1}{2}kx^2$$

Therefore Lagrange's equations

$$\frac{d}{dt}\left(\frac{\partial L}{\partial \dot{x}}\right) - \frac{\partial L}{\partial x} = 0$$

$$\frac{d}{dt}\left(\frac{\partial L}{\partial \dot{\theta}}\right) - \frac{\partial L}{\partial \theta} = 0$$

become

$$\frac{d}{dt}(M\dot{x} + m\dot{x} + ml\cos\theta\dot{\theta}) + kx = 0$$

$$\frac{d}{dt}(ml^2\dot{\theta} + m\dot{x}l\cos\theta) + m\dot{x}l\sin\theta\dot{\theta} + mgl\sin\theta = 0$$

or

$$\ddot{x} + \frac{m}{M+m}l\cos\theta\ddot{\theta} - \frac{m}{M+m}l\sin\theta\dot{\theta}^2 + \frac{k}{M+m}x = 0$$

$$\ddot{\theta} + \frac{1}{l}\ddot{x}\cos\theta + \frac{g}{l}\sin\theta = 0$$

The last two equations are the equations of motion for the system.

===

Rayleigh's dissipation function. In nonconservative systems (damped systems) energy is dissipated. Rayleigh developed a dissipation function D from which the damping force can be derived. Assuming that the system involves r viscous dampers, Rayleigh's dissipation function is defined by

$$D = \tfrac{1}{2}(b_1\dot{\delta}_1^2 + b_2\dot{\delta}_2^2 + \cdots + b_r\dot{\delta}_r^2)$$

where b_i is the coefficient of the ith viscous damper and $\dot{\delta}_i$ is the velocity difference across the ith viscous damper. (Thus $\dot{\delta}_i$ can be expressed as a function of the generalized velocities $\dot{q}_i$.)

By using Rayleigh's dissipation function, Lagrange's equations for nonconservative systems become

$$\frac{d}{dt}\left(\frac{\partial L}{\partial \dot{q}_i}\right) - \frac{\partial L}{\partial q_i} + \frac{\partial D}{\partial \dot{q}_i} = 0 \qquad (i = 1, 2, \ldots, n) \qquad \text{(C-6)}$$

Example C-5. **Spring-mass-damper system.** In the spring-mass-damper system shown in Fig. C-5 the only generalized coordinate is the displacement x, which is measured from the equilibrium position.

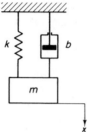

Fig. C-5. Spring-mass-damper system.

The kinetic energy T of the system is

$$T = \tfrac{1}{2}m\dot{x}^2$$

The potential energy U is

$$U = \tfrac{1}{2}kx^2$$

where the potential energy at the equilibrium position is taken as zero. (Note that although the instantaneous potential energy is the instantaneous potential of the weight of the mass plus the instantaneous elastic energy stored in the spring, the increase in the total potential energy of the system is due to the increase in the energy of the spring because of its deformation from the equilibrium position. See Sec. 2-5).

The Lagrangian L of the system is

$$L = T - U = \tfrac{1}{2}m\dot{x}^2 - \tfrac{1}{2}kx^2$$

The Rayleigh's dissipation function D is

$$D = \tfrac{1}{2}b\dot{x}^2$$

So on substituting in Lagrange's equation for nonconservative systems

$$\frac{d}{dt}\left(\frac{\partial L}{\partial \dot{x}}\right) - \frac{\partial L}{\partial x} + \frac{\partial D}{\partial \dot{x}} = 0$$

we obtain

$$m\ddot{x} + b\dot{x} + kx = 0$$

which is the equation of motion for the system.

Example C-6. **LRC system.** In the LRC system shown in Fig. C-6 capacitor C is initially charged to q_0 and switch S is closed at $t = 0$. For this system, charge q is the only generalized coordinate.

The kinetic energy T of the system is

$$T = \tfrac{1}{2}Li^2 = \tfrac{1}{2}L\dot{q}^2$$

The potential energy U is

$$U = \frac{1}{2}Ce^2 = \frac{1}{2C}q^2$$

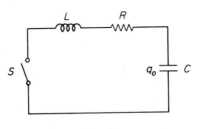

Fig. C-6. *LRC* system.

The Rayleigh's dissipation function *D* for the system is

$$D = \tfrac{1}{2}R\dot{q}^2$$

The Lagrangian *L* for the system is

$$L = T - U = \frac{1}{2}L\dot{q}^2 - \frac{1}{2C}q^2$$

Upon substitution of *L* and *D* into Lagrange's equation for nonconservative systems

$$\frac{d}{dt}\left(\frac{\partial L}{\partial \dot{q}}\right) - \frac{\partial L}{\partial q} + \frac{\partial D}{\partial \dot{q}} = 0$$

we have

$$L\ddot{q} + \frac{1}{C}q + R\dot{q} = 0$$

or

$$L\frac{d^2q}{dt^2} + R\frac{dq}{dt} + \frac{1}{C}q = 0$$

where the initial conditions are $q(0) = q_0$ and $\dot{q}(0) = 0$.

Lagrange's equations for systems with input forces. If the system is subjected to an input force (generalized force), then Lagrange's equations become

$$\frac{d}{dt}\left(\frac{\partial L}{\partial \dot{q}_i}\right) - \frac{\partial L}{\partial q_i} + \frac{\partial D}{\partial \dot{q}_i} = Q_i \qquad (i = 1, 2, \ldots, n) \qquad \text{(C-7)}$$

where Q_i is the input force corresponding to the *i*th generalized coordinate.

Example C-7. **Dynamic vibration absorber.** The mechanical vibratory system with a dynamic vibration absorber shown in Fig. C-7, where $p(t)$ is the input force, is a two-degrees-of-freedom system. The generalized coordinates are displacements *x* and *y*, where they are measured from their respective equilibrium positions in the absence of the input force $p(t)$.

The kinetic energy *T* of the system is

$$T = \tfrac{1}{2}M\dot{x}^2 + \tfrac{1}{2}m_a\dot{y}^2$$

The potential energy *U* of the system is

$$U = \tfrac{1}{2}kx^2 + \tfrac{1}{2}k_a(y - x)^2$$

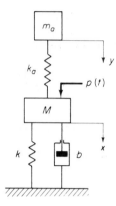

Fig. C-7. Mechanical vibratory system with a dynamic vibration absorber.

where the potential energy when $x = 0$ and $y = 0$ is taken as zero. The Rayleigh's dissipation function D is

$$D = \tfrac{1}{2}b\dot{x}^2$$

The generalized force corresponding to the x coordinate is $p(t)$.

The Lagrangian L for the system is

$$L = T - U = \tfrac{1}{2}M\dot{x}^2 + \tfrac{1}{2}m_a\dot{y}^2 - \tfrac{1}{2}kx^2 - \tfrac{1}{2}k_a(y - x)^2$$

Lagrange's equations

$$\frac{d}{dt}\left(\frac{\partial L}{\partial \dot{x}}\right) - \frac{\partial L}{\partial x} + \frac{\partial D}{\partial \dot{x}} = Q_1 = p(t)$$

$$\frac{d}{dt}\left(\frac{\partial L}{\partial \dot{y}}\right) - \frac{\partial L}{\partial y} + \frac{\partial D}{\partial \dot{y}} = Q_2 = 0$$

become

$$\frac{d}{dt}(M\dot{x}) + kx - k_a(y - x) + b\dot{x} = p(t)$$

$$\frac{d}{dt}(m_a\dot{y}) + k_a(y - x) = 0$$

or

$$M\ddot{x} + b\dot{x} + kx + k_a(x - y) = p(t)$$

$$m_a\ddot{y} + k_a(y - x) = 0$$

The last two equations are the equations of motion for the system.

Concluding comments. As seen from the preceding examples, once the energy expressions of the system are derived, Lagrange's method will give as many equations as the number of degrees of freedom of the system. These equations are the equations of motion for the system and describe completely the dynamics of the system.

For complicated systems, in order to ensure the correctness of the equations of motion, it is preferable to obtain these equations independently by using (a) Lagrange's method and (b) Newton's second law (or Kirchhoff's laws, etc.). The two approaches should yield the same result.

INDEX

B

Balanced vane pump, 163–164
Barna, P. S., 200
Barometer pressure, 157
Basic logic functions, 273
Basic quantity, 564
Basic unit, 564
Bayley, F. J., 200
Bellows-type pneumatic cylinder, 231
Bernoulli's equation, 182–183
BES units, 9, 565
Bistable device, 267
Bistable fluidic amplifier, 267
Bleed-type pneumatic relay, 496–497
Block, 480
  and tackle, 58–59
Block diagram, 480–489
  algebra, 484, 486–487
  reduction, 483–489
Blowers, 228–229
Bohn, E. V., 336
Boolean algebra, 272
Brake, 67, 90
Branch point, 481–482
Bridge circuit, 153
British Engineering System of units, 9, 565
British systems of units, 9
British thermal unit, 244
Btu, 125, 244, 574
Bulk modulus, 173, 200, 218–219

C

Camp, D. T., 287
Candela, 566–568
Cannon, R. H., Jr., 63, 536
Capacitance, 101, 123, 188, 260–262
Capacitor, 101–102
Capacity, 188
Capillary tube, 207–208
Cavitation, 171–172
Celsius, 575
Centipoise, 174
Centistoke, 174
Centrifugal compressors, 228–229
Centrifugal force, 394–395, 446
Centripetal force, 394–395, 447
cgs system of units, 565
Chain hoist, 59–60
Change of state:
  adiabatic, 247–248
  at constant pressure, 247
  at constant temperature, 247
  at constant volume, 246–247
  isentropic, 247, 249, 290–291
  isothermal, 247
  for perfect gas, 246–248
  polytropic, 248
  reversible adiabatic, 247
Chapman, W. P., 287
Characteristic equation, 25
Charge, electric, 98
Check valve, 168, 242
Churchill, R. V., 336
Circuit, 103–111
  analysis, 111–121

Closed-loop control system, 478–479
Closed-loop transfer function, 483
Commutative law, 274
Complement, 273
Complementary solution, 351
Complex algebra, 310–313
Complex conjugate, 309, 313
Complex function, 313
Complex impedance, 378–379, 441, 443–444
Complex numbers, 308
  addition of, 310
  division of, 311–312
  equality of, 310
  multiplication of, 310–311
  polar forms of, 308, 310
  powers of, 312
  rectangular forms of, 308, 310
  roots of, 312
  subtraction of, 310
Complex variable, 313
Compliance, 21
Component, 1
Compressibility, 173
  flow rate, 218
  modulus of, 173
Compressor aftercooler, 230
Compressors, 228–229
  axial, 228–229
  centrifugal, 228–229
  four-stage, 229
  positive displacement type, 228–229
  reciprocating, 228–229
  rotary, 228–229
  two-stage, 229
Conductance, 100
Conical pendulum, 446
Conservation of angular momentum, law of, 370
Conservation of energy, law of, 53, 82–86
Conservation of mass, principle of, 178–179
Conservation of momentum, law of, 369–371
Conservative systems, 53–56, 577–583
  Lagrange's equations of motion for, 577–583
Contact forces, 12
Continuity equations, 178
Contraction coefficient, 183, 186, 249
Control:
  derivative, 514
  integral, 512
  on-off, 491–493
  proportional, 493
  proportional-plus-derivative, 493–494
  proportional-plus-derivative-plus-integral, 493–494
  proportional-plus-integral, 493–494
  two-position, 491–493
Control systems, transient response analysis of, 509–516
Control volume, 178
Controllers, 491–509
Conversion table, 11, 570–575
  for area, 572
  for energy, 574
  for length, 571
  for mass, 570
  for power, 574
  for pressure, 573
  for temperature, 575
  for volume, 572

First-order system, 24
  proportional control of, 509–512
  ramp response of, 356
  step response of, 354–356
  transient response analysis of, 352–357
First variation, 578
Flapper valves, 167–168
Flip-flop, 267, 279–281
  electrical analog of, 280–281
Flow:
  control valves, 238
  through an orifice, 182–186
  in pipes, 176–177, 189–190
Flowboard amplifier, 282–283
Fluid amplifier, 265
Fluidic bistable device, 267
Fluidic control, 265
Fluidic control system, 265
Fluidic devices, 265, 295
  advantages of, 271
  disadvantages of, 271–272
Fluidic digital amplifier, 267
Fluidic impedance, 266
Fluidic logic gate, 272
Fluidic transducer, 266
Fluidics, 265
Force, 12–13
Force-current analogy, 127–128, 147–148
Force transmissibility, 398–401, 473
Force-voltage analogy, 126, 145–147
Forced response, 24, 351
Four-pulley hoist, 58–59
Four stage compressors, 229
Four-way spool valve, 166–167, 196–200
Four-way valve, sliding-spool, 166–167, 196–200
Free vibration, 26–29, 357–368
  of a spring-mass-damper system, 359–365, 367–368
  of a spring-mass system, 357, 359
Frequency, 28–29, 360–361
  damped natural, 361
  natural, 29
  undamped natural, 29, 360
Frequency response, 384–394
Friction:
  coulomb, 39
  dry, 39
  kinetic, 39–40
  linear, 23
  nonlinear, 23
  rolling, 41–42
  sliding, 39–40
  square-law, 23
  static, 39–41
Functional block, 480

G

Gage pressure, 157
Gas constant, 243, 245, 292–293
  of air, 243, 245
  universal, 245, 292–293
Gases, thermodynamic properties of, 243–248
Gas flow through an orifice, 248–258
Gate, 272

Gear:
  motor, 161, 166
  pump, 161, 163–164
  train, 61–63, 88–90
Generalized coordinates, 576–577
Generalized force, 585–586
Goldman, S., 336
Governor, 554–556
Gravitational acceleration constant, 12
Gravitational systems of units, 9, 565
Guillemin, E. A., 128

H

Hagen-Poiseuille formula, 187–188, 207
Half adder, 273
Hamilton's principle, 577
Hanging-mass system, 94, 96
Head, 186
Healey, M., 536
Heat, 244
  energy, 246
Henke, R. W., 200, 287
Henry, 102
Hertz, 28
Higdon, D. T., 536
Hohmann, C. J., 200
Hoist, 58–61, 88, 97
  chain, 59–61
  four-pulley, 58–59
  pneumatic three-pulley, 235–236
  six-pulley, 288–289
  two-pulley, 58–59, 88
Homogeneous cylinder, 14, 17, 41–47, 55–56, 67–68, 83–84, 94–95
  moments of inertia of, 14
Homogeneous disk, 71, 74–78
  moments of inertia of, 65–66
Houpis, C. J., 128
Hybrid systems, 242
Hydraulic actuators, 164
Hydraulic circuits, 158–159
Hydraulic control valve, 166–169
Hydraulic controllers, 505–507, 541–545
  jet pipe, 541–542
Hydraulic power units, 158–169, 202, 221
Hydraulic pumps, 159–164, 203
Hydraulic servo system, 196–200, 214, 216–219, 559
Hydraulic servomotor, 505–506
Hydraulic systems, 157–172
  advantages and disadvantages of, 169–170
Hydraulic tracing system, 560

I

Ideal damper, 23
Ideal spring, 22
Impulse function, 321–322
  Laplace transform of, 321
Impulse input, 369, 438–440
Impulse response, 369, 371–375
  of mechanical system, 371–375
INCLUSIVE OR, 273
Inductance, 102–103, 124, 185, 189
Inductors, 118, 123–124
  energy stored in, 123–124

Pumps (*continued*)
    radial piston, 161–162
    unbalanced vane, 163
    vane, 161–163

Q

Quality factor, 103

R

Radial piston motor, 161, 165–166
Radial piston pump, 161–162
Radian, 68, 566, 568
Radius of gyration, 15–16
Ramp function, Laplace transform of, 316
Ramp response:
    of first-order system, 356
    steady-state error in, 527–529
Rayleigh's dissipation function, 583–586
Reciprocating compressors, 228–229
Relative acceleration, 18
Relative angular acceleration, 18
Relative angular velocity, 18
Relative stability, 479
Relative velocity, 18
Relay:
    bleed-type pneumatic, 496–497
    nonbleed-type, 496–497
    pneumatic, 496–497
    reverse acting, 496
Relief valve, 168–169
    direct-acting, 237–238
    pilot-acting, 238
Residue, 331
Resistance, 100, 122, 186–188, 258–260
    average, 207–209, 293
    combined, 131–132, 154
    laminar flow, 187
    liquid flow, 187
    turbulent flow, 188
Resistance elements:
    electrical, 100
    mechanical, 23
Resistors:
    power dissipated by, 124–125, 154
    series and parallel, 103–106
Reswick, J. B., 63, 129
Reverse acting relay, 496
Reversible adiabatic change of state, 247–248
Reversible process, 246
Reynolds number, 176–177
Richardson, H. H., 64
Rigid body, 13
Rise time, 517–518, 521–522
Rolling friction, 41–42
    coefficient of, 42
Rolling motion, 201–202
Rotary actuators, 165–166
Rotary compressors, 228–229
Rotating unbalance, 396
    vibration due to, 396–397
Rotational damper, 22–23
Rotational system, 24–25

S

Safety valve, 290
Scotch yoke, 57
Second, 566, 568
Second-order system, 26, 357–375
    step response of, 365–368
    transient response analysis of, 357–375
Seely, S., 64, 129
Seismograph, 403–405
Self-inductance, 102, 119
Series circuit, 103–104
Servomechanism, 519–520
Settling time, 365, 517–518, 523
Sharp-edged orifice, 186
Shearer, J. L., 64
Shuttle valve, 242
SI, 565–569
SI units, 9–11, 564–569
    abbreviated prefixes of, 10
    abbreviations of, 10
    prefixes of, 10
Siemens, 100
Sign inverter, 411–413
Simple harmonic motion, 28
Simple pendulum, 69–70, 579
Simple pole, 313
Simulation, analog computer, 431
Sinusoidal transfer function, 387
Six-pulley hoist, 288–289
Slide valves, 239
Sliding friction, 39–40
    coefficient of, 40
Sliding-spool four-way valve, 166–167
Sliding-spool three-way valve, 167
Sliding spool valves, 166–167
Slug, 12
Smith, R. J., 129
Sound, 291
Space booster, attitude control of, 532–536
Specific gravity, 173
Specific heat, 243, 246, 353
    at constant pressure, 246
    at constant volume, 246
    ratio, 243
Specific volume, 172
    of air, 245
Specific weight, 172
Specifications:
    engineering, 6
    transient response, 516–525
Speed, 17
Speed-control system, 563
Speed of sound, 253, 291–292, 303
Spool valve, 166–167, 214–217, 239
    four-way, 166–167, 196–200
    linear mathematical model of, 216
    overlapped, 203–204
    three-way, 167
    underlapped, 203–204
Spring, 20
    ideal, 22
Spring constant, 20
    equivalent, 68–69, 71, 90–91
    torsional, 20
Spring-mass system, 25–29, 87–88, 357–359,
        384–387, 449–450

Spring-mass-damper system, 30–33, 359–365, 376, 561–562, 584
Spring-mass-pulley system, 72, 84–85
Spring-pulley system, 71
Square-law friction, 23–24
Stability, 479
  absolute, 479
  relative, 479
Standard atmospheric pressure, 172
Standard pressure, 243
Standard temperature, 243
Static friction, 39–40
  coefficient of, 40
Static system, 1
Steady flow, 178
  Euler's equation of motion for, 181
Steady-state error, 356, 512
  in ramp response, 527–530
Steady-state response, 351
Step function, Laplace transform of, 315
Step response:
  of electrical system, 365–367
  of first-order system, 354–356
  of mechanical system, 367–368
  of second-order system, 365–368
Steradian, 566, 569
Stoke, 174
Streamline, 178
Stream tube, 178–179
Streeter, V. L., 200
Summer, 413–414
Summing point, 481
Superposition, principle of, 2
Surge tank, 171, 220–221
Synthesis, 5
System, 1
  analysis, 5
  design, 5
  dynamic, 1
  static, 1
Systematic units, 564
Systems of units, 8–9, 564–569
  absolute, 565
  British, 9
  British engineering, 565
  cgs, 565
  gravitational, 565
  International, 565–569
  metric engineering, 565
  mks, 565

T

Tachometer feedback, 525
Taft, C. K., 63, 129
Taylor series, 195–196, 212–214, 578
Temperature, conversion table for, 575
Test signals, 350
Thermal capacitance, 353
Thermal energy, 125
Thermal equivalent of heat, 246
Thermal resistance, 353
Thermal system, 352–356, 470
  mathematical modeling of, 352–354
Thermodynamic properties of gases, 243–248
Thermometer system, 352–356, 560–561

Thin-plate orifice, 186
Thomas, G. M., 200
Three-port pneumatic pilot valve, 239, 241
Three-port, two-position, direct-acting magnetic valve, 239–240
Three-way valve, sliding-spool, 167
Time constant, 25, 193, 207, 383
Time-scale factor, 420–423, 459
Time scaling, 420
Toggle-joint, 302–303
Torque, 13
Torsional damper, 22
Torsional spring constant, 20
Torsional viscous friction coefficient, 23
Torsional viscous friction constant, 23
Traffic control, 479
Transducer, electric-pneumatic, 557
Transfer function, 375–383
  of loading elements in series, 381–382
  of nonloading elements in series, 379–381
  sinusoidal, 387
Transformers:
  energy, 57
  motion, 57
  power, 57
Transient response, 351
  specifications, 516–525
Transient response analysis:
  of control systems, 509–516
  of first-order systems, 352–357
  of second-order systems, 357–375
Translated function, Laplace transform of, 317
Translational damper, 22
Transmissibility, 398–403
  for force excitation, 398–401
  for motion excitation, 402–403
Truth table, 274
Turbulence amplifier, 267–268
Turbulent flow, 177
  resistance, 188
Two-degrees-of-freedom system, 35–38, 456–459, 580–583, 585–586
Two-port, direct-acting magnetic valve, 239–240
Two-position control, 491–493
Two-position controllers, 492
  pneumatic, 492, 499
Two-pulley hoist, 58, 88
Two-stage compressors, 229
Two-stage servovalve, 167–168
Two-stage valve, 168
Two-terminal circuit, 378–379

U

U-shaped manometer, 92
Unbalanced vane pump, 163
Undamped natural frequency, 29, 360
Underdamped system, 30
  free vibration of, 360–361
Underlapped spool valve, 166, 203–204
Unit-impulse function, 322
Unit-ramp response, 549–550
Unit-step function, Laplace transform of, 315
Unit step response curves, 521, 523, 530
Unit systems, 8–9, 564–569
Units, 8–9, 564–569

239

1200